Principles of Paleoclimatology

THOMAS M. CRONIN

COLUMBIA UNIVERSITY PRESS
NEW YORK

Columbia University Press
Publishers Since 1893
New York Chichester, West Sussex

Library of Congress Cataloging-in-Publication Data
Cronin, Thomas M.
Principles of paleoclimatology / Thomas M. Cronin.
p. cm. — (Perspectives in paleobiology and earth history series)
Includes bibliographical references and index.
ISBN 0-231-10954-7 (cl : alk. paper). — ISBN 0-231-10955-5 (pbk. : alk. paper)
1. Paleoclimatology. I. Title. II. Series.
QC884.C74 1999
551.69'09'01—dc21 98-48495

Casebound editions of Columbia University Press books are printed on permanent and durable acid-free paper.

Printed in the United States of America
c 10 9 8 7 6 5 4 3 2 1
p 10 9 8 7 6 5 4 3 2 1

TO MARGARITA

JASON, NICOLAS, AND ANTHONY

CONTENTS

PREFACE

MOST SCIENTISTS can recall an incident, probably insignificant at the time it happened, when their interest in natural history was stirred. I had two such moments. First, when I was six or seven, I asked my dad how the immense rocks had found their way into my Connecticut back yard. He explained that they had been carried there by great glaciers thousands of years ago. I never gave a second thought to the rocks and glaciers, nor to the climate changes that ultimately were responsible for them, until a course on the Pleistocene and glacial geology with Dan Miller at Colgate University resurrected my interest. The second event was a childhood trip to New York's American Museum of Natural History, where the dinosaur exhibit fostered a keen interest in paleontology. Quaternary geology and paleobiology are the body and soul of this book, and I am deeply indebted to my mentors, Bob Linsley and Steven Jay Gould, for encouraging me to pursue these endeavors.

Today, climate changes of the past have taken on a much greater importance than they had during my childhood. Concerns that humans may be altering the natural course of Earth's climate on a global scale through fossil fuel consumption, land use, and other activities have necessitated a better understanding of how the climate system responds to various types of forcings. Indeed, climate change has become a cause célèbre for environmentalists seeking to prevent a global catastrophe. In opposition, others are skeptical about the level of concern, advocating at least a wait-and-see approach, or even claiming that all the concerns are much ado about nothing.

I suspect that most of the exiting discoveries about paleoclimate described in this book come from researchers who, like myself, entered this field out of an interest in natural history and processes. To be sure, most scientists are conservative when it comes to speculating about the ramifications of past climate changes for future climate trends. Though climate change may now be in the public limelight, paleoclimatologists are motivated more by the intrinsically fascinating history told to them by glacial tills, pollen profiles, and geochemistry of protists than by any other factors.

When I developed a course entitled "Geological and Biological Records of Climate Change" at George Mason University, I believed that, given the myth and misunderstanding about climate history, a text describing how and why earth's climate has changed in the past and how climate changes are reconstructed was needed. Paleoclimatology is among the most interdisciplinary of sciences, melding earth, biological, chemical, and physical disciplines as well as field and laboratory investigations. It is also a pervasively historical pursuit requiring a keen appreciation of historical events. I was fortunate that Rick Diecchio, Chris Jones, and Bob Jonas at GMU endorsed this sentiment and supported me in this course and the writing of this book. I am extremely grateful to them.

I have also been fortunate to enjoy field experiences with many of North America's accomplished Quaternary geologists—Art Bloom, Hal Borns, Alex Dreimanis, Dick Goldthwait, Claude Hillaire-Marcel, Serge Occhietti, Vic Prest, and others. Over the years, I have benefited from stimulating exchanges with David Bowen, Harry Dowsett, Gary Dwyer, Joe Hazel, Nori Ikeya, Julio Rodriguez-Lazaro, Maureen Raymo, Bill Ruddiman, Deb Willard, and Ike Winograd. The scope of paleoclimatology is enormous and for those chapters in which I delved into unfamiliar territory, I thank Paul Baker, Joan Bernhard, Peter Clark, Rick Diecchio, Frans Hilgen, Lloyd Keigwin, Amy Leventer, Judy Lean, Brad Linsley, Ellen Mosley-Thompson, Terry Quinn, Rob Ross, Dave Scott, Todd Sowers, Lonnie Thompson, A. Tsukagoshi, and John Wehmiller for guidance and material, as well as Pat Megonigal for comments on the text itself. If the following pages succeed in conveying an appreciation of Earth's climate history, it is thanks to these and many other colleagues. If I falter, through missteps, mistakes, and misinterpretation, it is through my own shortcomings. I offer many thanks to these and other colleagues and hope that their ongoing research makes this book soon obsolete. This book would not have been possible without the patience and support of my editor Ed Lugenbeel, the copyeditor Amanda Suver, Ron Harris, and the staff at Columbia University Press, as well as the help of Erwin Vil-lager with graphics at GMU and Peter Trick with the text. In addition, the following scientists kindly granted permission for the use of illus-trations: Richard Alley, Larry Benson, Wolf Berger, Bill Berggren, Ed Cook, R. Delmas, Rob Dunbar, Pieter Grootes, Henry Hooghiemstra, Brian Huber, John Imbrie, Jean Jouzel, Scott Lehman, Paul Mayewski, Mike MacPhaden, Jim McManus, Alan Mix, Nat Rutter, R. Scherer, and Minze Stuiver. Last but not least, I owe my greatest debt to Mar-garita for her insight, enduring patience, and Dominican joie de vivre.

PRINCIPLES OF PALEOCLIMATOLOGY

I

Contemporary Issues in Climate Change: The Role of Paleoclimatology

We attach the utmost importance to a full understanding of the physical, chemical, and biological processes by which subtle changes in insolation are amplified to induce long-term changes in global climate. For this, a knowledge of the sequences of events and the exact timing of forcings and of the climate responses in various parts of the earth system is essential.

C. LORIUS ET AL. 1990

MODELS AND OBSERVATIONS

AS THE ROYAL MATHEMATICIAN living near Prague, in what is now the Czech Republic, Johannes Kepler's job in the early seventeenth century was to forecast the weather and cast horoscopes for the empire. But Kepler was involved with more than weather. As related by Ivars Peterson (1993) in his book *Newton's Clock: Chaos in the Solar System,* Kepler was engaged in a personal search to explain the solar system. He wished not merely to predict the motions of planets but to develop and test theories about the physical laws of the solar system. In 1543 Kepler's predecessor Copernicus had shaken humankind's notion of an earth-centered universe by proposing a heliocentric model to explain the relationship between the sun and the planets. But Kepler wanted to determine how predictable the orbits of the other planets were, in particular the troublesome Mars. Were they circular, a notion held by Aristotle, Hipparchus, Copernicus, and their contemporaries for centuries, or were they elliptical? Did a unifying mathematical relationship explain the orbits of the planets, reflecting an underlying harmony to the universe? Could a model of elliptical orbits sufficiently account for our natural universe?

Kepler knew that support for his model of the solar system would come after his predictions about planetary motions, made through his observations, were confirmed. With his assistants he undertook the daunting task of model validation through observation of planetary bodies and through the use of extensive data compiled over the prior 40 years by his predecessor, Tycho Brache. Through persistence and genius, Kepler showed in a series of publications an organization to the solar system hitherto unknown, embodied in his three laws of planetary motions. His achievements stand as a benchmark in model testing by means of careful data gathering and analysis.

Currently many other earth models, like Keplerian models of the solar system centuries ago, are under scrutiny for their predictive capabilities of earth's climate. Models called coupled general circulation models (GCMs) (McGuffie and Henderson-Sellers 1997; Trenberth 1994) are based on fundamental principles of physics and energy balance and the fluid flow of the earth's atmosphere and oceans. GCMs are designed to simulate earth's climate and test theories about climate change. Climate modeling research includes experiments aimed at better understanding of an array of key processes involved with climate change; some of these are listed in table 1-1.

Each of the studies in table 1-1 and scores of others were designed to investigate different aspects of the climate system; several excellent works on the basics of climate modeling are available (e.g. McGuffie and Henderson-Sellers 1997; Semtner 1995; Taylor 1994; Trenberth 1994).

Much of the current impetus behind modeling climate change is to answer the question, how is earth's climate responding to the large anthropogenic influence on its atmosphere through the emission of carbon dioxide (CO_2), methane (CH_4), nitrous oxide (N_2O) and other gases that trap radiation reemitted by the earth? Few scientists today dispute the idea that human activities—industrial, cultural, and land use—have led to increased concentrations of these gases in our atmosphere. Atmospheric CO_2 concentrations, for example, have increased from preindustrial levels of 280 parts per million volume (ppmv) to the current 360 ppmv (Houghton et al. 1990, 1996). Nor would many dispute the fact that these gases are radiatively active; that is, they absorb long-wavelength radiation, in theory trapping some of the heat that would otherwise be reemitted to space. Furthermore, land-use changes on a global scale such as deforestation have led to changes in surface albedo and disturbed the natural carbon cycle, as well as global biogeochemical cycles of elements such as nitrogen and phosphorous (Vitousek 1994; Schlesinger 1997). Models play a critical role in evalu-

TABLE 1-1. Some climate model studies.

Goal	*Reference*
Effects of atmospheric aerosols	Charlson et al. 1987
Influence of clouds, atmospheric radiation budgets, and climate	Ramanathan et al. 1995
Ocean circulation and climate	Mikolajewicz and Meier-Reimer 1990
Ocean circulation and carbon sequestration	Manabe and Stouffer 1994, Sarmiento and Le Quére 1996, Washington et al. 1994
Ocean heat and salt budgets	McCann et al. 1994
Past 60 years of coupled atmosphere-ocean general circulation history	Cubasch et al. 1995
Albedo feedbacks from land ice	Dickinson et al. 1987
Albedo feedbacks from snow	Cess et al. 1991
Feedbacks from sea ice	Meehl and Washington 1990, Washington and Meehl 1996
Antarctic ice sheet behavior	Verbitsky and Saltzman 1995
Sea-level change	Wigley 1995
Impact of elevated CO_2 levels on El Niño–Southern Oscillation	Knutson and Manabe 1994
Regional climate changes	Grotch and MacCracken 1991
Recent global temperature trends	Graham 1995
Carbon cycling and the role of terrestrial and oceanic processes	Keeling et al. 1989, Tans et al. 1990, Levin 1994, Dixon et al. 1994, Ciais et al. 1995
Intermodel comparisons of different GCMs	Cess et al. 1993
Role of carbon and other elements in climate	Oeschger et al. 1975, Sundquist and Broecker 1985, Siegenthaler and Sarmiento 1993, Schimel et al. 1995

ating the potential impact of these global and regional environmental problems.

Climate modeling research has expanded in tandem with a second component of the study of earth's climate: the extensive worldwide efforts to measure, observe, and monitor climatic variables using land- and satellite-based methods. Global and regional atmospheric temperatures (Jones 1994; Nicholls et al. 1996), carbon budgets (Murray et al. 1994; Keeling et al. 1989), and sea level (Gornitz 1995a,b) are just a few of the many types of data used by climatologists, atmospheric scientists, and oceanographers to examine secular trends and variability in climate. Observational trends in climate-related parameters form a powerful means of evaluating the output from computer model simulations, just as Kepler's astronomical observations were ultimately used to test his theory of planetary motions.

PALEOCLIMATOLOGY: THE TIME DIMENSION

A THIRD COMPONENT to the study of earth's climate is another essential ingredient in the quest to understand climate change: the field of paleoclimatology, the subject of this book. Paleoclimatology is the study of past climate changes that can be reconstructed from a plethora of geological and biological archives such as ocean and lake sediments, ice sheets, tropical corals, tree rings, and other sources.

Why is the history of climate changes that have taken place over years to millennia to millions of years important? The one overarching premise to the study of paleoclimatology is that the study of historical records of climate of the past decades and centuries and current climate trends is insufficient to fully understand how and why earth's climate changes. Whereas thermometers, tide gauges, and the like provide climatic trends for the past few decades or a century at most, the long-term history of earth's climate reconstructed from geological and biological sources adds a vital dimension—time—to the study of climate change.

More specifically, the need for paleoclimate reconstructions has at least four interrelated parts that I call the "uniformitarianism issue," "fingerprinting," "climate sensitivity," and "model verification."

The uniformitarianism issue involves the application of uniformitarianism to the study of climate change. Uniformitarianism, has deep roots in the geological sciences. In its simplest form, it is the notion that "the present is the key to the past." The uniformitarian philosophy of earth history, developed by Scotland's James Hutton in the late eighteenth century, championed by England's Charles Lyell (1830–1833), and accepted by many geologists since, holds that active geological processes, such as mountain building or sedimentation, occurring in the modern world have operated similarly throughout earth history in a gradual almost imperceptible way.

In the past few decades uniformitarianism has given way to a fundamentally different view of earth history. Earth scientists have rediscovered that catastrophes, evident in such diverse events as the 1993 Mississippi floods and the Cretaceous-Tertiary meteor impact and mass extinctions (Berggren and Van Couvering 1984), punctuate geological history. In essence, the geological record tells us that the present is not always the best analog for the past or the future.

Just as uniformitarianism has been questioned as the dominant approach to earth history in the geological sciences, so too has the idea that climate changes of the past have occurred at a slow and steady pace or in a cyclic and predictable manner. The concept that earth's climate system holds surprises—periods when the climate changes abruptly (Berger and Labeyrie 1987), sometimes in an unpredictable manner—has been advocated by several leading paleoclimatologists such as Herbert Flohn and Wallace Broecker. The paleoclimate record has become the primary source of evidence that abrupt climate events, unrivaled in human history and unrecorded by human measurement, occur frequently. Evidence gathered over the past few years has indicated that rapid climate changes, resulting from various mechanisms, characterize the earth's history. This idea has gained almost universal acceptance and has fostered a rethinking about the stability of earth's climate.

Paleoclimatology can also help resolve the contentious fingerprint problem surrounding historical climatic trends (Schneider 1994). Current evidence shows that mean annual temperature and sea level have risen over the past century. By *fingerprint,* I refer to this dilemma, which has led some climatologists to believe that at least some of this signal represents the "fingerprint" of human activities on global climate. But just how much of the past century's temperature and sea-level trends is due to natural climate variability caused by solar, volcanic, or other processes, and how much is due to human activities is unknown. Are scientists seeing the impact on earth's atmosphere of human-induced increased concentrations of radiatively active gases that are both massive in scale and anomalously rapid by our planet's past standards? Is the timing and scale of regional patterns of observed temperature and sea-level rise that which is expected from first principles of atmospheric circulation and chemistry? Or are these historical climate changes due to "natural" forces that would have occurred regardless of man's activities?

Schneider (1994) estimated that 80–90% of the observed 0.5 ± 0.2°C mean annual global warming of the past century is not a "wholly natural climatic fluctuation." Although there is by no means universal agreement on this issue, few active climate researchers would question that continued greenhouse gas emissions will eventually impact future climates. Paleoclimatologists can help identify the scale of the fingerprint of anthropogenic influence on climate by reconstructing climate changes that occurred during the past few thousand years of interglacial climate and during prior interglacial periods, thereby establishing the natural baseline of climate variability (Martinson et al. 1995). Climate history also exposes the role of the terrestrial biosphere and the oceans in the uptake of anthropogenic CO_2 through the study of past changes in the global carbon budget (Tans et al. 1990; Levin 1994; Sarmiento and Le Quére 1996). Finally, climate history can play an important role not only in determining patterns of climate change, but also in identifying the processes involved and the attribution of causal factors to explain the processes.

A third merit to paleoclimate study revolves around the sensitivity question (Covey et al. 1996). How sensitive is earth's climate to changes in various factors that are external and internal to the climate system? Internal processes that cause climate change include explosive volcanic emissions, changes in atmospheric and ocean circulation and chemistry, biological cycling of carbon and other elements, and tectonic processes. External factors include those stemming from processes outside the earth, such as solar variability and changes in earth's orbit. Paleoclimatologic reconstruction provides a means to examine earth's climate sensitivity to these various "forcing" factors. Natural climate changes of the past embody all naturally occurring feedbacks within the climate system, which cannot easily be incorporated into computer models (e.g., Hoffert and Covey 1992). In this sense, climate changes of the past serve as natural experiments conducted at spatial and temporal scales that cannot be carried out in a laboratory. Research on periods of global climatic warmth (Webb et al. 1993) and on the self-regulatory role of biotic processes such as photosynthesis and respiration in amplifying and dampening climate changes (Prentice and Sarnthein 1993) are examples of such natural processes. The temporal record of climate change derived from geology thus constitutes a means of testing climate theory and a valid line of hypothetical-deductive scientific inquiry.

A fourth advantage of paleoclimatology is the way paleoclimatic data are used to complement climate modeling. Climate models are types of experiments and paleoclimatology provides an independent check on whether or not the experimental results can be verified from the paleoclimate record. This point is particularly important because of disagreement about the rate and scale of the global climate system response to elevated greenhouse gas levels and the ability of climate models to successfully predict the impact of radiative gases on earth's temperature, precipitation, ecosystems, and polar ice caps. To cite one example, mean annual temperature is sometimes used as a significant measure of earth's climate sensitivity to perturbation such as elevated atmospheric levels of CO_2. Lindzen and Pan (1994) pointed out, however, that equator-to-pole gradients are more appropriate than mean annual temperature to explain glacial-interglacial cycles of the past million years. Covey (1995), in contrast, contended that this approach ignores the amplitude of the climate signal, which is better measured through mean annual temperature. Paleoclimate reconstructions can also shed light on feedbacks that occur in the earth's oceans and terrestrial ecosystems that either enhance or ameliorate the effects of external climate forcing. Conversely, climate models can help identify mechanisms that explain observed paleoclimatic reconstructions. Paleoclimate data—climate model comparisons are becoming standard fare in the investigations of climate change (Rind et al. 1986; Rind 1996; COHMAP 1988; Sloan et al. 1995).

In sum, climate surprises, identifying the human fingerprint on twentieth century climate, testing earth's climate sensitivity, and paleoclimate data—model comparisons are key subjects for paleoclimate research.

EARLY THOUGHTS AND CONTEMPORARY SOURCES

THE EARTH CONTAINS a treasure of information about climate history that adds the unique dimension of time to the study of earth's climate. Indeed, climate change has permeated humans' perception of their place in the universe for centuries. The presence of marine fossils in sediments lying above sea level, which today provide evidence for relative sea-level and tectonic changes, was for centuries linked to the Noachian flood. The concept of climatic cycles, an integral part of modern paleoclimatology, is also deeply rooted in Biblical and historical accounts of earth's climatic processes. For example, cyclic changes in earth's axial tilt influence climate by redistributing solar radiation, a notion embodied in today's orbital (Milankovitch) theory of climate change (Imbrie and Imbrie 1979). Stephen Jay Gould points out in *Time's Arrow, Time's Cycle,* a reference in chapter 1:5–9 Ecclesiastes to solar and hydrological cycles, even though the primary metaphor in biblical accounts was a view of time as an arrow, a unidirectional and irreversible sequence of events. Modern paleoclimatology is replete with cycles occurring at frequencies ranging in periodicity from 400,000 years for earth's orbital eccentricity, to 11 years for solar sunspots, to 3–7 years for the El Niño—Southern Oscillation (ENSO).

One of the earliest climate theorists was Sir Charles Lyell. In *Principles of Geology* (1830–1833), Lyell devoted chapter 7 to the "causes of vicissitudes of climate" from the standpoint of geology. He described theories of climate related to earth's axial position in relation to the plane of its ecliptic and the connection between climate and earth's cooling from a hot fluid, as well as the effects of tectonic uplift and the Antarctic continent on climate. Lyell also recognized the value of paleoclimate data in the broader question of climate: "A theory of climate can be subjected to the *experimentum crucis.*" He also offered the following comments about catastrophes, climate, and geological evidence from past sea-level changes: "In speculating on catastrophes by water, we may certainly anticipate great floods in the future, and we may therefore presume that they have happened again and again in past times" (p. 101). Lyell wrote extensively on fossil evidence for past climate changes with uniformitarianism as his guiding principle. Although geologists have abandoned strict uniformitarian concepts, Lyell must nonetheless be given credit for recognizing important paleoclimate concepts, such as the contrast between the "greenhouse" world of the Cretaceous and the "icehouse" of the Tertiary (Fischer 1981). More generally, Lyell must be credited with recognizing the value of the geological record for paleoclimate reconstruction.

Before the twentieth century, paleoclimatology did not exist as a unified scientific field. Instead, each area of modern paleoclimate research has its own distinct and fascinating history that parallels scientific inquiry into the various factors that cause climate change. The orbital theory of climate change (chapter 4) developed in tandem with advances in early nineteenth century glacial geology and later paleontology and marine geology (Imbrie and Imbrie 1979). The solar-climate relationship (chapter 6) blossomed with Galileo's first use of the telescope to link historical weather patterns to changes in sunspots (Hoyt and Schatten 1997). Interannual climate variability, which is most obvious in ENSO (chapter 7), has roots deep in atmospheric and oceanic sciences (Philander 1990). Explaining sea-level change (chapter 8) has historically attracted the attention of eminent scientists such as Celsius and Lyell (Morner 1979a). The evolution of earth's atmosphere before and after human influence has concerned geologists and atmospheric scientists alike since the nineteenth century (Arrhenius 1896; Revelle 1985; see chapter 9). Although it is impossible and premature to attempt to write the history of paleoclimatology here, each of the chapters of this book traces some of the more important historical aspects of the study of climate change.

Today paleoclimatology is blossoming, transformed from a group of disparate areas into a mature and united field. In the twentieth century paleoclimatology has become more than simply an offshoot of other disciplines such as climatology, atmospheric sciences, stratigraphy, marine geology, or glaciology. As the field expands at a dizzying pace, a seemingly endless array of new methods of climate reconstruction are available to study climate change. These methods examine such features as growth bands in corals, long-chain alkenone chemistry of the minute phytoplankton, tree rings, air bubbles trapped in polar ice and alpine glaciers, the calcite precipitated on the walls of subterranean caves, and the chemistry of marine shells, among others described in the following pages. With such a vast store of information, it is important to mention, albeit briefly, paleoclimate texts by my predecessors.

Much of paleoclimatology today remains closely allied with the earth sciences (Nairn 1961). Frakes (1979) and Frakes et al. (1992) are two sources of long-term geological records of climate changes covering the 4.5 billion years of earth history. Parrish (1998) also focuses on the geological evidence used to reconstruct past climate changes.

Berger (1981), Morner and Karlen (1984) and Berger and Crowell (1982) were among the first multiauthored volumes explicitly covering such important paleoclimate events as the Younger Dryas cooling. These books also are important sources of information on climate variation and variability. Berger et al. (1984) is still the primary single reference source on the Milankovitch theory of orbital influence on climate changes, although many concepts related to orbital theory have since been revised (see Imbrie et al. 1992, 1993a, and chapter 4).

Sundquist and Broecker (1985) contains numerous landmark papers on the role of CO_2 in global climate and spans geological history from the Precambrian to the most recent trends. This volume also describes pioneering efforts to develop models that explain natural CO_2 variability, especially the 30% glacial-age reduction of atmospheric CO_2 concentrations. How and why atmospheric CO_2 concentrations dropped during ice ages and rose during successive deglacial periods remains a perplexing and central problem in paleoclimatology. More recently, important volumes have appeared on the role of oceans in regulating the global carbon budget (Berger et al. 1989), on glacial-age carbon cycling (Zahn et al. 1994), and on the response of the terrestrial biosphere (Koch and Mooney 1996) and of populations and communities to elevated CO_2 levels (Korner and Bazzaz 1996).

Bradley (1985) provides a comprehensive text on Quaternary paleoecology, with emphasis on the multitude of paleoclimate dating and proxy methods used to reconstruct climate history over the past 2 million years. Bradley (1989) also edited a volume containing brief but lucid papers on selected aspects of Quaternary climate change. Bradley and Jones (1992) and Jones et al. (1996) provide state-of-the-art volumes on climate changes of the past 500 and 2000 years, respectively.

Berger and Labeyrie (1987) edited what was very likely the first major volume designed to specifically address the question of abrupt climate changes. As such, its chapters contain many examples of paleoclimatologists' newfound recognition that climate can change with startling rapidity.

The marriage of empirically reconstructed climate history to computer modeling of climate change has been the focus of several texts, including Hecht (1985), Duplessy and Spyridakis (1994), and Crowley and North (1991). The latter volume is an excellent overview of the use of climate models to evaluate climate history over short and long geological time scales.

Volumes produced by Russian paleoclimatologists include the pioneering work of Budyko (1982) and colleagues (e.g., Zubakov and Borzenkova 1990). These authors place great emphasis on the use of paleoclimate analogs as indicators of future climate with particular reference to the control of past and future climates by atmospheric CO_2 concentrations (see Webb et al. 1993). Dragan and Airinei (1989) wrote a comprehensive book, originally published in Romanian, entitled *Geoclimate and History.*

Hubert Lamb (1977; 1995) has written several major texts on recent climate history and human civilization. Lamb's work must be consulted for an introduction into historical observations of climate, especially in Europe. Ladurie (1971) and Grove (1988) also provide historical accounts of climate change.

Excellent texts also cover the fundamental principles and practice of reconstructing terrestrial environments within the dual context of the climatic factors and ecological processes that control the vegetation history of continents (Birks and Birks 1980; Delcourt and Delcourt 1990).

Eddy and Oeschger (1993) devoted a comprehensive volume to evaluating the strengths and limitations of paleoclimate data for assessing future climate changes and the contentious issue of "global warming." Considerable discussion has centered on the role of paleoclimate reconstructions in efforts to predict future climate changes. Their book is an excellent introduction to the opinions of leading climate theorists about this issue. It also contains excellent chapters on paleoclimate methodology and periods of global warmth in the past.

Finally, the book by Graedel and Crutzen (1994) entitled *Atmospheric Change* contains abundant citations of the paleoclimate record. *Atmospheric Change* exemplifies the recognition by nonpaleoclimatologists of the importance of climate history for addressing contemporary climate issues and for a fuller understanding of how earth's climate system operates.

A BRIEF OUTLINE OF THIS BOOK

PALEOCLIMATOLOGY is an important and inherently fascinating field of modern science. There is, however, a need to bring together the large and scattered literature and to synthesize recent advances in paleoclimate reconstruction within a set of principles. The text is divided into three parts. The first (chapters 2 and 3) describes the principles of paleoclimatology. Chapter 2 includes discussion of fundamental components of the climate system, the development of chronology for the study of past climate changes, and the use of climate proxies as surrogates of climate parameters. This chapter deals with the nuts and bolts that hold paleoclimatology together as a cohesive, albeit interdisciplinary, field aimed largely at the empirical reconstruction of past climate. Accurate chronology and reliable climate proxies are the sine qua non upon which paleoclimate research proceeds.

Chapter 3 focuses on the biological principles and concepts that underlie the use of biotic evidence to document climate change and the metabolic processes that form a major aspect of the causes of climate change. Biological processes are increasingly recognized as critical segments of many avenues of paleoclimate research. However, the biological principles that have evolved independently in the fields of ecology and evolutionary biology have seldom been applied to paleoclimatology.

The second part of this book (chapters 4 through 7) describes climate changes of the past over various time scales, such as tens of thousands to hundreds of thousands of years (chapter 4), millennia (chapter 5), decades to centuries (chapter 6), and single years (chapter 7). These temporal subdivisions are not arbitrary; they permit us to focus on the various forcing factors that cause climate change at different time scales.

The third part (chapters 8 and 9) focuses on two special aspects of what are often viewed as global climate changes—the geological record of sea-level changes and the paleo-atmospheric record obtained from polar and alpine ice cores. Although sea level and atmospheric evolution often reflect large-scale climatic changes, there are many regional and local aspects to the paleo–sea-level and ice-core records that can complicate the record. These chapters explain the complexities of reconstructing sea level and atmospheric changes from coastal geology, isotope geochemistry, and the trapped gases in polar and alpine ice caps and glaciers.

In all the chapters, I try to convey several themes to readers. First, a plurality of mechanisms, both external and internal to the earth, affect earth's climate. No single mechanism—such as atmospheric CO_2 or solar insolation changes due to orbital geometry—explains climate change. Instead, multiple causes and complex feedbacks must be a necessary part of climate change over all time scales. Delineating climate history through paleoclimatology is one means of sorting out these causes.

Second, in paleoclimatological practice, biological principles are, as compared with physical and chemical principles, sometimes relegated to a secondary role. Explanations of empirical climatic trends derived from paleoclimatological sources sometimes seem to approach organisms as physical entities, not as complex biological systems. Indeed, the idea that organisms are passive archives of climate-related physical and chemical processes recalls the fascinating history of competing philosophies about organisms themselves—historical debates between proponents of physicalism and vitalism that permeated the philosophy of science for centuries (see Sober 1993).

However, many processes inherent to those biotic systems that are routinely used to reconstruct climate history—ecological processes such as competition and predation, as well as evolutionary processes such as adaptation—are extremely complex. Some biological processes have elements of stochasticity that often go unappreciated in research on climate change aimed at determining causes of observed phenomena. Yet ecological and evolutionary concepts are not only germane but are intricately linked to interpretations of past climatic and environmental change (e.g., Bennett 1997). I further discuss these issues and the biological aspects of paleoclimatology in chapter 3.

Third, there is much to be gained from an appreciation of the historical development of the many branches of paleoclimatology. As mentioned above, whereas paleoclimatology is a very new field compared to physics, chemistry, and geology, virtually every avenue of paleoclimate research today rests on the foundations built by some of the greatest scientific thinkers of the nineteenth and early twentieth century. Consequently, I devote considerable time in each chapter to describing the historical background for each segment of climate history. Although they may not meet the standards of science historians, I believe these historical overviews convey important thinking and discoveries in widely dispersed fields of modern paleoclimatology.

In closing, I find the words of Thomas Kuhn on the structure of scientific revolutions enlightening: "The proliferation of competing articulations, the willingness to try anything, the expression of discontent, the recourse to philosophy and to debate over fundamentals, all are symptoms of a transition from normal to extraordinary research"

(Kuhn 1962:92). Although identifying a Kuhnian scientific revolution is no easy matter (for example see Ruse's 1981 essay on the geological revolution of the 1960s), many of Kuhn's criteria ring true for today's research on climate change. Whether the level of excitement and activity and the recognition of the dynamic nature of climate in paleoclimatology heralds a revolution or a major paradigm shift cannot be said. Yet from the perspective of the principles and applications of paleoclimatology described in the following chapters, the existing ferment and tumult raises just such a possibility.

II

Principles of Paleoclimatology

...theories too do not evolve piecemeal to fit facts that were there all the time. Rather, they emerge together with the facts they fit from a revolutionary reformulation of the preceding scientific tradition, a tradition within which the knowledge-mediated relationship between the scientist and nature was not quite the same.

THOMAS S. KUHN, 1962

INTRODUCTION: A MULTIDISCIPLINARY APPROACH IN PALEOCLIMATOLOGY

THE OUTBACK OF western Australia has been home for millions of years to endemic flightless birds such as the emu and ostrich that evolved in isolation since the Australian plate separated from other continents. During the last glacial period, about 50,000–20,000 yr ago, the emu genus *Dromaius* and its cousin the giant, extinct mihirung (*Genyornis*) were prominent members of Australia's terrestrial fauna, their eggshells often becoming fossilized in the dry playa sediments now exposed in deflation hollows in the continent's interior.

Instead of becoming a footnote of avian paleontology, emu and mihirung eggshells are helping to answer one of paleoclimatology's most perplexing riddles, one that cuts at the core of how well scientists understand earth's climate and how well climate modelers can simulate ice age climatic conditions. For more than a century and a half, scientists have known that earth's climate has periodically plunged into glacial periods, when large mid-latitude ice sheets formed and oceanic and atmospheric temperatures dropped 10°C or more in middle and high latitudes (figure 2-1). However, the consensus was that equatorial regions did not cool during glacial periods like the higher latitudes. Glacial-age foraminiferal assemblages recovered from deep-sea sediments from equatorial regions are composed of species similar to modern tropical assemblages. Quantitative analyses of foraminiferal assemblages indicated glacial sea-surface temperatures (SSTs) in the tropical oceans were no more than 1–2°C cooler than interglacial temperatures (e.g., CLIMAP 1976, 1981).

The hypothesis that the tropics remained warm during glacial periods, however, was not supported by general circulation model (GCM) simulations of glacial climate (see Rind and Peteet 1985; Manabe and Broccoli 1985). GCM simulations suggested that tropical temperatures, like those in higher latitudes, should fall during periods of global cooling. If GCM models unsuccessfully simulated glacial climatic conditions reconstructed from hard evidence, their utility in simulating earth's future climates might be much diminished. One obvious means to test GCM simulation output by using additional paleoclimate data from continental and oceanic tropical regions.

Fossil emu and mihirung shells are helping to solve the paleo-

FIGURE 2-1 Large erratic boulder of "Millstone Grit" quart-conglomerate carried by ice from the hills on the skyline. Boulder lies at limit of Late Wisconsinan ice in Gower, South Wales, and has been dated by ^{36}Cl at 23.8 ka. Courtesy of D. Q. Bowen and B. Pillans.

climate-model dilemma of the ice age tropics. On the basis of biogeochemical analyses of emu shells they dated by radiocarbon analysis, Miller et al. (1997) recently argued that tropical Australia did indeed cool during the last glacial period, by as much as 9°C or more. They based their conclusion on the chemistry of fossil eggshells. Eggshells are essentially biomineralized archives of the past temperature regimes of tropical Australia. Like their ostrich cousins, emu shells preserve the proteinaceous shell residues with little or no postmortem chemical change; that is, they remain for the most part geochemically intact. Amino acids, the building blocks of proteins, are thus preserved in the eggshells. However, amino acids undergo a process after death called racemization, whereby L (levo, or left-handed) amino acids convert into a mixture of D (dextro, or right-handed) and L amino acids. The D/L ratio is a function of both the time expired since death and the temperatures to which the fossil material is subjected (Wehmiller 1982). This change is observed in a variety of fossils, including eggshells and mollusks among others.

Because the rate of racemization is a function of temperature, the emu shells preserve an integrated thermal history of the region since the time they were buried. Miller et al. (1997) analyzed dozens of eggshell fragments from the Lake Eyre Playa and Lake Victoria regions to reconstruct glacial-age paleotemperatures. Amino acid ratios from deeply buried shells should represent glacial mean annual temperature, whereas those shells buried less than 2 m deep and exposed to daily and seasonal changes over the past few centuries provide a less accurate but still valid range of temperatures. Using radiocarbon dating and stratigraphic evidence, Miller and colleagues concluded that glacial temperatures had dropped a minimum of 6°C and were an average of 9°C colder for the period 45,000–16,000 yr ago. Glacial cooling of this magnitude is discordant with some micropaleontological evidence from seas off northern Australia, which suggests oceanic cooling of less than 2°C, but it supports GCM simulations showing cooler glacial-age tropics.

Emu racemization paleothermometry is one of a growing, somewhat eclectic group of paleoclimate tools that are now used to test the hypothesis that the tropics were cooler during the last glacial period. Other notable examples suggesting cooler tropical regions include palynological evidence for tree line elevation changes in South America (Hooghiemstra et al. 1993; Colinvaux 1996), glacial geological data for lowered snow lines in the Andes Mountains (Broecker and Denton 1989), atmospheric temperature estimates derived from noble gases trapped in "fossil" ice age groundwater (Plummer 1992; Stute and

Schlosser 1993), deep-sea isotopic analyses of pore water from abyssal sediments below the sea floor (Schrag et al. 1996), and SST estimates from coral reef geochemistry (Guilderson et al. 1994). Large-scale glacial-age cooling in the tropics is not supported, however, by paleoclimate evidence from long-chain alkenone "biomarkers" measured in deep-sea sediments (Brassell et al. 1986; Sikes and Keigwin 1994) or by planktonic foraminiferal assemblages.

DEFINING THE PRINCIPLES OF PALEOCLIMATOLOGY

THE MULTIDISCIPLINARY approaches now used to investigate the dilemma of glacial-age climates introduces the subject of this chapter: the principles that form the foundations of the field of paleoclimatology. First, a definition of the field of paleoclimatology is useful. Paleoclimatology is the study of earth's climatic history over all time scales. Paleoclimatology derives its unique identity from the marriage of many fields. Climatology, the study of the earth's modern climate system, is obviously a close cousin. In recent years, the two fields have become virtually inseparable, each benefiting from the strengths of the other. Second, paleoclimatology also is allied with traditional subdisciplines in the geological sciences, notably earth history, stratigraphy, geochemistry, glaciology, as well as the newer subdisciplines ice core paleo-atmospheric studies, dendroclimatology (which uses tree rings), and scleroclimatology (which uses tropical corals). Third, organic remains and biological processes play a major role both in the reconstruction of past climates and in regulating earth climate. Paleoclimatology, therefore, also incorporates aspects of organic evolution, ecology, and metabolic processes. As a result of its hybrid nature, paleoclimatology is of necessity a patchwork, made up of a panoply of methods and specialties, each with its own strengths and weaknesses, nuances, and scattered literature. Are there unifying principles that guide paleoclimatology?

Some might argue that the study of past climate changes needs no guiding principles other than those derived from other disciplines. A case might be made, for example, that climate history is based mainly on first principles of radiation physics and energy balance. Energy balance models are among the most useful for the study of earth's climate (Crowley and North 1991). Similarly, P.A. Sheppard (1966:1) expressed a meteorological viewpoint of climate history at the International Conference on World Climate 8,000–0 B.C. "while dynamical oceanography, glaciology, geochemistry and other disciplines are inevitably involved . . . the general problem is the problem of the general circulation of the atmosphere and its evolution on the many time scales." Physical oceanographers likewise might argue that the laws governing fluid dynamics are central to paleoceanography because the oceans store and transport heat and carbon, two critical factors in studying climate change over longer time scales. Geologists who practice paleoclimatology might point out that stratigraphic principles underpin almost all studies of climate history. Geochemists might suggest that the thermodynamics and kinetics of the chemical reactions that govern the geochemical composition of shells, wood, sediments, or trapped air bubbles in ice cores are essential to paleoclimatology. All these viewpoints would, of course, be valid.

Nonetheless, the historical development of modern paleoclimatology suggests that certain standards and conventions—call them principles—have evolved that have molded the field into a mature discipline over the past few decades. Despite the interdisciplinary nature of paleoclimatology, three overriding principles govern most paleoclimatological research. These are the *climate system* itself and the dynamic interrelationships among its component parts; *chronology,* the dimension of time that is a prerequisite to trace changes in climate; and climate *proxies,* the surrogates of climate-related parameters, such as emu eggshell chemistry serving as a proxy for tropical temperatures. All three are essential for investigating climate history, whether tracing interannual variability of the El Niño–Southern Oscillation (ENSO) in the tropical Pacific Ocean or reconstructing global warmth and paleo-atmospheric carbon dioxide (CO_2) levels during the middle Cretaceous. One goal of this book is to convey how past climates are reconstructed using these three principles, with a focus on the strategies—the modus operandi—that paleoclimatologists adopt to study climate change.

A few comments about the scope of this chapter are in store. First, principles of climatology are described in several texts, including some with excellent coverage of earth's climate history (e.g., Graedel and Crutzen 1993). My treatment of climatology is necessarily cursory, aimed at summarizing key aspects of the climate system relevant to later chapters. I refer the reader to these texts for more thorough and quantitative coverage of climatology.

Second, the number and accuracy of chronological methods available to paleoclimatologists is enormous and rapidly growing. For example, consider that 20 years ago, radiocarbon (^{14}C) dating, the staple method to date late Quaternary material younger than about 40,000 yr, required a large amount of material to obtain a ^{14}C age. Today, the technique of accelerator mass spectrometry (AMS) radiocarbon dating allows the age of as little as a few grams of carbon to be determined. AMS dating has allowed the calibration of the ^{14}C chronology using annual-layered sediments and tree rings to produce an "absolute" age, or calibrated year chronology, of climate events extending back more than 10,000 yr. Likewise, uranium-series (U-series) radioactive isotopic decay by "alpha" counting provides a means of dating sea-level history from ~100,000-yr-old emerged coral reefs with errors of several thousand years. Currently, U-series thermal ionization mass spectrometry (TIMS) dating of corals now yields ages with errors of about 1000 yr, spawning new controversy about the theories of climate change. Other dating and correlation methods too numerous to discuss in detail have been developed in support of paleoclimate research.

Third, the number and variety of climate proxy indicators used to estimate past climates has also expanded exponentially. For example, a single deep-sea sediment core can contain a dozen or more different climate-related indices (e.g., stable isotopes; trace elements such as magnesium, strontium, vanadium, and barium; microfossil assemblages, magnetic and physical properties). Polar ice coring has also expanded the scope of paleoclimatology into new directions (Grootes 1995). The GISP2 ice core from Greenland, for example, has more than 40 different climate-related properties measured from the ice and the air trapped within it. Moreover, a rapidly growing number of long continental climate records, many using new proxy techniques on lake sediment cores, are becoming available (Swart et al. 1993). Virtually every fossilized group of organisms is used in one way or another for either age dating or as a proxy of climate or both.

Given the scope of modern paleoclimatology, a comprehensive list of the limitations and accuracy of all available dating and proxy methods could not be covered in a single volume, and such a compendium would be beyond one person's expertise. Moreover, it would not convey the *principles* of paleoclimatology. Instead, the goal of this chapter is to offer an entree into the literature on paleoclimate methods to illustrate the three principles of climate, chronology, and proxies. I will also show the strategies paleoclimatologists adopt to solve specific problems, using two case studies as examples. One example portrays the development of a near calendar-year accuracy, that is, a high-resolution chronology, to study rapid climate change during the Younger Dryas. The other is the development of new and innovative proxy methods over the past 150 yr that are designed to solve the mystery of glacial-age climates. These examples capture the essence of paleoclimatology more than a listing of methods.

Because biological processes are so germane to both the reconstruction of past climates and the regulation of climate itself, I devote chapter 3 to selected principles and processes governing organism-organism and organism-environment interactions that are woven through the fabric of paleoclimatology.

CAUSES OF CLIMATE CHANGE: THE CLIMATE SYSTEM

Components of the Climate System

Earth's climate system can be divided into five main components—the lithosphere (some refer to the geosphere, regolith, or simply land surface), the hydrosphere (oceans, lakes, rivers, and groundwater), the cryosphere (glaciers, sea ice, and ice sheets), the atmosphere, and the biosphere (figure 2-2). Each component interacts with the others over time scales ranging from millions of years to seconds. Land plants are part of the terrestrial biosphere that interact with the lithosphere, through plant-soil physical and chemical processes, and with the atmosphere, through the exchange of trace gases such as carbon dioxide (CO_2) and methane (CH_4). The cryosphere plays a critical role in storing water (H_2O) in its solid form in ice, which upon melting can lower

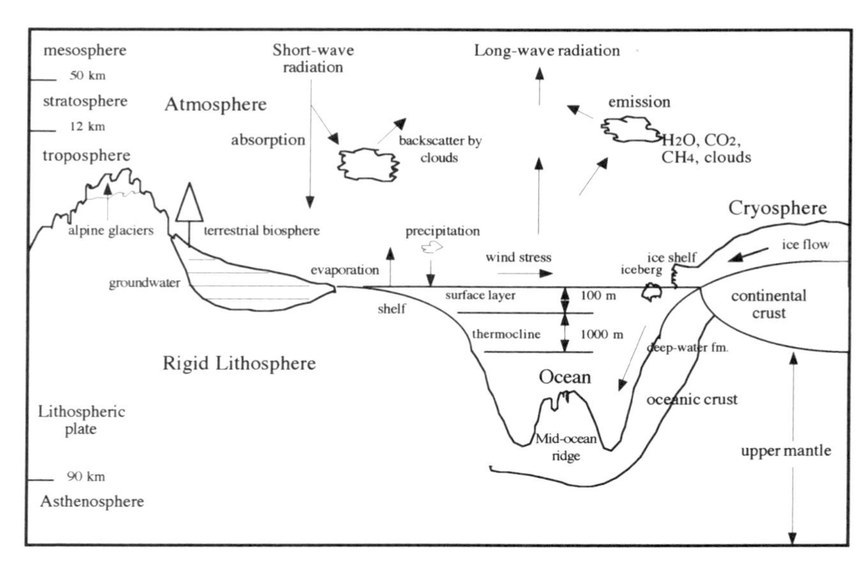

FIGURE 2-2 Major components of earth's climate system related to paleoclimatology.

the salinity and temperature of the surface oceans, changing global oceanic circulation. To understanding how climate has changed in the past, we must first consider certain basic processes that operate today within each component.

Lithosphere

The solid earth consists of an outermost lithosphere, about 100 km thick, an aesthenosphere, a mantle, and a core, with inner solid and outer liquid layers (figure 2-2) The crust is the outermost part of the lithosphere and is composed of 8 major plates and about 26 smaller plates. Continental plates are constantly shifting position, mainly as a result of unequally distributed subcrustal heat within the earth. Convergent plate boundaries are characterized by active tectonic processes such as mountain building and plate subduction; divergent boundaries by plates moving apart, mantle upwelling, and sea-floor spreading; and transform boundaries by plates sliding past each other. Relatively stable intraplate regions are usually characterized by slow uplift and subsidence due to processes such as isostasy, the force of gravity acting upon crustal materials of different densities.

The long-term evolution of earth's climate is linked to the effects of tectonic processes for several reasons. Changes in topography, for example, are believed to be an important cause of long-term climate cooling during the Cenozoic (Ruddiman and Kutzbach 1989; Hay 1992; Berner 1994). One effect of mountain building on climate is that extremely high rates of uplift can alter global weathering patterns and geochemical fluxes from continents to oceans (Ruddiman and Raymo 1988; Raymo et al. 1988), ultimately changing atmospheric levels of CO_2 and contributing to global cooling (Raymo 1991; Berner 1994). Raymo (1991) postulated that a 50% increase in global weathering rates since the Miocene is recorded in strontium isotopic changes recorded in deep-sea sediment cores. Cenozoic uplift of the Tibetan Plateau may also have contributed to global cooling over the past 40 million years by changing atmospheric circulation (Ruddiman and Kutzbach 1989; Raymo 1994). High mountains would block west-to-east airflow and surface and upper level jet stream flow in mid latitudes, as well as alter seasonal surface and atmospheric heating and regional monsoon patterns and, more generally, global atmospheric circulation and climate.

The location of a study area within or near a plate can significantly influence the interpretation of paleoclimatologic events. An illustrative example is the study of Quaternary sea-level change. In some tectonically active zones, emerged tracts of fossil coral reefs formed during past high stands of sea level during warm interglacial periods are uplifted, providing an excellent record of sea-level history. However, unless the rate of uplift is known, separating the effects of tectonics on the reef sea-level record from those of true changes in continental ice volume can pose problems. In contrast to tectonically active areas, sea-level history determined from relatively stable regions provides a partial solution, because the local sea-level record is unencumbered by major tectonic movements over the past few hundred thousand years.

Changing continental positions also impose major constraints on oceanic circulation, which in turn can have an impact on the global redistribution of solar energy by ocean currents and deep-water circulation (Rind and Chandler 1991). Oceanic circulation changes, for example, are considered a contributing factor, in addition to elevated atmospheric CO_2 concentrations, in causing warm global climates of the Eocene, about 50 million years ago (Sloan and Rea 1995; Sloan et al. 1995). These same long-term changes in ocean basin geometry driven by sea-floor spreading also lead to global sea-level changes (Hallam 1992).

In sum, many geophysical processes operate in earth's lithosphere, playing various roles in the reconstruction and interpretation of climate history. We will encounter many of these throughout this book.

The Hydrosphere: The Water Cycle

The hydrosphere and cryosphere include water contained in the world's oceans (97.3%), ice sheets and glaciers (2.1%), groundwater in aquifers (0.6%), and surface water, soil moisture, water vapor in the atmosphere, and water locked up in the biosphere (together $<< 0.1\%$). The study of processes governing water stored within each reservoir includes many separate disciplines such as groundwater hydrology (Freeze and Cherry 1979), physical oceanography (Tchernia 1980; Tomczak and Godfrey 1994), atmospheric science (Graedel and Crutzen 1993), and glaciology (Paterson 1978; Peltier 1993).

Although some texts on hydrology specialize in one part of the climate system, a more holistic approach to the earth's hydrological system is favored by others. For example, Berner and Berner (1996) make the global water cycle the nucleus of the global climate system. The hydrological cycle refers to the flux of water through all five components of the climate system. In its simplest form, the hydrological cycle includes evaporation, mostly from the ocean surface but also from lakes and rivers; condensation of water vapor in the atmosphere into clouds; and precipitation back to the land and ocean surface. During the cycle some water becomes locked in snow and glacial ice, in biological systems, and in soil and rocks. Water is also circulated through tectonic processes, being outgassed through volcanoes and hydrothermal vents and locked into the earth's crust via plate tectonic processes. Water is also the medium through which major and minor chemical species are circulated and transported from one part of the climate system to the next.

Paleoclimatology is inherently concerned with changes in the hydrologic cycle over time scales of years to millions of years. Changes in the hydrological cycle can either reflect a response to climate change or constitute a causal factor in climate change. For example, recent discoveries from polar ice cores reveal that atmospheric circulation can change from one state to another abruptly, within decades and centuries, in both polar and tropical regions. While the causes of rapid atmospheric shifts are still uncertain, synchronous climatic warming in polar and equatorial regions may very likely involve changes in atmospheric water vapor content, because other processes such as glacial melting and ocean circulation changes are too slow to account for the abruptness of the climatic shift. In contrast, the 100-m sea-level rise between 20,000 and 6000 yr ago was clearly due to slower physical processes that govern ice-sheet melting.

The following qualitative discussion emphasizes aspects of the hydrosphere and cryosphere—the world's oceans, ice sheets and glaciers, and atmosphere—that pertain to climate changes taken up in the following chapters.

The Oceans

The oceans are the world's largest reservoir of heat derived from solar insolation. Ocean circulation is the most important process transporting solar energy from low to high latitude areas and, through deep-oceanic circulation, around the world. Oceanic circulation is driven mainly by wind stress, heating and cooling, and evaporation and precipitation, all ultimately controlled by solar radiation and linked to atmospheric circulation. Major surface ocean currents are wind driven, created by pressure gradients that cause fluids and gases to flow from high- to low-pressure regions. Pressure gradients are balanced by the Coriolis "force," which is not a force *sensu strictu,* but rather is the result of the eastward rotation of the earth. In simplistic terms, Coriolis force serves to "deflect" currents moving poleward. In the Southern

Hemisphere, Coriolis force deflects currents toward the left, or westward, and in the Northern Hemisphere toward the right, or eastward. The net product of pressure gradients and Coriolis force is called *geostrophic flow,* in which ocean circulation proceeds along isobars, or lines of equal pressure. Geostrophic flow predominates in ocean interior regions away from the equator and leads to the primary surface-ocean currents often the subject of paleoceanographic research.

Ekman layer flow refers to wind-driven oceanic flow in the surface ("Ekman") layer that is perpendicular to the main wind direction. Ekman flow occurs because momentum from wind is transferred to the ocean surface layer by friction; it thus contrasts with the density-driven oceanic flow and the Coriolis force that drive surface currents and deep-ocean convection. The nature of Ekman flow varies throughout the oceans depending on latitude and other factors. Two major types of Ekman flow are upwelling (Ekman pumping) and downwelling. Upwelling is most important to climate reconstruction in some coastal and equatorial regions where colder nutrient-rich waters rise into the surface layers to replace the wind-driven surface waters. For example, in the Pacific Ocean near the equator, strong Ekman flow northward and southward creates a zone of equatorial divergence, or upwelling, that is linked with climate changes occurring over various time scales. Oceanic divergence and strong upwelling zones along coastal margins, such as those off Peru and Ecuador, result in extremely high biological productivity. These areas are among the world's most biologically productive regions. Upwelling and associated biological activity in semi-isolated silled ocean basins provide detailed temporal oceanic climate records.

Some oceanic features arising in later discussions are shown in the schematic cross section of the eastern portions of the Atlantic Ocean (figure 2-3). The figure shows the North Atlantic Current (the northeastward extension of the Gulf Stream); the equatorial divergence; the subtropical gyres, labeled the North and South Atlantic Current water (NACW, SACW); the location of deep-water formation and the major deep-water masses, North Atlantic deep water (NADW, not shown is Labrador Sea deep water) and Antarctic bottom water (AABW); and the intermediate water masses, Mediterranean deep water (MEDW) and Antarctic intermediate water (AAIW). Two important surface ocean zones are the North Atlantic's Polar Front, a convergence zone located at the boundary between the North Atlantic Current and NACW water and the South Atlantic's Antarctic divergence zone, at about 59–60°S latitude, where strong upwelling leads to high productivity.

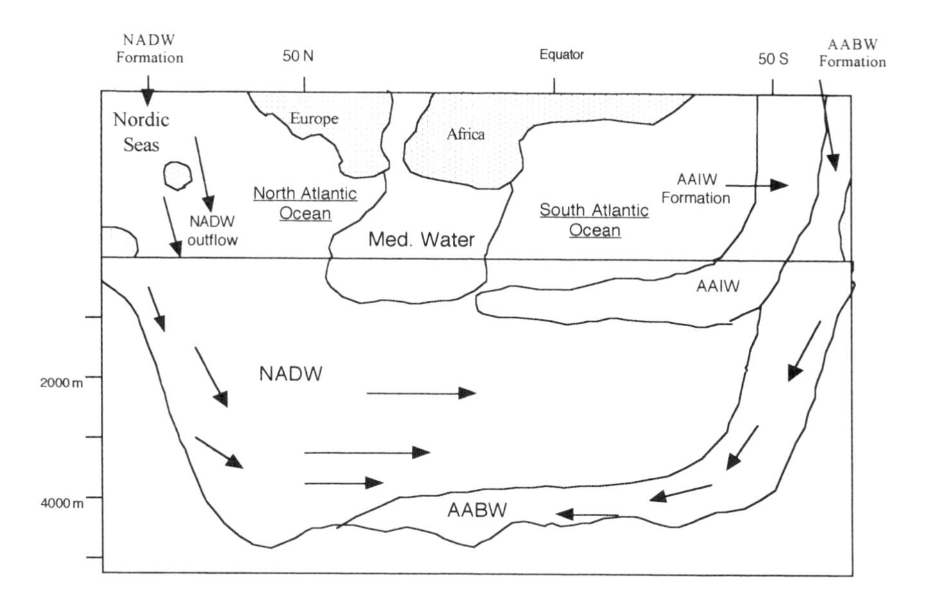

FIGURE 2-3 Schematic view of Atlantic Ocean deep-water circulation. Major deep-water masses are Antarctic Intermediate Water (AAIW), Antarctic Deep Water (AABW), North Atlantic Deep Water (NADW), and Mediterranean Water (Med Water).

Western Boundary Surface Currents

On the western sides of Northern Hemisphere Oceans, northeast-ward-flowing western boundary currents—the Pacific's Kuroshio and the Atlantic's Gulf Stream System—are major exporters of warm water from low to high latitudes. Poleward heat transport via these currents is the reason that northeastern Atlantic and Pacific coastal regions in northern Europe and the Gulf of Alaska currently have relatively warm climates despite their relatively high latitude. Changing strength of western boundary currents can have significant impacts on the climate history of high latitude regions (Dowsett et al. 1992).

Eastern Boundary Currents

In the Southern Hemisphere, the northward-flowing Benguela and Peru-Chile Currents flow along the western coasts off Africa and South America, respectively. These Southern Hemisphere boundary currents are associated with regions of the strongest coastal upwelling in the world (Suess and Thiede 1981). Coastal upwelling oceanographic processes are a complex subject in their own right, but upwelling of cold nutrient waters generally occurs within the broader context of the large-scale equatorward flow of the surface ocean in the eastern margin of the subtropical gyre.

Mid-ocean Gyres

In the central subtropical regions of the Atlantic and Pacific Oceans, between about 20–40°N and 20–40°S latitude, the surface oceans are characterized by subtropical gyres circulating clockwise in the Northern and counterclockwise in the Southern Hemisphere. Mid-ocean gyres are bounded by equatorial currents and western and eastern boundary currents. The oceanographic boundary between the North Atlantic subtropical gyre and subpolar North Atlantic water is called the Polar Front, a critically sensitive zone in terms of glacial-interglacial and shorter-term climate changes.

Equatorial Oceanic Upwelling

In equatorial regions, trade wind-driven oceanic divergence characterizes large regions where significant upwelling of cooler, nutrient-rich waters occur. Regions of equatorial upwelling are especially relevant to glacial-interglacial cycles caused by orbital changes due to earth's precessional cycles. Enhanced biological productivity in equatorial upwelling zones is believed by some researchers to be responsible for decreased atmospheric CO_2 levels during glacial periods, because enhanced productivity would lead to greater export of carbon to deep-ocean reservoirs. Chavez and Barber (1987) calculated that eastern Pacific Ocean upwelling accounts for as much as 18–56% of the ocean's total productivity and thus represents a major part of the biological pump and a means of carbon sequestration into the oceans. Mix (1989) argued changes in biological upwelling were a major cause of reduced atmospheric CO_2 levels during the last glacial period.

Thermocline

The oceanic features discussed above are all located in the uppermost 50–150 m, called the *surface mixed layer.* Below the mixed layer, water temperature decreases in a zone called the *thermocline.* The seasonal thermocline is located just at the bottom of the mixed layer and forms when oceanic overturning results from seasonal warming and cooling. The seasonal thermocline is usually shallow in spring and summer, deep in fall, and absent during winter cooling.

Below the seasonal thermocline is the permanent oceanic thermocline that represents the steep temperature gradient from the surface mixed layer to deep-sea abyssal regions. The temperature transition in the permanent thermocline is roughly 20–5°C in tropical regions and 15–5°C in higher latitudes. Thermocline processes are important in understanding nutrient and ocean circulation changes related to CO_2 storage during climate changes and changes in vertical ocean temperature structure.

Deep-Oceanic Circulation

Deep-ocean circulation is referred to as *thermohaline* or *density-driven circulation.* The ocean's deep water masses form in high latitudes, mainly the Greenland-Iceland-Norwegian (Nordic) Seas and the Labrador Sea in the North Atlantic regions, called North Atlantic deep water (NADW), and around the periphery of Antarctica, called Antarctic bottom water (AABW). Note in figure 2-3 that relatively warm (2.5–4°C) NADW is underlain by colder (0–2°C) AABW. Two major processes are responsible for density-driven deep-water formation. First, open-ocean convection occurs as warm salty water reaches high latitudes, experiences seasonal wintertime cooling, and sinks owing to its decreased buoyancy. A second process known as *brine rejection* occurs in regions of large-scale sea-ice formation such as the margins of the Arctic Ocean and around Antarctica. When ocean water freezes it "rejects" salt, which also increases the density of upper ocean water, causing it to sink.

Once formed, deep-water masses carry with them distinctive temperature and salinity signatures that are slowly altered by entrainment with other water masses as they circulate globally (Dickson and Brown 1994). For example, McCartney (1992) traces the flow of deep-water boundary currents in the North Atlantic Basin, where it is strongly influenced by submarine ridges and basins. He shows that the transport of deep water in the North Atlantic Ocean originates south of the Icelandic Faeroes in the eastern North Atlantic Basin, where Nordic Sea overflow joins with deep Antarctic water. The deep boundary current then snakes westward around southern Greenland into the Labrador Sea, where entrainment of LSDW occurs before it flows southward along eastern North America as a deep western boundary current. Deep-ocean circulation via deep boundary currents are a means of interhemispheric heat transport such that changes in thermohaline circulation can influence global climate changes over millennial time scales.

The most important message in this brief overview is the dynamic nature of oceanic processes in relation to climate change. Surface oceanic conditions obviously change over short (seasonal, interannual) time scales in response to seasonal changes; deep-water circulation is also dynamic and unstable over various time scales. For example, observations gathered over several decades show that centers of North Atlantic deep water formation in the Labrador and Nordic Seas vary decadally in their temperature, density, and depth of penetrations (e.g., Dickson et al. 1996) and circulates across the North Atlantic at rates of 1.5–2.0 cm/s (Sy et al. 1997). Short- and long-term variability in surface and deep circulation and linkages with atmospheric circulation will be covered in later chapters.

The oceanic features just described have distinct biotic characteristics, comprising the earth's primary oceanic ecosystems. Some marine ecosystems are distributed more or less ocean wide. For example, along latitudinal gradients of the surface mixed zone, distinct foraminiferal assemblages inhabit equatorial oceans, subtropical gyres, subpolar water masses, and upwelling regions. Other marine ecosystems, such as benthic communities found on continental shelves, show high degrees of endemism.

At smaller spatial scales, vertical oceanic structure controls the distribution of phytoplankton requiring light for photosynthesis, whereas some zooplankton, and even some phytoplankton (diatom mat-forming genera), migrate vertically within the water column on a daily basis. The interlocking nexus of distinct biotas and major physical oceanographic features at various spatial and temporal scales is an important theme in marine ecology (Mann and Lazier 1996), and paleoclimatology is infused with marine ecosystem dynamics.

Fresh Water and Groundwater, Soil Moisture

Those portions of earth's water stored on the continents' surface (lakes and rivers), as groundwater, and in soils comprise an important hydrological reservoir intimately associated with biological processes (respiration and photosynthesis), physical processes, (erosion), and chemical processes (dissolution, chemical weathering). Hydrological reservoirs on the continents are linked closely with climate change over all time scales. Over millions of years, for example, chemical weathering plays a critical role in global biogeochemical fluxes of key chemical elements like carbon, silicon and strontium. Over thousands of years, changes in regional or global hydrological balance are often reflected in lake levels that act like giant rain gauges to record large-scale changes in precipitation. Past African lake levels, for example, were relatively high during periods of enhanced mean precipitation during the early-middle Holocene (e.g., Street-Perrot and Harrison 1985).

The Cryosphere

The earth's cryosphere—that part of the global water stored in ice—is a dynamic and critical part of the entire climate system. Like other subsystems of earth's climate, the cryosphere is constantly in a state of flux, interacting with the oceans and the atmosphere. Here I begin by briefly describing the scope of earth's modern ice inventory.

Among the various inventories compiled for the world's ice, the recent papers edited by Williams and Ferrigno (1988, 1993), Satellite Atlas of Glaciers of the World, uses Landsat imagery to document the location of glaciers, ice fields, and ice sheets. The chapter by Swithinbank (1988) estimates that the Antarctic Ice Sheet, by far the largest volume of ice on earth, contains 60% of the world's fresh water and 91% of its ice, equivalent to 13.9×10^6 km^2 in area and 30.1×10^6 km^3 of volume. The Antarctica Ice Sheet is divided into the larger, continental East Antarctic Ice Sheet and the smaller, marine-based West Antarctic Ice Sheet. Examining Antarctica in more detail, Antarctic ice can be divided into five parts: the region including the Transantarctic mountains to the Ross Sea, the Indian Ocean sector, including the Amery Ice Shelf system with tidal glaciers moving at a 2.5 km/year; the Atlantic sector fringed with ice shelves fed by large ice streams (the Filchner and Ronne Ice Shelves are the size of Texas), the Pacific sector including the eastern Ross Sea margin, the Ellsworth Mountains, with five major ice streams draining into the Ross Ice Shelf, and the Antarctic Peninsula, with fjords, ice shelves, and the recently disintegrated Wordie and Larsen Ice Shelves (Vaughan and Doake 1995).

The grounded ice portion of the Antarctic Ice Sheet—that portion not floating around the margin of Antarctica—contains the water equivalent to a sea-level rise of about 73 m, covering a land area of 12.1 million km^2. Substantial regions of ice are not grounded and consist of large floating ice shelves, such as the Ross and Filchner Ice Shelves. Ice shelves are an important part of the world's cryosphere. Should they melt, they would not contribute directly to a sea-level rise because they already displace ocean water, but disintegration of ice shelves may nonetheless draw down ice from interior regions, which would lead to a sea-level rise (Scherer et al. 1998).

The smaller Greenland Ice Sheet is one tenth the size of the Antarctic Ice Sheet, about 1.71 million km^2. The equivalent sea-level rise, should the Greenland Ice Sheet melt, would be 7.4 m. Greenland ice also differs from Antarctic ice in the lack of large floating ice shelves and in the fact that it loses ice mass through both calving and surface melting during summer.

The world's mountain glaciers and small ice caps, such as those in Iceland and the Tibetan Plateau, and including small glaciers on Greenland, cover only 640,000 km^2. If they melted, they would cause the sea level to rise 0.3 m.

Earth's cryosphere holds special significance for many aspects of paleoclimatology covered throughout this book. The sheer elevation of Pleistocene ice sheets influenced hemispheric atmospheric circulation. The influence of glacial meltwater on ocean surface salinity and temperature regimes may have caused rapid global climate change. Sea-level history reflects processes that govern the behavior of ice sheets and glaciers, and modern ice sheets serve as archives of earth's atmospheric history preserved in ice cores.

The Atmosphere

The lower three divisions of earth's atmosphere are called the troposphere, stratosphere, and mesosphere (figure 2-2). In the troposphere, daily and seasonal temperature variability controls to a large extent vertical circulation patterns whereas the stratosphere is relatively stable over shorter time scales. Atmosphere circulation, like that of the oceans, is driven by a combination of factors including pressure gradients caused by heating from solar radiation and the Coriolis force from earth's rotation. Atmosphere circulation in a north-south direction is called *meridional flow,* whereas east-west flow is referred to as *zonal flow.* Shifts from meridional to zonal flow have occurred during rapid climatic reorganizations of the late Quaternary deglaciations (chapter 5).

Major features of atmospheric circulation are driven by strong solar insolation at the equator, which causes warm air to rise, then to sink, as it cools at about 30° latitude in both hemispheres (figure 2-2). These rising and sinking air masses form a Hadley cell (Hadley circulation). The rising limb of the Hadley cell is located at the Intertropical Convergence Zone (ITCZ), a narrow zone of maximum rainfall in oceanic equatorial regions that forms at the convergence of northern and southern trade winds and results from warm moist air rising and releasing moisture to the atmosphere. The sinking limb of the cell near 30°N and 30°S latitude is characterized by weak winds and the high pressure system known as the subtropical high. The world's modern desert areas are located in regions of sinking dry air and low precipitation.

Planetary circulation in low latitude subtropical regions is characterized by northeast and southeast trade winds in the Northern and Southern Hemispheres, respectively. These winds form where a portion of the descending air mass in Hadley cells moves equatorward. Another portion moves poleward to form the westerlies, which in mid latitudes are the prevailing winds characterized by strong cyclonic (low pressure) cells. Large-scale undulations in the westerlies causing shifts in storm tracts are known as *Rossby waves,* similar to oceanic Rossby waves that characterize the surface of the oceans.

Ferrel cells occupy the mid latitudes of both hemispheres and include the regions of strong westerly winds associated with western boundary oceanic currents. Polar cells occupy high latitudes, flow in a fashion similar to Hadley cells, and produce high-pressure systems in polar regions.

Many key atmospheric features are encountered in paleoclimate studies discussed in later chapters. The seasonal rainfall anomalies associated with the ITCZ are particularly important in paleoclimate research on interannual climate change related to the ENSO. Over tens of thousands years, the ITCZ varies in relation to equatorial oceanic currents. This variability is related to climate cycles associated with orbital variations.

Atmospheric circulation changes are also critical in past climate changes. Tropospheric and stratospheric circulation changes are particularly relevant in climate studies that use geological records of airborne particles (dust). These particles may either be blown off continents or consist of sea salt and may be found in oceanic sediments or in polar and high-elevation ice cores. Periods of elevated eolian sedimentation in deep-sea cores, such as occurs during glacial episodes, can reflect greater continental aridity. Although large-scale reorganization of atmospheric circulation and changes in wind strength and direction also occur during past climate change, greater eolian deposition in deep-sea sediments does not necessarily imply stronger wind strength, a misconception pointed out by Rea (1994). Volcanic particulate material from volcanic eruptions is also transported in the troposphere and stratosphere and is used as an important age-dating tool in ice-core studies.

Atmospheric trace gas chemistry is a critical and rapidly growing sector of paleo-atmospheric research because the radiative properties of water vapor, CO_2, and nitrous oxide (N_2O) make them potential causes of climate change. Paleo-atmospheric chemistry is studied mainly through polar and low-latitude, high-elevation ice cores. Atmospheric concentrations of "greenhouse" gases such as CO_2, CH_4, N_2O, and water vapor vary over time scales ranging from millions to thousands of years and influence short-term (e.g., Lorius et al. 1990; Raynaud et al. 1993; Chappellaz et al. 1993) and long-term climate change (Crowley and North 1991). Moisture transport between the atmosphere and ground also plays an important role in climate change.

The atmosphere also contains many chemically active species, sometimes lumped under the heading of glacio-chemical species. Glacio-chemical paleoclimate records come mainly from ice cores and include nitrates, sulfates, chlorine, and complex aerosols. This burgeoning field of investigation is particularly relevant both to natural atmospheric and climatic variability and to anthropogenic effects on the atmosphere (e.g., Delmas 1993).

The Biosphere

The earth's biosphere occupies a unique niche in paleoclimatology. The biosphere, for example, is a key component in global biogeochemical cycling of carbon, nutrients, and other chemicals, which all influence climate. William Schlesinger's excellent text *Biogeochemistry: A Study of Global Change* (1997) is an authoritative treatment of this field.

Earth's biosphere is, however, subject to processes very different from those that govern atmospheric circulation or glacier flow because it is comprised of living organisms. Processes governing organisms pertain to the fields of organic evolution and ecology and involve both organism-organism and organism-environment interactions. The unique qualities bestowed upon living organisms and the growing importance of many biological processes in paleoclimatological reconstruction have led me to devote an entire chapter to the biological aspects of paleoclimatology. A discussion of the biospheric component of the climate system is postponed until chapter 3.

A Systems Approach

Earth's climate and its interlocking components are often studied using a systems approach (e.g., Lockwood 1979). Systems can be viewed as existing in a stable or unstable state (figure 2-4). A stable system is unlikely to be knocked out of equilibrium easily, whereas an unstable system may be poised for rapid change when triggered by a perturbation. Systems can also oscillate in cyclic or quasicyclic fashion owing to external or internal factors.

Three fundamental concepts about the climate system permeate most of paleoclimatology: climate forcing mechanisms, linear versus nonlinear climatic response to perturbation, and climate feedback loops.

Major Forcing Mechanisms

Two central goals of paleoclimatology are to document patterns of past climate change—that is, the *how* and *when* of climate change; and to infer *why* climate changes. It is therefore important to conceptualize the distinction between reconstructed *patterns* of climate change, which may be exceptionally well-documented by multiple-proxy data in several regions, and the *processes* that may have produced those patterns.

The processes that cause climate to change are both external and

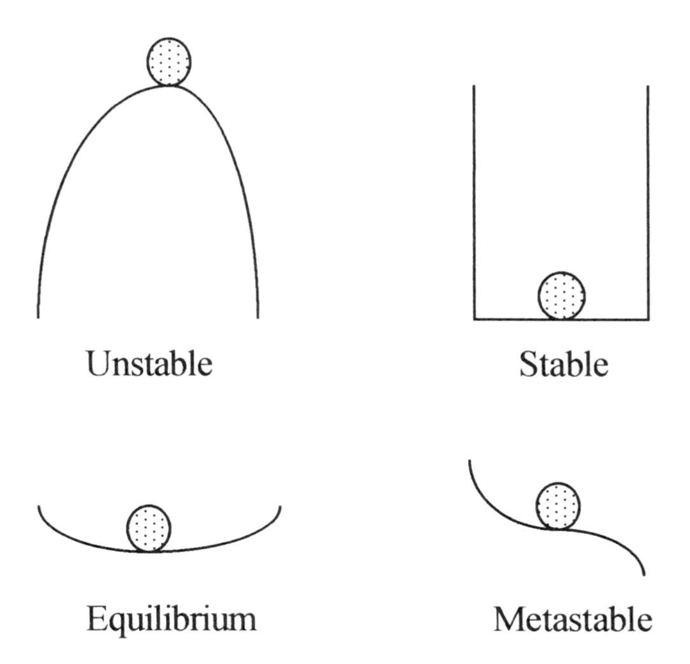

FIGURE 2-4 Schematic depiction of the concepts of stability, instability, and equilibrium as they pertain to the climate system.

internal to the earth (table 2-1). The term *forcing* is often used to refer to these processes. Climate forcing mechanisms external to the earth include solar radiation, changes in earth's orbit, and extraterrestrial phenomena; those internal to the earth include volcanic activity, ocean-atmosphere coupling, and atmospheric greenhouse gas concentrations, among others.

Many forcing mechanisms act simultaneously upon earth's climate

TABLE 2-1 Major Climatic Forcing Mechanisms

- External
  - Solar radiation
    - Sunspot variation and irradiance changes
    - Solar diameter
    - Solar ultraviolet wavelength variability
    - Magnetic variation
  - Earth's orbital changes
    - Eccentricity
    - Obliquity (tilt)
    - Precession of equinoxes
    - Axial precession
  - Asteroid impacts
- Internal
  - Plate tectonics
    - Mountain building
    - Ocean-volume changes
  - Volcanism
    - Aerosols
    - Gases
  - Ice sheets and sea ice
    - Albedo
    - Meltwater
    - Atmospheric circulation
  - Ocean-atmosphere feedbacks
    - Thermohaline circulation
    - Internal dynamics (El Niño–Southern Oscillation)
  - Biosphere-atmosphere gas exhanges
    - Dimethylsulfide (DMS–cloud condensation nuclei)
    - Carbon dioxide (biological pump)
    - Methane (wetlands)
  - Dust (nonvolcanic)
  - Anthropogenic impacts*
    - Carbon dioxide (CO_2), methane (CH_4), nitrous oxide (N_2O), halocarbons, tropospheric nitrogen oxides, carbon monoxide (CO), sulfate aerosols

*See Houghton et al. 1996.

systems, and the challenge is to separate the effects of one factor from the effects of others. Solar energy is the predominant source of energy. It drives much of earth's climate system through its influence on atmospheric circulation and evaporation and precipitation. But the other processes listed in table 2-1 modify the effects of solar energy. Paleoclimatologists often try to isolate the dominant causes of reconstructed climate change by looking for characteristic climate patterns—called "fingerprints" by Schneider (1994) and "footprints" by Rind (1996)—that should, in theory, characterize climate change caused by a particular mechanism. A fingerprint might be a particular spatial or temporal pattern of climate change. Distinguishing between potential causes via fingerprints is illustrated by the following example of three climate forcing mechanisms acting in opposing ways to warm and cool climate.

Earth's mean annual temperature has risen about 0.5°C during the past century (Houghton et al. 1996). Many scientists attribute at least part of this temperature rise to anthropogenic influence of greenhouse gas forcing. Early climate models based on fundamental physical principles called for increasing atmospheric temperatures and greater atmospheric moisture due to the radiative forcing of CO_2, CH_4, and N_2O. However, during the 1980s, the mean annual temperature rise of the past century was shown to be not as great as that predicted from some climate models, nor did the timing of the rise in temperature correlate with the rise of anthropogenic greenhouse gas emissions. In other words, the twentieth century's temperature rise did not have the fingerprint expected from elevated atmospheric CO_2 concentrations. Inconsistencies such as these posed a serious challenge to the use of models for projecting future climate conditions.

One way to explain the subdued twentieth century temperature rise is to hypothesize a second forcing mechanism—another type of anthropogenic pollution, sulfate aerosols—might be counteracting the warming effects of greenhouse gases. Humans have injected considerable quantities of particulate aerosol material into the atmosphere during the past few decades, especially in heavily populated regions. Atmospheric particulates can have a negative forcing on atmospheric temperatures because they shield the surface layers from solar radiation. Twentieth-century aerosols probably have counterbalanced the effects of greenhouse gases, partially explaining the less-than-expected rise in surface temperatures.

To complicate matters, a third forcing mechanism, twentieth-century solar variability, long written off as a viable mechanism of large-scale climate change, has now regained favor as a candidate to explain at least some decadal and centennial climate change. Researchers are now actively seeking to identify the characteristic fingerprints of solar influence on climate. The net change in total solar irradiance over an 11-year sunspot cycle, however, is only about 0.1% (daily mean irradiance ranged from about 1367 to 1369 W/m^2 during solar cycle 21 between 1980 and 1991 [Willson 1997]), and its climatic fingerprint is difficult to identify. Consequently, enormous controversy surrounds the degree to which humans have altered earth's climate with greenhouse gases and sulfate aerosols, as well as how much historical climate change is due to natural solar variability.

Linear–Nonlinear System Response

Climate systems can respond in either a linear or a nonlinear fashion to a particular forcing mechanism. In a linear response, the rate of the climate-system response at any one instant is directly proportional to the magnitude of the input signal. For example, earth's axial tilt oscillates at angles between 22° and 25° over a period of about 41,000 yr, causing changes in the amount of solar radiation striking different latitudes. Other factors being equal, the greater the tilt, the more insolation strikes polar regions and the larger the climatic response. Orbital cycles of tilt are believed to initiate high-latitude climate response in a linear fashion during much of the Pliocene (5 to 1.6 million years ago) (Imbrie et al. 1992).

Conversely, a nonlinear climatic response to a perturbation is reflected in a lagged response. In this case, the climate-system response may be amplified or dampened by additional feedback controls. Nonlinear response to orbital forcing is one hypothesis to explain the predominant 100,000-year climate cycles of the past 600,000 yr, which cannot be explained solely on the basis of radiative forcing from changes in solar input due to orbital factors (Imbrie et al. 1993a). Pisias et al. (1990) and Hagelberg and Pisias (1990), for example, inferred a nonlinear response to orbital changes on the basis of Quaternary radiolarian assemblages from the eastern Equatorial Pacific Ocean. Their studies supported the idea of a separation of high- and low-latitude climatic response to orbital changes.

Distinguishing between linear and nonlinear types of climate change requires information about the response time of each component of the climate system. Table 2-2 lists how long different earth systems take to regain equilibrium after a perturbation (McGuffie and Henderson-Sellers 1997). Climate modelers construct their models using these response times. Paleoclimatologists apply them to inter-

TABLE 2-2 Equilibrium Times for Climate Component Subsystems

Subsystem	*Time*
Atmosphere (free)	11 days
Atmosphere (boundary)	24 days
Oceans—mixed layer	7–8 years
Oceans—deep	300 years
Sea ice	Days to centuries
Continents—lakes and rivers	11 days
Soil vegetation	11 days
Snow, surface ice	24 hours
Mountain glaciers	300 years
Ice-sheet decay	1000–10,000 years
Ice-sheet growth	10,000 to 100,000 years
Crustal isostatic adjustment	2000 to 10,000 years
Mantle convection	30,000,000 years

See McGuffie and Henderson-Sellers 1997.

preting reconstructed climatic trends. In an excellent review of climate models, Taylor (1994) lumps these subsystems into two categories, those with fast physics (i.e., atmospheric circulation) and those with slow physics (deep ocean circulation, ice sheet growth). The time scales given in table 2-2 point to why paleoclimatologists require accurate chronological tools to identify the response of different parts of the climate system.

Climate Feedback Loops

The concept of feedbacks within the climate system reflects the idea that certain processes can either a dampen or amplify climate change initiated by an initial event. Positive feedback occurs when the system is pushed in the same direction as the initial forcing, negative feedback when the shift is dampened, returned toward the original state, or both (Santer et al. 1993).

One example of positive feedback is the ice-albedo feedback. Albedo is the ratio of reflected to incident radiation. Earth's surfaces have different albedos ranging from the very reflective ice-sheet and sea-ice surfaces to the radiation-absorbing forested-land surfaces. Because of its high albedo, ice reflects back to space a greater proportion of incoming solar radiation than most earth surfaces. A decrease in solar radiation initiated by orbital changes may initiate a possible feedback in which more snow remains late in the spring, reflecting more radiation. Other factors being equal, the more radiation reflected back to space, the less is absorbed by earth's surface and the colder a region becomes, in theory allowing more extensive snow cover and eventually ice-sheet growth. Cooling may also be amplified by land-surface vegetation changes associated with a climate-induced shift from taiga to the more reflective surface of a tundra environment.

A second example of feedback is the water vapor–greenhouse feedback loop. Here, as atmospheric temperatures rise during climatic warming, evaporation from oceans and surface water increases and the amount of atmospheric water vapor increases. Water vapor is the largest natural greenhouse gas because its molecules absorb long-wave radiation reemitted from earth's surface. In the absence of other stabilizing controls, a climate would progressively become warmer if this feedback continued. Likewise, if climatic cooling were initiated, atmospheric water vapor would decrease, leading to diminished absorption of reemitted radiation and further climate cooling.

Feedback loops do not operate in a vacuum; they coexist with other complex processes occurring in the climate system and tend to counterbalance any runaway climatic cooling or warming. Recognizing the interdependence of different processes operating within or upon the climate system is a basic tenet of applied paleoclimatology.

"TIME IS OF THE ESSENCE"

THE DIMENSION OF time is the second unifying principle underlying the reconstruction of past climates. Of the slogans about time and earth history dear to geologists and paleontologists, the uniformitarian mantra—"the present is the key to the past"—is perhaps the most famous. As discussed in chapter 1, a uniformitarian view of earth history holds that changes occur in a gradual, steady series of steps, by "evolution" rather than by "revolution," governed by natural processes, such as volcanic activity, that we observe today.

Uniformitarianism, however, is no longer the sole unifying theme in geology. Many phenomena preserved in the geologic record—from evolutionary radiations and speciation to asteroid impacts (Berggren and Van Couvering 1984; Berggren in press)—imbue the geologic record with a more catastrophic character.

As have geoscientists, paleoclimatologists have abandoned a strict uniformitarianism view of climate change as evidence accumulates from ice cores, ocean, and lake sediments showing the catastrophic nature of rapid climate change. Evidence from calendar-year chronologies and decadal-scale paleoclimatic events provides a whole new dimension to "neocatastrophism" and a stark portrait of the dynamic and sometimes fickle nature of climate change. Webb et al. (1993:52) offer a "revised" version of uniformitarianism to express paleoclimatology's role in the study of future climate: "The past is *a,* but not necessarily *the,* key to understanding the present and the future." Wallace S. Broecker has called rapid climate changes of the past "surprises" in the climate system, and they are the subject of intense study. These characterizations carry an important and very authentic message about the practical contribution paleoclimate data offers to understanding future climate change.

One basic premise for reconstructing climate change from the geological record is that the record of climate derived from historical accounts and instrumental records is insufficient for a full understanding of climate change. Why is this so?

Humans have been on earth a very short time in geological terms, and they have kept records of climate-related phenomena a brief few centuries or so. How far back do historical climate records take us? The answer to this question depends on the type of record. Hubert Lamb, the acknowledged forebear of the study of climate change and cultural history (1977, 1995) draws on a wealth of historical accounts to identify many cultural events, tracing them as far back as several millennia, that he associates with climate changes. Ladurie's (1971) monograph on Europe's 1000-year history of climate and cultural evolution is a testament to human observation of natural climate-related events. Grove's (1988) treatise on the Little Ice Age provides extensive historical and glaciological documentation of glacier advances throughout the world during the thirteenth through nineteenth centuries. Quinn et al. (1987) mine the documentary record of colonial South America to reconstruct a multicentury chronology of ENSO events in the eastern Pacific Ocean, later extending the El Niño chronology to records of Nile floods kept by ancient Egyptian cultures (Quinn and Neal 1992). Other recent volumes (e.g., Bradley and Jones 1992; Jones et al. 1996) cite an immense amount of climate data derived from historical records preserved especially in Europe and Asia for the last five centuries and occasionally farther back in time.

Paleoclimatologists, however, have the unique tools to complement measured and observed records of the past few millennia with paleoclimate data from regions where no records were kept and to extend historical records far beyond the records of satellite measurement, surface thermometers, and historical archives. Moreover, during the past decade or so, the need to understand twentieth-century climate change has spurred a conscious effort to splice historical and instrumental records with paleoclimate records reconstructed from corals, tree rings, and high rapidly accumulating sediments (e.g., Hughes and Diaz 1994). As with paleoclimate studies of longer time scales, these efforts require a firm basis in chronology, which begins with earth's geological time scale.

Dating and Correlation of Climate Change

The Geological Time Scale

All paleoclimatology is firmly rooted in the standard geologic time scale accepted by most geologists who study earth history. Table 2-3 lists the major subdivisions of the geological stratigraphic column. Throughout most of this book, the focus is on climates of the Cenozoic—the last 66 million years of earth history—especially the past 5 million years of Pliocene and Quaternary history. References will also be made to long-term climate history, which are described in general surveys of pre-Cenozoic climate history (Crowley and North 1991; Frakes et al. 1992).

Our discussion of chronology begins by defining the conventions used to designate the age of climate events. The literature contains many, often confusing, ways to cite age, and there is no universally accepted terminology. Most researchers, however, use one form or another of the abbreviations and conventions listed in table 2-4. Henceforth, I will use *ka* to refer to thousands of years before present (B.P.) and thousands of years duration, *Ma* for millions of years ago and millions of years duration, and *yr* for years, unless otherwise stated.

The accuracy of a geological age depends on the quality of the dated material and method used to date it. The convention in the primary literature is to give the analytical errors associated with dates obtained through various radiometric dating techniques. The case of radiocarbon dating, a primary method to date material less than about 40 ka in age, illustrates the complexities of geological age-dating. A "raw," uncorrected radiocarbon age obtained on carbon-bearing material (e.g., fossil marine shells or peat) of 11,250 ± 250 yr means the dated material is somewhere between 11,500 and 11,000 radiocarbon years old, the uncertainty stemming from errors in counting random events. The error is usually equivalent to 1 or 2 standard deviations (σ, or sigma).

TABLE 2-3 Geological Time Scale

Eon	*Era*	*Period*	*Epoch*	*Age (in millions of years)*
Phanerozoic				
	Cenozoic			
		Quaternary	Holocene	0.01–0
			Pleistocene	1.6–0.01
		Tertiary	Pliocene	5.3–1.6
			Miocene	23.7–5.3
			Oligocene	36.6–23.7
			Eocene	57.8–36.6
			Paleocene	66.4–57.8
	Mesozoic			
		Cretaceous		144–66.4
		Jurassic		208–144
		Triassic		245–208
	Paleozoic			
		Permian		286–245
		Carboniferous (Mississippian & Pennsylvanian)		360–286
		Devonian		408–360
		Silurian		438–408
		Ordovician		505–438
		Cambrian		570–505
Precambrian				4600–570

Note: See Berggren (1998) for discussion of history of Cenozoic chronostratigraphy.
Source: Skinner and Porter 1995.

An uncorrected radiocarbon age, however, does not record calendar years before present because it does not take into account factors such as correction for oceanic reservoir, changes in cosmogenic production of radiocarbon, or changes in carbon cycling (see Stuiver and Reimer 1993 and later). Thus, it is critical to understand the sources of age uncertainty in radiometric dating in order to establish whether climate trends from different regions are synchronous or asynchronous, and, ultimately, to infer causes and mechanisms of these climate change.

A second point about absolute calendar years versus radiometric years needs clarification. Advances in late Quaternary chronology described later allow near–calendar year chronology back to climate events more than 10,000 yr ago. Some authors cite age in ^{14}C ka in

TABLE 2-4 Chronological Conventions and Abbreviations Used in Paleoclimatology

yr = years
kyr = kiloyears = thousands of calendar (sidereal) years duration
kyr BP (or B.P.) kiloyears before present*
^{14}C ka = radiocarbon kiloyears before present (not calibrated)*
ka† = kiloannum = thousands of years before present
Ma = millions of years before present

*Before present usually refers to years before 1950 A.D.
†Used throughout this book (e.g., 5 ka means 5000 years before present), except where precision is required, in which cases yr BP is used (e.g., 4,678 yr BP).

their papers; others use standard calibration equations (Stuiver 1993) to convert ^{14}C ages into calibrated ages. Because of rapidly improving chronologies for late Quaternary climate studies, "ka" will be used to refer to "calibrated" year ages in later discussions of climate events occurring over the past 15,000 yr, unless otherwise indicated by ^{14}C ka.

Paleoclimatology uses many techniques to date and correlate climate history: radiometric dating, magnetostratigraphy and biostratigraphy, astronomical "tuning," annual-layer counting, among others. In this overview, I synthesize those major dating and correlation tools that pertain most to paleoclimatology.

Radiometric Dating

Several naturally occurring, unstable isotopes of certain elements are frequently used to determine the age of climate-related phenomena. The age of geological material can be computed from the ratio of parent to daughter isotopes using the radioisotopes' half-life, the time it takes for half the original number of atoms to decay. Radioactive isotopic decay occurs when an isotope of an element undergoes spontaneous emission of either electromagnetic radiation or particles. Two radioactive decay processes are *alpha particle emission,* in which a nucleus having two protons and two neutrons is emitted (reducing the atomic number by 2), and *beta emission,* in which a neutron is converted to a proton by electron emission (increasing the atomic number by 1).

Uranium-series age dating involves one of the most important radioactive decay series used in paleoclimatology: the progressive decay of uranium to thorium to lead (see Schwarcz 1989). This decay series results in the conversion of ^{238}U to ^{206}Pb (lead), which is useful over time scales of 10 Ma to 4600 Ma. U-series dating involving the intermediate step in the uranium decay series in which ^{234}U is converted to ^{230}Th is a commonly used technique to date Quaternary corals younger than about 400 ka. Coral take up uranium from seawater when they build their skeletons. Mollusks, teeth, carbonates, and phosphates are also used for uranium-series dating. The criteria used to determine if material is suitable for dating by the uranium-series method vary among researchers. Generally three main criteria are involved: the material formed at a discrete time; the radioisotope parent-daughter ratio was "reset" at the time of formation; and the system has remained a closed chemical system, or at least any alteration processes are well understood.

Reef-building hermatypic corals meet these criteria more often than most other material, and U-series–dated corals are used extensively in climate and sea-level research. The corals should retain 100% of their original aragonitic skeleton; even a small percentage of calcite would suggest that the coral has been altered and the uranium-series age might be suspect. Also important are tests, such as measuring the concentration of the isotope protactinium, to determine if postmortem chemical uptake of impurities has occurred.

Potassium-argon dating ($^{40}K/^{40}Ar$ is one type) is another common radiometric method used to date older events in the range of 50 ka to several billion years old. One important application of K/Ar dating has been to calibrate paleomagnetic polarity reversals (see below), which are part of the foundation of the standard geological time scale used in paleoclimate studies.

Radioactive isotopes produced through cosmogenic bombardment of earth's atmosphere are also key dating tools in paleoclimatology. The three most important cosmogenic isotopes are those of the elements carbon (^{14}C, radiocarbon), beryllium (^{10}Be), and chlorine (^{36}Cl). Primordial carbon occurs in two isotopic forms ^{12}C and ^{13}C, with a ratio of $^{13}C/^{12}C$ atoms of about 1.12×10^{-2}. Isotopes of carbon are also produced in the atmosphere by the reaction of nitrogen with cosmogenically formed neutrons in the reaction ^{14}N + neutron = ^{14}C. According to Stuiver (1994), about 100 kg of ^{14}C is produced annually, and the $^{14}C/^{12}C$ ratio is about 10^{-12}. These carbon atoms form carbon monoxide, then CO_2 molecules, and eventually are taken up into the global carbon cycle through various pathways, including biological routes such as photosynthesis, precipitation by carbonate-secreting organisms, and incorporation into other organic forms. Biogenic material is often the source material for radiocarbon dating, which is the most widely used method to date climate events back to about 40 ka. Organisms take up $^{14}C/^{12}C$ in proportion to the ratio in the atmosphere; upon death, the ^{14}C begins to decay to ^{14}N and the ratio of $^{14}C/^{12}C$ in the dated material is used to compute its age, relative to 1950 A.D., using the half-life of 5730 ka (some researchers use a slightly different half-life for radiocarbon).

Radiocarbon dates are usually corrected first for the isotopic composition of the source material (Stuiver and Polach 1977). Then another correction is performed by applying the model of radiocarbon production and atmospheric and oceanic carbon cycling over the past 15 ka developed by Stuiver and Braziunas (1993). For marine shells, the radiocarbon age can be > 400 yr older than the actual calendar year age owing to the reservoir effect of carbon storage in the deep sea.

Beryllium production in the atmosphere is influenced mainly by solar wind and is a radioisotopic tool used in dating ice cores. ^{10}Be differs from cosmogenic carbon in that it is not taken up by biological systems, but its pathways are complicated by the way it cycles through the earth's atmosphere. One formed, beryllium has a short residence time of only 1–2 yr, during which it is subjected to modulation by climate factors. For example, because ^{10}Be attaches to aerosol particles, it is subject to atmospheric mixing and transport processes. Beer et al. (1988, 1996) studied these factors controlling ^{10}Be deposition by examining ^{10}Be trends in several polar ice cores. They determined that atmospheric production of beryllium was the dominant factor in controlling ^{10}Be concentrations in polar ice. Thus, beryllium concentrations in polar ice or lacustrine sediments can be a source of information on solar activity going back tens of thousands of years.

Cosmogenic radioactive isotopes do not necessarily form at continuous rates in the atmosphere over extended periods of time. Solar and earth geophysical processes, some of which represent forcing agents of climate change, influence the rate of formation. This obviously complicates the use of such isotopes as a dating tool. Indeed, the study of cosmogenic isotopes serves as both a dating and correlation method, while the isotopes themselves are a proxy of solar activity, two inseparable aspects of this field of research (e.g., Stuiver and Braziunas 1993). Recently, solar variability has entered the limelight as a potential cause of high-frequency climate change over the past millennium (e.g., Nesme-Ribes 1994; Lean et al. 1992; Hoyt and Schatten 1997; see also chapter 6).

Magnetostratigraphy and Biostratigraphy

Magnetostratigraphy is the application of reversals in the earth's magnetic field to the correlation of sediments and other rocks. Because the earth's magnetic polarity occasionally reverses itself quickly, the magnetic polarity of minerals in rock sequences preserves magnetic polarity reversals as near-instantaneous time horizons that can be used to correlate rocks—and climate history—globally. A standard magnetostratigraphic scheme is used in most paleoclimate studies of Cenozoic sediments. The period since about 780 ka is called the Brunhes (normal polarity) Chron, a period when earth's magnetic inclination was as it is today. Progressively older magnetic epochs are dated approximately as follows (see Lourens et al. 1996): Matuyama Chron, reversed polarity, 2.58 Ma–780 ka; Gauss Chron, normal polarity, 3.60–2.58 Ma; Gilbert, reversed, 5.9–3.6 Ma). Brief magnetic reversals occur within the Matuyama (Jaramillo, 1.07–0.99 Ma; Olduvai, 1.94–1.79 Ma, Reunion, 2.15–2.13 Ma), in the Gauss (Kaena, 3.12–3.03 Ma, Mammoth, 3.33–3.21 Ma), and in the Gilbert (Cochiti, 4.30–4.19 Ma, Nunivak, 4.63–4.49 Ma, Sidujfjall, 4.9–4.8 Ma, Thvera, 5.24–5.0 Ma).

The ages of magnetic reversals are usually estimated on the basis of radiometric (e.g., K/Ar method) dating of rocks containing the magnetic signal and sea-floor spreading magnetic anomalies (Cande and Kent 1992). More recently, the ages of magnetic boundaries have been revised using the astronomical tuning method described later. Berggren et al. (1995) describe the fascinating historical development of the current standard astronomical-geomagnetic time scale for the Cenozoic Era.

Other magnetic properties of rocks aid in correlating climate events and identifying patterns of climate change. The earth's magnetic pole wanders, and paleomagnetic studies document this by measuring changing inclination and declination. Magnetic susceptibility refers to the degree to which various sedimentary lithologies take up magnetic "imprints," which is a function of the physical properties of the sediment. Oscillations in magnetic susceptibility provide a convenient means to measure cyclic changes in the physical nature of sediment that often can reflect climatic factors.

Biostratigraphy, the use of distinct fossil species or assemblages characteristic of different time intervals to date and correlate sediments, is usually united with magnetostratigraphy to produce an integrated time scale for paleoclimate research. The age resolution obtainable using biostratigraphic data depends on the fossil group, the climate zone, sedimentation and other factors. The first (evolutionary) or last (extinction) stratigraphic appearances of many marine microfossil species have been carefully calibrated to magnetic and astronomically tuned time scales for the Cenozoic and, in a continuous sedimentary sequence, can yield highly precise age estimates within a few tens of thousands of years for a particular sedimentary horizon.

Astronomical "Tuning"

"Tuning" represents a novel and somewhat controversial method to date and correlate paleoclimatic events related to the orbital theory of climate change. This theory holds that gravitational influences on earth's orbital eccentricity, tilt, and precession affect seasonal and geographical distribution of solar radiation and global climate. Geologists have for more than a century tried to establish whether the timing of orbital cycles as determined by celestial mechanics (e.g., Berger et al. 1984, Berger and Loutre 1994) coincide with events in the geological record of climate change, such as the waxing and waning of ice sheets. Geological dating of climate events through radiometric means, however, is not as accurate as the calculations by astronomers of the history of earth's orbital cycles of eccentricity, tilt, and precession. For example, Berger (1984) calculated that the accuracy of the astronomical calculations of the frequency of orbital cycles over the past 5 Ma ranges from 1% for the precession cycle, < 0.01% for the tilt cycle, and about 3% for the eccentricity cycle of earth's orbit. Such precision cannot be achieved with conventional radiometric techniques.

Tuning the geological record of climate change to the astronomical time scale of orbital cycles evolved as a formal step in a general procedure to test the orbital theory of climate. This procedure, pioneered by John Imbrie and colleagues (Imbrie et al. 1984, 1992), made use of deep-sea oxygen isotope records. After careful stratigraphic analysis of deep-sea cores, Imbrie and colleagues linked the continuous oxygen isotope record to radiometrically dated tie points (e.g., the ^{14}C-dated last glacial period, 18–21 ka; the last interglacial maximum, 127 ka; and the Brunhes-Matuyama and Jaramillo paleomagnetic boundaries). They then "tuned" the oxygen isotopic fluctuations in an iterative fashion using the frequencies of the two orbital parameters, precession and tilt (22 and 41 ka, respectively). When they observed a phase difference in the isotope curve, they made an adjustment so the curve "fit" the astronomical time scale of orbital changes. Martinson et al. (1987) extended the tuning of the deep-sea oxygen isotope record to establish the SPECMAP (spectral mapping) time scale, which is the standard time scale used in most late Quaternary climate studies.

Why is tuning the geological time scale so useful? Shackleton et al. (1995b) gave three reasons for its utility as pertains to paleoclimate research: (1) tuning allows the study of leads and lags among various climate proxies; (2) it yields an estimate of the amount of time represented within a sedimentary section; and (3) it provides a more accurate geologic time scale than that provided by traditional radiometric dating methods.

Tuning involves certain assumptions, however. For a linear climate response, one assumes that the phase relationship between the climate forcing (i.e., obliquity) and the system response (i.e., the proxy record) is both constant and known (see Hagelberg and Pisias 1990). Tuning is also complicated by sedimentation rates that do not remain constant through the studied interval, which may impart a serious bias into the chronology. Another basic assumption is the that observed climate cycles are indeed the products of orbitally induced changes in solar insolation. This assumption gives tuning an air of methodological circularity and raises the issue of whether tuning might eliminate some of the real climate signal. Tuning also poses problems when radiometric ages are obtained that conflict with astronomically tuned ages. Such is the case with the Devils Hole vein calcite isotope record (Winograd et al. 1988; Ludwig et al. 1993).

Despite these assumptions, tuning has provided a means to test the accuracy of radiometric ages of climatic sequences. For example, Hilgen and Langereis (1989) and Hilgen (1987, 1991a,b) dated Pliocene sedimentary cycles in southern Italy using biostratigraphy and the standard magnetic time scale. But by using the standard ages for magnetic boundaries, they discovered that the periodicity of climatic cycles were at odds with those hypothesized to be the result of orbital forcing. By "tuning" the age to the astronomical time scale, they achieved an excellent match and a new age model for the Pliocene.

The astronomically tuned time scale has since been extended back to about 3.0 Ma by Ruddiman et al. (1986; 1989), Raymo et al. (1989) and Shackleton et al. (1990) and through the early Pliocene to 5.0 Ma by Shackleton et al. (1995a) in the Pacific; Hilgen and Langereis (1989), Hilgen (1991a,b), and Lourens et al. (1996) in the Mediterranean; and Tiedemann et al. (1994) in the Atlantic. Figure 2-5 shows an integrated time scale for the past 6 Ma based on multiple geochronological data and astronomical tuning. Astronomical tuning of some Miocene sequences has also begun (e.g., Krijgsman et al. 1995). These and other

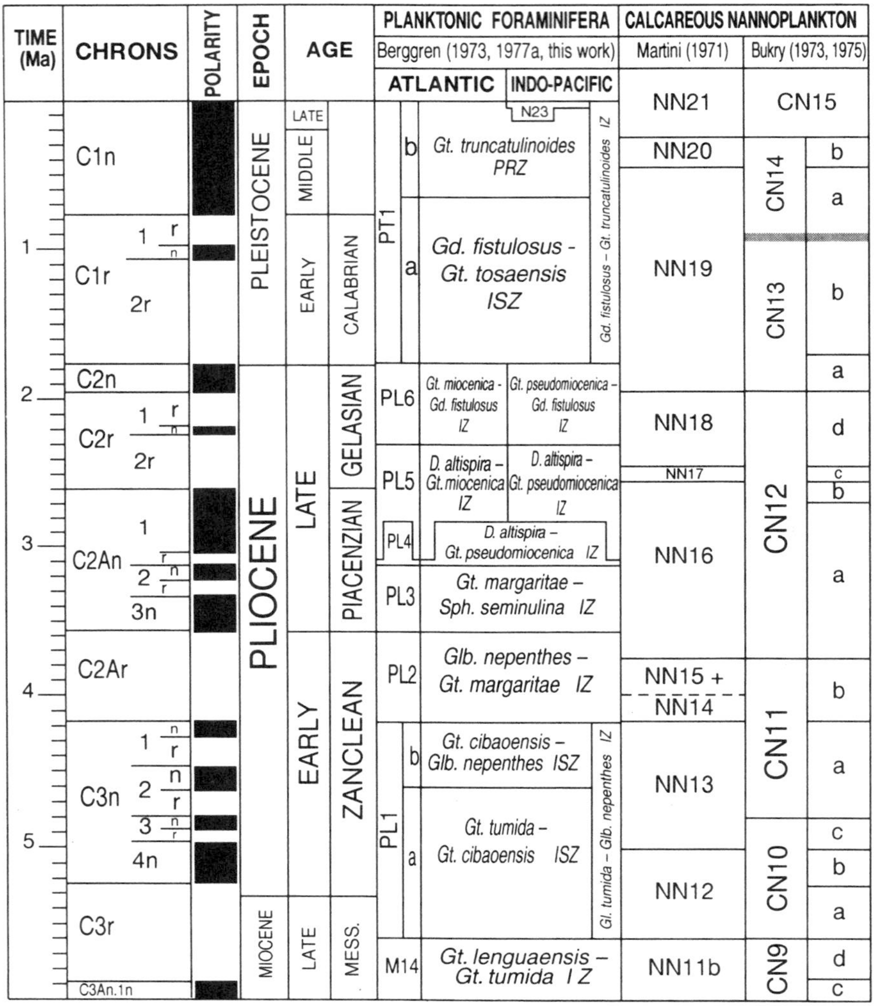

FIGURE 2-5 Integrated time scale for the Pliocene and Pleistocene developed from stratigraphic, paleomagnetic, and astronomical tuning of the geological record based on the ages of past changes in solar insolation due to changes in orbital geometry. K, R, O signify the Kaena, Olduvai, and Reunion paleomagnetic events. From Berggren et al. (1995). Courtesy of Bill Berggren and Geological Society of America.

studies have culminated in an integrated time scale for the Neocene, astronomically tuned back > 5 Ma (Berggren et al. 1995). Future revisions are likely.

In sum, tuning should be viewed like other geological age schemes, as a model subject to revision and refinement. As improvements in radiometric dating are made, it will be necessary to continually reevaluate astronomically tuned time scales.

Annual-Layer Counting

In tuning, cycles of sedimentary, micropaleontological, or isotopic data are counted, and then the number of cycles are helpful in dating and correlating the climatic sequence. So too can simple counts of annually formed features be used to date climate events. The most useful annually resolved climate archives are tree rings (in dendrochronology), coral growth bands (sclerochronology; see figure 2-6), mollusk shell growth, layers of ice (cryochronology), and varved sediments. Varved sediments include three main types: annual glacial varves deposited near receding glaciers, biogenic laminated lake sediments formed from seasonal variations in productivity and sedimentation, and laminated marine sediments often deposited in anoxic basins. The layer-counting approach is simple—count annual layers back from the present, or from a known event (i.e., the death of a coral colony or a tree). To extend the record beyond the living tree, or reef, annual layers can sometimes be spliced together using marker events—i.e., unique varve or tree ring layers—to tie one record to another and obtain a longer, a continuous record.

Other Dating Methods

Other dating techniques applied to date soils, loess, and tephras are described in many books and in the pages of journals such as *Quaternary Chronology.* Methods such as amino acid racemization, optically stimulated luminescence (OSL), thermoluminescence (TL), and electron spin resonance (ESR) are usually applied to the Quaternary dating of material older than 35 ka, and younger than about 1 Ma, an interval when radiocarbon or U-series methods are inapplicable. TL dating of beach sands and loess, for example, is relatively accurate back to about 800 ka (Wintle 1990; Berger 1995). These methods are often useful in geological situations where there is a relatively incomplete stratigraphic record, such as sediments deposited in caves or isolated exposures.

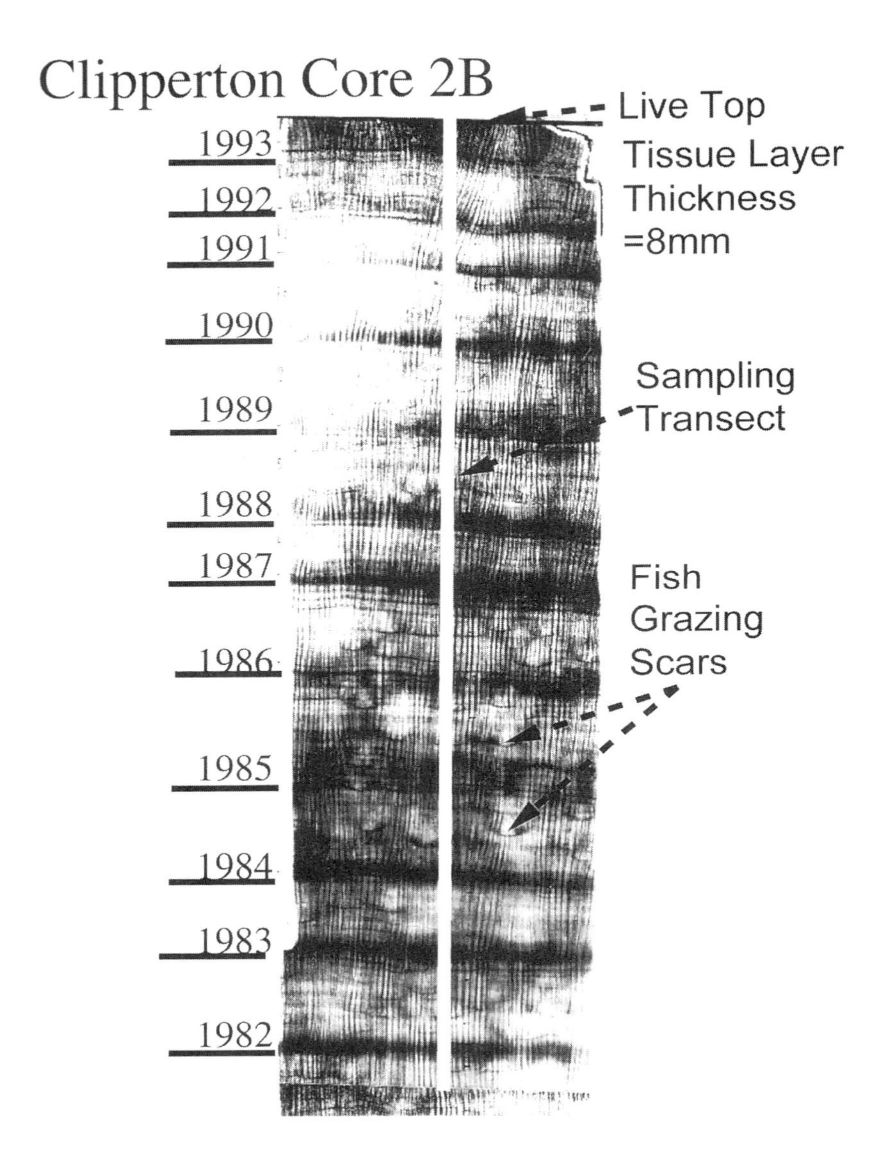

FIGURE 2-6 Annual growth bands in the skeleton of tropical coral colonies reflect seasonal growth variability and are one means to obtain annual and subannual chronological resolution in paleoclimatology. Courtesy of Braddock Linsley.

Floating Time Scales

A floating time scale refers to situations in paleoclimatology in which a high-resolution, often annually resolved climate history can be obtained from a period for which the absolute and/or calendar year age remains uncertain. A good example of a floating time scale is found in the study of early Holocene coral geochemistry from the island of Vanuatu in the southwest Pacific Ocean by Beck et al. (1992, 1997). These authors studied strontium-calcium (Sr:Ca) ratios in five coral colonies dated using a thermal ionization mass spectrometry (TIMS) uranium series method at 10,344, 9688, 9509, and 4166 yr B.P., as well as modern coral, to establish the evolution of tropical SSTs during the Holocene. For each period, they used coral growth banding to study five or six annual cycles of temperature oscillations. The precise age of each coral was limited to the age uncertainty associated with the TIMS dates, usually several hundred years, and therefore the coral could not be given a calendar year age. However, the annual cycles of climate history for each period of coral growth could still be determined.

Although Beck et al. could not reconstruct Pacific tropical climate history for the entire period covering the past 10,500 yr, their floating paleoclimate record is still useful to see changes in the annual range of SST variation during several distinct periods of the Holocene. Roulier and Quinn (1995) also determined interannual climate variability about 3 Ma using a floating time scale of a Pliocene coral from Florida. In addition to coral records, floating time scales can be obtained from continental climate records such as annual layers in tree rings (Kromer et al. 1994) and varved lake sediments (Müller 1974).

Age Models

Many paleoclimate studies proceed by first developing a chronology, usually referred to as an age model (figure 2-7). An age model is just that, a model constructed from the various dating and correlation tools described above. Age models for some of the more important paleoclimate records seem to take on lives of their own, evolving as more data accumulate or dating and correlation methods improve. One notable example is the age model for the Vostok ice core from Antarctica, a keystone climate record spanning the past 400 ka. Early studies of Vostok combined isotopic and chemical stratigraphy of the ice to correlate to other climatic records (Lorius et al. 1979). Lorius et

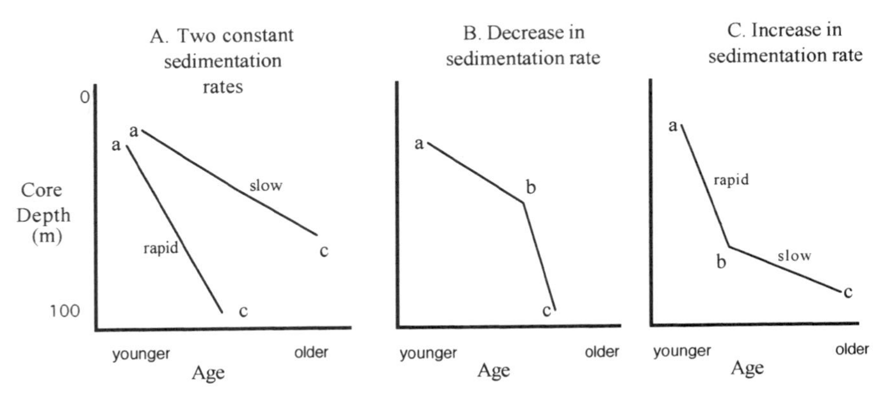

FIGURE 2-7 Sedimentation rate and age-model development in sedimentary sequences having different sedimentation rates. The *y* axis represents depth in a core or an outcrop section; the *x* axis represents age in years. Left: lines a-b-c depict two different constant sedimentation rates. Middle and right: the break in the slope of the line at point b signifies a decrease or increase in sedimentation rate, respectively.

al. (1985) later produced a major revision of the Vostok record back to 150 ka using mainly a two-dimensional glaciological model of how Antarctic ice flowed from its central regions toward the Vostok ice core site. Since then, several revised age models have been proposed based on additional chronological tools (see Barnola et al. 1987, 1991), such as the oxygen isotope stratigraphy of molecular oxygen trapped in air bubbles (Sowers et al. 1991; Bender et al. 1994; Sowers and Bender 1995). The Vostok paleoclimate record is critical for understanding interhemispheric climate change and the role of greenhouse gases in past climate change; its chronology is likely to see continued improvement.

STRATEGIES TO STUDY CLIMATE HISTORY

HOW DO paleoclimatologists capitalize on the fact that their discipline is embodied in the time dimension? Temporal aspects of climate change research can be divided into three distinct, but not mutually exclusive, approaches: *synoptic (time-slice) studies, climate change in the frequency domain (climatic cycles)*, and *time-transient climate change*.

Synoptic (Time-Slice) Studies

Although climate is always changing, paleoclimatology seeks to compare earth's climate as it existed during different extreme endmember climate states. Paleoclimatologists may wish to compare a glacial with an interglacial period, for example. Or, over shorter time scales, they may want to examine the difference between global climate anomalies during an El Niño versus a La Niña event. Assuming that the climate system attains near equilibrium conditions during such climatic extremes, synoptic paleoclimate studies reconstruct climatic boundary conditions during a "time-slice," or a synoptic, period. A synoptic paleoclimate study is then a condensation or abridgment of the climate of a particular time when it was in a distinct state. Later, I use the example of the last glacial maximum (LGM), about 20 ka when most continental ice sheets were near their maximum extent, mean global temperature was coldest, and sea level was lowest, to illustrate the progressive development of proxy techniques.

For any time-slice study, an error margin always stems from dating and correlation uncertainties; as a result, paleoclimatologists use the term *synchronous* to mean "within the limits of dating and correlation." The consequence of dating uncertainty in synoptic studies is that high-frequency climate oscillations that occur within the reconstructed time slice can become obscured and lead to a misinterpretation.

Several multidisciplinary synoptic paleoclimate studies of regional and global climate are benchmark contributions to earth history. Among the most notable studies are those on the last glacial (CLIMAP 1981, Denton and Hughes 1981), the last interglacial (CLIMAP 1984), several intervals during the Holocene (reconstructions at 3,000-yr increments, COHMAP 1988), the Cretaceous (Barron and Washington 1985; Barron et al. 1995), and the Pliocene (Dowsett et al. 1994, 1996; PRISM 1995). Each of these studies involved the compilation of new and existing data from many sources; each provides a glimpse of unique climate regimes that existed in the past.

A final comment is necessary about the use of synoptic paleoclimate reconstructions as analogs of future climates. Budyko (1982) and Zubakov and Borzenkova (1990) have made direct comparisons between periods of global climate warmth in the past, such as the Eocene and the early Pliocene (about 50 and 5–3 Ma, respectively), and potential future climates in terms of elevated atmospheric CO_2 levels. While recognizing the plurality of mechanisms that cause global change, Budyko and colleagues attribute periods of global warmth of the past 100 Ma mainly to levels of atmospheric CO_2 inferred to be higher than current levels. The main thrust of Budyko's analog approach is that rising CO_2 levels will similarly lead to future global warming.

The use of past climates as *direct* analogs of the future has been roundly criticized by Webb et al. (1993) and others (see Schneider 1993). One major criticism is that long-term CO_2 has varied in the earth's atmosphere (Berner 1994; Mora et al. 1996; Cerling and Quade 1993), but at much slower rates than the rapid postindustrial build-up. A second problem is that boundary conditions such as continental positions, continental elevations, and the location of oceanic gateways and deep-water sills were vastly different from what they are today. A third weakness is that stringent age control and qualitative proxy methods were used. In regard to the analog approach, several leading paleoclimatologists (Webb et al. 1993:52) stated: "Our group unanimously rejects such a simplistic application of paleoclimate data but strongly favors the use of such data to check the predictions of climate models, to learn about key climatic processes, and to characterize the behavior of the climate system and its major components during periods of rapid global change."

A more sanguine outlook of paleoclimate reconstructions is taken by Covey (1995), who suggests that, despite the uncertainty associated with age and proxy climate estimations, the signal-to-noise ratio for past global warmth is so great as to make them extremely valuable tools in evaluating processes in the climate system. Regardless of what one calls them, or how one views the value of synoptic climate reconstructions for the study of future climates, they should be carried out under the guiding principles described here.

Climate Change in the Frequency Domain: Climatic Cycles

Earth's climate can also oscillate cyclically between two extreme states. Climatic cycles are identified in paleoclimatology by one or more proxies fluctuating in a wavelike pattern characterized by a distinct frequency. On the crest of the wave, climate is in one state; in the trough it exists in the opposite extreme. Identifying the periodicity in proxy variables helps to establish the causes of climate change. Some of the more widely recognized periodicities are earth's orbital cycles of precession, tilt, and eccentricity (mainly 23, 41, 100, and 400 ka); internal ice sheet dynamics (about 6 ka); various solar cycles (1450, 200, 80–90, 11 yr); and ocean-atmospheric cycles of ENSO (7–3 yr).

In nature, however, climatic cycles rarely if ever progress with perfect regularity. For example, Pacific Ocean SSTs associated with ENSO vary quasicyclically at periods that range from about 3 to 7 yr. The term *quasicyclic climate change* is often applied to situations in which endmember climatic states, such as the El Niño warm phase and the La Niña cool phase of ENSO, occur over irregular cycle lengths. Since its discovery in the nineteenth century by Schwabe, quasicyclic variability has also been known for the "11-yr" sunspot cycle (Hoyt and Schatten 1997).

Studies of the frequency domain of climate change often involve the statistical technique of time series analysis. Time series techniques help identify periodicities in proxy trends. They also help to establish the lead and lag relationships among climate variables, as well as their relationships with forcing mechanisms internal or external to the climate system suspected as the cause of the cycles. Mathematically, time series analysis extends the statistical concepts of mean, variance, and correlation of random variables to a sequence of climate-related variables ordered into a temporal series (Crowley and North 1991). Frequency bands can be identified by plotting power spectra on a peridogram, in which frequencies that explain various amounts of variance peak above random climate noise.

Two questions pertain to the application of time-series analyses to climate change. First, do two distinct climate variables vary coherently, that is at about the same frequency, or at different frequencies? Cross-spectral analysis, which compares the covariance of two variables at a certain frequency band in a fashion analogous to the correlation between two variables, can show whether two variables are coherent. Second, is there a lead or a lag between two variables or between a variable and the initial forcing that might reflect the delayed response of some part of the climate system? Establishing the phasing between two variables yields clues into the temporal relationships between the variables within the framework of a single climate cycle. The phase relationships between climate variables are sometimes expressed using the "phase wheel" convention (Imbrie et al. 1984). A phase wheel depicts a single climatic cycle as a 360° circle; in the case of a 41-ka obliquity (tilt) cycle, 90° represents one quarter of the cycle, or 10,250 yr. In studies of orbital climate change, the phase wheel usually sets global ice volume at 0°, with specific climate events identified at points around the periphery corresponding to their lead ahead or lag behind global ice volume.

Clemens et al. (1996) analyzed the evolution of the Asian monsoon using time series to study orbital influence on tropical climate. They compared the frequency of four climate variables—foraminiferal oxygen isotopes (a measure of global ice volume), percent opal (an indicator of biological productivity), percent *Globigerina bulloides* (a foraminiferal species indicating oceanic upwelling), and lithogenic flux (a measure of continental aridity)—from Ocean Drilling Program deep-sea sediments deposited in the Arabian sea over the past 3.2 Ma. They examined the relationship between these climate variables and low-latitude solar insolation that varied owing to changes in the earth's orbital precession. Precessional insolation variability is calculated independently by astronomical calculations (Berger and Loutre 1991, 1994).

Clemens and colleagues examined the records for coherence, the linear measure of correlation between two time series—i.e., opal and oxygen isotopes—over a given frequency band when the phase difference between the two variables is zero. Then they determined the phasing for the two frequency bands suspected of influencing monsoonal climate, the 41-ka obliquity and 23-ka precessional bands. They found that maxima in African continental aridity (measured by lithogenic grains) were both coherent and in phase with global ice volume (measured by foraminiferal oxygen isotopes). Conversely, monsoonal strength was offset from aridity and ice volume indicators; that is, there was phasing among these different proxies. Moreover, the phasing among climate variables changed several times, at 2.6, 1.2, and 0.6 Ma. These shifts represented periods of global climate transitions associated with progressively larger-amplitude glacial-interglacial cycles, more intense glacial maxima, and larger continental ice volume. The upshot is that as the Asian monsoon evolved over the past few million years, it weakened progressively as Northern Hemispheric ice sheets expanded during each successive glacial period.

Time series analysis has become a common analytical method in paleoclimatology, and my treatment of the subject has been elementary and qualitative. Further detailed discussion of the general methodology can be consulted in Jenkins and Watts (1968) and, as applied to paleoclimate research, in Imbrie et al. (1984) and Crowley and North (1991). Three points regarding time series are important. First, time-series analyses of paleoclimate data involves many choices. As Crowley and North warn, one should first have a hypothesized physical mechanism suspected as the cause of a cyclic climate pattern. Second, other professionals who use time-series analyses, such as financial analysts interested in periodic changes in the Dow Jones Industrial Average, know the precise "age" of the series of events under study. Paleoclimate data obtained from the geological record do not have such a precise chronology. Thus, time series should only be attempted when the sampling density is adequate, the age model is reasonable, and the forcing mechanism and response are suspected. Third, some researchers eschew time series analyses and compare climate records by simple curve matching, basing climate inferences more on their experience and judgment than on statistics.

Time-Transient Climate Change

When climate changes from one equilibrium state to another, due to an external or internal forcing, the resulting climate change often occurs over a geologically short period. Rapid, or abrupt, climate change is a type of *time-transient* climate change that represents one of the most intensely researched areas in paleoclimatology. The goal in these types of study is usually to investigate the transition from one climate state to another, often during a period when the climate system appears to be out of equilibrium.

The need to document and understand rapid time-transient climate change during the late Quaternary has led to the development of increasingly accurate chronological age models. One such rapid climatic event is known as the Younger Dryas climatic reversal and is described in the following section.

THE DEVELOPMENT OF AN ANNUAL-RESOLUTION DEGLACIAL CHRONOLOGY

DURING THE PAST decade the need to understand the nature and causes of abrupt climate change has led to aggressive efforts to develop absolute time scales that would allow land, sea, and atmospheric (ice core) climatic records to be dated by calendar years and thus correlated to each other with great precision. The development of an absolute calendar-year paleoclimate chronology is exemplified by efforts to understand the last time earth's climate shifted from a glacial to an interglacial state, between 21 and 10 ka (in this section, ka refers to calibrated radiocarbon years and ^{14}C ka to uncalibrated radiocarbon years). Paleoclimatologists have taken on the ambitious goal of obtaining calendar-year temporal resolution of climate events that took place thousands of years ago to answer vexing questions about where and at what rate did the climate change and, ultimately, what triggered the changes?

The deglacial history of northern Europe and the Nordic Seas between 15 and 10 ka is an intensely studied interval of climatic change. Early European fossil pollen zones (Jensen 1935; Iversen 1954) represented changes in both vegetation and climate during deglaciation and were used in early studies for correlation of deglacial climatic events. These pollen zones led to a standard chronostratigraphic scheme of climate stages for the period 15–10 ka. The ages of pollen zones were originally determined by the conventional ^{14}C method. The climatic zones were as follows: Oldest Dryas (cool), Bølling (warm), Older Dryas (cool), Allerød (warm), Younger Dryas (cold), and Preboreal (warm early Holocene) (Mangerud et al. 1974). In this chronostratigraphy, the Younger Dryas was a brief, cool stadial climate event that lasted approximately 1000 yr from about 11 to 10 ^{14}C ka. For a generation, this scheme served the paleoclimate community well. However, problems arose as researchers tried to correlate the Younger Dryas pollen sequence of Europe with paleoceanographic changes in the North Atlantic Ocean, pollen sequences in North America, and deglacial climatic changes in the Southern Hemisphere.

As mentioned above, several complications make conventional radiocarbon techniques inadequate for dating climate events with the precision needed to document and correlate rapid climate transitions such as the Younger Dryas. First, a relatively large quantity of material is required, which means that a large interval of a sediment must be "homogenized," blurring the chronology during the key transition. The development of radiocarbon dating by means of accelerator mass spectrometry (AMS) (Bard et al. 1990a,b) solved this problem by allowing paleoclimatologists to date only a few grams of material.

A second complication involves possible changes in the production of ^{14}C in the atmosphere by cosmic rays. Natural changes in ^{14}C production would in theory affect the ratio of ^{14}C in biogenic material such as trees and marine carbonate. Oscillations of ^{14}C due to variable production are especially important in understanding decadal-to-centennial scale climate change.

In the case of millennial-scale changes of the deglacial period, changes in atmospheric ^{14}C concentration due to climate effects on carbon cycling are significant. Several investigations have shown that radiocarbon age does not match absolute, or calendar-year, age in material older than about 9 ka. First, calibration of ^{14}C dates by uranium-thorium TIMS dates on corals (Bard et al. 1990a,b, 1993) showed that, before about 9 ka, U-series and radiocarbon ages in corals did not match. The ^{14}C–U/Th age discrepancy is about 1000 yr near 10 ka, and as much as 3000–3500 yr near 15 ^{14}C ka.

Second, anomalously young ^{14}C dates were obtained from lake sediments in Europe dated independently by counting annually deposited varves (Zbinden et al. 1989). In addition, Swiss lake sequences were dated with AMS by Ammann and Lotter (1989). They recognized two ^{14}C plateaus, one in the later Younger Dryas zone and the other near the Oldest Dryas–Bølling Boundary, which signified a mismatch between radiocarbon and absolute ages.

The immediate causes of the ^{14}C–U/Th coral and varve-^{14}C age discrepancies are very likely changes in global carbon cycling caused by a weakening of ocean thermohaline circulation (Zbinden et al. 1989; Ammann and Lotter 1989; Stuiver and Reimer 1993). Slower deepwater formation in high latitudes reduces the amount of atmospheric ^{14}C taken up and stored in the deep ocean, increasing its concentration in the atmosphere, and hence leading to relatively young ^{14}C dates in plants and corals that uptake carbon (e.g., Broecker and Peng 1982; Stuiver and Reimer 1993).

These complex processes explain why many climate events correlated by conventional radiocarbon dates led to debates in the literature about the synchroneity in various regions of events like the Younger Dryas cooling. Uncorrected ^{14}C dates cannot be used as the sole basis for correlating rapid climate events of the last deglacial interval; independent chronology was needed to calibrate the radiocarbon dates.

Breakthroughs in the development of a revised chronology of climate events have come through the calibration of ^{14}C-dated macrofossil plant material to calendar years using four main sources capable of yielding near-annual resolution: tree rings, lake sediments, glacial varves, and ice cores (table 2-5). German oak tree rings (counted back

TABLE **2-5** Dating the Younger Dryas Climate Reversal

Chronology	*Beginning*	*End*	*Reference*
German Lake Holzmaar	11,940	11,490	Hajdas et al. (1995)
Swiss Lake Soppensee	12,125	10,986	Hajdas et al. (1993)
Polish Lake Gosciaz	12,520	11,440	Goslar et al. (1995)
Swedish varved lakes	12,100	11,440	Stromberg (1994)
German dendrochronology		11,045	Kromer and Becker (1992)
GISP2 Greenland ice core	12,900	11,460	Alley et al. (1993)
GRIP Greenland ice core		11,550	Johnsen et al. (1992)

Sources: Bard et al. 1993; Alley et al. 1993; Wohlfarth 1996.

to about 9971 yr) (Becker 1993) and pine tree rings (counted between about 9800 and 11,597 yr) (Kromer et al. 1994) provided an annual chronology back to about 11.6 ka. Two types of trees were used because European vegetation shifted from pine to oak forests about 9 ka. The tree ring records of oak and pine are spliced together—about 295 tree rings overlap the two records—based on the recognition of a clear climate-induced oscillation near 8.8 ^{14}C ka (Kromer and Becker 1992, 1993).

The second annual chronology comes from lake sediments in Poland and Germany and in the ice-dammed Baltic ice lake. Annually laminated sediments (varves) provide an exceptional means for paleoclimatologists to attain annual resolution. Hajdas et al. (1993, 1995) dated terrestrial macrofossils from Lake Soppensee in Switzerland and Lake Holzmaar in Germany, providing a chronology back to 13.8 ka. Hajdas et al. (1993) dated the Younger Dryas at 12,125–10,986 yr, or a duration totaling 1139 yr. For Lake Holzmaar, they were able to date the beginning and end of the Younger Dryas at 11,940 and 11,490 yr, respectively. In Lake Gosciaz in Poland, the Younger Dryas episode is dated at 12,520–11,440 yr (Goslar et al. 1995). Thus, there is generally a good match at two sites but a young age at the third.

To augment annual layer counting, marker volcanic ash layers (tephras) were used to correlate the European deglacial sequence. At least two well-dated ashes are useful: the Vedde Ash, generally considered to be near the Younger Dryas age of 12.2–11.8 ka, and the Lacher See Ash, deposited at the end of the Allerød warm period about 12.2 ka (Hajdas et al. 1995).

A third annual-resolution chronology is derived from Greenland ice cores, which preserve a climate record of atmospheric temperatures and circulation over the North Atlantic region. Two primary ice cores, the GRIP (Johnsen et al 1992) and GISP2 (Alley et al. 1993) pertain to the Younger Dryas–late glacial chronology, providing absolute age revisions for this key climatic event.

The European deglacial-age relationships are central to understanding the rapid time-transient climatic transition of the last deglaciation, and they are discussed at length by Wohlfarth (1996), who concluded that some Younger Dryas age discrepancies within Europe may simply be due to the way the pollen zones are characterized at the different sites. This means the Lake Soppensee pollen transition into the Younger Dryas may not actually represent the same pollen transition that is recorded elsewhere. She recommends adopting a layer-counting calibration of AMS radiocarbon-dated material and the abandonment of pollen zones at least as a means of regional correlation. Further work will certainly improve the correlations.

From the procedural standpoint of how one develops a paleoclimate chronology, the integrated late glacial chronology obtained from tree-ring, lake, varve, and ice core records yields an unprecedented age model for the Younger Dryas climate reversal. This model is a critical step in establishing the reversal's cause. The cold reversal began about 12.1–12.9 ka and ended 11.4—11.5 ka. These ages are about 1000–1200 yr older than those based on early ^{14}C dates. Although additional research will certainly refine the age estimates in table 2-5, understanding the climate implications and possible causes of the now well-dated Younger Dryas episode, at least in Europe and the North Atlantic, is possible (Björck et al. 1996a).

The development of a late glacial chronology for Europe and the North Atlantic region is a scene played out in most paleoclimate studies. The time scales range from tens to hundreds of thousands of years, millennia, centuries, decades, years, and seasons. As will become clear throughout the later chapters in this book, each time scale requires a different level of temporal resolution and age-model refinement before the causes of climate variability can be evaluated.

PROXIES OF CLIMATE

THE THIRD MAJOR theme of paleoclimatology is the use of proxies of environmental parameters to reconstruct climates of the past. Because paleoclimatologists cannot revisit the past to observe or measure climate parameters, they have developed a diverse arsenal of proxy tools—such as emu eggshell geochemistry—to attack complex problems of climate change. Before examining the classes of proxy tools, I discuss briefly a few of the more important sources of climate data from historical and instrumental records.

Humans have recorded climate history over the past few centuries. Modern climatologists struggle to piece together a continuous and reliable climate history from sources that can be divided into instrumental and phenomenological types of data. Instrumental records are self-evident. The earliest thermometer recordings of temperatures, mainly in the urban centers of Europe and North America, are one example of instrumental records. A few instrumental observations are several centuries old, including telescope observations of solar activity (Schove 1983; Hoyt and Schatten 1997) and the tide gauge measurements of sea-level change available from eighteenth century Europe and, to a lesser extent, from the Americas and a few other isolated places (Emery and Aubrey 1991; Pirazzoli 1991). Periodic flooding of the Nile River was a central part of Egyptian society, and Nile flood records recorded by the "Nilometer" provide a means of evaluating long-term precipitation trends for the Nile Valley (Quinn and Neal 1992). However, the farther back one looks, the murkier the instrumental record becomes and the less reliable the data for interpreting climate trends and establishing causality.

Phenomenological records are slightly different from instrumental records. They are derived mainly from historical accounts of human activities, such as farm records, and of human interactions with the environment. The late-nineteenth century advance of glaciers in the European Alps documented with photographs and etchings is a prime example of historical documentation of climatic events (Grove 1988; Lamb 1995). Sea-ice records around Iceland or fishing records of South American villagers are other examples. Early cultures were most likely unaware that their observations would be used in future climate research. Yet as incomplete and qualitative as phenomenological records may be, they provide an invaluable means to cross-check paleoclimate trends of the past few millennia derived from natural systems such as ice cores or tree rings (Jones et al. 1996). Like instrumental records, however, phenomenological climate records have severe spatial and temporal limitations that impede an understanding of long-term climate history.

The many climate proxies used in paleoclimatology can be classified in several ways: into the components of the climate system they reconstruct (i.e., hydrosphere, lithosphere, atmosphere, etc.), into oceanic versus nonmarine realms (Webb et al. 1993), by taxonomic group (Parrish 1998), or into biological versus nonbiological (e.g., lithological, geochemical) methods (Webb et al. 1993). Another approach is to focus on proxies that delimit climate-related processes such as biological productivity, sea-level change, nutrient fluxes, or terrestrial-atmosphere carbon exchange via photosynthesis.

In table 2-6, climate proxies are listed by the component of the system from which they are derived (continents, oceans, ice, atmosphere) and the climate parameters that are reconstructed from them. This classification highlights the interlocking subsystems of climate and has the advantage of showing the maximum time resolution achievable. It emphasizes the places that researchers go to find climate history.

In table 2-7, proxies are listed by scientific specialty, for example, paleontological or geochemical. This arrangement shows more about

TABLE 2-6 Geological and Biological Sources of Paleoclimate Proxy Data

Source	*Best resolution (yr)*	*Parameters measured**
Continents		
Lake sediments	1–20	Hydrology, *P*, *T*, *S*
Loess (windblown silt)	100	Wind
Groundwater	> 1000	T_a, *P*
Paleosols	100	T_a, *P*
Peats	100–1000	C
Precipitated calcite (caves, speleothems)	1000–2000	T_a, *P*, hydrology
Packrat middens	500–1000	Vegetation, *P*, T_a
Tree rings, cellulose	<1	Seasonal *P*, T_a
Geomorphological features	100–1000	Ice volume, sea level
Oceans		
Ocean sediments	< 100–1000	*S*, T_o, Prod, Nut, aridity, Circ (ocean and atmospheric)
Corals, mollusks	<1	*S*, T_o, Nut
Coral reefs	1000	SL-Ice vol
Marine peats (marsh)	< 1000	SL-Ice Vol
Ice Sheets and Glaciers	< 1–1000	T_a, *P*, wind, Circ, trace gases and biological activity, cosmogenic (solar, geomagnetic) activity, volcanic activity (see chapter 9), atmospheric chemistry, biomass burning, anthropogenic activity

*P = precipitation; T_a = atmospheric temperature; T_o = oceanic temperature; *S* = salinity; C = carbon cycle; hydrology = regional changes in evaporatioin and precipitation; Circ = ocean or atmospheric circulation; SL-Ice Vol = sea-level and/or ice-volume change; Prod = biological productivity; Nut = nutrients.
Source: Bradley 1989, White et al. 1993)

what paleoclimatologists measure, such as the proportions of species in fossil assemblages (census data) and isotopic ratios, for example. The emphasis in this classification is on first principles of chemistry or biology applied to paleoclimatology rather than on components of the climate system.

Many crosscutting relationships exist between the two systems, and neither is preferred over the other. Isotopic fractionation, for

TABLE 2-7 A Selected Taxonomy of Paleoclimate Proxy Methods

Category	*Proxy*	*References**
Paleontological		
	Planktonic foraminifers	Be 1977; Erez and Honjo 1981; Curry et al. 1993; Spero et al. 1991; Tolderlund and Be 1971
	Benthic foraminifers	Corliss 1985, 1991; Gooday 1986, 1988, 1993; Gooday et al. 1992; Gooday and Turley 1990; Mackensen et al. 1994; Scott 1978 (Marsh Foraminifera)
	Radiolaria	Lozano and Hays 1976; Morley 1989; Moore 1973
	Diatoms	*Marine:* Fry 1996; Koç and Jansen 1994; Leventer et al. 1996; Sancetta 1992
		Nonmarine: Laird et al. 1996
	Dinoflagellates	Edwards et al. 1991; Versteeg 1994, 1996
	Ostracodes	*Marine:* Cronin 1988; Cronin and Dowsett 1991; Dwyer et al. 1995; Hazel 1970
		Nonmarine: Chivas et al. 1986, 1993; Curtis and Hodell 1993; De Deckker et al. 1988; **Holmes 1994**
	Coccolithophorids	Molfino et al. 1982; Molfino and McIntyre 1990
	Beetles	Atkinson et al. 1987; Coope 1977; **Elias 1994**
Geochemical		
	Stable isotopes	
	Carbon	**Curry et al. 1988**; Rau et al. 1989
	Oxygen	**Berger 1979;** Duplessy et al. 1981; Emiliani 1955; Labeyrie et al. 1987; Shackleton 1967; Shackleton and Opdyke 1973; Urey et al. 1951
	Deuterium	**Grootes 1995**
	Nitrogen	Verardo et al. 1990
	Trace elements—tracers	
	Magnesium	Cronin et al. 1996; Dwyer et al. 2995
	Strontium	Beck et al. 1992; **Guilderson et al. 1994**
	Cadmium	Boyle 1988a, **1992**
	Barium	Lea and Boyle 19990

continued

TABLE 2-7 *(continued)*

Category	*Proxy*	*References**
	Carbon dioxide (atmospheric)	Barnola et al. 1991; Paytan et al. 1993; Raynaud et al. 1993; **Schwander 1989**
	Methane (terrestrial)	Chappellaz et al. 1993
	Alkenone biomarkers	Brassell et al. 1986; Prahl and Wakeham 1987
	Stromatolites	Casanova and Hillaire-Marcel 1993
	Geochemistry (see chapter 9)	Delmas 1995; **Hammer 1989;** Mayewski et al.
	Tree cellulose	Stuiver and Braziunas 1987
	Cosmogenic isotopes	
	^{10}Be	Raisbeck et al. 1987
	^{14}C	Stuiver 1965; Jirikowic et al. 1993 (tree rings)
Sedimentological		
	Dust in deep-sea sediments	**Leinen and Sarnthien 1989,** Pye 1987, Rea 1994; Rea and Janacek 1985
	Glacially derived sediments	
	Tree ring dendro-climatology	**Briffa et al. 1996;** Fritts 1976; Schweingruber 1988
	Coral scleroclimatology	Beck et al. 1992; Patzold 1984

*Boldface indicates papers that represent initial applications, review papers, or volumes of the proxy method as it pertains to paleoclimate research and serve as an entree into the literature. This list does not include the multitude of applied studies that followed initial technique development. Additional references are given in later chapter discussions.

example, applies to most parts of the climate system, as an indicator of temperature, salinity, changes in the global ice volume (ice-sheet growth and decay), or regional changes in the hydrological cycle.

The following summary highlights some general principles underlying the use of faunal, floral, and geochemical proxy methods.

Indicator Species and Fossil Assemblages

The species and assemblages of about a dozen major groups are used routinely in paleoclimatology, and many more minor fossil groups are used less commonly (figure 2-8, table 2-7). What makes a species a good proxy of climate parameters? Although the answer to this question varies widely from group to group, Coope (1977) provides five attributes of species of Coleoptera (beetles) that also apply to most other groups: evolutionary stability, morphological complexity (species-specific fossilizable characters), abundance in sediments (allowing quantitative assemblage analyses), physiological constancy (difficult to ascertain directly), and rapid response to climate shifts. Coope

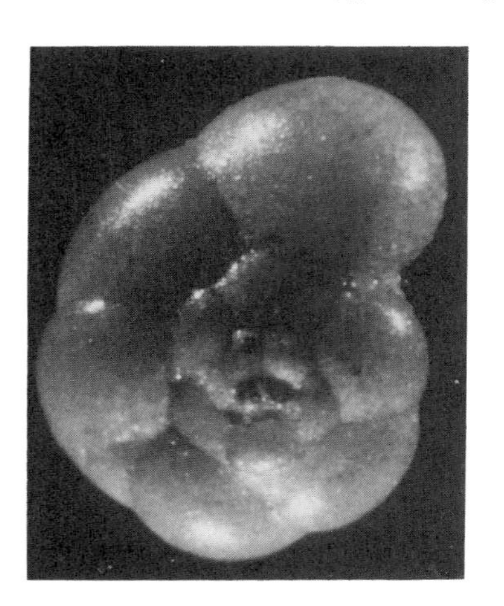

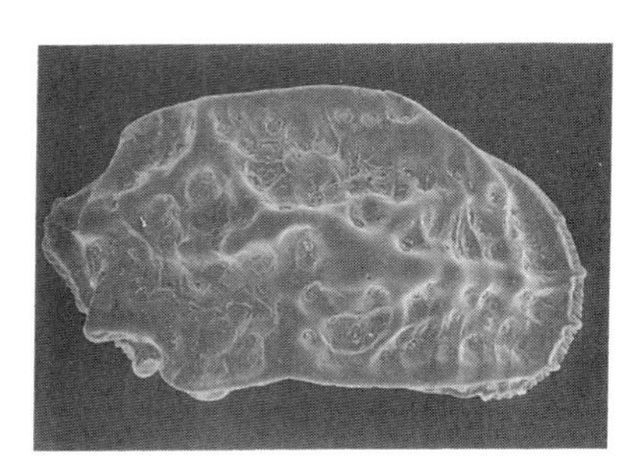

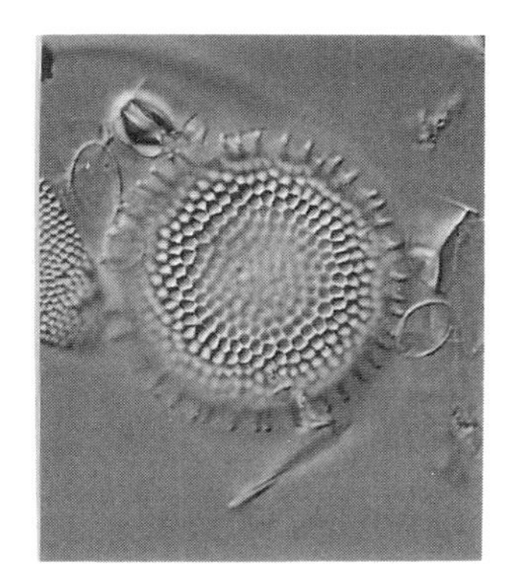

FIGURE 2-8 Several micropaleontological groups used as proxies in paleoclimatology. From upper left: high marsh benthic foraminifer (*Trochammina inflata*), ostracode (*Schizocythere kishinouyei*), planktonic foraminifer (*Globigerinoides ruber*), marine diatom (*Thalassiosira antarctica*). Courtesy of D. Scott, A. Tsukagoshi, H. Dowsett, A. Leventer, and R. Scherer.

adopted an actualistic approach to indicator species and communities in reconstructing past climate whereby the modern ecological requirements of species are used to directly interpret the significance of fossil assemblages and infer past climates. Coope did not mince words in expressing his confidence in his beetles as indicators of climate change: "fossil assemblages when treated in this way make ecological sense; [they are] the best possible justification for a hypothesis, namely that it works" (p. 317). In late Quaternary continental deposits of England, Europe, Canada, and elsewhere, fossil Coleoptera have been used extensively to document rapid climate change.

An actualistic approach is used by many other paleoclimatologists who use fossil evidence to reconstruct climatic parameters. For example, any study in which the statistical relationship between the modern distribution of species and a climate-related parameter like temperature or salinity be established takes an actualistic approach. Imbrie and Kipp (1971), in their seminal study of planktonic foraminiferal assemblages and Atlantic Ocean surface water masses, relied heavily on foraminiferal species biology (e.g., Be and Hamlin 1967, Be 1977) to develop the transfer function paleoecological method of SST estimation. They stated clearly that this approach involves ecological assumptions regarding the use of core-top foraminiferal assemblages to estimate SSTs. One of the most common assumptions is that the maximum abundance of a species in a sample indicates that it was inhabiting its preferred ecological niche. This assumption also implies that the species is evolutionarily "best adapted" to a particular set of environmental conditions.

Establishing the modern ecological preferences of foraminiferal species can be accomplished by mapping species distributions using dead specimens from core-top material (Imbrie and Kipp 1971), living plankton tow samples, living and dead bottom samples (Cronin and Dowsett 1990), or sediment trap samples. These sampling methods can result in useful databases that can serve as a baseline for quantitative analyses of past assemblages, especially over large spatial scales. Biogeographic databases are even more valuable when additional experimental or ecological data are available on species' habitat preferences (e.g., Spero et al. 1991).

Although the ecological limits of species can be estimated from their modern distributions, two complications arise in the application of statistical associations of species to reconstruct past changes in climate-related parameters. First, modification of the original faunal or floral assemblage occurs by taphonomic processes. Taphonomy, a large subdiscipline of the geosciences, involves the study of processes that alter a living community of organisms after death. The proper interpretation of taphonomic effects of paleocommunities is not a trivial matter. Even if the ecological tolerances of species are well known, the transition from life to death to fossil involves a series of complex processes, each of which can further complicate paleoclimatological interpretation. In the example of modern pollen rain, many biases are introduced between the time plants pollinate, the time pollen settles into lakes and bogs, and the time it is recovered from lake sediments (Birks and Birks 1980). These biases include biological factors such as differential pollen production, dispersal capabilities from pollen grain aerodynamics, and preservation within the sediments. In the case of marine microfossils, such as foraminifers, ostracodes, and diatoms, processes of transport, burial, and differential dissolution must similarly be taken into account before an assemblage can be properly analyzed. A firm grasp of taphonomic processes is essential for the successful reconstruction of climate history using fossil organisms over all spatial and temporal scales.

The second complication is whether ecological or climatic factors are responsible for observed temporal variability in species' proportions. Some researchers, especially those investigating vegetation changes reconstructed from pollen spectra, have attributed changes in pollen assemblages to plant-type life histories, dispersal, or competition, rather than to climatic factors. Community disequilibrium among species in forest communities has been cited as the cause for the differential rate and dispersal of plant taxa after deglaciation (e.g., Davis 1981). Population dynamics and interspecific interactions have also been cited as processes more important than climate change to explain patterns of pollen spectra since the last glacial period (see Prentice et al. 1991). Because biological processes can influence paleoclimatological interpretations so heavily, they are discussed at length in chapter 3.

During the past few decades, research on many fossil groups has reached a point where the geographic and/or bathymetic distributions of species are well known, the biology of many key species is understood, and quantitative methods are routinely applied to fossil assemblages. One database of eastern North American surface pollen samples, for example, includes pollen census data from about 1000 modern pollen samples (Prentice et al. 1991); the European pollen database of Guiot (1990) consisted of 655 samples. The planktonic foraminiferal core-top data have grown to thousands of ocean core-top samples. Other regional faunal and floral data sets often include tens to hundreds or more modern samples (e.g., ostracodes [Cronin and

Dowsett 1990; Ikeya and Cronin 1993], coccoliths, benthic foraminifers [Polyak et al. 1995], diatoms [Leventer et al. 1996]).

Many statistical methods have also been applied to translate faunal and floral census data into paleoclimatic reconstructions. Of the dozens of techniques available, the most commonly used are the transfer function (TF) (Imbrie and Kipp 1971), the modern analog technique (MAT) (Overpeck et al. 1985), and the response surface (Bartlein et al. 1986) methods. The TF method first uses factor analyses to establish a small number of "factors," or assemblages, based on the relative proportions of different species over an environmental gradient. For example, Imbrie and Kipp (1971) defined six major Atlantic Ocean assemblages: polar, subpolar, subtropical, equatorial, gyre, and ocean margin. Once the factors are defined, multiple regression analysis is used to formulate equations that relate the factors to the climate parameter, usually temperature or salinity. These equations are used to calculate paleotemperatures from downcore assemblages.

In MAT, species census data are also used to reconstruct past conditions, but instead of factor analysis, the statistical measures known as *coefficients of dissimilarity* are used to compare two assemblages on the basis of the relative proportions of key taxa. Overpeck et al. (1985) explored MAT using a number of dissimilarity coefficients in a study of North American deglacial pollen records. In Europe, Prentice et al. (1991) compared last interglacial (Eemian) pollen profiles to a modern database on pollen distribution for 28 plant taxa in surface samples from across Europe and argued that plant types exhibit a nonlinear response to several climate variables.

The response-surface method (Bartlein et al. 1986) measures the response of common tree taxa to climate parameters such as precipitation and temperature, identifying the ideal habitat conditions for key plant taxa. The many excellent reviews of these and other methods include the papers by Guiot (1990) and Prentice et al. (1991).

Tracers

The term tracer is used to refer to geochemical proxies that indicate oceanic processes or conditions. Elderfield (1990) lists three types of ocean tracers. The first and most commonly used is the indirect measurement of seawater composition by a tracer that is incorporated into the parent material in proportion to its composition in seawater. These tracers include stable isotopic ratios such as $\delta^{13}C$ and elemental ratios such as cadmium-calcium and germanium-silicon ratios in marine microfossils. A second type of tracer includes those that define a flux of an important climate parameter, such as estimates of carbon obtained from measurements of mass accumulation. A third type of tracer measures particular oceanic processes, such as the long-chain alkenone chemistry of marine phytoplankton (notably Prymnesiophyceae) (Brassell et al. 1986, Sikes et al. 1991), or barium-calcium ratios in foraminfera to estimate changes in oceanic organic productivity. Several oceanic tracers are listed in table 2-7.

Paleoclimatology likewise has a host of direct and indirect methods to estimate past atmospheric CO_2. These techniques include measures of carbon isotopes of oceanic foraminifera (Shackleton and Pisias 1985), nonmarine peat carbon isotopes (White et al. 1994), leaf stomata (van der Burgh et al. 1993), and—the most direct method—fossil CO_2 trapped in air bubbles in glacial ice (Oeschger et al. 1985, Schwander 1989). The section on glacial-age climatology covers the development of proxies to reconstruct past atmospheric CO_2 concentrations for the last glacial period and related parameters that influence the global carbon cycle, such as oceanic nutrients, circulation, and biological productivity.

Stable Isotopes

Stable isotope ratios of oxygen ($^{18}O/^{16}O$), carbon ($^{13}C/^{12}C$), hydrogen (deuterium, $^{2}H/^{1}H$, in ice cores), and to a lesser extent nitrogen ($^{15}N/^{14}N$) probably comprise the single most important family of paleoclimate tools used to reconstruct past climates. A proper treatment of stable isotope paleoclimatology would require volumes; the following brief section on the terminology of stable isotopes gives the basic tenets of the stable isotope method and a foundation for many applications described in later chapters.

The ratio of two isotopes of a particular element is expressed using the following convention, taking carbon as an example:

$$\frac{\delta^{13}C = (^{13}C/^{12}C)_{\text{sample}} - (^{13}C/^{12}C)_{\text{standard}}}{(^{13}C/^{12}C)} \times 1000.$$

The value δ is referred to as "per mil" (‰) and is the standard convention used in paleoclimatology for other elements. The standard against which a sample is compared is a fossil Belemnite from the Cretaceous Pee Dee Formation of South Carolina. This fossil, known as the PDB (for Pee Dee Belemnite) standard, was used in the pioneering study of Urey et al. (1951), who first applied oxygen isotopes for paleotemperature estimation. The PDB standard has been adopted ever since.

The principal rationale behind stable isotopes as proxies is that the ratio of heavy to light isotopes of an element in any material is a function of many variables, among them, climate-related variables such as SST, salinity, ice volume, atmospheric temperature, and moisture source. By identifying the dominant factors that control isotopic fractionation in a particular material, paleoclimatologists obtain a proxy that can provide a quantitative measure of secular and geographic variability in key parameters. Samples analyzed isotopically on a mass spectrometer include a wide array of geological and biological material containing the elements oxygen and carbon: foraminiferal shells, precipitated calcite, ice, tree wood, corals, speleothems, and organic rich sediments, to mention a few.

The use of oxygen isotope ratios in oceanic planktonic foraminifera to reconstruct Quaternary SSTs and global ice volume is an illustrative case. Temperature (Urey 1947, Craig 1965) and salinity (Craig and Gordon 1965) of seawater affect the isotopic composition of precipitated biogenic (e.g., foraminiferal shells) and inorganic calcite ($CaCO_3$). For example, if a species secretes its shell in thermodynamic equilibrium with the seawater in which it lives (some species do, others do not), the shell is enriched in $\delta^{18}O$ relative to the seawater. However, the $\delta^{18}O$ is temperature dependent such that the greater the temperature of the seawater, the less enriched the foraminiferal $\delta^{18}O$.

Salinity also affects isotopic ratios because nonmarine aqueous $\delta^{18}O$ varies widely, ranging from 0.0‰ for freshwater to less than –30‰ for glacial ice (Dansgaard and Tauber 1969). As a result, the $\delta^{18}O$ of seawater varies, ranging from about –0.3‰ (e.g., Weddell Sea water, Antarctica) to +1.0‰ (North Atlantic surface water) (Craig and Gordon 1965; Berger and Gardner 1975), mainly because of regional differences in evaporation and precipitation and in seawater salinity, which reflect the source of precipitation.

The volume of continental ice also affects foraminiferal $\delta^{18}O$ because, during glacial periods, a large quantity of freshwater with very negative δ values ($\leq$ 30‰) is removed from the oceans and stored on the continents (the equivalent of 100 m of sea-level drop), such that the isotopic composition of seawater changes. This change is reflected in foraminiferal shells that secreted their shells in seawater. During interglacial periods when there is less continental ice, there is proportionally more light ^{16}O in seawater. Thus interglacial seawater is isotopically more negative (i.e., lighter). Conversely, during glacial periods, when continental ice preferentially stores the light isotope (^{16}O), seawater is preferentially enriched in ^{18}O and, other factors being equal, glacial foraminiferal calcite $\delta^{18}O$ will be less negative. This "glacial effect" (Olausson 1965, Shackleton 1967) is believed to dominate the isotopic signal in foraminifera living in regions with relatively small temperature and salinity variability, such as the tropical Pacific Ocean (Shackleton and Opdyke 1973; Matthews and Poore 1980) or deep-sea abyssal environments.

To sum up, the difference between a foraminifera that secreted its shell during a glacial and one that grew during an interglacial can be expressed as follows (Berger 1979):

$$\Delta\delta_{\text{foram}} = \Delta T + \Delta G + \Delta E/P + \Delta V,$$

where T, G, E/P, and V are the differences in glacial-interglacial foraminiferal calcite $\delta^{18}O$ due to temperature, ice volume, evaporation and precipitation (salinity), and "vital effects" (when certain species secrete shells out of isotopic equilibrium), respectively. Omitted from this equation are effects of seawater density and postmortem dissolution, which can alter the original isotopic composition. If, as reasoned by Shackleton and Opdyke (1973) and others (see Mix 1992), tropical SSTs remained relatively stable during glacial periods, and salinity effects were minor, the observed glacial-interglacial $\Delta\delta_{\text{foram}}$ for Pacific Ocean foraminifers of about 1.3‰ (glacial-interglacial $\Delta\delta_{\text{foram}}$ varies regionally) would be due mainly to ice volume changes, that is, the glacial effect. Fairbanks and Matthews (1978) calibrated the $\Delta\delta_{\text{foram}}$ to sea-level equivalent and estimated that a 0.1‰ change is the equivalent of an ~10-m drop in sea level, or a little less than one tenth the total glacial-interglacial change in ice volume.

The application of isotopic research in continental paleoclimatology (see Swart et al. 1993) revolves around an understanding of the atmospheric factors that control the fractionation of oxygen and hydrogen in atmospheric precipitation. Rozanski et al. (1993) reviewed the development of the global monitoring of atmospheric $\delta^{18}O$ by the International Atomic Energy Agency (IAEA) and the World Meteorological Organization (WMO) since the 1950s. IAEA/WMO data show that there are latitudinal gradients in the $\delta^{18}O$ of precipitation in which long-term mean values are –2‰ to –8‰ in low latitudes but > 15‰ to 20‰ in high latitudes. Gradients in atmospheric $\delta^{18}O$ develop because the major source of water vapor is the tropical ocean. As atmospheric circulation transports water vapor poleward, precipitation occurs, reducing the total available precipitation and preferentially leading to more negative water vapor at high latitudes. Much of the scatter in precipitation $\delta^{18}O$ values is the result of altitudinal and continental effects related to the removal of heavy isotopes during condensation and other processes. Temperature variability over seasonal–to–long-term time scales is also an important factor controlling the $\delta^{18}O$ of precipitation. These factors must be sorted out in any study attempting to reconstruct paleo-atmospheric trends using the many sources of isotopic data from continental records.

In addition to the use of oxygen isotopes, references to carbon, nitrogen, and hydrogen isotopic methods in paleoclimatology will be encountered throughout the following chapters. In each case, one must remember that multiple factors influence isotopic composition of any material and thus the interpretation of paleoclimate records.

THE EVOLUTION OF GLACIAL-AGE CLIMATOLOGY

WITH THE BASIC concepts of faunal and isotopic proxy methods in hand, I wish to elaborate on the development and use of new and innovative proxy methods, such as the emu geochemistry introduced earlier in this chapter, to reconstruct glacial-age climates. New proxy development and application does not occur in a vacuum. Paleoclimatologists operate in a hypothetico-deductive mode, so that as new hypotheses develop to explain past climate changes, new tools are needed to test the hypotheses. A case history approach to global climate research on the last glacial maximum (LGM) is a convenient platform to introduce many proxy methods now used in contemporary paleoclimatology.

The LGM, about 21 ka, is probably the single most investigated period in the geological record. What confers the LGM such a lofty status? To answer this question, I divide the history of inquiry into five stages, during each of which new questions were posed about global climate. (1) Did continental ice sheets exist? (2) How cold was glacial climate? (3) How much ice was stored on the continents? (4) Were LGM tropical climates colder than tropical climates of today? (5) How did atmospheric chemistry, ocean circulation, and ocean chemistry differ during the LGM compared to the modern ocean? Each stage of inquiry was catalyzed by new discoveries and new ambiguities about the causes of glacial climates. Necessity has been the mother of invention, and paleoclimatologists were repeatedly compelled to develop new proxy tools to answer new questions.

Glacial-age paleoclimatology began in earnest when, in the early eighteenth century, Louis Agassiz (1840) developed his revolutionary theory of continental ice sheets from a suite of glacial geological proxies that are now standard tools of modern glacial geologists mapping formerly glaciated terrain. These proxies include glacial erratic boulders (figure 2-1), glacial moraines, and striations in bedrock carved by former ice sheets, among others. As Agassiz's glacial theory became widely accepted in Europe and North America, glacial geology was born (Imbrie and Imbrie 1979).

It was more than a century after Agassiz's glacial theory that a second stage of inquiry led to quantitative isotopic and faunal and floral techniques that produced temperature estimates of the glacial climates. These proxies included oxygen isotopes of foraminifera in oceanic sediments (Emiliani 1955) and glacial ice in continental environments (Dansgaard and Tauber 1969), as well as planktonic foraminiferal (Imbrie and Kipp 1971) and radiolarian (Moore 1973; Lozano and Hays 1976) faunal assemblages, and were used to reconstruct SSTs in the world's oceans. Methods like the factor analytic transfer function provided a first-order quantification of global mean annual temperature drop of about 5°C, with much greater glacial cooling in higher latitudes but only 1–2°C drop in tropical regions (e.g., CLIMAP 1981).

More recently, new micropaleontological proxy tools for oceanic surface conditions have been developed using phytoplankton groups (diatoms, dinoflagellates, and calcareous nannofossils). Diatoms are unicellular golden-brown algae that secrete a siliceous frustule preserved in oceanic sediments. They are the most important primary producers in the world's oceans. Dinoflagellates are protists, many of which are motile (using flagella) and have a free-swimming stage; many species also have chloroplasts and photosynthesize. Many dinoflagellate species form a cyst during their life cycle. This cyst is preserved in marine sediments. Dinoflagellates are second only to diatoms in their oceanic productivity. Calcareous nannofossils are minute algal protists that, during a phase in their life cycle, secrete a coccosphere that surrounds the cell. The coccosphere is composed of minute (about 8 μm) calcitic platelets called coccoliths that are an important component of oceanic sediments in tropical and subtropical regions, where they reach their maximum density.

Diatoms, dinoflagellates, and calcareous nannofossils have an advantage over foraminifers for surface paleoceanography in that some members require sunlight for photosynthetic activity. Thus, they must live in the uppermost euphotic zone of the oceans, and they provide an unambiguous surface-ocean signal. Diatom (Koç and Jansen 1994), dinoflagellate (e.g., de Vernal et al. 1996; Versteeg 1994, 1996), and calcareous nannofossil (Gard 1993) assemblages have proven extremely useful in reconstructing glacial and deglacial SST, sea-ice conditions, and temperature-related productivity changes in the North Atlantic, the Mediterranean, the Arctic Ocean, and the Nordic Seas. Studies of surface-dwelling phytoplankton groups are leading to major revisions in our understanding of glacial-age upper-ocean temperature and thermocline structure in the North Atlantic Ocean (de Vernal et al. 1996).

Another key question lingering from nineteenth century glacial geology is how much ice existed on the continents during peak glaciation? As mentioned above, the isotopic composition of foraminiferal shells is controlled by many climate-related factors, including ice volume. However, interpreting isotopic records is even more difficult when the biology of foraminiferal species is considered. For example, some species secrete their shells out of equilibrium with surrounding seawater, requiring calibration with or indexing to other species. Moreover, foraminiferal species have complex life cycles; different species live and secrete their shells at various depths in the upper layers of the ocean. These complicating factors led to extensive research on foraminiferal ecology and depth zonation using sediment traps instead of core-top samples (Erez and Honjo 1981; Deuser et al. 1981). Eventually, oxygen isotopes of planktonic and benthic foraminifers yielded excellent estimates of global ice volume (Shackleton and Opdyke 1973; Chappell and Shackleton 1986; Mix 1992). Research on planktonic foraminiferal ecology and isotopic composition and dissolution continue unabated, providing more realistic interpretation of foraminiferal isotopic records (Spero and Lea 1993; McCorkle et al. 1995).

Another round of proxy method development arose because the tropical glacial-age paleoclimatology posed special problems. Did the tropics cool during the LGM, as did high latitudes, or did they remain "tropical"? During the past few years, partly in response to modeling studies that suggested tropical areas should in fact cool during glacial climates (Rind et al. 1986; Manabe and Broccoli 1985), a suite of new proxy methods have been developed to measure past oceanic SSTs. These include the use of alkenone biomarkers (Brassell et al. 1986; Prahl and Wakeham 1987) and Sr:Ca ratios in tropical reef-building corals (Guilderson et al. 1994; Beck et al. 1992; de Villiers et al. 1995). The former technique suggests that LGM tropical oceans cooled only slightly, supporting planktonic foraminiferal SST estimates; the later technique indicates tropical SSTs fell as much as 5°C. New studies using planktonic foraminiferal oxygen isotopes suggest that the tropical Atlantic Ocean off Brazil cooled by about 2 or 3°C, twice the CLIMAP estimate (Wolff et al. 1998).

Low-latitude LGM continental climates have also received considerable attention in response to controversy surrounding tropical cooling. Paleo-vegetation studies have demonstrated elevation shifts in alpine tree lines in South America and Africa, indicating a major environmental change, very likely associated with decreased glacial-age cooling (see Colinvaux 1996). Snow line depression in low-latitude alpine regions is also considered strong evidence for LGM cooling in continental regions (Broecker and Denton 1989). Noble gases trapped in glacial-age groundwater suggested Holocene-LGM temperature differences were about 5°C for Texas and 9°C for Hungary (Stute and Schlosser 1993). Glacial-age groundwater geochemistry in the subtropical southeastern United States also showed a LGM temperature drop (Plummer 1992).

The discovery in trapped air from Antarctic ice cores that LGM atmospheric CO_2 concentrations were about 30% lower than those during the pre-industrial Holocene (LGM CO_2 = 190–200 ppmv, late Holocene = 280 ppmv [Neftel et al. 1982; see Raynaud et al. 1993]) sent reverberations through the scientific community and provided a strong impetus to reevaluate many aspects of LGM climate. If atmospheric CO_2 varied naturally on such a large scale, what did this portend for future climates, given anthropogenically elevated CO_2 levels? The most significant question for paleoclimatology was, What caused reduced atmospheric CO_2? The most likely candidate was enhanced carbon uptake by the oceans (Broecker 1982; see Sundquist and Broecker 1985). Increased oceanic carbon uptake would, in theory, draw down atmospheric CO_2. But was the CO_2 drawdown due to changes in ocean chemistry or circulation; biological productivity in coastal upwelling, equatorial, or polar regions; enhanced nutrient cycling from deep water; or iron fertilization from glacial continental aridity and greater quantities of dust blown into the oceans (Broecker 1982; Berger et al. 1989; Longhurst 1991)? Or was the cause a combination of these factors?

The response of the paleoclimate community to these new challenges was to develop new proxy methods for key parameters related to carbon cycling: oceanic nutrient levels, partial pressure of CO_2, biological productivity, oceanic circulation, and atmospheric dust transport and continental aridity. The cadmium-calcium ratio of benthic foraminifers, for example, was hypothesized to be a proxy for oceanic nutrients (Boyle 1988a, 1992). Barium-calcium ratios in benthic foraminifers also held promise for identifying trends in productivity (Lea and Boyle 1990).

However, before confidence could be established in trace elemental or isotopic composition of benthic foraminifers as tracers for productivity, significant advances in benthic foraminiferal ecology were required (e.g., Corliss 1985; Altenbach and Sarnthein 1989; Pederson et al. 1988). Some benthic species live infaunally and their shell chemistry reflects conditions of the pore water rather than the bottom water, which complicates their isotopic and trace elemental signals. Benthic foraminiferal ecology and geochemistry now constitutes a large and growing research area devoted to understanding glacial-interglacial productivity and ocean circulation changes (Corliss 1991; Gooday 1988, Gooday and Turley 1990; Mackensen et al. 1993; McCorkle et al. 1995).

Because primary productivity in the highly productive high-latitude oceans was considered a potential mechanism to explain the increased transport of carbon from the atmosphere to the oceans during the LGM (e.g., Sarmiento and Toggweiler 1984; Wenk and Siegenthaler 1985), new paleo-productivity proxies were developed for polar regions. Among the more innovative tools were the use of germanium-silicon ratios (Shemesh et al. 1989; Froelich et al. 1989), biogenic opal in marine sediments (Charles et al. 1991; Shemesh et al. 1992), and diatom species ecology (Leventer et al. 1996).

Additional advances in foraminiferal carbon isotope research has addressed the question, to what extent can foraminiferal carbon isotopic ratios be used as a proxy for paleo-CO_2 levels (Shackleton and Pisias 1985; Rau et al. 1991)? Through field and laboratory study of living individuals, Spero et al. (1991), among others, specifically addressed the question of carbon isotopes in terms of planktonic foraminiferal ecology, the role of symbiotic algae in isotopic composition, and foraminiferal ontogeny and life history.

The literature on paleoproductivity proxy tools has grown enormously, and several excellent sources are available on these methods and the more general topic of oceanic productivity during the LGM (e.g., Berger et al. 1989; Berger and Wefer 1991; Zahn et al. 1994).

Oceanic circulation changes related to LGM CO_2 drawdown have also been reconstructed using several new proxy methods. These methods include foraminiferal oxygen isotopic reconstruction of sea-surface salinity (Duplessy et al. 1991), foraminiferal carbon isotopes for deep-water formation (Curry et al. 1988), and deep-sea bottom-water temperature estimates derived from abyssal benthic ostracodes (Dwyer et al. 1995; Cronin et al. 1996).

Finally, paleo-atmospheric circulation during glacial episodes has received considerable attention. Rea (1994) reviews the large literature on paleo-atmospheric circulation and wind transport shown by eolian sediments preserved in deep-sea sediment cores. When a distinction is made between eolian- and water-borne sources (hemipelagic silt and clays derived from continental margins), glacial-age eolian dust records from deep-sea sediments show that three to five times more dust was transported from northern hemisphere continents during late Quaternary glacial episodes than during typical interglacial periods. This large difference was due to much greater continental aridity during glacial ages, especially in semiarid regions of Africa and Asia. Rea cautions that increased eolian flux does not necessarily mean that LGM wind strength and intensity was greater.

To sum up, the proxies used in a century and a half of glacial-age climate study have progressed from glacial geology, stable isotope geochemistry, phytoplankton ecology, foraminiferal life histories, organic molecular biomarkers, trace elemental composition, geochemical indicators of ocean productivity, and eolian sediment transport and deposition. Novel climate proxy development and application, facilitated by technological breakthroughs such as AMS and TIMS dating and other geochronological methods, comprise the fundamental principles of paleoclimatology. The essence of paleoclimatology involves the ingenious development of climate proxies—such as the emu eggshells and oxygen isotopes—and improved chronology to answer questions about earth's climate history. It is with these underlying principles that we now proceed.

III

Vital Effects: Biological Aspects of Paleoclimatology

[Vital effects:] A name which hides the fact that we don't understand them.
WALLACE BROECKER 1982

BIOLOGICAL PRINCIPLES IN PALEOCLIMATOLOGY

AUSTRALIAN EMUS, prymnesiophyte algae, *Porites* coral reef colonies, and South American trees may seem to have little in common. Actually, all are biological groups adapted to live in the modern tropical climatic zones characterized by relatively warm year-round temperatures. All four also have organic parts that fossilize easily. A third commonality is that emu eggshell, prymnesiophyte, and coral skeleton chemistry, as well as pollen from South American trees are all paleoclimatic proxies used to understand the climate history of tropical regions.

In the previous chapter, a rich array of biological sources of paleoclimate data was cited both as a means to develop the chronology of climate change and as proxies to reconstruct climate parameters. The study of climate change, of course, involves understanding physical processes governing the circulation of earth's atmosphere and ocean and its energy balance. Climate change also involves understanding earth's biotic systems. Indeed, most branches of paleoclimatology examine directly the remnants of biological entities such as fossil pollen or foraminiferal shells to document patterns of climate change or measure the outcome of biogeochemical and physiological processes that regulate climate through complex feedbacks with earth's oceans and atmosphere.

With both direct fossil and indirect biogeochemical evidence, paleoclimatology enters the domain of the biological sciences, most notably such fields as ecology, evolutionary biology, and physiology. Many biological processes, the subject a large experimental and theoretical literature, govern the history of organisms, just as physical and chemical processes govern atmospheric and oceanic circulation and global chemical cycling. Consequently, the theoretical foundations of evolution and ecology that underpin biology deserve consideration in paleoclimatology for the simple reason that reconstructed patterns of climate change often reflect the outcome of complex biological processes. In this chapter, my goal is to provide a general overview of key biological processes and principles that either are relevant for the reconstruction of paleoclimates through the use of biological proxies or contribute to climate change itself through biogeochemical processes.

The Fossil Record Sensu Latu

Why stress a biological focus in paleoclimatology, a field traditionally viewed as a discipline of the physical sciences? There are four overarching reasons to bring a biological focus into our attempts to understand climate change.

The first reason relates to the prior discussion of tropical glacial-age paleoclimatology. Much if not most evidence for climate change is based on information derived from biological sources preserved in the fossil record. In paleoclimatology, *fossil record* refers to direct biotic sources of paleoclimate data obtained from fossilized shells, pollen, bone, and other fossils preserved in sediments (figure 3-1). From fossils, paleoclimatologists obtain a wide spectrum of concrete ecological, morphological, and geochemical evidence for past climate change. The record also includes some organisms that are not strictly speaking fossils—living trees and coral colonies—from which climate history of the last few centuries and millennia can be obtained by analyzing annual growth bands.

Indirect biological evidence for climate change also comes from the biogeochemistry of microfossils, sediments, and polar ice sheets and high-elevation ice caps. These archives capture a record of metabolic activity of organisms and often go under the rubric *global biogeochemistry* or *geophysiology*. They are manifestations of important processes such as photosynthesis (which has an impact on carbon cy-

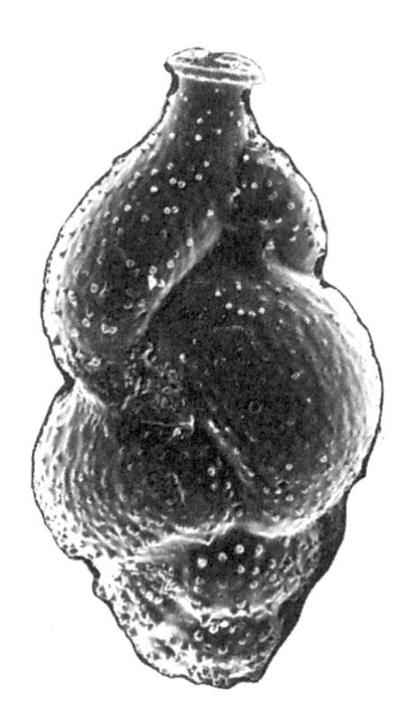

FIGURE 3-1 A foraminiferal shell, a symbol of vital effects in paleoclimatology.

cling), respiration (which affects atmospheric molecular oxygen), intracellular dimethylsulfonioproprionate (DMSP) production (DMSP is a precursor to dimethylsulfide [DMS], which influences atmospheric chemistry), and anaerobic bacterial respiration (methane production). Taken together, these tissue- and cellular-level processes play important regulatory roles in climate through feedbacks with the atmosphere and oceans. Ecological processes of competition and predation may also have an impact on large-scale ecosystem and biome functioning and thus, indirectly, climate.

Hierarchy is an important concept in modern biology, and biological sources of climate data encompass several hierarchical levels of biological organization, from cells to entire biomes. These levels are listed in table 3-1 as they pertain to topics described in later chapters. Intracellular DMSP production by different groups of marine phytoplankton is an example of a lower level process that contributes to global sulfur cycling through complex chemical pathways, sometimes mediated by bacterial enzymatic processes, producing DMS. DMSP production is a species-specific phenomenon carried out in modern oceans by coccolithophores, dinoflagellates, and prymnesiophyte algae. DMS is in turn removed from oceans via photochemical oxidation,

TABLE 3-1 Selected Biological Sources of Climate Data

Chapter	*Data source covered*
Chapter 4	Planktonic foraminiferal communities, world oceans
	Coral reef communities, tropical oceans
	Terrestrial plant communities, Africa
	Equatorial marine phyto- and zooplankton communities
	Deep-sea benthic communities
Chapter 5	North Atlantic foraminiferal species ecology
	Arctic plant species ecology
	Benthic foraminiferal species ecology
	European terrestrial plant communities
Chapter 6	Tree growth and physiology
	Marine planktonic communities (diatoms, productivity)
	South American vegetation (dust)
Chapter 7	Colonial coral-skeleton growth and ecology
	Tree growth and physiology
Chapter 8	Coral species ecology
	Coastal salt marsh community ecology
	Intertidal and subtidal mollusks and foraminifers
Chapter 9	Global terrestrial vegetation, continental aridity, and atmospheric dust
	Tropical and high-latitude wetlands' methane production
	Marine algae and dimethylsulfide (DMS)
	Marine surface ecosystems, nutrients, and carbon cycling

consumption by bacteria, and air-sea exchanges when it becomes a source of acidic aerosols, altering atmospheric chemistry. Thus Charleson et al. (1987) hypothesized that dimethylsulfide production by marine organisms affects clouds and global climate.

Other metabolic activities operating at lower hierarchical biotic levels have an impact on climate. These include the isotopic fractionation of molecular oxygen, known as the Dole effect (e.g., Bender et al. 1985; Sowers and Bender 1995); the production of terrestrial biogenic methane and its export to the atmosphere (Chappellaz et al. 1990); and photosynthesis by marine organisms, which affects the carbon export from the atmosphere into the deep-sea reservoir via the biological pump (e.g., Sundquist and Broecker 1985; Longhurst 1991; Zahn et al. 1994). These metabolic biological processes can influence climate on hemispheric and global scales.

The biology of an individual organism, such as a tree, or a single colony, such as a group of corals also can be crucial to a paleoclimate record. The growth, physiology, and morphology of organisms can yield information about their responses to climatic disturbance. For example, during ontogeny, trees, mollusks, and coral develop annual growth bands at different rates depending on environmental conditions. Growth bands can preserve distinct records of climate change. Other examples of organism-level processes include foraminiferal chamber growth and isotopic geochemistry (e.g., Erez and Honjo 1981; Mackensen et al. 1993) and crustacean molting and shell trace metal chemistry (Chivas et al. 1986; De Deckker et al. 1988), two processes affected by the ambient environment. Understanding the ontogeny of organisms is a prerequisite for interpreting geochemical signals captured in their shells.

A population of an ecologically sensitive species is another level of biological organization used to record climate history. Populations are subsets of species, geographically and (potentially) reproductively isolated sets of organisms. An indicator species is one whose ecological tolerances—as measured by its changing abundances—can be used to establish past temperature, precipitation, sea level, and other parameters. The fossil populations of hundreds of species of protists, mollusks, crustacea, insects, trees, vertebrates, corals, and other taxa have been used to track climate change in virtually all habitable environments. To cite one example, the planktonic foraminiferal species *Neogloboquadrina pachyderma* is an invaluable paleoceanographic tool in both Northern and Southern Hemisphere polar regions because variability in its populations is a reliable indicator of changing sea-surface conditions during rapid climate transitions.

Multispecies groups are a yet higher level of organization. Multispecies groups are usually referred to in paleoclimatology as assemblages, biofacies, or associations. They often characterize a modern climatic zone, an ecosystem, or a particular habitat that has climatic significance. In ecological parlance, these groups are referred to as communities—sets of interacting but geneologically unrelated species. Paleocommunities often show complex, dynamic shifts that are routinely used to track climate change. Forest communities are a common, albeit complex, tool used in paleoclimatology. North American pollen assemblages used to record the evolution of forest communities during the glacial-deglacial transition 20—10 ka represent one of the many well-documented examples of multispecies groups (COHMAP 1988; Wright et al. 1993).

Our understanding of past patterns of climate change would be incomplete if we relied solely on physical evidence (obtained through sedimentology, glacial geology, or geomorphology) or chemical evidence (e.g., isotopic fractionation) to detect climate change.

Reciprocity: Climate Change and Biological Processes

A second reason to focus on biology is the reciprocity between climate change and biological processes. Just as climate change can have severe impacts on biotic systems, biotic activity likewise influences climate. The study of climate history thus requires an understanding of the inherent role that organisms play in biogeochemical processes that link the atmosphere and ocean in the flow of energy and matter. An obvious example of how biotic activity regulates climate is the role in the storage and transfer of carbon—a key element in atmospheric carbon dioxide (CO_2) and methane gases. Changes in atmospheric CO_2 concentration are vital to the planet's long-term (tens of millions of years) (Crowley 1991; Berner 1994) and short-term (tens of thousands of years) (Barnola et al. 1987, Raynaud et al. 1993) climatic evolution. Oceanic and terrestrial biotic-atmospheric fluxes of carbon are tightly integrated into all modern theories about climate change.

Paleoclimatologists' views on the topic of organism-climate reciprocity vary widely. These views run the gamut from those promoting active biological regulation of global climate (e.g., Gaia hypothesis) to those supporting a more passive role for organisms as part of the larger, complex, geophysical system (Prentice and Sarnthein 1993).

The Gaia hypothesis was developed by James Lovelock (1972; see Lovelock and Margulis 1974) and holds that the earth's biota plays an *active* role in climate regulation through feedbacks with the atmosphere, ocean, and crustal rocks (via weathering rates). This field of "geophysiology" (Lovelock 1989; Lovelock and Kump 1994) emphasizes the global scale of biotic interactions and climate, in particular the surface distribution of two major biotic components—photosynthetic marine algae and terrestrial plants. The planetary-scale distributions of both marine plankton and terrestrial plants change substantially during large-scale climate change, such as glacial-interglacial cycles. These changes are sufficient to modify the global cycling of carbon and influence atmospheric concentrations of CO_2.

Lovelock and Kump (1994) expanded upon early notions of Gaia, integrating the geophysiology of climate within the context of climate changes that accompanied the last deglaciation, about 15–10 ka. Global ocean productivity was almost certainly higher during the last glacial maximum (LGM) compared with that during such interglacial periods as now (Berger et al. 1989; Herguera and Berger 1989, 1994). Lovelock and Kump posited that an extension of the ocean's thermocline occurred during the temperature rise of the last deglaciation and that, because marine plankton grow most efficiently at temperatures of 5–10°C, this extension decreased the ability of marine photosynthetic algae to sequester carbon into the world's oceans. This aspect of the Lovelock-Kump hypothesis offers the premise that regulation of climate is a function of organic activity. They state that "if living organisms participate in climate regulation in an active and responsive way, they do so most probably as part of a tightly coupled system, which includes the biota, the atmosphere, the oceans and the crustal rocks" (Lovelock and Kump 1994:732). The Gaia viewpoint may approach the teleological limit for some (see discussion in Schneider and Boston 1991), but it focuses attention on biological reciprocity in climate.

An alternate approach to biogeochemical processes and feedbacks is simply to treat organisms as a passive component in the overall transfer of energy and matter within the integrated climate system that includes the oceans, atmosphere, ice, and lithosphere. The equatorial Pacific Ocean surface ecosystem provides an example. The important link between ocean primary production and carbon cycling has recently fostered efforts to understand large-scale changes in earth's paleoproductivity stemming from the restructuring of major organic ecosystems (figure 3-2). The equatorial Pacific may be the largest single natural source of carbon to the earth's atmosphere because this region is characterized by strong oceanic upwelling of CO_2-rich deep waters (Berger et al. 1989). Geochemical and micropaleontological evidence indicates that during the last glacial period, oceanic productivity was probably much higher than during the present interglacial period, contributing to drawdown of atmospheric CO_2 (e.g., Mix 1989; Berger et al. 1989). Paytan et al. (1996) reconstructed a 450-ka paleoceanographic record of multiple glacial-interglacial cycles in the equatorial Pacific Ocean to study paleoproductivity. They used marine barite records from sediment cores as a productivity proxy because mineral particulate barite is commonly found in sediments in regions of high productivity. Paytan and colleagues discovered that productivity in the central and eastern equatorial Pacific during all glacial periods of the past 450 ka was twice the level of the intervening interglacial periods. Moreover, the twofold increase in glacial productivity was of such a scale as to play a significant, though still incompletely understood, role in regulating global climate through its influence on atmospheric CO_2 levels.

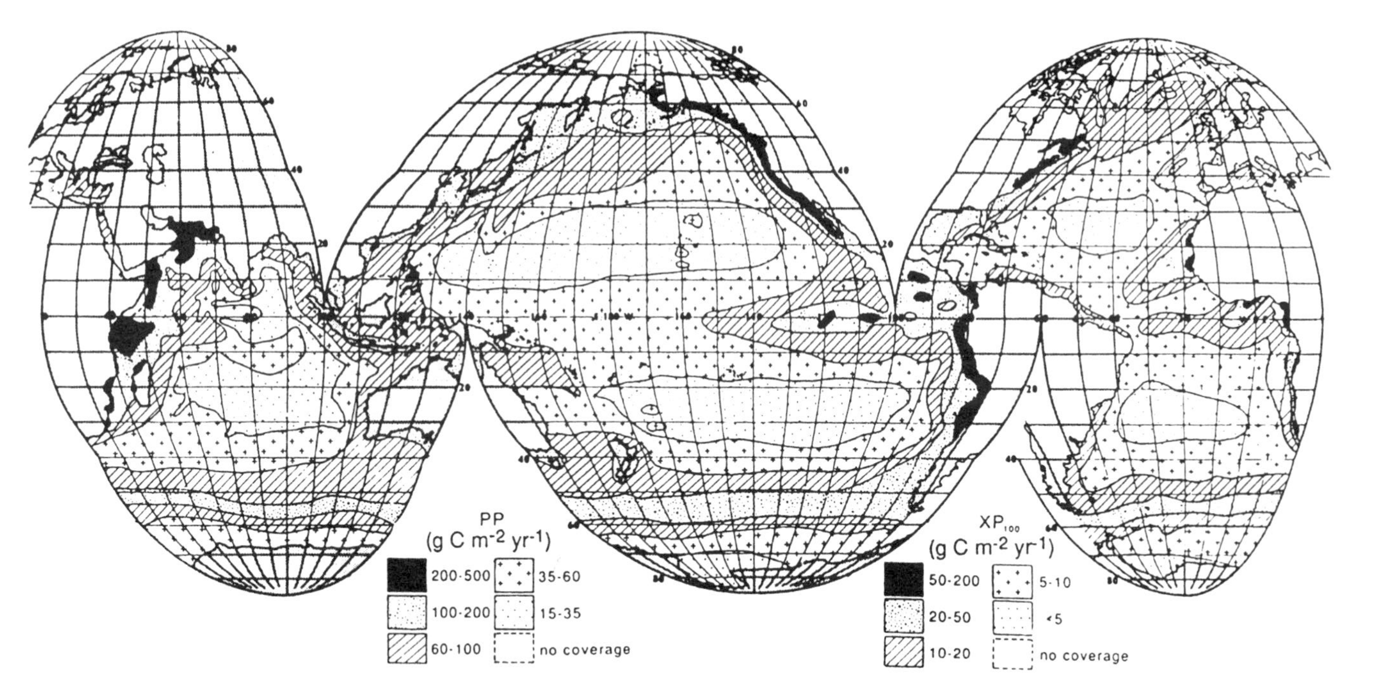

FIGURE 3-2 Map of primary productivity (PP) and export productivity (XP), at 100-m water depth of the world's oceanic regions, reflecting biome-scale processes in the climate system and the reciprocity between climate and biotic activity. Courtesy of the World Meteorological Organization (WMO) and W. Berger.

Several theories have been proposed to explain enhanced glacial productivity. Most revolve around changes in nutrient fluxes (iron, nitrogen, phosphorous) or in atmospheric and/or oceanic circulation (Broecker 1982; Sundquist and Broecker 1985; Boyle 1988b). Other researchers, however, theorize that ecosystem dynamics are also critical in controlling modern oceanic productivity. Frost (1991), for example, cites the importance of zooplankton grazing on phytoplankton as a primary control of marine productivity. This top-down ecological control of ecosystem functioning contrasts with the view that nutrient concentrations limit productivity and implies that temporal changes in various trophic levels can severely affect carbon sequestration. Whatever the exact mechanisms controlling global atmospheric CO_2 and ocean productivity, ecosystem functioning—expressed in the day-to-day activities of organisms—must somehow be involved.

Hypothetico-Deductive Testing of Biological Theories

We have seen that biological sources of climate history span various hierarchical levels, from the cell, whole organism, population, species, community, ecosystem, or biome. A third reason to focus on the biotic aspects of climate change is that the temporal dimension provided by paleoclimatology provides a means to test ecological and evolutionary theories about these entities, just as it offers a means to test theories of climate change.

Most ecological and evolutionary research, like paleoclimatological research, proceeds from a hypothetico-deductive approach, often using experimentation to test hypotheses. In ecology experimental field studies are conducted to test concepts of community structure and interspecific competition. Controlled biological experiments are, however, obviously limited in the temporal and spatial dimensions in which they can operate; scientists simply cannot examine large-scale and long-term phenomena with accuracy obtained in small-scale ecological experiments. For example, anthropogenic emission of greenhouse gases has been termed the great geophysical experiment because, regrettably, there is no control experiment against which to compare the impacts on global climate under elevated greenhouse-gas concentrations. Similarly, continent-scale geographic shifts in entire forest communities over thousands of years cannot be studied in a laboratory.

These phenomena can, however, be studied using the paleoecological record. Oceanic and continental paleoclimate records provide in many cases the spatial and temporal resolution to test ecological and evolutionary theories and processes where the environment—i.e., climate—is an inherent part (table 3-2). Delcourt and Delcourt (1990) argued fossil pollen and plant records can be used to test classical ideas about ecological succession. Jablonski and Sepkoski (1996) described the contribution of paleobiological data to understanding ecological processes such as community ecology, species invasions, within-habitat and global species diversity, and biogeographic response to disturbance and predation. Tests of the theories of island biogeography, community structure and succession, niche partitioning, and patchiness all can be derived from long-term temporal records, especially those with annual, decadal, or century-scale sampling resolution, although the potential has yet to be realized.

Similarly, the paleoclimate record provides a means to examine the role of the environment in evolutionary processes. Evolutionary theories predict certain spatial and temporal patterns over long periods of time that can be tested in natural "experiments" during climate change (Bennett 1997). Intraspecific stasis—the concept that species are unique entities with spatial and temporal boundaries, inherently stable over long periods of time—can be examined with the high-density temporal and spatial sampling now available for deep-sea and lacustrine sediments. Theories to explain speciation and the controls on biodiversity can also benefit from the paleoclimate record. Cronin and

TABLE 3-2 Evolutionary and Ecological Theories and Processes Studied Through Climate History

Ecological
Community structure
Succession
Biological diversity and island biogeography
Niche theory
Dispersal versus vicariant biogeography
Competitive exclusion principle within a trophic level
Character displacement
Predator-prey relationships between trophic levels
Evolutionary
Natural selection (individual)
Geographical modes of speciation
Red Queen versus Stationary Mode of evolution
Evolutionary rate
Extinction and macroeveolutionary processes
Adaptation and adaptedness

Raymo (1997), for example, found that the time-stability hypothesis of biodiversity, which holds that long-term deep-sea environmental stability has led to relatively species-rich biotas, was inadequate to explain long-term diversity trends in deep-sea environments. The many paleoclimate records cited throughout this book reveal a rich biotic record that may hold answers about some long-standing biological enigmas.

Reductionism and Holism

The fourth reason for devoting attention to biological processes involves a discussion of the underlying philosophical basis of paleoclimatology. Paleoclimatology is essentially a hybrid science. It is at once an observational and a historical science that involves natural climatic "experiments" of the past, often measuring biological and biochemical parameters as evidence, to test theories about the causes of climate change. At the same time, it is a quantitative predictive science devoted to understanding physical processes related to atmospheric and oceanic dynamics, energy balance, and the like.

Despite the importance of observational, historical, and biological aspects of paleoclimatology, there is a tendency among some researchers to adopt a reductionist approach to the study of past climate changes, as if climate change were a solely an extension of the fundamental laws of physics. Reductionism is a complex, sometimes ambiguous concept that historically has had a number of meanings. A full discussion of reductionism is beyond the scope of this text (but see Mayr 1982). Nevertheless, in the history of science, reductionism in its simplest form has referred to the philosophy that all theories of biology can be reduced to theories of physics. One extreme example of a reductionist view would be the idea that all biological processes can be reduced to the molecular interaction operating at the level of the gene and that the behavior of molecules themselves can be explained solely by physical and chemical laws. Reductionism thus bestows a unity upon all the sciences.

Unfortunately, this approach can lead to oversimplification of complex systems and erroneous interpretations. The physical and biological sciences often proceed with different methodologies and underlying philosophies. Physics, for example, seeks universal laws to explain observed phenomena. Chemistry and functional biology (e.g., physiology) use experiments to measure natural processes. In contrast to seeking laws and conducting experiments, many scientists interested in the earth and organismic biology "observe" historical events of the past. These fields depend heavily on historical events as natural experiments.

In paleoclimatology, a reductionist view emerges in two areas. The first involves the issue of causality in explaining past climate events. For example, researchers sometimes consider only physical mechanisms as potential causes of reconstructed patterns of climate change, but biological mechanisms may also be valid causes. The problem of tropical glacial-age sea-surface temperatures discussed previously is a case in point. The similarity between glacial and interglacial planktonic foraminiferal assemblages suggests the tropical ocean did not cool during ice ages; if it had, cooler-water species would have migrated equatorward and been discovered in glacial-age fossil assemblages. This enigma has engendered much debate about the physical explanation of glacial-age climate, yet there is an ecological explanation for similar glacial and interglacial assemblages. An ecological explanation would invoke interspecific competition as a mechanism that prevented cooler water planktonic foraminiferal species from inhabiting tropical regions. Tropical species under this scenario essentially outcompeted their more northerly counterparts for habitation of tropical oceans.

The reductionist approach is also evident when organisms used as proxies of climate parameters are viewed as passive recorders of climate-related parameters such as temperature and salinity. Such a view is most prevalent when paleoclimatologists try to calibrate a chemical or physical parameter measured in organisms to a set of values measured in the organism's environment. The case of stable oxygen isotope ratios in foraminiferal shells, which vary as a function of the water temperature in which the shell was secreted, is an example discussed later. Research has shown, however, that foraminiferal shell chemistry is a function of many complex, still partially understood factors, including so-called vital effects. The term *vital effects* refers to biological characteristics unique to each foraminiferal species that can influence isotopic fractionation in their shells. The existence of vital effects means organisms are not merely passive recorders of the biogeochemical environment but complex living things. Vital effects are discussed more fully later in the chapter.

These issues stemming from a reductionist view of living things point out that biological systems are inherently different from physical systems. Mayr (1982) lists eight distinct characteristics of living things that together make the case for autonomy of biology: organized complexity (DNA molecules, tissues, species, etc.), chemical uniqueness (enzymes, phosphates, lipids, etc.), "quality" (as opposed to quantification), uniqueness and variability, possession of a genetic program, historical nature, natural selection, and indeterminacy. All eight characteristics are important in biology; five elements are especially germane to paleoclimatology—quality, uniqueness and variability, historical nature, the indeterminacy of biological events, and the complexity of living systems.

Quality refers to the fact that many features of biological systems are best described qualitatively and in a relational context, rather than through rigid quantification. Species and ecosystems generally defy the rigorous quantification that befits the Newtonian physical world. As Mayr points out, overly ambitious efforts in the past to quantify biological systems have almost invariably met with failure. Likewise, efforts to quantify aspects of biological systems, such as the temperature tolerance of an individual species, and apply them to the reconstruction of climate change should be tempered by the fact that living systems are distinct from physical systems and are not easily subjected to quantification.

Biological systems are also unique and highly variable, posing additional complications for paleoclimatology. The evolution of a new species, for instance, is certainly a unique event. Moreover, the evolution of a new species can have significant impact on a community and an ecosystem. Take the example of the evolution (or extinction) of a species that plays a key role in a climate-relevant biogeochemical process, such as methanogenesis, photosynthesis, or nitrogen fixation. Pollock (1997) suggests that photosynthetic activity of key Antarctic diatom species is in part responsible for drawdown of atmospheric CO_2 and, indirectly at least, is a causal factor in Pleistocene glaciations. Speciation or extinction of species could in theory affect climate in unknown ways.

Some biological systems, such as a species' populations, also possess a much higher statistical variance that do most physical phenomena. High variability in the population structure of an indicator species, for example, would limit its paleoecological use in tracking environmental changes over time. Because paleoclimatologists often attempt to quantify past changes in physical parameters using indicator species (or assemblages), recognizing the inherent variability in these biological tools is critical when interpreting past trends.

Like paleoclimatology, the study biological entities such as individual organisms and species to reconstruct climate is a historical science. Organisms and species are the products of genetic programs, and they have evolved through common descent. These attributes clearly distinguish living things from inanimate objects.

Prediction is also a key component in the study of climate change. Climatologists ultimately seek to predict events of the future through the development of theories of climate and computer modeling of global and regional climate change. Paleoclimatologists likewise seek to "retrodict" past climate events on the basis of their ideas about the forcing factors that may have caused climate to change. Yet the indeterminacy of many biological events, such as major extinctions and random genetic mutations, renders prediction (and retrodiction) of living systems a difficult task. As Mayr (1982:57) states: "Temporal predictions are much more rarely possible in the biological sciences" than in the physical sciences. Who would have predicted the evolution of land plants 350 to 400 Ma and their subsequent influence on the global carbon cycle and climate? Given the fact that biological systems such as land plants regulate climate-related processes in a reciprocal relationship with abiotic systems, climate prediction will remain an extremely difficult task.

The final aspect of living things germane to climate prediction is their complexity. Complexity is an inherent part of biological systems that also complicates the study of climate change. O'Neill et al. (1989) point out that ecologists have ignored complexity, circumvented it by working with simple systems, or just invented mathematical models as surrogates for real ecosystem functioning. They argue that reductionism in the field of ecology ignores the complexity of real natural systems. Similarly, reductionism permeates the history of evolutionary biology (Mayr 1982). The role of the species, the population, and the individual organism have at times been subsumed beneath laws and processes governing genetics (e.g., Dawkins 1976). Joel Cohen summed this issue up by claiming that "physics envy is the curse of biology," a pun denoting an apparent desire to find regularity and predictability in a very unpredictable world in which chance plays a bigger role than is wished (see Oster and Wilson 1978). Mayr (1982), Grene (1984), and Sober (1984) discuss the intriguing literature on reductionism in biology, to a large extent denouncing reductionist views. In its place, many biologists advocate a holistic view of the organism that treats it as a cohesive entity rather than an amalgamation of separate parts (see the discussion of adaptation at the end of this chapter).

From a practical standpoint, the interplay between complex biological processes and physical aspects of earth's climate greatly complicates paleoclimate interpretation. The concept of vital effects in paleoclimatology illustrates the concerns raised in the preceding paragraph. The term *vital effect* was first used by Urey et al. (1951) in their classic paper on the development of oxygen isotopes as paleothermometers. Urey and colleagues made reference to the fact that two coexisting species can secrete their calcareous shells out of chemical equilibrium with the seawater in which they live, imparting a "vital effect" on the isotopic signal. As discussed in chapter 2, the calibration of isotopic ratios in shells to seawater temperature (or other parameters) is central to paleoclimatology. Specifically, the disequilibrium in oxygen isotopic ratios of $^{18}O/^{16}O$ derived from shells of different foraminiferal species poses problems when using isotopic ratios derived from two species as a proxy for oceanic temperature or for global ice volume.

The isotopic disequilibrium in different species was originally attributed to unknown biological reasons. Since then, the term vital effect has evolved into a catch-all phrase in paleoclimatology for biologically mediated processes that are poorly understood. Broecker (1982:1698) summed up vital effects as "a name which hides the fact that we don't understand them." Many other types of vital effects have been uncovered in recent years, including those for other foraminiferal species and their carbon isotopic ratios and ostracode coprecipitation of shells (Chivas et al. 1986, 1993), among other proxy methods. In some cases, the causes of vital effects are gradually becoming clearer, as in the case of symbiotic algae and foraminiferal isotopic chemistry (Spero et al. 1991) and other aspects of foraminiferal shell chemistry and life history (e.g., Boyle 1994). In others, a full appreciation of the underlying biological causes to explain vital effects is still lacking.

The recognition of vital effects adds complexity to the interpretation of climate change because paleoclimatology seeks to establish direct linkages between physical and biological processes in order to use measured parameters to track climate change. As lofty a goal as this may be, striving for it can foster reductionist approaches to complex biological systems. Because biological data are the basis used to document past climate changes, and many biological events are random, unique, and unpredictable, a dimension of uncertainty and unpredictability will always impair climatic reconstruction. Biological uncertainty should not impede the search to improve proxy tools or reconstruct climate history, but it does impose complications if an overly reductionist approach is adhered to. Furthermore, because biological processes regulate climate through feedback loops, then one must also recognize that these biological forcing mechanisms themselves may involve a degree of stochasticity.

The topics of vital effects in particular, and reductionism in general, introduce thorny issues in an already complex field. I raise them in part because I have not seem them discussed explicitly in the paleoclimate literature and because they merit fuller discussion. Clearly it is difficult to know the degree to which the randomness inherent in some biotic systems complicates climate history. One philosophical challenge facing paleoclimatology is how to bridge the physical-biological gap without falling into a deterministic view of climate change that might otherwise obscure potentially important chance events. As paleoclimatology matures into a major, integrated field of research, climate should not be reduced to physical laws and deterministic processes. Rather, a more holistic view of both the climate system and the biological systems used to study past climate changes is needed, and this requires a conceptual framework from which to proceed.

A CONCEPTUAL FRAMEWORK OF BIOLOGY AND CLIMATE CHANGE

THE TWO FIELDS OF organism-level biology are ecology and evolution. Ecology encompasses all aspects of organisms in their interaction with members of other species and with the physical and chemical environment. To a large degree, this organism-environment interaction constitutes the economics of biology—the transfer of energy and matter through biotic systems. Evolutionary biology is the branch of biology that deals with the organic evolution of species and of higher and lower taxonomic units within the context of the Darwinian struggle for existence. The unifying theme of evolutionary biology is the genealogical continuity of phylogenetic lineages derived from the differential reproductive success of individuals.

While these twin pillars of organism-level biology often overlap, they have over the past century evolved as discrete disciplines, to a large degree in isolation from one another. Each field has developed its own theoretical and conceptual frameworks in which organisms are studied. Conceptual linkages between distinct branches of biology recently emerged that provide a platform from which to examine biological processes that influence earth's climate. In this section I draw on efforts to unite ecology and evolution within a common hierarchical framework as it applies to climate change.

Hierarchical Notions in Ecology and Evolution

Hierarchy permeates aspects of biology that seek to explain how organisms evolve and how they interact with each other and with abiotic elements of their environments (O'Neill et al. 1989). Hierarchies are often discussed under various theoretical frameworks such as general systems theory, neural networks, hierarchy, and chaos theory. Most of these are systems designed to explain and to predict naturally occurring phenomena. In the general sense, a conceptual hierarchy of biotic systems resembles our conceptual hierarchy of the climate system in that both are designed at least in part to obtain a better predictive capability for natural phenomena.

Table 3-3 presents a hierarchical framework for paleoclimatology. With minor modifications, this scheme follows closely the hierarchy defined by Eldredge (1992) and Eldredge and Grene (1992). It distinguishes between ecological and evolutionary systems, consists of a workable number of levels within each system, and allows for cross-level or cross-system interaction. In the system of ecological levels, the economic side of organismic biology is represented by a progressively decreasing spatial scale: the global biosphere, biomes, ecosystems, communities, populations (or avatars, sensu Damuth 1985), organisms, and soma (e.g., the cellular level). The evolutionary side lists conventional levels of genealogical hierarchy, from higher taxonomic groups to the germ line. This evolutionary system stresses the point that within any one level, all members are unified by the fact that they share a common ancestor—that is, the group is a monophyletic group that shares a genetic history. Species and, more properly, populations within species, are common genealogical levels chosen to study climate change. The term higher taxa refers to the standard nomenclature of phylum, class, order, family, and genus encountered in paleoclimatology.

TABLE 3-3 Conceptual Hierarchy of Biological Systems for Paleoclimatology

Ecological system	*Evolutionary system*	*Paleoclimate applications**
Biosphere	Higher taxa (corals, algae)	Global biogeochemical cycles
Biomes	Higher taxa (grasslands)	Hemispheric, continental-scale change
Ecosystem	Species (key species)	Orbital, suborbital climatic cycles
Community	Species	Orbital, suborbital climatic cycles
Populations (avatars)	populations (demes)	Regional-scale events
Organisms	Organisms	Adaptive traits
Soma	Germ lines	Physiology, morphology

*These are general categories; many other applications exist

Several authors offer hierarchical taxonomies similar in many respects to this scheme. Brown (1995) offers a scheme that includes the following levels: genes and enzymes, cells, colonies of cells, organism, population, species, monophyletic lineage, biosphere. Allen and Hoekstra (1992) propose an eight-level hierarchy of criteria similar to the ecological system in table 3-3 with the additional level—landscape—placed between the ecosystem and community. Huston's (1994) explanations of biodiversity focus on functional biotic groups, leaving aside the genealogical aspects. DiMichele (1994) distinguishes three separate sets of terminology used by ecologists, plant paleoecologists, and animal paleocologicists. DiMichele's ecological levels include the guild, community, landscape, biome, and province. His paleoecological schemes are generally similar to mine except the fossil equivalents of the modern community are called plant assemblages and animal biofacies.

Paleoclimatology uses all levels of biological hierarchy and patterns of organisms, populations, species, and communities preserved in the geologic record, because response to climate can stem from ecological or evolutionary processes or both. Also, as discussed by Levin (1992) and Allen and Hoekstra (1992), there is no correct level from which to view ecological or evolutionary interactions—the level depends on the goal of the viewer in terms of the natural phenomenon under study, as well as logistic limits. In a similar fashion, each level can have its own utility for the study of specific paleoclimate studies. For example, one does not need to examine the individual morphological complexity of each individual ostracode carapace if the goal is to examine species' geographic distributions (i.e., the biome level) and their importance for marine paleoclimatology (Hazel 1970; Cronin 1987). Similarly, lower-level processes of growth and physiology operating within a coral colony are sufficient to study El Niño–Southern Oscillation (ENSO) variability using coral growth bands and skeletal geochemistry; large-scale, higher-level biogeography of tropical biomes are not of concern. From a practical standpoint, this means it is not necessary to incorporate processes operating at higher or lower levels.

Scale in Processes

Scale is critical in both paleoclimatology and biology. Ecological research agendas have traditionally operated at two scales. Population ecologists, often studying genetics of populations, work at fine spatial and temporal scales, whereas ecosystem ecologists, often in applied field research, conduct their research at the much larger "landscape" scale (Levin 1992; Kareiva et al. 1993).

Brown (1995) divides ecologists into two camps. One group, comprising the majority, includes hypothetico-deductive, experimentally oriented ecologists who examine specific behavioral and physiological aspects of populations and individuals. These ecologists, who might conduct experiments on a small patch of landscape, provide important information on the interactions of species within controlled habitats. Their experiments often have paleoclimate significance when processes are extrapolated over large regions or entire continents at the level of the biome. Brown (1995) puts in the other group biogeographers, paleobiologists, macroevolutionists, and "macroecologists." Members of this group tend to be more inductive in their approach, use mainly species as the unit of study, adopt a more holistic approach, and focus on higher-level spatial and temporal patterns and processes. There is a clear alliance between Brown's macroecological camp and a segment of the paleoclimate community that deals with large-scale patterns of organisms and communities.

Other prominent ecologists have evaluated scale in ecology. O'Neill et al. (1989) criticized a reductionist approach in ecology and called for a hierarchical approach to sort out ecological complexity. Levin (1992) claimed that the problem of pattern and scale is the central problem in ecology, unifying population biology and ecosystem science, and marrying basic and applied ecology. Vitousek (1994) appeals for "scaling up" in ecological thinking in addressing questions of climate and land-use changes. Kareiva et al. (1993) call for a marriage between the formerly disjunct fields of population ecology and ecosystem science to address broad questions of global climate change and its impact on biotic systems.

To a certain degree, this trend toward a new ecological perspective on large-scale spatiotemporal properties of organisms may be due to a renewed interest among ecologists in the topic of climate change (McGowen 1990). Brown (1995) opens his book *Macroecology* with a discourse on global climate change and then devotes much of his discussion to processes and patterns that operate above the population level, that is, those scales important to many paleoclimate studies. This interest in climate change may stem from concern over anthropogenic influence on climate (Peters and Lovejoy 1992). It also may reflect the fact that long-term time-series of data are beginning to yield significant results about ecosystem functioning and climatic factors previously unappreciated in ecological studies limited to sampling over a few years and a limited area. Whatever the reasons, many ecologists recognize that larger scale phenomena such as climate change play important roles in the functioning ecosystem and even on a global scale, sometimes exerting a greater influence on organisms than do biotic interactions among species.

The concept of scale in evolutionary paleobiology has also been widely discussed in recent years. For example, Stanley (1979) and Vrba and Eldredge (1984) discussed the concept of species selection—a level of selection operating at the level of species instead of classic Darwinian natural selection operating at the level of the individual. Gould (1985) postulated a hierarchy of three tiers representing different levels of macroevolutionary theory—natural selection (acting on individuals), species selection (acting on species), and mass extinctions (acting on whole biotas). Bennett (1990) added a fourth level corresponding to the evolutionary phenomena occurring between speciation and natural selection. In Bennett's scheme, ecological processes occurring over thousands of years are distinct from Milankovitch climatic cycles occurring over 20–100 ka cycles. Bennett (1990) and Cronin and Ikeya (1992) point out that Milankovitch climate cycles provide a mechanism for geographic isolation of populations and thus for incipient speciation to occur, but that more often than not, the duration of the climate event is too short for a new species to evolve. Bennett's scheme does not account for long-term community and ecosystem changes unrelated to evolutionary events and mixes ecological and evolutionary processes and entities. Roy et al. (1996) also discuss problems and implications of scale in understanding Quaternary biotic changes in light of abundant evidence for relatively rapid climatic changes. Although evolutionists do not all agree that a hierarchical scheme is necessary to deal with evolutionary processes, the concepts of species selection and long-term species stability are important when using species as paleoclimate proxy tools.

SCALE IN PALEOCLIMATOLOGY

SCALE ALSO LIES at the heart of applied paleoclimatology and therefore becomes a central organizing principle behind the biology-climate framework in table 3-3. A scaled hierarchy of organization is useful for several reasons. First, scaling lets paleoclimatologists consider the theories and principles that are relevant to any particular biotic level. If one is using pollen assemblages to reconstruct changes in forest communities, one should be cognizant of the principle of ecological succession, as well as species ecology. Second, scaling requires paleoclimatologists to consider the impact of processes operating on a higher or lower level than the level of interest. For example, in the aforementioned efforts to understand the role of oceanic equatorial phytoplankton in glacial-interglacial carbon cycling, nutrient concentrations are considered a major limiting factor in CO_2 uptake through photosynthetic activity (Berger et al. 1989; Codispoti 1989). But predation, operating through the process of grazing by zooplankton (mainly copepods), represents an alternative hypothesis to explain variation in carbon and nutrient cycling (Chisolm and Morel 1991; Frost 1991; Longhurst 1991). Cellular and population-level processes provide competing explanations for the same observed patterns.

A third important aspect of scale in paleoclimatology and ecology is that one gains improved predictability as one moves up the hierarchy. For example, consider orbital and millennial-scale climate change, which is discussed in chapters 4 and 5. Changes in geographic and seasonal solar insolation of earth's upper atmosphere, which are due to gravitational effects on the earth's precession, tilt, and orbital eccentricity, result in glacial-interglacial cycles occurring at 21-, 41-, and 100-ka frequencies, respectively. Relatively coarse temporal-scale sampling (e.g., one sample every 2000–3000 yr) and spectral analyses commonly show that orbital frequencies are characteristic in many biologically based climate proxies (e.g., isotope chemistry of microfossils, faunal and floral assemblages) and provide convincing evidence that astronomical factors influence global climate (chapter 4). If one is interested in testing orbital theory, this temporal and spatial scale appears quite appropriate.

If one looks at the record from the same region with a finer temporal sampling resolution, however, one finds a different paleoclimate story. Paleoclimatologists have discovered climatic patterns in high sedimentation–rate sequences representing millennial scale events that cannot be explained solely by orbitally induced changes in solar insolation. As one moves down the temporal scale of observation, climate history that was apparent at a higher resolution becomes less predictable and more difficult to understand. Temporal patterns of faunal and floral communities that were "predictable" based on orbital theory at the scale of tens of thousands of years become very confused at shorter time scales and behave in a less predictable manner in relation to insolation changes. Thus, other mechanisms are required to explain suborbital-scale climate events (chapter 5).

In ecological parlance, Allen and Hoekstra (1992) refer to this problem of ecological integration in terms of the "grain" and the "extent" of the study. These concepts, although lacking exact parallels in paleoclimatology, are still conceptually valuable. The grain—the resolving power of measurement—limits the organizational level one can study. The grain of paleoclimatology is akin to the level of precision in the climate proxies. Allen and Hoekstra's (1992) term "extent" refers to the spatiotemporal constraints imposed on the study to examine higher-level phenomena, such as the biome or global biosphere. Extent in paleoclimatology is the temporal and spatial resolution attainable in the particular study, which constrains the level of climate phenomena that can be examined.

From a practical standpoint, a definition of scale is essential in studying climate change because in each study an age model is developed to describe patterns of change, to study trends, and if cycles exist, to establish their frequencies. The *time scale* one chooses depends on many factors that will determine the level of climate phenomena one can study. Chapters 4–7 move progressively through different time scales, from tens and hundreds of thousands of years to annual and seasonal climate phenomena.

REAL BIOLOGICAL ENTITIES OR HEURISTIC CONSTRUCTS?

UP TO THIS POINT, we have assumed the organizational units in table 3-3 have some basis in reality. That is, to be real, entities such as a population, an organism, or a community should have definable spatiotemporal boundaries. This concept is simple to understand for the individual organism, which with certain exceptions (colonial organisms and some plants), is born, lives for a certain time, and then dies. That species are real entities is not so clear and has long been debated in evolutionary circles. Nominalists view species as mere human constructs, but currently many evolutionists generally accept that species are entities (see Ghiselin 1969; Hull 1973). Species evolve (quickly over geological time scales), exist for usually 1–10 Ma over limited geographic areas (Stanley 1979), and go extinct. Indeed, the documentation of species' longevity in the face of constantly shifting environmental extremes in paleoclimatology is itself a testament to the reality of species as discrete units in time and space (Cronin 1985). Higher level taxonomic units are more difficult to define, especially in terms of their origin and spatial boundaries, but they are still united by their common heritage.

In ecology, the reality of ecosystems and communities has also been much discussed. In brief, Eldredge and Grene (1992) concluded that the reality of ecosystems is quite well established in the spatial dimension. Though ecosystems are usually well defined entities, ecosystems "leak"; that is, their boundaries are not always so distinct that all members of the ecosystem remain within those limits.

The temporal nature of ecosystems has until recently been based on fairly spotty data. This deficiency, however, is remedied in part by the extensive quantitative, well-dated paleoclimate data available for terrestrial and marine ecosystems. The reality of many ecosystems and communities is demonstrated by their recurrence over long periods (see later). This is not to suggest that ecosystems and communities are somehow fixed, permanent entities; transient climate changes frequently perturb and temporarily disaggregate communities. Nevertheless, communities often reassemble into a remarkably similar structure in terms of the types and relative abundances of their constituent species.

Within-Level Functioning

How do different elements of a single hierarchical level interact? Levin's (1992) example of the extrinsic and intrinsic control of two interrelated communities that comprise the pelagic Antarctic marine ecosystem illustrates such interaction. The Antarctic zooplankton community consists of a monospecific assemblage of Antarctic krill that is associated with the primary producing phytoplankton community. Levin noted that large-scale climatically driven shifts in oceanic circulation dictate the patterns of the Antarctic phytoplankton community. Conversely, Antarctic krill are subject to finer spatial and temporal scale variability such that other mechanisms control their population variability. Intrinsic biotic processes such as predation and swimming may override the effects of physical processes such as oceanic currents.

The contrast between within-community functioning of krill and control of phytoplankton by physical processes anticipates situations in paleoclimatology in which a multiproxy approach to marine ecosystems is taken to reconstruct climate history. In paleoceanography, trends determined from phytoplankton (e.g., diatoms, dinoflagellates) can reveal patterns different from those of zooplankton (foraminifera) groups during periods of climatic transition (de Vernal et al 1996). Likewise, in continental ecosystems, long-term orbital climate cycles invoke different responses in distinct east African terrestrial ecosystems. Some plant communities respond at precessional frequencies; others tend to show patterns that reflect the influence of earth's variations in obliquity. The degree to which one can track climate changes depends heavily on the identification of the primary types of climate change that influence different types of marine and terrestrial communities.

Cross-Level Interactions

Although hierarchical levels of organization are convenient concepts, they are in no way immutable. In ecology, for example, energy and matter flow up and down different hierarchical levels. In evolution, interchange likewise occurs between different levels, such as when a population becomes extinct but conspecific populations survive elsewhere. Such events are frequently documented in paleoclimatology when populations become locally extinct owing to climatic factors.

Another question arises with respect to cross-level interactions: Do the dynamics of one ecological level of organization dictate, or at least influence, higher levels through some sort of collective behavior? Or are the levels discrete in terms of interaction up and down the hierarchy? This issue often surfaces in paleoclimatology. In particular, in terrestrial plant and marine plankton communities, the term *nonanalog* flora or fauna is used to designate those past communities reconstructed from fossil evidence that bear little or no resemblance to any known modern assemblage. Thus the fossil assemblage, in terms of species composition, has no apparent counterpart in the modern database. This aspect of paleoclimatology is a bane to the researcher seeking a simple tracking of climate by the extension of a modern assemblage back through time, but it reflects the fundamental point that populations of different species respond differently from others to climate change. The net effect can be the evolution of a community comprised of a totally alien mix of species as compared with any existing currently. Paleoclimatology's "no analog" problem brings out the importance of fundamental ecological principles related to scale dynamics.

In evolutionary circles, the topic of cross-level interactions involves the extrapolation of microevolutionary processes operating at the individual and population levels to macroevolutionary processes at the species level and above. Paleobiologists (Stanley 1979; Gould and Eldredge 1972; Vrba and Eldredge 1984) have argued for treating species and higher taxa as subject to evolutionary processes distinct from those operating at the population-genetics level. They hold that emergent properties of species, such as biogeographic ranges, are important from an evolutionary standpoint. Although others have argued against this view, the concept that species possess emergent qualities is significant in the application of species ecology to climate reconstruction.

Cross-System Relationships

An important aspect of any hierarchical system is that of membership. Some entities are members in both ecological and evolutionary groupings, with a different role and confronted by different processes in each system. Thus, evolutionary levels are defined by their genealogy, not by their role in the ecosystem or by their interactions with other unrelated species. Populations of the eastern North American species of oyster, *Crassostrea virginica,* are members of distinct communities living in Pamlico Sound and Chesapeake Bay. Both the Pamlico and Chesapeake populations belong to the genealogical unit called the species *C. virginica* and also to the various cross-genealogical communities living in the two semi-isolated waters of Pamlico and Chesapeake. In contrast, two totally different mollusk species may occupy the same functional role in an ecosystem or community in terms of the transfer of energy and matter without regard for genealogy. The distinction between membership in an ecological or in an evolutionary unit pertains when applying the species in paleoecology.

BIOLOGICAL CONCEPTS IN PALEOCLIMATOLOGY

THIS SECTION illustrates the role of biology in paleoclimatology by describing four interrelated concepts that permeate ecological and evolutionary biology: the niche, the community, the biome, and the evolutionary concept of adaptation to the environment. Obviously, these concepts are neither all-inclusive nor totally distinct from one another. They incorporate, however, a large amount of theory and principles useful in climate studies. The first three concepts relate to successively higher levels of organization and climate variables: fossil populations and the reconstruction of environmental variables, species assemblages and their relationship to paleocommunities and ecosystems, and large-scale biomes and their role in hemispheric and global climate change. The last concept, evolutionary adaptation, pertains to assumptions about organisms and their environment that pervade much of paleoclimatology.

Niche

In his classic paper "Homage to Santa Rosalia," G.E. Hutchinson (1959) asked the simple question "Why are there so many species?" Hutchinson's papers spawned an explosion in theoretical ecology that reached a zenith in seminal publications by Robert MacArthur (1972; MacArthur and Wilson 1967), Robert May (1981), and many other ecologists (see Cody and Diamond 1975). Much of this effort was devoted to expressing mathematically the relationships between competing species and populations and to conducting field investigations of organisms in nature to verify the mathematical relationships. There was a drive to deduce ecological "laws" that could be applied across a spectrum of organisms. Now, scientists of many persuasions still ask what drives organic diversity (Huston 1994), but it is widely recognized that biological systems are not governed by strict laws like those in the physical sciences. Instead, the behavior of biological systems follows generalizations, each of which has many exceptions.

The roots of modern theoretical ecology found in the works of Hutchinson, MacArthur, and colleagues introduce the first ecological concept pertinent to paleoclimatology: the niche. The concept of the niche is familiar to most of us. Organisms in all communities play specific roles in terms of the daily interaction with other biotic elements. Darwin (1859) certainly had the notion of a role for each species in his *Origin of Species.* Each species has a certain function in the overall economy of energy and matter fluxes within the ecosystem.

In addition to the function or role played by a species, the concept of niche can also connote location. Grinnell (1914) was among the earliest ecologists who explicitly referred to the niche concept as representing a species' habitat and geographical range. In his studies of birds and mammals of western North America, Grinnell focused more on the spatial aspects of organisms' interactions with abiotic physical and chemical parts of the ecosystem.

Hutchinson's (1957, 1959) formal concept of the niche has several components that are germane to paleoclimatology. First, Hutchinson conceived of the niche in terms a multidimensional space in which the axes represented variables, that is, abiotic and biotic factors, that defined the conditions a population needed to reproduce, survive, and persist (figure 3-3). The variables included biological factors such as food resources, prey, and competitor species, but they also included physical parameters of perhaps greatest interest in studies of climate, such as temperature and salinity. Significantly, both biotic and abiotic

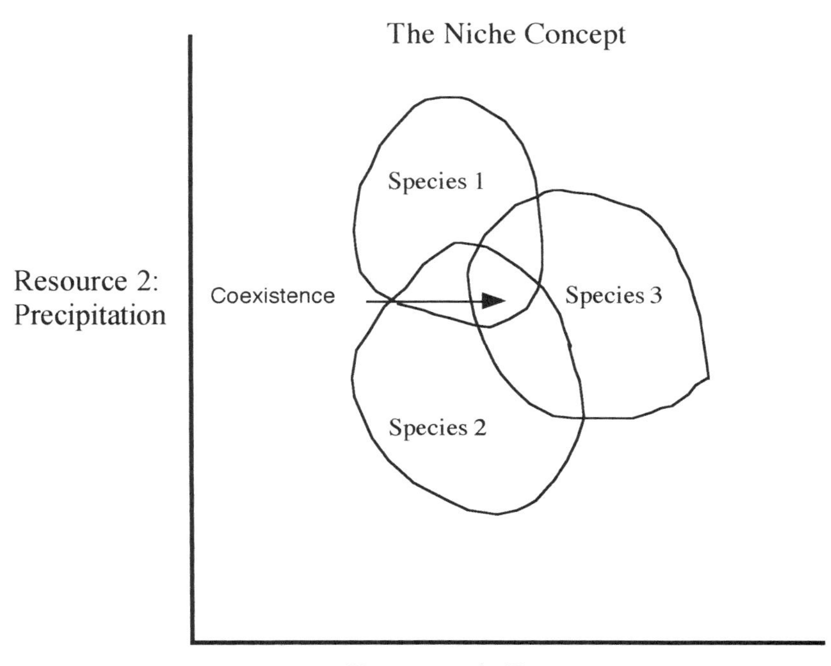

FIGURE 3-3 The concept of the niche. In paleoclimatology, a species' niche defines the resource limitations that render the species a useful indicator of climate-related parameters such as temperature and precipitation.

factors were recognized as important, although many ecologists place more emphasis on the biotic aspects of a niche.

Second, the niche concept engendered the idea that competition occurred between species. This imbued the niche with clear Darwinian overtones. Interspecific competition was an important process that might exclude weaker species from the community.

Third, early concepts of the niche implied a uniqueness for each species. The niche was an attribute of the species, not an attribute of the habitat in which its populations lived. Thus, members of a species occupy a "hypervolume" in which the axes are resources (food, prey) or physical factors (e.g., temperature and precipitation [figure 3-3]).

Since Hutchinson's early studies, measuring and quantifying the parameters that define the niche have occupied a large part of population ecology (MacArthur 1972; Cody and Diamond 1975; Pianka

1975). Much of this effort went into developing mathematical expressions to explain how so many species can coexist under certain environmental conditions and in the presence of other species. Generally, ecologists wished to improve their predictive abilities to account for the distribution and abundance of species in nature.

Concepts of the niche are still quite varied among different ecologists. Schoener (1988) concluded that the modern niche concept emphasizes the factors food, space, and time as the key elements. Brown (1995) argued that abiotic factors should also be included in any effort to explain the structure and dynamics of populations. Pianka (1975) offered a utilitarian view that the niche can be divided into three main elements: trophic resources (what organisms eat), space (where they live and reproduce), and time (when they are active). Plant and animal ecologists will tend to view niches differently—the former think of the niche as more a geographic area occupied by plants, the latter consider a more active role for organisms interacting within the habitat. In sum, definitions of niche bear a large degree of subjectivity.

Niche in Paleoclimatology

In paleoclimatology, the niche concept is manifested as one of the discipline's most fundamental principles embodied in the concept of an indicator species. A useful indicator species is one whose abundance and distribution (or sometimes simply its presence or absence) in time and space are limited by physical parameters related to climate. The niche variables that, to the evolutionary ecologist, were agents of natural selection among individuals to the paleoclimatologist become the critical parameters for tracking climate change. The questions posed by the ecologist and paleoclimatologist are similar. Why does a particular species vary in abundance and distribution? The answers, however, are distinct. Some ecologists would answer that prey availability, social behavior, absence of competitors, and food resources are factors that control the abundance and distribution of its populations. Many paleoclimatologists, on the other hand, would concentrate their efforts on identifying the role of temperature, salinity, dissolved oxygen, sedimentary substrate, or other physical and chemical environmental factors.

I advocate a broad concept of niche that includes three major aspects. First, the niche must be viewed in terms of the potential biotic as well as abiotic factors interacting among species. Paleoclimatologists should consider that species' temporal and spatial patterns might be unrelated to physical parameters and instead may reflect biological processes between species. To ignore biological processes may cloud the interpretation of paleoclimate patterns, as is the case with glacial-age planktonic foraminifers mentioned earlier.

Second, this niche concept is evolutionary in that it views species as adapted to environments. (The concept of adaptiveness is addressed separately below.)

Third, the niche is temporally dynamic and transient. This statement becomes obvious from any examination of a temporal sequence of fossil pollen or radiolarian assemblages. Paleoclimatological evidence shows that the biotic and abiotic niche variables that control the populations of a species are extremely unstable in space and time. The transient nature of niches is obvious from both long-term paleoclimatological records, spanning geological or evolutionary time scales (hundreds of thousands to millions of years), and also from short-term records of high-frequency climate change (annual, decadal, and centennial time scales). These short-term climate records overlap with the traditional domain of ecological time scales and thus provide important mechanisms for understanding niche dynamics in relation to climate (chapters 6, 7).

A corollary to the transient nature of niches is that they are recurrent over multiple climate cycles that occur at very different frequencies. For example, the 3- to 7-yr quasicyclic ENSO (see chapter 7) exerts a major influence on tropical environments, substantially altering niche relationships in pelagic marine or coral reef ecosystems. Indeed ENSO cycles are an excellent way to examine marine-ecosystem response to perturbation (Peterson et al. 1991; Glynn 1990). Conversely, low-frequency climate cycles present an entirely different perspective from which to view the niche.

Applying the niche concept to climate change requires the researcher to ask, How do two or more species coexist sympatrically within the context of interspecific interactions, limiting resources, and physical parameters related to climate? The following example illustrates this idea.

Planktonic foraminifera consitute a commonly used paleoclimate tool that illustrates the applicability of the niche concept in climate studies. About 30 species of planktonic foraminifers inhabit the upper 400–500 meters of the world's ocean. Shells of foraminifers settling through the water column now blanket the bottom sediments of much of the Atlantic and the shallower parts of the Pacific and Indian Oceans. Early foraminiferal studies established a correspondence between death assemblages of shells obtained from the uppermost sediment in cores and life assemblages of living foraminifera collected in plankton tows in the vicinity of the core site (see Be 1977). However, assemblages preserved in the sediment at the ocean floor represented spatially and temporally averaged assemblages that blurred the fine scale daily and seasonal variation that characterized foraminiferal populations living within the upper layers of the water column. Despite this homogenization process, the broad geographic distributions of species were clearly temperature-limited, and planktonic foraminifers were successfully used as the preferred tool for sea-surface temperature reconstruction at the ocean basin scale (Imbrie and Kipp 1971).

As foraminiferal assemblages and their shell isotope chemistry became an integral segment of paleoceanography, it became necessary to obtain a better understanding of the ecology of species. Several questions were especially important for interpreting the isotopic signal preserved in the shell. Which species actually lived and secreted their shell near the surface, close to the atmosphere-ocean interface, and which lived deeper in the water column was of primary importance to paleoclimatologists. Did certain species migrate seasonally or daily in the water column, providing a more complex signal than average annual sea-surface temperature (Curry et al. 1983)?

Efforts were mounted to better define each species' niche, that is, what factors allowed many species to coexist in the same region. Improved understanding of planktonic foraminiferal ecology came about through plankton traps, culturing, and other methodological advances, leading to a clearer picture of species' distributions at finer spatial and temporal (seasonal, interannual) scales (Berger et al. 1981). Deuser et al. (1981), for example, demonstrated seasonal variation in the size, weight, abundance, and isotopic composition of 11 species obtained from sediment traps placed in the Sargasso Sea in the North Atlantic Ocean over a 14-month sampling period. Erez and Honjo (1981) used a series of descending sediment traps to discover shallow-dwelling (0–50 m), intermediate-dwelling (50–255 m), and deep-dwelling (255–600 m) species. They found that some species secreted their shell below the photic zone, others above it, and still others at several levels during their ontogeny. Biological factors such as the presence of symbiotic algae and behavioral and ontogenetic aspects of diurnal migrations were found to be related to preferred depth habitats and were essential to the application of foraminiferal isotopic geochemistry in paleoceanography (see Hemleben et al. 1989).

It soon became apparent that many species of planktonic foraminifera coexist sympatrically in the geographic sense because they partition microhabitats temporally and spatially in the upper layers of any particular region of the ocean. While the ecological niche of foraminiferal species is still a subject of intense research, the fact that foraminiferal species have evolved to partition the limited resources in the upper water column is an important application of niche theory to paleoclimatology.

Communities and Ecosystems

Communities of plants and animals are defined as groups of species that have evolved to coexist and interact in a variety of ways. Communities represent a higher level of biotic integration than populations and species. The concept of community in ecology incorporates the biological elements (i.e., species) but usually excludes the physical parameters that influence individuals and populations living in the community. A community is therefore not solely defined by the spatial dimensions of the habitat in which a group of species live, but rather as an evolutionarily accommodated group of species. Allen and Hoekstra (1992:44) call communities "the integration of the complex behavior of the biota in a given area so as to produce a cohesive and multifaceted whole." A forest is thus a typical terrestrial community; the phytoplankton and zooplankton in the ocean surface comprise a marine community. These communities are not merely a group of isolated populations that can be studied independently but are also an integrated group of biotic elements competing with and preying on each other.

To apply the community concept to climate change, the distinction between the community and the ecosystem must be clarified. The community is the association of biotic elements interacting with each other. In this sense, the community is a somewhat intangible entity, because of the difficultly in observing and measuring all animal and plants and their interactions with each other in any particular area.

The ecosystem concept since its early formulation (Tansley 1935) has usually connoted an entity that is more concrete than the community (see Golley 1993 for a review of the history of the ecosystem concept). This is mainly because the ecosystem includes both the community of organisms (including their interactions with each other) and also the physical attributes of the region in which the community lives. Thus, the ecosystem includes soil, atmosphere, precipitation, temperature, and seasonality and other parameters. In the broadest sense, the ecosystem incorporates the influences that abiotic processes, such as changes in solar radiation, volcanic activity, fire, among others, exert upon organisms.

The application of fossil paleocommunities to derive information about past climatic parameters is so central to paleoclimatology that it is worth tracing the historical development of the community concept. Early opposing views of community evolved from studies of terrestrial plant communities by Frederic Clements (1916) and Henry Gleason (1922, 1926, 1939). Clements studied communities of the Nebraskan Plains, Gleason the forest-prairie interface. McIntosh (1981) provides a excellent analysis of contrasting Clementsian and Gleasonian views of communities, which I draw upon here. Clements believed in a universal law that bare regions gave rise, through the process of ecological succession, to a climate-controlled, mature climax community. So long as resources (food, light, water, proper soil conditions) were adequate, Clements thought that environmental disturbances such as fire or drought would be followed by a succession of plant communities eventually leading to the climax community. He thus viewed the community as a discrete entity with its own emergent properties (i.e., properties that are characteristic of a higher level of organization) that were distinct from the properties of individual parts (i.e., the populations within the community). Clements's approach is aligned with the modern concept of the ecosystem and the application of systems theory to explain its functioning (Allen and Hoekstra 1992).

Clements's holistic view of the community strongly contrasts with the individualistic view of Gleason. Gleason emphasized the importance of the life history and biotic characteristics of the populations as the major factors that held the community together. Population characteristics were derived from an evolutionary approach that held that natural selection determined the community structure. Thus, Gleason's community can be viewed in a reductionist manner in that he did not view the community as a distinct, higher biotic organizational level. Although the views of Clements and Gleason were not totally at odds with each other in terms of community structure and functioning, a simplified distinction is that Clements's view generally emphasized plant communities from the standpoint of the landscape and the broad spatial context, whereas Gleason's emphasis was on individual organism-level processes.

Community ecology grew from a qualitative descriptive field into a deductive science aimed at developing quantitative laws to explain and predict species diversity and relative abundance, competitive strategies and resource allocation, and community structure and the assembly of species (see Cody and Diamond 1975). Many classic experimental studies of competition in the field (e.g., Connell 1961;

Paine 1966) were designed to study predation and competition between species, with limited focus on the impact of physical and chemical factors.

Although biotic interactions were often paramount in community ecology, certain concepts emerged with potential application to the study of climate change. One example is the competitive exclusion principle. Hutchinson (1978) developed this concept from classical Lotka-Volterra equations describing predator-prey relationships within a community in the absence of environmental change. This principle holds that if two species have identical physical tolerances or resource needs, when they come in contact with one another, one species will out-compete the other and exclude it from the region. Population dynamics thus calls for the exclusion of one species from a community when two species have similar niche requirements. While the lack of an environmental context renders the Lotka-Volterra equations not directly applicable to paleoclimatology, competitive exclusion in nature might be the cause of some observed faunal and floral patterns. The absence of a particular species in fossil assemblages may at one time be related to environmental factors but at other times be the result of biological competition.

Modern views on community structure vary, but the idea that communities involve primarily biological processes has been modified in recent years. There is now an acute awareness that physical parts of the environment limit the distribution of species within a community, and ecologists today recognize different types of community structure. Brown (1995) lists several community structures that vary owing to uniform or aggregated dispersion of species traits, rules of species assembly, core-satellite distribution patterns, and nested subsets of different species diversities. Each type of community structure may have a different signature, depending on whether competition, predation, or abiotic factors are controlling species' distributions.

Can one then objectively define the characteristics of a community? Some early studies on community behavior under natural circumstances represented attempts to derive statistical methods to measure community structure and distinguish elements of one community from the next. The reasoning went, if communities are real entities, then there should be a means to define them. Among the more important methods were similarity indices using ordination techniques (e.g., Bray and Curtis 1957), which treated species of a community as variables. Multivariate studies using either species' presence or absence or species' proportions to define communities were the precursors of similar types of studies of fossil communities in paleoclimatology (Imbrie and Kipp 1971; Prentice et al. 1991) and paleoecology (Birks and Birks 1980). As discussed later, multivariate analyses of species assemblages support the concept of communities as real entities consisting of integrated groups of species.

In sum, the nature of communities is subject to many viewpoints, as are interacting organisms in nature. Nevertheless, a few generalizations are relevant to climate change. Many community ecologists (usually animal ecologists) focus on populations and the processes of competition, mutualism, predator-prey relationships, and the symbiotic interactions that influence species. Other ecologists, usually terrestrial ecologists, approach the community from the standpoint of individuals within the community. Animal populations are inherently easier to observe and define; trees reproduce and disperse in a very different manner. Population ecologists usually come from evolutionary backgrounds, community ecologists from ecological backgrounds. Both cross paths at the level of the community. Applying the principles derived from both camps aids in the interpretation of climate change, as illustrated in the following example.

Paleocommunities and Paleoclimate

Although the term *community* itself is rarely encountered in paleoclimatology, the idea of an integrated group of species is implicit in paleoecological studies that use species assemblages to reconstruct climate. Indeed, along with species, no biotic entity is more germane to paleoclimatology than fossil associations of plant and animal species that change in their relative proportions over time. Paleoclimatology is heavily dependent upon paleocommunities to answer the most basic questions about climate change. Paleoclimatologists use terms like assemblages, associations, biofacies, communities, and biomes to refer to multispecies communities of organisms from a particular region or habitat. These terms refer to groupings of species that may be members of a higher taxonomic group, such as a planktonic foraminiferal assemblage, or taxonomically unrelated species, such as a pollen assemblage from a forest. In both cases, however, these terms refer to a cross-genealogical group of species that coexist in a region; all of these terms fall into the category of the paleocommunity.

The paleoecological fossil assemblage is, of course, an incomplete subset of the former community. The foraminiferal assemblage recovered from marine sediments does not represent the entire surface ocean community; many parts of the community are not preserved as fossils, and other parts experience postmortem alteration. Likewise, fossil pollen assemblages provide a biased picture of a forest community; pollen tells us little direct information about coexisting forest vertebrates, insects, soil microbes, and so on. Still, the use of thanatocoeneses, or death assemblages, in paleoclimatology to reconstruct environments is not unlike a modern ecological field study in that obtaining a complete census of all organisms in a natural community at any time and place is virtually impossible. Thus, paleoecologists and ecologists alike accept to varying degrees that an assemblage is a representative subset of the entire community.

Many basic questions about communities can be addressed through the use of paleoclimate data. Two major concepts central to the link between climate change and communities illustrate this point: The reality of community structure over extended time scales and multiple climate cycles and the dynamic disequilibrium of communities during transient climate change.

Reality of Communities

Are groups of species called communities real entities or heuristic constructs of convenience? Stated another way, can we define the spatiotemporal boundaries to communities? If assemblages of species disassemble when faced with a large-amplitude, "permanent" climatic change and then reassemble into roughly the same abundances as a climatic cycle completes itself, then the answer is very likely yes. One might infer that the same competitive and predator-prey interactions that existed during the prior cycle, prior to disruption, reemerge intact. Moreover, in this case, there is a much firmer basis to infer climate-related data from paleocommunities. Conversely, if species reassemble in totally different groupings, even though the same or similar environmental conditions return to a region, then communities probably do not exist as integrated cohesive units consisting of species. In this case, one would have difficulty using communities to reconstruct distinct environments.

Community structure can be studied during the recurrent habitat disturbances that accompanied cyclic climate and sea-level changes of the late Quaternary. As described in chapter 4, the past few million years of earth's climate history have been dominated by climatic cycles related to variation in solar radiation due to gravitational effects on earth's axial tilt, precession, and orbital eccentricity. The orbital (also called the Milankovitch) theory of climate holds that over frequencies of tens to hundreds of thousands of years, seasonal and geographical variations in solar insolation constitute the "pacemaker of the ice ages" (Hays et al. 1976). During glacial ice ages, large continental ice sheets cause sea-level lowering of about 125 m (chapter 8); interglacial periods have relatively high sea levels, as we have now.

Along uplifting tropical coasts such as the Huon Peninsula, an exceptional record of past high interglacial sea levels associated with periods of high solar insolation and warm global climate are preserved in fossil coral reef tracts. Intensive mapping and radiometric dating of Quaternary Huon reefs provided firm stratigraphic and chronological understanding of reef development and its relation to orbitally induced global climate change. This understanding was necessary to examine reef community evolution (Chappell et al. 1996).

Huon reef tracts offered Pandolfi (1996) a chance to test the concept of community unity by investigating coral reef community structure over 130 ka of climate history. Distinct coral reef tracts formed on the Huon Peninsula about every 20 ka—coincident with the frequency of the precessional cycle. Pandolfi asked the question, Are ecological communities "open" assemblages with each species responding to climate and sea-level change in a singular individualistic fashion (a Gleasonian view), or do they behave like integrated units, multispecies assemblages, reacting in concert to large-scale environmental change (an Eltonian model)? Using quantitative analyses of coral assemblages, Pandolfi compared the taxonomic composition of reefs during each high stand of sea level. Then Pandolfi asked, Did each reef assemblage have the same species composition, and if not, then were the observed differences expected from a random sampling of the available within-habitat species pool?

For the first question, Pandolfi found that species diversity and composition remained constant from 95 ka to the present through nine full cycles of sea-level regression and subsequent transgression. Each assemblage showed what Pandolfi called "limited membership"; that is, Pacific Quaternary coral reef assemblages did not appear to be open communities. In fact, the spatial differences associated with different reef microhabitats (or biofacies in geological terms) were found to be greater than the temporal differences among assemblages from different high stands.

For the second question, Pandolfi found a lack of differentiation among the reef assemblages from reef tracts of different ages; the similarity among them was greater than chance sampling would produce. Reef assemblage structure apparently did not consist of random associations. This result implied a level of biotic integration that was repeated over multiple climate cycles due to nonrandom factors. This use of reefs to track en masse Quaternary environmental change caused by glacio-eustatic sea-level oscillations raises important, still-unanswered questions about the role interspecific biotic interactions play within these speciose reef communities in maintaining community structure once interglacial oceanic conditions are attained.

The cyclic pattern of Pacific coral reef community evolution during the Quaternary is one example of many dozens of similar faunal and floral communities that oscillate freely over orbital time scales. Indeed, Pandolfi's conclusions are in many ways routine fare for micropaleontologists who use quantitative microfaunal or microfloral assemblage data to study Quaternary climate change instead of community dynamics. Several of these investigations are discussed in detail in the next chapter's coverage of orbital climate change. From both ecological and climatic perspectives, the overwhelming evidence for such long-term community stability, punctuated by frequent, periodic environmental disruption, lends credence both to the reality of many animal and plant communities and to the applicability of paleocommunities for tracking long-term climate and environmental change.

Communities During Transient Climate Change

The second aspect of paleocommunities and climate change pertains to rapid climate transitions that occur over short time scales of decades to millennia. At certain times of relatively rapid, large-scale climate and environmental change, communities tend to be dynamic and often unstable, consisting of assemblies of species unlike those encountered during periods of climate equilibrium.

Transient climate change raises the complex question of ecological succession—the theory to explain the progressive development of plant and animal communities after an environmental disturbance. Originally formulated to explain spatial, not temporal, relationships among plants in heterogeneous environments, succession allows us to directly focus on how well plant and animal communities track a directional trend in climate rather than cyclic climate change.

The most intensely studied period of transient climate change is the last deglaciation, from about 20–10 ka. As earth's climate shifted from a glacial to an interglacial climate extreme, overall global warming did not proceed in unidirectional fashion. Instead, it was characterized by several fits and starts when brief but intense periods of cooling interrupted the more general warming trend. These rapid climate reversals lasted centuries to a millennium, and some ecosystems experienced temporary shifts half the magnitude or greater than those of a full glacial-interglacial cycle.

Rapid climate transitions have a severe impact on marine and terrestrial communities. Abundant faunal and floral data show that the assemblages of species developed in some plant and animal communities bear little or no resemblance to those that exist anywhere in the modern interglacial period (chapter 5). These nonanalog assemblages reflect the environmental instability communities face during rapid climate change. They are also characteristic of rapid climate change.

The existence of nonanalog assemblages raises an important point about community disequilibrium that has bearing on paleoclimatological reconstruction. Over relatively short time intervals of centuries to millennia, communities that become disrupted are of diminished utility as climate indicators because the species that comprise them disassemble as a cohesive unit. Instead of communities, individual species, which often have a well-defined response to climate forcing of specific niche characteristics, become a better tool for reconstructing paleoclimate trends during rapid climate change.

The use of species-level proxies to document rapid transient climate change and community level assemblages to track long-term cycles is not universal. The degree to which both communities and individual species respond to perturbation depends on the complexity of the community and species dynamics within it and the individual life-history traits of the species. Some relatively simple communities may in fact shift quickly enough in response to rapid changes to be extremely useful. Likewise, some species may have such broad ecological tolerances as to render them unfit in paleoclimatology. Still, a perusal of the multitude of paleoclimate studies using quantitative fossil assemblages tends to confirm what ecological studies on modern species show (Bazzaz 1990): that each individual species will exhibit a distinct response to rapid climate change.

In sum, communities are a level of biotic organization that by their nature are more difficult to define than an individual organism or a species. The blurring of community boundaries occurs prominently at both their spatial boundaries and also at times of rapid climate disequilibrium. If one views the transient state of terrestrial plant communities during the rapid climate change of deglaciation, one finds evidence for disequilibrium along the advancing margins of species distributional ranges. Species dynamics within a natural community do not operate within a vacuum but under constantly shifting climatic and environmental conditions. Allen and Hoekstra (1992:44) state: "Competition in communities is set in a variable environmental context that does not allow population competition to come to a simple resolution." Communities sometimes seem to be artificial constructs with no basis in reality.

Conversely, if one extends the paleocommunity record back 500 ka, through several full glacial-interglacial cycles, one finds recurrent vegetation associations that are virtually indistinguishable from one interglacial to the next. The interglacial reef community takes on a dimension of reality, albeit intermittently. In this case, Quaternary paleocommunities of species exist not merely as statistical associations but as real entities that can be explained by both biological and physical mechanisms and traced through time. Thus, communities can be real, cohesive biotic entities under certain conditions, even though communities sometimes "leak" (Eldredge and Grene 1992).

If we accept that there is in principle a natural cohesion to many communities (due to both biotic interactions and physical constraints on morphology and physiology of their species), and that communities periodically disassemble and reassemble, then we can more readily accept the application of paleocommunities to paleoclimatology.

Biomes

There is a natural progression from the level of species and populations, to communities and ecosystems, and finally to the larger spatial and temporal scales in which we study faunal (zoogeographic) and floral (phytogeographic) provinces and the biomes that inhabit them. Biogeography has traditionally examined the broad distribution of animals and plants over thousands of kilometers and continent- and ocean-basin distances. Early naturalists such as Charles Darwin and Alfred Wallace established biogeography as a traditionally descriptive field emanating from nineteenth and early twentieth century discoveries of new faunas and floras in unexplored (by Europeans) lands. In recent years, a hypothetico-deductive approach to biogeography has appeared in a school called *vicariance biogeography.* However, the accumulation of biogeographic data on species in the field of paleoclimatology remains to some degree descriptive and inductive.

Many scientists currently view the large-scale features of the earth's faunas and floras in terms of the biome. The biome, like the ecosystem, includes both the biotic elements of a large region and the physical environment in which organisms live. A grassland biome, for example, includes the dominant plant types and the soil, the atmospheric conditions, the physiography of the land, the region's hydrology, and so on. The biome is thus defined both by its flora and fauna and its climatic characteristics.

Table 3-4 lists the major biomes of the world. Several classification schemes are available, depending on one's purpose and viewpoint. For example, general circulation climate models classify the world's land surface from the standpoint of physical characteristics (albedo, evapotranspiration, etc.) to simulate those aspects of global climate change that involve large-scale biogeochemical processes. These models divide the earth into a finite number of discrete categories having well defined spatial boundaries. In the Goddard Institute of Space Studies (GISS) model, more than 20 major biomes characterize the world. Global and regional biome models, a separate class, simulate terrestrial biotic response to changing atmospheric and climatic conditions (e.g., Nielson and Marks 1995). Biome models are a fundamental tool in attempts to predict future plant response and biogeophysical feedbacks to elevated atmospheric CO_2 concentrations.

There are subtle distinctions between biomes and ecosystems and communities. Allen and Hoekstra (1992) suggest that biomes are not simply big ecosystems but rather biomes are climate-mediated entities in which the physical environment defines the context of the system. The soil represents the accommodation between the vegetation, precipitation, the underlying regolith (bedrock), physical and chemical weathering, and organic and inorganic processes. Drawing a distinction between biomes and ecosystems may seem like splitting hairs, as ecosystems are comprised of biotic plus abiotic elements, and at least with respect to the spatial dimensions of biomes and ecosystems, there is no definitive scale to distinguish between the two. Nonetheless, the

TABLE 3-4 Some Biomes of the World

Terrestrial	*Aquatic*
Tropical rainforest	Open surface ocean
Tropical seasonal forest	Coastal upwelling zone
Tropical evergreen forest	Continental shelf, littoral zone, and estuary
Tropical deciduous forest	Algal bed
Boreal forest	Coral reef
Wood-shrubland	Deep-sea thermal vent
Savanna	Equatorial upwelling zone
Temperate grassland	Swamp and marshland
Tundra and alpine meadow	Lake and river*
Desert-shrubland	Continental slope and thermocline*
Continental ice sheet	Abyssal plain

Cultivated land & urban areas are not included.
*More ecosystem level but grouped as biomes.

concept that a biome characterizes a region equivalent to a major climatic zone is conceptually important in paleoclimate research.

The biome-community distinction is more clear-cut. Biomes are not simply scaled-up communities. Communities are characterized by accommodation between biotic elements only; biomes show accommodation between the whole biota of a large region and its physical environment. Thus, the most important point stressed by many ecologists is that the biome concept connotes the scale of a climate zone, a physical-biotic hybrid that forms an entity that is bounded by major physical gradients in temperature, precipitation, and other parameters.

From the evolutionary viewpoint, another key element regarding biomes and biogeographical-scale processes emerges. Large-scale plant provinces on one continent might be defined by species that have distinct physiognomy and morphology that reflect adaptation to a particular precipitation and temperature regime. However, convergent evolution tends to produce similar plant morphologies independently on different continents in regions with similar climates. Animal physiological adaptation to climate factors might also be reflected in species' distributions. At the level of the biome, processes of interspecific competition and predation are relegated to minor roles, superseded by the large-scale evolutionary processes governing the limits of species' physiological tolerances of extreme climatic conditions or their dispersal capabilities in response to major environmental changes.

The importance of the biome concept in paleoclimatology can be illustrated in the context of biotic-climatic reciprocity of the global carbon cycle described earlier in this chapter. Both temperate and tropical forests contribute CO_2 to the atmosphere and store carbon as biomass at rates that depend on their primary productivity. The interaction between terrestrial biomes and climate via carbon sequestration and CO_2 utilization is in fact one of the most active areas of climate research because of concern about increasing modern CO_2 concentrations (e.g., Korner and Bazzaz 1996; Koch and Mooney 1996). Although lush tropical forests may seem to have high rates of biological productivity, consistent with their relatively high species diversity and forest density, in fact, current understanding of tropical lowland terrestrial biome forest suggests that the *rate* of biological productivity in tropical forests is lower than in many temperate forests (Huston 1994). Relatively low rates of tropical forest productivity are a function of two factors: the predominance of weathered acidic soils (a result of high temperature and precipitation), and typically high day and nighttime temperatures, which lead to high respiration rates and a large carbon—fixation by forest plants. High gross annual productivity of tropical forests is due to the year-round growing season, not to high growth rates. On a monthly basis however, tropical productivity is lower, or at most equal, to temperate forest production.

In cases of tropical and temperate forest biomes, climate weighs heavily in overall chemical and energy cycling through its impact on carbon storage. Ultimately, the functioning of these biomes has an impact on the global atmospheric CO_2 budget. The same is true of high-productivity marine equatorial ecosystems, which help sequester carbon into the world's ocean from the atmosphere, affecting global carbon budgets over thousands of years. Forest and ocean ecosystems are prime examples of large-scale biome-climate interactions. Although their boundaries are not as discreet as those of individual organisms, species, or even communities, biomes lend themselves to global paleoclimatology through a steadily improving understanding of large-scale biotic-climatic processes.

Biome Boundaries and Functioning and Paleoclimatology

The dynamic nature of biomes during periods of climate change is frequently the subject of research in paleoclimatology because modern biome boundaries, where large-scale species turnover occurs over relatively small distances, correspond to critical climatic or environmental gradients. Some of the more important biome-climatic boundaries are the temperature gradients across the Polar Front, nutrient levels across the Equatorial Convergence Zone, temperature and/or dissolved oxygen through the oceanic thermocline and oxygen minimum layer, and atmospheric temperature and precipitation across the forest-tundra boundary. In fact, it is axiomatic when studying climate history to develop a strategy that focuses sampling near modern climate and biotic gradients. A strong climatic gradient should, in theory, be most sensitive to forcing of climate.

The complexity of terrestrial biome dynamics during rapid climate change emerges from studies of the spatial and temporal response of North American mid-continent vegetation to the warming of the last deglaciation (Davis et al. 1986). The work of Davis and colleagues was unique and valuable because they distinguished between patterns of vegetation that were caused by the physical influence of climate change on different plant species that had well-defined ecological tolerances. They contrasted the impact of climate change with that of migrational barriers such as Lake Michigan, at 1000-yr intervals, on the migration of two tree species, American beech and eastern hemlock. The former species has heavier seeds and is dispersed mainly by birds and mammals; the latter has light-winged seeds and is wind-dispersed. Both species migrated into the region 5000–7000 yr ago, several thousand years after deglaciation had occurred and considerably later than when other forest species had reached equilibrium in the region. Did this delay result from poor dispersal capabilities, continuing climate change, or physical barriers?

Shifting species borders such as those encountered at the Great Lakes forest biome edge represent states of disequilibrium that are due either to biological or environmental (climatic) factors, or both. The edge of the biome as defined by the northern limit of its dominant species could in theory be limited by four factors: the effectiveness of the dispersal mechanism, the rate of environmental change, the frequency of the disturbance in providing suitable habitat, and interspecific biotic interactions (Delcourt and Delcourt 1987). Davis's studies discovered that both climatic and biotic factors mattered in the invasion and expansion of northern forest species, depending on the particular conditions at any given time and place. During the late glacial and early Holocene, climate change was more rapid than the dispersal capabilities of the plant species. In other cases, during the mid and late Holocene, dispersal was fast enough for plants to keep pace. Invasion was complicated by the fact that precipitation variability and changes in late glacial outlets caused Great Lake levels to oscillate, constantly shifting potential dispersal pathways. In general, because of its dispersal mechanisms, eastern hemlock was more limited during the Holocene in both its initial establishment as an invading species and its subsequent expansion into new territory. Biological characteristics, not climate change, accounted for many aspects of the hemlock invasion patterns.

Delcourt and Delcourt (1991) cited the study of Davis and colleagues as a prototypical example of climatic versus biotic (dispersal) factors controlling the invasion of species at the forest edge. This particular example highlights the complexity of the fine-scale events occurring near a biome margin. Similar events were played out across North America along the retreating Laurentide Ice margin, in other segments of the biome boundary, and at other biome boundaries in North America (Prentice et al. 1991; Webb et al. 1993) and Europe (Huntley and Prentice 1988).

Forest biome dynamics and climate change are the subject of a large literature reviewed by Webb (1992). As in the case of Michigan forest response to the last deglaciation, plant response to climate change over 10- to 10,000-yr time scales has been found to be quite variable. Some change in local plant populations and genotypes over 30–150 yr represents the results of competitive interaction among species. Other shifts in plants' geographical ranges at rates varying from 50 to 2000 m/yr depending on the species represent physiological limits to dispersal and survival. Climate change acts in concert with biological factors to control the dispersal and distribution of species and communities. Ultimately these large-scale biome shifts can impact regional and global biogeochemical cycling. In chapter 9, we will return to the topic of large-scale biome processes and biogeochemical fluxes of carbon and other elements in discussions about polar and alpine ice core paleoclimate records.

Adaptation

The final biological concept that is an essential plank in the theoretical foundation of paleoclimatology is that of evolutionary adaptation. Although paleoclimatologists accept a role for ecological principles in the interpretation of paleoclimatological data, evolutionary principles and processes in general, and adaptation in particular, are rarely mentioned in paleoclimatology. If, however, we accept that species are real entities, consisting of organisms whose phenotype (their morphology, physiology, behavior, and geochemistry) is used to study climate change, then a logical question is *why* are species so useful as indicators of the environment? How do species come to be so (apparently) "adapted" to their environment, and under what circumstances is the organism's phenotype not necessarily linked to the physical environment?

Let us take the example of the planktonic foraminiferal species *Neogloboquadrina pachyderma.* The abundance of *N. pachyderma* is inversely correlated with sea-surface temperature, making it an excellent indicator of frigid climates in high latitudes (e.g., Lehman and Keigwin 1992; Bond et al 1992, 1993). *Neogloboquadrina pachyderma* is considered to be physiologically *adapted* to frigid ocean temperatures such that its populations disperse into oceanic regions when temperatures fall below a certain point. Although *N. pachyderma* exhibits complex behavior in the water column at smaller spatial and temporal scales (e.g., Kohfeld et al. 1996), its physiology dictates its large-scale geographical distributions. *N. pachyderma* has served paleoclimatology well as a proxy for measuring the relative cooling of North Atlantic Sea-surface temperatures because it is well adapted to such conditions.

A more complex situation arises when one considers the chemical composition of the calcium carbonate shells of deep-sea benthic foraminifers, an important tool in understanding deep-ocean circulation changes. The ratio of coprecipitated cadmium to the primary ion of the shell, calcium, in *Cibicidoides wuellerstorfi* correlates positively in most of the world's oceans with the nutrient content of seawater (see Boyle 1988a, 1992, 1994). Because seawater cadmium concentration correlates positively with that of phosphorous, an important nutrient in the oceans, researchers use the concentration of cadmium in *C. wuellerstorfi* as a proxy for paleonutrient concentrations over glacial-interglacial time scales. Is the temporal and spatial variability in foraminiferal shell trace elements such as cadmium a reflection of physicochemical kinetics of foraminiferal calcite precipitation in seawater unrelated to the organism's biological functioning? Or is variable trace elemental concentration an evolutionary adaptation, somehow related to the biology of the foraminifer's shell, perhaps involved with its protective role?

The question as to whether or not there is an adaptive value to shell chemical composition pertains to virtually all phenotypic variables of fossil organisms that are used in paleoclimatology to reconstruct climatic parameters. Consequently, the concept of adaptation of plant and animal species to their environment, with its deep roots in evolutionary theory, lies hidden below the surface of applied paleoclimatology.

The idea that species are adapted to their environment—physiologically, morphologically, behaviorally, and otherwise—may at first seem so obvious that it does not need mentioning. There nevertheless may be cases where an assumption of rigid evolutionary adaptation of morphology or physiology to the environment is unwarranted. Paleoclimatic interpretations will be affected significantly, for instance, if a species can inhabit a wider range of temperatures than that in which it currently lives. Similarly, paleoclimatologists who focus on a single climate variable such as temperature risk an oversimplified interpretation of faunal or floral patterns of change because species are complex entities coadapted to a variety of physical and biological parameters.

Another complication can arise if an evolutionary change occurs that improves a species's "adaptedness" to changing environmental conditions. One example would be alterations in the physiology of marine phytoplankton, such as the chain-forming Antarctic diatoms *Delphineis* or *Chaetoceros*, that could enhance their ability to adapt to changing nutrient levels, affecting carbon sequestration in the world's oceans (Pollock 1997).

As alluded to earlier, paleoclimatology sometimes (and certainly unwittingly) suffers from an "adaptationist" approach to organisms and their application to climatic reconstruction. Organisms are often treated as abiotic entities, as tools to measure physical parameters related to climate. Their biology is pushed aside, tossed into the black box of "vital effects." This practice is unfortunate because species do not exist, nor did they evolve, to serve the needs of paleoclimatology. Given the high level of variability in biological systems, physical and biological parameters will probably be rarely correlated at levels found in physical systems. Stated another way, there is no a priori justification to the underlying assumption that all aspects of a species' phenotype expressed in its form and function, morphology, physiology, and behavior is evolutionarily adapted to fit lock-step–like into its environment. So it is worthwhile to briefly examine the concept of adaptation in evolutionary biology.

The concept of adaptation of organisms to their environment has a long, engaging history. In certain segments of evolutionary biology, however, researchers tended to overextend the concept to explain every morphological feature as an evolutionary adaptation resulting from the process of natural selection. A critique of the adaptationist view in evolutionary biology, in which natural selection is viewed as an "optimizing agent," appears in the classic paper on the spandrels of the Cathedral of San Marco in Venice by Gould and Lewontin (1979) and is useful in the context of the present discussion. The following quote from Gould and Lewontin gives their arguments that organisms must be viewed as integrated wholes, not as a set of disconnected phenotypic parts (p. 581):

> We fault the adaptationist programme for its failure to distinguish current utility from reasons for origin. . . , for its unwillingness to consider alternatives to adaptive stories; for its reliance upon plausibility alone as a criterion for accepting speculative tales; and for its failure to consider adequately such competing themes as random fixation of alleles, production of non-adaptive structures by developmental correlation with selected features (allometry, pleiotropy, material compensation, mechanically forced compensation), the separability of adaptation and selection, multiple adaptive peaks, and current utility as an epiphenomenon of non-adaptive structures.

Gould and Lewontin were essentially saying that each organic structure, each phenotypic trait, each complex pattern of reproductive behavior, does not necessarily require the explanation that it evolved via natural selection to optimize the organism's place in the environment. They use the metaphor of the ornately decorated spandrels of the Cathedral of San Marco to make their point. The spandrels were not designed to exhibit Christian iconography per se, nor to "house the evangelists" painted on them, but rather to adhere to architectural constraints related to the construction of the entire cathedral. Likewise, biological structures did not evolve in isolation, but in the context of the whole organism.

The extension of this point to paleoclimatology should be evident. Phenotypic traits of organisms that are used as proxies of climatic factors are not evolutionary isolates designed solely to meet organisms' needs in terms of singular climate variables. They are parts of an integrated whole with its own properties.

Paleoclimatology, like evolutionary biology, warrants a greater appreciation for the nonadaptive aspects of organisms. To expand upon this idea, the following brief review of the theory of natural selection and adaptation draws on a large modern literature on adaptation. Among the more authoritative papers are Brandon (1978), Burian (1984), Lewontin (1978), Gould and Lewontin (1979), Gould and Vrba (1982), the volume edited by Rose and Lauder (1996), and the excellent book by Brandon (1990) entitled *Adaptation and Environment.* Though I cannot do justice to the concept of adaptation and its place in the field of evolutionary theory, this brief outline of some basic concepts of adaptation may lead the reader to pursue these ideas in the literature.

Brandon (1990:11–12) defines natural selection with an emphasis on the adaptedness of a species: "Natural selection is not just differential reproduction, but rather it is the differential reproduction that is *due* to differential adaptedness, that is, due to the adaptive superiority of those who leave more offspring." Evolution through natural selection thus has several requirements: inherent variation in the genotype, hereditability of traits, and selection in the context of the environment. In brief, this means that the evolutionary processes involve the intraspecific natural selection of individuals selected on the basis of their *relative adaptedness* in the face of environmental change. Environmental change for our purposes involves climate-related changes induced by various forcings, both internal and external to the climate system.

Adaptation of a species is then generally defined in terms of the differential fitness of offspring produced. Fitness is correlated with morphological, physiological, and behavioral traits that embody the phenotype, that allow an organism to survive and reproduce. The term fitness is often used interchangeably with adaptedness (but see Brandon 1990 for detailed discussion of the philosophical aspects of this relationship). The environment is pertinent in that more offspring are produced than can survive because of the inherent resource limitations. In the Malthusian view of population growth, a population will not expand beyond the carrying capacity of the environment. Some organisms will out-compete others, reproduce, and survive, and these will be those individuals that are better adapted. In this sense, the classical Darwinian concept of adaptation permeates paleoclimatology when species abundances or shell chemistry are linked directly to climate-related parameters such as temperature, dissolved oxygen, or nutrients. The main thesis of Brandon's work, however, is that relative adaptedness should be viewed as a "relation between genotype (or organism) and environment" (p. 39). Adaptedness varies across different environments—it is not "environmentally invariant."

I wish to raise four points with respect to adaptation in general, and its relationship to the environment in particular, as pertinent to paleoclimatology. First, and perhaps most important, is the fact that not all phenotypic expressions are adaptations. Random processes such as genetic drift and historical events sometimes determine patterns of organic change. To take an example, most theories of speciation hold that new species arise most often through allopatry, whereby small populations become wholly or partially isolated from parental populations through geographic barriers to reproduction. Indeed, climate change is itself a viable mechanism by which small populations can become isolated by environmental barriers such as temperature gradients or unsuitable habitats. The isolation of populations, however, involves stochastic processes in that a population of individuals with advantageous genetic characteristics are not more likely to become isolated than members of the parental stock without those characteristics. Historical events such as the isolation of populations necessarily implies a degree of randomness and, as much of paleoclimatology is devoted to understanding historical events, it provides a fertile field in which to examine evolution and environmental change.

Moreover, as argued by Gould and Vrba (1982) many phenotypic structures are not adaptations at all. They introduced the term "exaptations" to characterize such nonadapted features. The notion that some phenotypic characters may not have originated as an adaptation to meet some specific requirement has not been studied in terms of biotic factors used in climate reconstruction.

A second aspect of the adaptation-environment linkage, described in depth by Brandon (1990), is the idea that the environment that matters in terms of natural selection is the *selective* environment. The selective environment includes both biotic (e.g., competition, symbiotic relationships) and abiotic (e.g., nutrients, temperature) and is clearly dynamic and variable in time and space. However, external environmental heterogeneity alone is neither a necessary nor a sufficient condition to produce selective evolution through natural selection. Indeed, most temporal environmental changes do not result in evolutionary changes (i.e., speciation, see Cronin 1985; Cronin and Ikeya 1992). Conversely, internal biotic factors unrelated to environmental heterogeneity can produce a selective environment. This separation of external environmental variability from the actual selective environment makes the problem of identifying the origin of true adaptations even more difficult than would otherwise be apparent. The adaptation and environment issue also recalls the ecological concept of the niche discussed earlier in the chapter. A niche is defined broadly in terms of parameters both external (environmental, physical, climatic) and internal to the biological population under consideration.

A third general aspect of evolutionary adaptation is that an organism should be viewed as a whole entity, not atomized into distinct, unrelated parts. In evolutionary biology, this issue pertains to individual phenotypic traits that are often interpreted to have evolved on their own as adaptations rather than within the context of the whole organism (Gould and Lewontin 1979). In paleoclimatology, this issue pertains most directly to the development, through field and laboratory (experimental) studies, of proxies of specific climate-related parameters to meet the needs of the particular climate problem at hand. The discussion in the previous chapter of the history of paleoclimatology's use of proxies to reconstruct glacial-age climatic parameters of sea-surface temperature, aridity, ocean productivity, nutrients, and so on, reflects the practical need to find new methods. In the development of each proxy, however, virtually all other aspects of the organism are subsumed within the search for single-parameter correlations between a particular aspect of the species' phenotype and the desired environmental parameter. Even when multiparameter control of organisms is considered, such as when temperature and precipitation optima for individual tree species are identified (Bartlein et al. 1986), many aspects of the organism's biology must necessarily be left out. Thus, the standard approach to proxy development using biological entities necessarily precludes a more holistic view of the organism, neglecting the fact that an organism's overall phenotype, and its inherent variability, reflects a more complex integrated entity.

Paleoclimatology is full of examples in which complications arise during proxy development as a result of unknown biological or physical factors. The most common general category is that of vital effects, which were also called "biologically mediated" effects by Dunbar et al. (1996). As mentioned earlier, an example is shell or skeleton secretion by different species of foraminifers and coral out of isotopic equilibrium with the seawater they live in. Another category is inter-population phenotypic variability, whereby the statistical correlations between an environmental parameter (e.g., nutrients) and a phenotypic variable (e.g., shell chemistry) vary considerably in different populations of the same species (Boyle 1994; Hemleben and Bijma 1994). In such situations, the life histories of the entire organism may reflect a response to a host of external factors and to internal biological constraints. These situations naturally pose challenges to the use of proxies in climatic reconstruction.

A fourth consideration for paleoclimatology is the theoretical and empirical basis of various levels of selection that have been discussed in the evolutionary literature. Whereas the individual organism is the focus of classical Darwinian natural selection, other levels of biological hierarchy have been viewed as subject to selection. These include the smallest organizational level—the concept of the "selfish gene" (Dawkins 1976, 1996)—and selection of entire species (Vrba and Eldredge 1984). The idea that different levels of biological hierarchy might be selected for is important because of the widely varying temporal and spatial time scales at which paleoclimatology operates (see Jablonski and Sepkoski 1996). Chapters 4 through 7 successively treat temporal scales of climatic variability ranging from millions of years to interannual and seasonal variability, with greatly different biological processes operating within organisms, populations, species, and communities.

In sum, evolution by natural selection and perhaps other levels of selection shape the organism to survive in a changing environment. The principles of classical Darwinian natural selection lie at the heart of much of paleoclimatology, but the idea that organisms' traits reflect a strict adaptation to their environment—that species are simultaneously a reproductively isolated unit (with respect to other species) and a genetically suited entity that produces a phenotype able to withstand climatically related resource limitations—is an extremely complex topic. Historical and stochastic factors and multiple causalities in the form of biological processes engender competing hypotheses to explain the temporal and spatial patterns of variability of any biotic proxy.

SUMMARY

AT THE OUTSET of this chapter, I raised the subject of reductionism and the tendency of paleoclimatologists to oversimplify the interpretation of complex patterns caused by both physical and biological processes. The integration of biological principles within a physically leaning discipline such as the study of climate change is a daunting task because such a wide variety of processes are relevant to the study of climate change: competition, predation, dispersal, migration, growth, social and reproductive behavior, speciation, adaptation, and natural (and other) selection. Still, theoretical aspects of such concepts as the niche, community structure, population dynamics, biogeographic distribution and biome functioning, and evolutionary adaptation are inherently embedded in the discipline of paleoclimatology. Likewise, organism-environment relationships derived from paleoclimatic investigations should be a fertile research field in the disciplines of ecology and evolutionary biology. These concepts should become more evident in the following chapters describing the geological and biological records of climate change.

IV

Orbital Climate Change

When these ice sheets, forced by precession and obliquity, exceed a critical size, they cease responding as linear Milankovitch slaves and drive atmospheric and oceanic responses that mimic the externally forced response.

JOHN IMBRIE ET AL. 1993A

THE JIA-YI MONUMENT

IN 1908 CHINESE ARCHITECTS built the Jia-Yi Monument in Taiwan to commemorate the newly constructed railroad located at the Tropic of Cancer. The Tropic of Cancer is the reference point that marks the highest latitudinal point reached by the subsolar point each year. According to Chao (1996), architects inscribed the precise latitude and longitude on the monument. Over the years, the monument has been rebuilt after repeated destruction by earthquakes, wars, and storms. Jia-Yi was always rebuilt next to the site of the original monument, and today the Jia-Yi Monument still lies at the "original" site, determined in 1908 as the Tropic of Cancer. However, the true Tropic of Cancer had moved 1.27 km southward. This movement resulted from the changing position of earth due to its precessional wobble and tilt in space, and it will continue to move a total of 90 km southward until the obliquity cycle reverses itself thousands of years from now. The tilt of the earth's axis, gradually decreasing each year by 0.5', can be measured by the slow shifting of the tropics southward at a rate of 14.4 m/yr—a kilometer per century.

The Jia-Yi Monument serves as an excellent reference point by which we can see earth's geometry within the solar system changing over historical time scales. Jia-Yi's movement is also a convenient point of departure for a chapter describing astronomical processes that influence climate changes during glacial-interglacial cycles spanning tens to hundreds of thousands of years. The waxing and waning of earth's great ice sheets during the past few million years—sometimes referred to as "the mystery of the ice ages" — has, more than any topic in paleoclimatology, captured the imagination of scientist and layman alike since the mid nineteenth century when the Swiss naturalist Louis Agassiz proposed that massive glaciers had covered his native Switzerland, much of the rest of Europe, and much of North America (Imbrie and Imbrie 1979). Since Agassiz's original glacial theory, we now know that during Quaternary glacial periods, large mid-latitude ice sheets reaching a thickness of 3 km covered more than 15% of the earth's land surface. During intervening interglacial periods such as the current one, continental glaciers like those in Greenland and Antarctica, and smaller ice caps in the Canadian Arctic, Iceland, and elsewhere cover only about 7.6% of the earth's surface (Denton and Hughes 1981; Dyke and Prest 1987; Sibrava et al. 1986). Among the causes proposed to explain Quaternary ice ages, none has been more widely discussed, and especially in the past few decades, none has gained widespread (though not universal) acceptance as the orbital theory of climate change.

The orbital theory of climate is the prevailing theory of glacial-interglacial climate change over tens of thousands to hundreds of thousands of years. This theory is also called the astronomical theory and the Milankovitch theory, after the early twentieth century Serbian astronomer Milatun Milankovitch, who popularized one version. In its basic form, it holds that variations of earth's orbit lead to changes in seasonal and geographic distribution of incoming solar radiation (insolation) reaching the earth's upper atmosphere. These changes in turn catalyze the growth and decay of huge mid-latitude ice sheets and major changes in terrestrial and oceanic ecosystems, atmospheric circulation and chemistry, oceanic circulation, and other perturbations of earth's climate system. The shape of earth's ellipse, the tilt of its axis, the axial precession (wobble), and the precession of the equinoxes (shifting of the earth's orbital rotation around the sun) oscillate at 100-ka, 41-ka, and 23- and 19-ka frequencies, respectively, owing to gravitational forces within the solar system. Resulting insolation changes at different latitudes are a major factor influencing climate changes and produce spectacular cyclic environmental change in many geologic records (figure 4-1). In essence, orbital insolation changes are the metronome and the earth's climatic responses, as measured by geological and biological proxies, constitute the metronomic record (Martinson et al. 1987).

Orbital theory spans many enormously complex and still partially understood topics in astronomy, geology, oceanography, glaciology, climate modeling, and atmospheric sciences. For example, what

FIGURE 4-1 A. Loess section at Weinan, China. Courtesy of N. Rutter and T. Liu. B. (*opposite*) Rhythmic sedimentary layers from the Pliocene Trubi Marl, Punta de Maiata, Sicily. Light and dark layers are caused by changes in biological productivity, nutrients, and rainfall related to changes in orbital geometry. Courtesy of F. Hilgen.

FIGURE 4-1 (*continued*)

processes link insolation changes occurring at the top of earth's atmosphere to climate change? How are small changes in energy input and distribution amplified into massive perturbations of the global hydrological cycle, which in turn lead to increases in the net accumulation of snow and reduced ablation during winter seasons, thus forming ice sheets? What was the role of atmospheric greenhouse gases and ice sheets in affecting earth's radiation budget and atmospheric circulation, respectively? What internal factors (ice-sheet albedo effects, ice-sheet bedrock interaction, precipitation and temperature feedback loops) amplify or dampen externally forced insolation changes?

To simplify its study, Andre Berger et al. (1992) broke orbital theory down into several components: (1) the theoretical computation of long-term orbital and geometric changes of insolation, (2) the design of climate models to transform insolation changes into climate, (3) the collection of geological and biological data to test models, and (4) modeling orbitally induced climate and the comparison of paleoclimatic data to simulated climate variables. Orbital theory thus melds quantitative astronomical "retrodictions" about earth's orbit together with patterns of climate change derived from the geological record. The third component and much of the fourth are the domain of paleoclimatology. Paleoclimatology has extracted from the geologic record extensive evidence about climatic cycles attributed to orbital changes. In this chapter I describe evidence for orbital climate change preserved in the geological record with focus on several of earth's major ecosystems. Because the literature on orbital theory of climate is so extensive, it is important to outline the scope and objectives of this chapter.

First, I present a brief historical development of orbital theory, which has developed over more than 150 years of scientific inquiry. Then, I discuss the basic tenets of the theory. The rest of the chapter is devoted to paleoclimate records—especially several high-resolution, long time-series studies of major ecosystems carried out over the past few decades—that constitute compelling evidence that orbital frequencies are a basic feature of earth's climate for at least the past few million years. Evidence presented for orbital climate change builds upon the principles introduced in chapters 2 and 3, revolving around long-term secular stability of climatically sensitive communities and species periodically disrupted by physical and chemical changes to their environment. Reliable chronology and numerous proxies of climate-related parameters are central to testing the theory.

Orbital variations impact earth's ecosystems in unique ways. Imbrie et al. (1992) reached this important conclusion in their evaluation of 16 different proxy records of climate changes of the past 400 ka. Thus, for example, tropical terrestrial ecosystems are affected very differently from deep-sea benthic ecosystems. The ecosystem-specific response to orbital forcing reflects many internal mechanisms operating in earth's climate subsystems that produce important feedbacks. Feedbacks that either amplify or dampen the response of an ecosystem to initial insolation forcing include processes related to ice sheet dynamics, atmospheric carbon dioxide (CO_2) forcing, and deep-ocean circulation changes, among others. Challenges to orbital theory are also discussed in the context of internal factors influencing climate.

Considerable effort has been devoted to modeling the physical processes and feedbacks that translate orbitally induced changes occurring at the top of earth's atmosphere into energy redistribution and global climate change at the earth's surface (Imbrie and Imbrie 1980; Berger et al. 1984). Although modeling is not a major focus here, in the final section I briefly discuss modeling of orbital scale climate change and a comprehensive process model of orbital climate change put forth by Imbrie et al. (1992, 1993a).

The orbital influence on pre-Cenozoic climates has a storied history (Gilbert 1895; Wanless and Shepard 1936); discussions can be found in volumes edited by Berger et al. (1984) and de Boer and Smith

(1994). Buoyed by the success of paleoclimatologists documenting orbital cycles in the Quaternary, many researchers have expanded efforts to determine whether or not cyclic sedimentary sequences in pre-Cenozoic deposits exhibit orbital-scale frequencies. Some have found the evidence to be compelling for orbital cycles throughout the Phanerozoic (e.g., Schwarzacher and Fischer 1982; Fischer et al. 1991; de Boer and Smith 1994). Others, such as Dott (1992), issue a firm caution about invoking orbital forcing of pre-Cenozoic sea-level cycles and cyclic sedimentation unless deposits are dated well enough to prove orbital forcing.

Pre-Cenozoic evidence for orbital climate change is intrinsically captivating, but it is beyond the scope of this text. I focus mainly on late Cenozoic climate cycles; the principles applied here also hold throughout the geological column. Four aspects of the late Cenozoic climate record make it especially suitable for evaluating orbital influence on climate. First, clear evidence from glacial geology for cyclic growth and decay of major polar and mid-latitude ice sheets can be traced back at least 7 Ma in the Northern Hemisphere (Jansen and Sjøholm 1991) and at least 35 Ma in Antarctica (Miller et al. 1996). Second, as described in chapter 2, late-Cenozoic geochronology is exceptionally well developed, and correlation of climatic events is based on an integrated time scale derived from radiometric ages, magnetostratigraphy, and biostratigraphy and on tuning to the orbital time scale (Berggren et al. 1995). Third, the astronomical and celestial mechanics that provide calculations of earth's orbital configuration and associated solar radiation at various latitudes and seasons are also well known (Berger and Loutre 1991, 1994). Solar insolation cycles going back several million years can be compared directly to paleoclimate records. Berger et al. (1992) showed that the periods of orbital cycles have changed over the last 500 Ma but that the eccentricity period changed little. Finally, most species applied to late Cenozoic paleoclimatology are extant, and their ecology and physiology are known to varying degrees. The interpretation of their fossil and isotopic records is not a major source of uncertainty as can sometimes be the case with paleoclimate inferences based on extinct taxa. These factors give paleoclimatologists a firm basis on which to test orbital theory for the past few million years.

EARLY DEVELOPMENT OF ORBITAL THEORY

THE RELATIONSHIP between earth's orbit and climate change is among the oldest links between astronomy and geology. The development of this theory encompasses many important breakthroughs in paleoclimatology, as recounted by Imbrie and Imbrie (1979) and Berger et al. (1992). References to variability in the astronomical factors that might influence earth's climate go back at least to the ancient Greek Hipparchus, who in 120 B.C. observed secular discrepancies in the motion of stars that were due to the shifting position of the earth known as precession. Rapid advances in celestial mechanics in the eighteenth and nineteenth centuries by mathematicians and astronomers, and parallel progress in geological sciences, set the stage for ideas about an astronomical-climate link (see Imbrie and Imbrie 1979).

The nineteenth century French mathematician J. Adhemar is credited by Imbrie and Imbrie (1979) for proposing that changes in earth's orbital parameters led to the ice ages. Adhemar was influenced by the French astronomer Jean le Rond d'Alembert, who had in 1754 calculated that the gravitational pull of the sun and moon on the earth's equatorial bulge caused the 26-ka precession of the equinoxes. Adhemar theorized that the two components causing the precession of the equinoxes—the 26-ka wobble and the 19-ka rotation of the elliptical orbit itself—shifted the positions of the equinoxes along earth's orbit with a 22-ka cyclicity that would cause an ice age in whichever hemisphere experienced winter. Ice ages would occur every half cycle—that is, every 11 ka and would grow in the Northern Hemisphere while decaying in the Southern Hemisphere and vice versa. Under this theory, the number of hours of daylight and darkness in each hemisphere was considered a key factor in explaining ice ages.

In 1837, about the same time as Adhemar's work, Louis Agassiz developed his landmark theory of the ice ages in Europe, which he later described in his book *Etudes sur les Glaciers* (Agassiz 1840). Agassiz discovered first in Switzerland, later in Scotland, North America, and elsewhere, widespread evidence that vast continental ice sheets occupied low latitudes. By the time Scottish scientist James Croll (1864, 1867a,b; 1875) began his study of the orbital influence on climate, Agassiz's ice age theory was accepted by many geologists. Like Agassiz, Croll integrated geological field evidence with the astronomical advances made by French astronomer Urbain Leverrier, who in 1843 determined that the earth's eccentricity changed about every 100 ka owing to gravitational pull of the planets. Croll postulated that high eccentricity (when earth has a relatively elongate orbit) led to ice ages because decreased winter sunlight created a positive feedback loop. As more radiation was reflected, the climate became cooler. Croll also reasoned that the precession of the equinoxes was a factor. When earth's Northern Hemisphere winter is close to the sun, as it is currently, we experience an interglacial period. Conversely, 11,000 yr, when the Northern Hemisphere was farther from the sun during its winter season, the potential existed for a glacial period because the eccentricity of the orbit was also elongate. Thus, Croll postulated the simultaneous interplay of two orbital parameters influencing climate: the elongate orbit and a winter solstice far from the sun. He graphically displayed a hypothetical paleoclimate history in which he estimated that the last glacial period lasted from 250 to 80 ka.

During the nineteenth century development of orbital theory, when astronomical discoveries were commonplace, the geological and biological sciences were also flourishing, providing empirical and theoretical statements about climate change. Sir Charles Lyell in his influential *Principles of Geology* (Lyell 1830–1833) developed his marine drift theory to explain glacial deposits, a notion at odds with Agassiz's theory that continental ice sheets had deposited them. In later writings, however, Lyell acknowledged the existence of the ice ages. Though Lyell was a firm advocate of the concept of cyclicity in geological time (Gould 1987), he nonetheless called upon the slow and steady subsidence and uplift of land areas as the primary mechanism to account for evidence that climate had changed. Charles Darwin also freely cited climate change as support for his theory of natural selection in later editions of *On the Origin of Species* (1859). In chapter 13 of *Origin,* for example, Darwin marshaled considerable faunal and floral evidence for species migration in response to the last glacial period, describing vegetation changes in Europe after the retreat of the last great ice sheets.

The modern version of orbital theory was developed in the early part of the twentieth century by Pilgrim (1904) and Milatun Milankovitch (1930, 1938, 1941). Milankovitch, a Serbian mathematician, developed through his lifelong work the major theoretical underpinnings explaining the influence of orbital changes on solar radiation in various seasons and latitudes. One of Milankovitch's major contributions was to provide a mathematical description of insolation changes and their influence on climates of the past and of how ice sheets would respond to insolation changes. His efforts to quantify the contributions of changes in earth's eccentricity, tilt (obliquity), and precession represent the birth of the modern orbital theory of the ice ages. Milankovitch calculated how the variations affecting insolation in high latitudes during northern summer were critical in the formation of Northern Hemispheric continental ice sheets. His argument was that when summer insolation was reduced, snow of the previous winter would be preserved. This preservation would enhance the albedo and more radiation would be reflected back to space, causing further cooling through a positive feedback (Pisias and Imbrie 1986/87). The interplay of solar radiation and the ice sheet itself became emblematic of Milankovitch's version of orbital theory, and Milankovitch himself remains an icon in paleoclimatology.

SUPPORT FOR ORBITAL THEORY FROM GEOCHRONOLOGY

DATING OF paleoclimate events using techniques of geochronology has alternatively been invoked to support and to refute the influence of solar insolation on climate. For example, the early development of radiocarbon and uranium-series (U-series) dating had a significant impact on testing of orbital theory (Imbrie and Imbrie 1979; Broecker 1968). Recent methodological advances in radiocarbon and U-series dating vis-à-vis accelerator mass spectrometry (AMS) and thermal ionization mass spectrometry (TIMS) are today central elements in efforts to test orbital theory.

One cogent example of disputed chronology involves the relationship between glacio-eustatic sea-level history and orbital climate change. Glacio-eustasy is a concept traced back at least as far as Maclaren (1842). It holds that sea level is low during glacial periods and high during intervening interglacial periods because ice sheets store huge volumes of water on continents. Whittlesey (1868) hypothesized that the storage of water in continental ice sheets caused global sea level to fall by more than 100 m. Coral reef terraces along tropical coastlines provided a direct test of orbitally induced glacio-eustasy because they represent snapshots of periods of climatic warmth and high sea level. During the 1960s, the U-series disequilibrium method was used to date emerged fossil coral reefs exposed on tropical coastlines. The ages of Quaternary emerged reefs formed during high sea level were found to match times of high insolation and climatic warmth predicted by orbital theory (Thurber et al. 1965; Broecker 1968; Broecker et al. 1968). Detailed mapping and U-series dating of coral terraces from Barbados and New Guinea and elsewhere by Mesolella et al. (1969), Bloom et al. (1974), Chappell et al. (1996), and others thus provided support for orbital theory.

The chronology of tropical reef development also led to modifications in the relative weight placed upon different elements of the three main orbital parameters. For example, Broecker et al. (1968) dated Barbados reef terraces, one of which yielded an age of 103 ka. This age suggested a high sea level at a time that did not correspond to a period of high insolation evident from the solar insolation curve from 65°N. Broecker et al. argued that less emphasis be placed on high-latitude insolation stemming from the 41-ka tilt component of earth's orbit. Instead, the low-latitude precessional influence on solar radiation would provide a better explanation for the sea-level record. Later in this chapter we will see how more recent dating of reefs from tropical coastlines has refueled debates about the timing of changes in sea level, ice volume, and insolation.

THE DEEP-SEA RECORD OF ORBITAL CLIMATE CHANGE

TESTING ORBITAL THEORY requires a more continuous temporal record of climate than that provided by the snapshot-like sea-level record of coral-reef tracts. The advent of deep-sea sediment coring, foraminiferal micropaleontology, and stable isotope geochemistry provided continuity of climate indicators. Continuity is a prerequisite for tests of periodicity, an inherent part of Milankovitch's theory. The fairly continuous record of ocean history afforded by micropaleontological data yielded two essential ingredients: a timeseries of climate history and a quantitative method to estimate ocean temperatures and serve as a proxy for climate change. Changing proportions of temperature-sensitive foraminiferal species during glacial and interglacial periods could be compared to astronomer's calculations for changes in solar insolation.

The roots of modern paleoceanography reside in classic stable isotopic, foraminiferal, and oceanographic studies of Schott (1935), Arrhenius (1952), Ericson and Wollin (1968), and others (see Phleger 1976). Emiliani's deep-sea paleotemperature curve derived from the oxygen isotope record of foraminifers (1955, 1993) explicitly supported orbital theory. The landmark paper by John Imbrie and Nilva Kipp (1971), published in the influential volume *Late Cenozoic Glacial Ages* (Turekian 1971), legitimized planktonic faunal assemblage analysis as a tool for estimating sea-surface temperatures (SSTs) and testing orbital theory in the ocean record. The Imbrie-Kipp transfer-function method became a standard tool to estimate past SST variability over orbital time scales (Imbrie et al. 1984, 1992; Ruddiman and McIntyre 1981). A generation of paleoclimatologists have applied this method to paleocommunities of diatoms (Koç and Jansen 1992), radiolaria (Morley 1989), ostracodes (Cronin and Dowsett 1990), and other groups.

Orbital theory has also derived support from the marine oxygen-isotope record of foraminifera. Emiliani (1955) showed that oxygen isotope variability in foraminifera, which he believed recorded oceanic temperature changes, oscillated at orbital frequencies (see Berger 1979; Mix 1992 for reviews). Applying the basic temperature-isotope relationships developed by Epstein et al. (1953), Emiliani postulated that light isotope ratios signaled warm SSTs and heavy ratios cooler temperatures, and that isotope values oscillated at orbital frequencies. Shackleton (1967) revised Emiliani's interpretation of the isotopic signal, suggesting instead that oxygen isotope ratios recorded mainly changes in continental ice volume. Shackleton reasoned that during ice ages, continental ice sheets preferentially uptake the light isotope ^{16}O, leaving seawater and shells secreted by foraminifers enriched in the heavier isotope ^{18}O. Glacial foraminiferal oxygen isotope ratios thus become heavier; that is, the $\delta^{18}O$ is more positive. Interglacial seawater and foraminiferal shell $\delta^{18}O$ is lighter, or more negative.

Many leading paleoclimatologists adopted oxygen isotope stratigraphy as a standard tool in deciphering the oceanic record of orbital cycles (e.g., Broecker and van Donk 1970; Shackleton and Opdyke 1973, 1976; Hays et al. 1976; Pisias et al. 1984; Pisias and Imbrie 1986/87). Several key aspects of orbital theory emerged from these studies. First, during the last 700 ka, the predominant frequency was 100 ka, suggesting a relationship to the eccentricity component of astronomical parameters. Second, the ages of the interglacial periods, defined by the lightest oxygen isotopic ratios, were estimated by interpolation between the Brunhes-Matuyama paleomagnetic boundary and the Recent, and these ages generally matched those for periods of high solar insolation and the U-series ages of coral reef tracts. Third, the isotope curve exhibited a sawtooth pattern characterized by rapid periods of ice decay (known as *glacial terminations*) and progressive, stepwise decline from interglacial to glacial climates (Broecker and van Donk 1970).

By the 1970s, the combination of U-series–dated coral reef tracts, the SST record from planktonic foraminifera and other groups, and the deep-sea oxygen isotope–ice volume curve formed a Grand Synthesis in support of the orbital theory. Much of this research was carried out under the auspices of the CLIMAP and SPECMAP Projects. CLIMAP represented a large-scale multidisciplinary research effort to understand the causes of Quaternary ice ages incorporating both empirical paleoreconstructions and modeling research. By mapping the ice ages and intervening interglacial periods in time and space, CLIMAP culminated in publications including "Variations in the Earth's Orbit: Pacemaker of the Ice Ages" (Hays et al. 1976) and *Milankovitch and Climate,* edited by Andre Berger, John Imbrie, and others (Berger et al. 1984), as well as publications on time-slice reconstructions of the last glacial maximum (LGM) (CLIMAP Project Members 1981) and the last interglacial climate (CLIMAP Project Members 1984), Quaternary paleoclimatology (Cline and Hays 1976), and reconstruction of global ice sheets of the LGM based on glacial geology and modeling (Denton and Hughes 1981).

SPECMAP took the testing of orbital theory a step further by investigating the *frequency domain,* or the periods of major Quaternary climatic cycles. SPECMAP explicitly accepted orbital variations as the driver of Quaternary ice volume changes and, in lieu of direct geochronological age control, tuned the deep-sea isotope curve and other proxy records to the retrodated astronomical time scale (Imbrie and Imbrie 1980; Pisias et al. 1984, see chapter 2). The SPECMAP time-scale model of Martinson et al. (1987) has become the standard against which many Quaternary paleoclimate records are dated and the foundation for the comprehensive model of orbital climate of Imbrie et al. (1992, 1993a) described later. Moreover, the approach pioneered by CLIMAP and SPECMAP has been expanded to older marine sedimentary sequences, deep-sea oxygen isotope records, and faunal and floral records predating the Brunhes-Matuyama boundary. Current evidence supports 6 major 100-ka cycles over the past 600 ka and as many 50 glacial-interglacial cycles of lower amplitude and of lower frequency (41-ka periods), going back about 2.5 Ma to the middle Pliocene (Ruddiman et al. 1989; Raymo et al. 1989, 1992) and to the Miocene (Hilgen 1991a,b; Hilgen and Langereis 1993; Shackleton et al. 1995a; Berggren et al. 1995). Orbital theory has become the prevailing explanation of earth's long-term climate history of ice sheets, sea level, and the oceans.

FUNDAMENTAL TENETS OF MODERN ORBITAL THEORY

THE SLOW BUT perceptible march of earth's orbital orientation conveyed in the Jia-Yi Monument has been formally described in many papers on the orbital theory of climate change (Imbrie et al. 1984; Pisias and Imbrie 1986/87; Berger et al. 1992). Berger (1984) and Berger and Loutre (1991) describe the celestial mechanics of earth's orbital geometry as it relates to climate change. In this section, I describe briefly the basic tenets of the theory derived from these papers.

Understanding how earth's orbit in the solar system varies over long time scales requires an exercise in three-dimensional thinking. The earth's orbital plane around the sun, called the ecliptic, serves as a frame of reference. The earth's axial tilt of 23° 26' 22" measures the angle between the ecliptic and earth's equatorial plane. The tilt leads to earth's seasons. The three components of orbital theory—precession, tilt, and eccentricity—can be described from this frame of reference.

Precession

Precession cycles reflect secular changes in the earth's perihelion—the time when earth moves closest to the sun. The precession of the equinoxes was first recognized in ancient Greece by the astronomer Hipparchus, who determined that minute, long-term secular changes in the position of the North Star caused the equinoxes to change over the centuries. In the sixteenth century, Johannes Kepler explained the precession of the equinoxes in terms of the elliptical orbit of the earth. Precession can be separated into two components—axial and elliptical precession. Axial precession results from torque of the sun and moon on the earth's equatorial bulge. The torque causes the earth's axis of rotation to wobble, which in turn causes the North Pole to prescribe a circle over a 26-ka period (figure 4-2). Elliptical precession re-

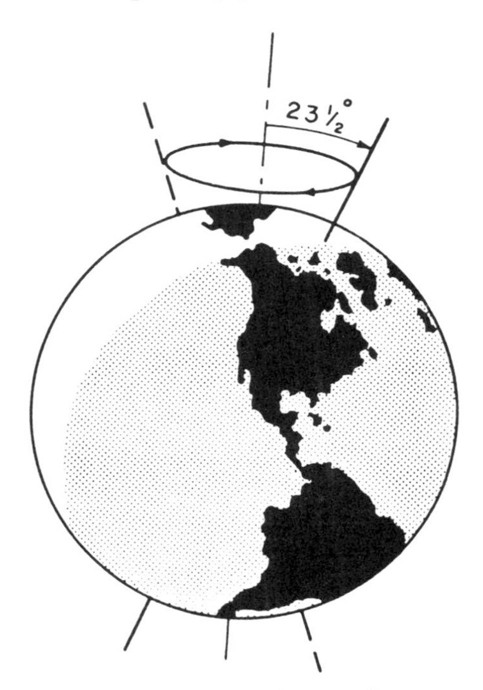

FIGURE 4-2 The earth's axial precession. Courtesy of John Imbrie and Enslow Publishers.

sults from planetary effects on the entire earth's mass and causes the earth's orbit itself to rotate independently about one focus (figure 4-3). Today, summer in the Northern Hemisphere occurs when the earth is farthest away from the sun; the perihelion occurs during the Northern

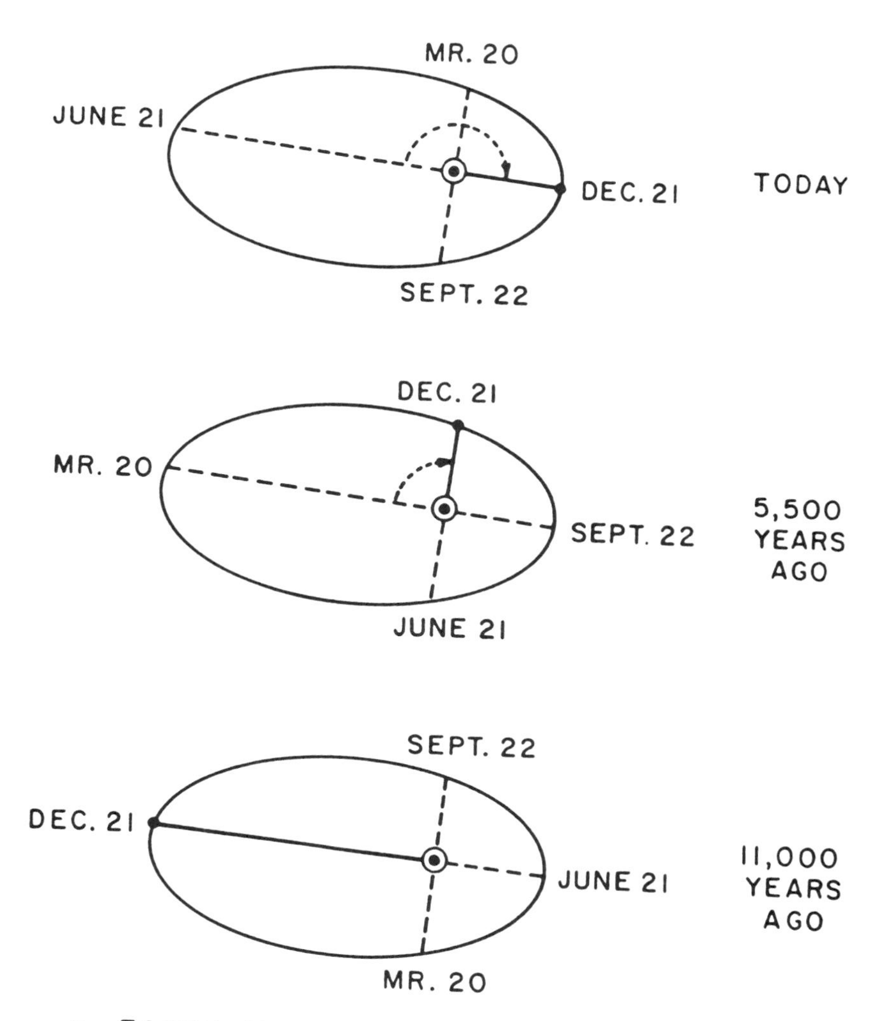

FIGURE 4-3 Precession of earth's equinoxes. Positions of the equinoxes and solstices shift around earth's eccentric orbit about every 22 ka. Courtesy of John Imbrie and Enslow Publishers.

Hemisphere winter, when 10% more radiation is received than will be received 11 ka from now during winter. Conversely, 12 ka ago, as great Northern Hemisphere ice sheets were melting, summer radiation was greater in the Northern Hemisphere than it is today.

The net effect is that the precession of equinoxes (March 20 and September 22) and solstices (June 21 and December 21) shift slowly around the earth's orbit with a period of 22 ka. This period is modulated by the eccentricity to produce 19-ka and 23-ka effects. The climatic impact of precession changes is greatest near the equator and is equal to about ±10% insolation change for a season. At summer solstice, the change is 8% around the mean (40 W/m^2). For comparison, this value is quite large compared with the climatic forcing of 4 W/m^2 that would occur with a doubling of atmospheric CO_2 concentrations. However, atmospheric greenhouse gas forcing occurs year round, whereas precessional changes are seasonal, so earth's climatic sensitivity to greenhouse gases is in theory much greater than it is to seasonal effects of precession. Paleoclimatologists use an index to measure the combined effects of axial and elliptical precession on solar insolation; in the last million years the precession index has varied from –6.9 to +3.7.

Obliquity or Tilt

Today the earth's axis of rotation is tilted about 23.5° relative to the ecliptic plane and is gradually declining at a rate of about 0.5"/yr. The inclination of the axis of Venus is, by contrast, tilted only 3°. When earth is tilted toward the sun, there is summer in one hemisphere and winter in the other. Earth's tilt has varied over long time scales between 22° to 25°, owing to planetary effects on the position of the ecliptic in space. One complete cycle of obliquity takes about 41 ka. The climatic effect of obliquity is that increased tilt tends to amplify seasonal cycles at high latitudes because greater tilt causes polar regions to receive relatively more radiation than reaches the tropics (figure 4-4). The net change in solar radiation reaching the top of the atmosphere in high latitudes during an entire 41-ka cycle is about 17 W/m^2.

Obliquity is therefore considered to be primarily a source of high-latitude climatic forcing, and insolation changes at 65°N latitude are often considered to be a standard measure of obliquity influence on climate. Empirical evidence for 41-ka obliquity cycles derived from the geologic record and modeling of the influence of seasonal tilt-

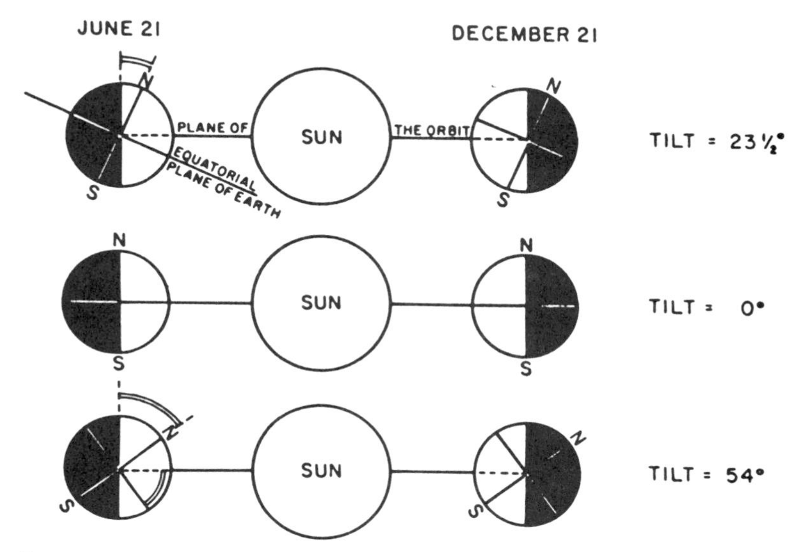

FIGURE 4-4 Effect of earth's tilt on the geographic distribution of sunlight striking various parts of the earth's surface. Courtesy of John Imbrie and Enslow Publishers.

related changes in radiation both suggest the tilt cycles prescribe direct, linear climatic forcing (Imbrie et al. 1993a).

Eccentricity

The earth's orbit is measured by two elliptical parameters—the semimajor axis and the focus of the ellipse. In the past the shape of earth's orbit has changed because of gravitational pull of the sun and planets. At one extreme it has approached a circle and at the other a more flattened ellipse. More elliptical orbits have greater climatic influence on seasons. The main periodicities for the changes in earth's orbit are about 400 ka and 100 ka. A reduction in elliptical orbit from its present value of 0.17 to about 0.04 translates into a very small net change in total solar radiation reaching the earth's upper atmosphere—only about 0.1% of total insolation, or 0.5 W/m^2. A change of this magnitude would cause a change in mean annual global temperature of only a few tenths of a degree Celsius. By comparison, a few tenths of a degree is an order of magnitude less than radiative forcing from glacial-interglacial changes in atmospheric CO_2 concentrations, which causes a mean global temperature change of 1–2°C due to feedback. Thus, the small net insolation due to a 100-ka eccentricity cycle cannot directly cause climate change of the magnitude of a glacial-interglacial cycle. This enigma has led many researchers to suggest that the main effects of changes in eccentricity are modulation of the amplitude of the precessional signal and production of high-amplitude glacial-interglacial cycles through feedbacks internal to the climate system (see later).

To examine the impact of insolation on climate over tens to hundreds of thousands of years, we begin with the assumption of constant solar output and thus equal net insolation over the course of orbital cycles. This assumption is valid over the last 5–10 Ma, the interval of most orbital research to date, although solar luminosity has increased over long geological time scales, during the past 600 Ma (Berger and Loutre 1994), and will oscillate over very short time scales of decades and centuries (chapter 6).

What is the major impact of each separate orbital component? First and foremost, the relative influence of each component does not remain constant in time and space. Over the past 5 Ma, precession, tilt, and eccentricity have affected different regions and ecosystems to varying degrees owing to factors internal to the climate system (see later). In general, however, the obliquity variable is mainly influential in changing the amount of solar radiation reaching the top of the atmosphere at high latitudes. At an eccentricity of 4%, a change in tilt of 1° causes a 2.5% increase in summer insolation at 65°N, a 1.2% increase at 45°N, and an overall average increase of only 0.8% for the entire Northern Hemisphere. Thus, an obliquity-forced climate signal should theoretically influence high-latitude climate processes such as deep-water formation and be evident in proxy indicators such as polar and subpolar planktonic faunas and floras and deep-sea faunal and geochemical records (Imbrie et al. 1992).

Precession, on the other hand, affects seasonal insolation at middle and low latitudes. The effect of precessional changes can be imagined if the eccentricity and obliquity were set at zero. The seasonal cycle would disappear and the latitudinal gradient would be accentuated. In paleoclimatology, the precession signal is generally identified by its impact on low-latitude terrestrial and oceanic processes. Precession-related variability also has been proposed as a modulator of eccentricity. Moreover, nonlinear climatic response to precessional forcing may explain suborbital-scale climatic variability occurring at a 10.3-ka frequency and at other high frequencies (Hagelberg et al. 1994; McIntyre and Molfino 1996). Precession cycles influence low-latitude climatic processes such as tropical wetland activity, reflected in fluctuations of atmospheric methane (CH_4) concentrations found in polar ice cores (Chappellaz et al. 1993; see chapter 9), and equatorial upwelling and tropical monsoons, evident in oceanic geochemical records (e.g., Prell and Kutzbach 1987).

Eccentricity changes introduce such small total changes in insolation, relative to those from precession and obliquity, that the degree to which eccentricity alters earth's climate has been a troublesome aspect of orbital theory. This is the case especially for the past 600 ka, when the 100-ka cycle was evident. Translating small insolation changes into huge glacial-interglacial climatic swings characteristic of earth's recent past has been difficult for paleoclimate modelers. Many workers generally accept that eccentricity modulates and amplifies the signal from precession and tilt and that large ice sheets have a major impact on amplifying the eccentricity signal (Imbrie et al. 1993a). Clemens and Tiedemann (1997) also argued based on the deep-sea oxygen isotope record from 1.2–5.2 Ma that 404-ka and 100-ka eccentricity cycles modulate high-frequency precessional cycles.

Paleoclimatologists and astronomers have devised an index that combines the three orbital parameters—eccentricity, obliquity (tilt), and precession (ETP) — into a single useful measure of insolation (Imbrie et al. 1984) (figure 4-5). There is, however, no "correct" insolation curve in paleoclimatology. The choice of latitude strongly influences whether one finds concordance or mismatching between astronomical cycles and paleoclimatic records and thus whether or not the geologic record supports or refutes orbital theory. In their study of Barbados sea–level record, Broecker et al. (1968) showed how important the choice of insolation curve can be. By plotting the insolation curve for 45°N rather than 65°N, they found a more realistic fit of insolation to the observed record. At a 4% eccentricity, the insolation during the perihelion, when earth is nearest to the sun, is 4.8% higher than it is when the perihelion is reached on the March 21 or September 21 equinoxes. Because a 1° increase in earth's tilt increases the insolation more at high latitudes than at the equator, the choice of an insolation curve for 45°N instead of 65°N leads to greater agreement between the Barbados sea-level record of the last 150 ka and astronomical cycles.

This observation leads to a final comment about chronology and dating geological evidence for orbital cycles. Deciphering the imprint of orbital variation upon earth's climate is largely a study of time. Testing orbital theory hinges on accurate geological correlations among different parts of the climate system (ice, oceans, land, atmosphere). With some exceptions, most studies that have uncovered paleoclimate evidence for orbital climate change have done so without much direct radiometric age dating. Instead, ages for deep-sea and continental climate records, where most orbital studies have been

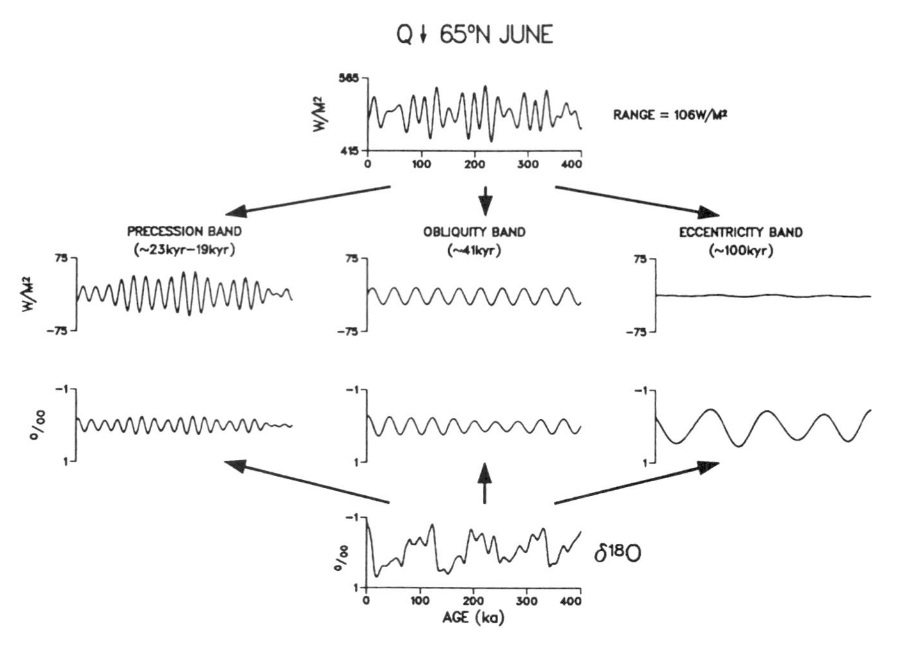

FIGURE 4-5 Partitioning of solar radiation (*Q*) striking earth in June at 65°N into precession, obliquity, and eccentricity bands to ultimately produce the 100-ka cycles of Quaternary glacial-interglacial climate that characterize the deep-sea oxygen isotope ($\delta^{18}O$) curve for the past 400 ka. Systemic response to orbital forcing over this interval is much larger that that produced as a direct linear response to solar radiation changes, suggesting nonlinear factors such as large temperate ice sheets amplify the signal. Courtesy of the American Geophysical Union and John Imbrie. From Imbrie et al. (1993a).

carried out, are usually obtained through relative dating methods used to establish an age model for the core based on correlation to the standard late Neogene time scale. As described in chapter 2, the late Neogene time scale, which itself was traditionally dated by radiometric techniques, is now astronomically tuned to produce an astrogeochronology supplemented by key radiometric tie points and paleomagnetic polarity reversals (Martinson et al. 1987; Berggren et al. 1995). The astrogeochronologic time scale applied to deep-sea paleoclimate studies uses the oxygen isotope curve for correlation from core to core, region to region, and ocean to ocean. The absolute age chronology of many continental climate records is also derived by linking curves of various climate proxy indicators to the deep-sea oxygen isotope curve (sometimes supplemented with additional dating methods). Because the isotope curve is dated between radiometric tie points primarily by tuning, and since most continental records (except ice cores, see chapter 9) do not have an isotope curve, there is an element of circularity in the evaluation of climate change. Interregional correlations are particularly difficult between oceanic and terrestrial records, posing problems related to leads, lags, and phasing between components of the climate system and, more generally, to testing the tenets of orbital theory.

Despite the assumptions of the tuning process and the lack of absolute age chronology for most paleoclimate sequences, high-quality paleoclimate records still provide evidence for orbital influence on climate. Orbital-band frequencies can be identified in proxy records simply by interpolating between polarity reversals under the assumption of constant sedimentation. Orbital forcing appears to drive the functioning and temporal rhythms of many marine and terrestrial biomes and ecosystems, even if the precise sequence of feedbacks and linkages to other regions are poorly understood or in dispute. As we proceed, it is important to keep in mind that improved radiometric dating will no doubt improve our understanding of the way insolation triggers, and internal processes transform, the initial insolation signal to the climate system.

GEOLOGIC AND BIOTIC EVIDENCE FOR ORBITAL CLIMATE CHANGE

IN THE FOLLOWING sections I describe how orbital climate change influences six major ecosystems of the world: (1) tropical coral reefs from the Pacific and Caribbean, whose growth is periodically disrupted by rapid sea-level change; (2) the North Atlantic marine pelagic ecosystem affected by SST variability; (3) the deep-sea benthic ecosystem, where cyclic ocean circulation and climate change has a profound effect on communities; (4) open-ocean equatorial upwelling and subtropical coastal ecosystems driven by low-latitude productivity and monsoonal processes related to precession; (5) African and South American terrestrial ecosystems where vegetation community structure oscillates—latitudinally in the former, altitudinally in the latter—during climatically induced perturbations of rainfall and temperature; and (6) the Mediterranean surface ocean ecosystem and its response to sea-surface salinity (SSS) and rainfall-induced productivity changes. These six topics are listed in table 4-1 with other selected papers showing evidence for orbital cycles.

This eclectic mix of paleoclimate records was chosen to illustrate

TABLE 4-1. A Selection of Papers on Late Neogene Orbital Climate Cycles

Record	*Region*	*Proxy*	*Selected references*
Coral reefs, sea level, and ice volume			
Sea level	New Guinea	Coral reefs	Bloom et al. 1974; Chappell and Shackleton 1986
Sea level	Barbados	Coral reefs	Edwards et al. 1997; Mesolella et al. 1969
Ice volume	Pacific	Isotopes	Shackleton and Opdyke 1973
Ice volume		Isotopes	Broecker and van Donk 1970
Ice volume	Pacific	Isotopes	Shackleton and Opdyke 1976
Sea level	New Zealand	Stratigraphy and foraminifera	Naish 1997
North Atlantic pelagic ecosystems and SST			
Ocean SST	Atlantic	Zooplankton	Ruddiman et al. 1986
Ocean SST	Global	Zooplankton	Imbrie et al. 1989
Deep-sea benthic ecosystems and oceanic circulation			
Ocean circulation	Global	Isotopes	Oppo et al. 1995; Raymo et al. 1989, 1990
Deep sea	Atlantic	Geochemical tracers	Boyle 1992, Boyle and Keigwin 1987
Deep sea	Atlantic	Shell chemistry	Dwyer et al. 1995
Deep sea	Atlantic	Benthic ostracodes	Cronin et al. 1996; Cronin and Raymo 1997
Terrestrial ecosystems, Africa, and South America			
Vegetation	Andes	Pollen	Hooghiemstra et al. 1993
Vegetation	Africa	Pollen	Dupont et al. 1989
Aridity	Africa	Dust, isotopes	deMenocal 1995; deMenocal et al. 1993; Pokras and Mix 1987
Equatorial oceanic ecosystems			
Ocean SST	Atlantic	Zooplankton and phytoplankton	McIntyre et al. 1989; McIntyre and Molfino 1996
Monsoon	Indian	SST, isotopes	Clemens and Press 1990; Prell 1984
Pelagic ocean	Pacific	Faunal isotopes	Farrell et al. 1995; McKenna et al. 1995
Mediterranean surface ocean			
Productivity	Mediterranean	Phytoplankton	Versteeg 1994
Productivity	Mediterranean	Multiple proxies	De Visser et al. 1989; Hilgen 1987; Hilgen and Langereis 1989; Zachariasse et al. 1989

continued

TABLE 4-1. *(continued)*

Record	*Region*	*Proxy*	*Selected references*
Other relevant studies of orbital cycles			
Vegetation	France	Pollen	Guiot et al. 1989
Phytoplankton	Pacific	Diatoms	Sancetta and Silvestri 1986
Aridity	Pacific	Sediments	Janacek and Rea 1985
Continental atmospheric circulation	China	Loess, paleosols	Ding et al. 1994; Kukla 1987; Kukla et al. 1988; Rutter et al. 1996
Continental	Europe	Loess	Kukla 1975
Ocean carbon	Pacific	Isotopes	Shackleton and Pisias 1985
Atmosphere	Devils Hole	Isotopes	Winograd et al. 1988, 1992
Atmosphere	Greenland	CH_4	Brook et al. 1996; Chappellaz et al. 1993
Atmosphere	Antarctica	Paleoatmosphere	Barnola et al. 1987; Genthon et al. 1987; Petit et al. 1997; Raynaud et al. 1993
Surface ice	Arctic Ocean	Plankton foraminifera	Poore et al. 1993
Monsoon	Lake Baikal	Productivity silica	Colman et al. 1995; Williams et al. 1997
Ocean	Indian Ocean	Dissolution	Chen et al. 1995b
Ocean	Pacific Ocean	SST	Hagelberg and Pisias 1990
Glacier	Nordic Seas	Ice-rafted debris	Jansen and Sjøholm 1991
Sea level	Japan Sea	Ostracodes	Cronin et al. 1994

orbital climatic change because it reflects the principles of chronology and proxy development described in chapter 2. First, each record is exceptionally well-documented in terms of its chronology and temporal continuity. In each case researchers obtained continuous, uninterrupted paleoclimate records covering 500 ka to more than a million years. Second, these studies show a relatively solid understanding of the processes that influence the regional ecosystem, including its biotic components. This knowledge permits inferences about climate change from temporal patterns of faunal, floral, and geochemical proxy indicators.

Coral Reefs, Sea Level, and Ice Volume

Sixteenth and seventeenth century explorers to the New World and Indo-Pacific region encountered emergent coral reefs along many tropical island coastlines. By the time he published his 1842 monograph of coral reefs, Charles Darwin had a rich collection of observations of his own from the 1832–1837 voyage of the *Beagle* and those of other explorers upon which to base his theory of coral-reef development. Darwin considered coral reefs evidence of the recent uplift of the land. He referred to fossil reefs in certain regions as "upraised organic remains of a modern date" (Darwin 1842:131), citing the presence of volcanoes near many emerged reefs as a mechanism to explain their elevation. In other regions, Darwin and later Dana (1853) surmised that reefs were undergoing submergence, which ultimately would take a coastline through the classic stages of fringing reef, barrier reef surrounding a central island, and finally an atoll ring of reef islands surrounding a central lagoon.

Darwin's ideas regarding "prolonged subsidence" and construction by reefs of thick accumulations of coral rock that cap atolls and guyots was eventually confirmed by deep drilling in the 1950s through the thousands of feet of coral rock that form the foundation of the Marshall Islands. Darwin was only partially correct, however, in his proposition that reefs now known to be late Quaternary in age were uplifted. Today we know that some fossil reefs formed above current sea level because of glacio-eustatic sea-level rise during periods of reduced continental ice volume, not because they are uplifted (see chapter 8). Moreover, the coral foundations under modern Pacific Ocean atolls are not the result of homogeneous, gradually accumulating reef growth but are composed of complex lagoonal and reefal sedimentary facies punctuated by erosional disconformities that form when the surface is exposed during periods of lowered sea level (e.g., Szabo et al. 1985; Quinn and Matthews 1990). Periods of subaerial exposure occurred because the sea level rose and fell at a rate much faster than tropical islands subside; that is, rapid glacio-eustatic sea-level oscillations due to orbital climate changes were superimposed on long-term subsidence.

During the past 30 yr, studies of Quaternary emerged coral reefs have generally found support for orbital control of sea-level fluctuations, although recent challenges to orbital theory based on new radiometric dating of reefs have emerged (see later). The reasoning for coupling coral reef dynamics to sea level is as follows. All other factors being equal, past sea-level positions along any particular coastline will be inversely proportional to the amount of ice stored in the poles. Less ice stored in polar regions means sea level would be higher than it currently is; with more ice, sea level would be lower. This is referred to as the glacio-eustatic component of sea level. Because tectonic uplift and subsidence occurs at widely varying rates on different coastlines, the original elevation at which fossil coral reefs formed can be altered. The amount and rate of subsidence and uplift will depend upon the regional tectonics, which depends mainly on the location of a region with respect to the shifting positions of continental plates. The location of a coast with respect to plate margin and tectonic activity determines how much uplift or subsidence has occurred over the past million years and therefore how much of the local sea-level record—measured by the elevation of the coral reef tract—is glacio-eustatic and how much is regional tectonic. Along uplifting coasts, rapid sea-level changes are superimposed on rising land, the opposite trend of that for subsiding atolls mentioned earlier. Stable coastlines, in contrast, can be viewed as regions of little or no net uplift or subsidence where emerged coral reefs will only record periods of sea level higher than present ones.

During periods of increasing solar insolation in high Northern Hemisphere latitudes due to the effects of eccentricity, obliquity, and precession, global ice volume should decrease and mean sea level should rise. The converse occurs during times of decreasing insolation and falling sea level. Coral reef development should reflect rising and falling sea level such that the ages of emerged coral reefs should in principle coincide with periods of maximum summer solar insolation and the ages of submerged reefs with minimum insolation.

Corals are a valuable paleoclimatic tool for studies of sea-level variability because of their unique biological attributes. Hermatypic corals grow symbiotically with zooxanthellate algae, which require sunlight to carry out photosynthetic activity. Thus many coral species live in the photic zone close to mean sea level. *Acropora palmata* is one such species; in the Caribbean it is limited to depths less than about 5–10 m, depending on location. Many corals also grow rapidly, as fast as 10 mm/yr. Rapid growth means that colonies can survive during periods of deglaciation and rapid sea-level rise (Blanchon and Shaw 1995; Montaggioni et al. 1997).

Another attribute of corals is the suitability of their skeletons for radiometric dating. During growth, corals take up radioactive elements from seawater, including radioactive elements used in U-series dating (^{234}U, ^{230}Th, and ^{231}Pa, with half lives of 244.5 ka, 74.4 ka, and 32.7 ka, respectively). Many coral skeletons are closed chemical systems; little or no exchange of uranium or its daughter products occurs after the skeleton has formed, and thus accurate ages can be obtained. The alpha counting method of ^{230}Th age dating used in many older studies of corals and Quaternary sea-level history has a large analytical age uncertainty of several thousand years. The TIMS method (Edwards et al.

1987; Bard et al. 1990a,b) vastly improved geochemists' ability to date corals. TIMS analytical age uncertainty is < 50 yr for corals from the last postglacial sea-level rise (< 15 ka) and < ±1000 yr for last interglacial corals (approximately 117–134 ka). TIMS has also recently been applied to measure protactinium-231 (^{231}Pa) concentrations in Quaternary coral reefs as a means of evaluating open-system chemical behavior and postmortem uptake by the coral skeleton of ^{230}Th that might compromise the age estimate (Edwards et al. 1997). Older coral reef tracts have been dated with the helium-uranium method (Bender et al. 1979), but the ages have large errors.

Researchers consider certain criteria as prerequisites for accurate age determinations from late Quaternary corals (table 4-2). Even in cases where these criteria are met, disagreement about whether dated reef tracts support or refute orbital theory can arise. Indeed, different opinions about the reliability of coral reef radiometric ages lie at the heart of debates about orbital theory.

The most intensive research on Quaternary sea level has been conducted on the stair-step emergent reef tracts of tectonically uplifting coasts and of stable coasts (table 4-3). Reef terraces along the Huon Peninsula, New Guinea (see figure 4-6) (Veeh 1966; Veeh and Chappell 1970; Bloom et al. 1974; Gallup et al. 1994) and the Caribbean Island of Barbados (Broecker et al. 1968; Mesolella et al. 1969; Fairbanks and Matthews 1978) have an almost revered aspect in paleoclimatology.

Due to the extremely high uplift rate on New Guinea, at least 20 separate reef tracts are exposed along the Huon Peninsula. Reef tract VII formed during the last interglacial, about 140–120 ka, lies approximately 260 m above sea level (ASL). Huon reefs record *relatively* high eustatic sea levels compared with the extreme low sea level during a glacial, although some New Guinea reef tracts actually formed when sea level was slightly lower than current levels, and they are now exposed because of the high uplift rate (Bloom et al. 1974). Early U-series ages of New Guinea reefs indicated high sea level occurred about

TABLE 4-2. Important Criteria for Dating Corals from Tropical Reefs and Temperate Terrace Deposits

Aragonitic shell composes 95–100% of skeleton
A ^{230}Th/^{232}Th ratio > 20 means no uptake of thorium-bearing mineral
Uranium concentration is 2–3 parts per million
Original ^{234}U/^{238}U ratio of 1.14–1.15 is consistent with ^{230}Th/^{232}Th ratio
There is no evidence of reworking of fossils from older deposits

TABLE 4-3. Selected Coral Reef Sea-Level Studies Pertaining to Orbital Theory

Region	*Reference*
Tectonically emerging areas	
Pacific, Indian Oceans	Veeh 1966
Huon, Peninsula, New Guinea	Veeh and Chappell 1970
Huon, Peninsula, New Guinea	Chappell et al. 1996
Huon, Peninsula, New Guinea	Bloom et al. 1974
New Guinea	Chen et al. 1991
New Guinea	Stein et al. 1993
Barbados	Mesolella et al. 1969
Barbados	Edwards et al. 1987
Barbados	Bard et al. 1990a
Barbados	Edwards et al. 1987
Barbados	Gallup et al. 1994
Barbados	Fairbanks and Matthews 1978
Haiti	Dodge et al. 1983
California	Muhs et al. 1994
California	Muhs 1992
Stable areas: minimal vertical movement	
Florida	Osmond et al. 1965
Florida	Broecker and Thurber 1965
Yucatan Peninsula, Mexico	Szabo 1979
Bermuda	Harmon et al. 1983
Bahamas	Neuman and Moore 1975
Bahamas	Chen et al. 1991
Oahu, Hawaii	Ku et al. 1974
Oahu, Hawaii	Muhs and Szabo 1994
Oahu, Hawaii	Szabo et al. 1994
Western Australia	Zhu et al. 1993
Salmon Bay, Australia	Szabo 1979
Baleartic Islands, Mediterranean	Hearty 1987
Subsiding areas	
Enewetak, Marshall Islands	Szabo et al. 1987; Thurber et al. 1965

Listed studies represent a subset of Quaternary reef studies; see also Muhs (1992), Szabo et al. (1994), Stirling et al. (1995), and Hillaire-Marcel et al. 1996).

every 20 ka: reef tracts VIIa, VIIb, VI, V, IV, and III were dated at about 140, 124, 103, 82, 60, and 40 ka, respectively, showing an apparent precessional periodicity.

On Barbados, Mesolella et al. (1969) dated the Rendezvous Hill, Ventnor, and Worthing terraces at 125, 103, and 82 ka. These represent

FIGURE 4-6 Flight of Quaternary coral reef tracts on the Huon Peninsula, New Guinea. Uranium-series dating of reefs has provided support for cyclic oscillations in global sea level corresponding to orbitally induced climate changes. Courtesy of A. L. Bloom and Prentice Hall. From Bloom (1991).

the tripartite subdivision of the last interglacial corresponding to marine oxygen isotope stages 5e, 5c, and 5a (see table 4-4 in the next section). A younger terrace dated at 60 ka lies just above the present surf zone. Mesolella et al. (1969) proposed that the ages of high coral reef tracts in Barbados supported orbital theory, a conclusion reached by Fairbanks and Matthews (1978) and others.

Bloom et al. (1974) compared the Huon and Barbados sea–level records and provided a strong argument for glacio-eustatic control of reef development over the past 150 ka. Despite the higher uplift rate at Huon, they established a firm correlation between the New Guinea and Barbados reef sea-level record in terms of their ages. By calibrating the Huon record to the Barbados reef record and pinning the 125-ka-old reef to a +6 m eustatic sea level estimated from reefs from stable areas (see next section), Bloom et al. (1974) constructed a sea-level curve showing the highest sea level near 125 ka and successively younger high stands at progressively lower elevations. Thus, the New Guinea curve had elements of a stepwise progression of cooling between the last interglacial near 125 ka and the LGM near 20 ka. This sawtooth pattern is characteristic of the deep-sea oxygen isotope curve (Broecker and van Donk 1970, Shackleton and Opdyke 1973).

In contrast to uplifted coasts such as those on New Guinea and Barbados, emerged fossil reefs along stable coasts where uplift and subsidence are minimal signify only periods when global (eustatic) sea level was higher than modern sea level and continental ice volume was less than it is now. Most investigators agree that sea level was about 4–6 m ASL during the last interglacial (LIG) period when polar ice volume (probably in Antarctica or Greenland) was lower than it currently is. Emergent reefs from the LIG formed in Florida, the Bahamas, and elsewhere (Osmond et al. 1965; Broecker and Thurber 1965) were among the first dated by the U-series method.

Many more LIG reefs in stable regions have been dated since these early studies, confirming a global sea level several meters above the present one (Cronin 1982; Muhs 1992; Neumann and Hearty 1996). Despite agreement that the last interglacial sea level was higher than at present, however, the exact age duration of this interglacial high stand of sea level is in dispute. One school holds that LIG sea level peaked near 125–127 ka, coincident with the northern hemisphere insolation maximum, and lasted about 6–10 ka. Another school holds that LIG sea level rose before 130–135 ka and remained at or near the present level until about 118–117 ka (Szabo et al. 1996; Neumann and Hearty 1996; Winograd et al. 1997).

Coral reef tracts older than those of the LIG exposed on Barbados were dated using the U-series and helium-uranium methods. Bender et al. (1973) found evidence for relatively high sea level about 350 ka and 500 ka; later Bender et al. (1979) produced a tentative correlation between the reefs in St. George's Valley, Barbados, and the deep-sea isotope record showing three reefs formed during isotope stage 7 (dated at 180, 200, 220 ka; see table 4-4, next section, for isotope stages), three during isotope stage 9 (280, 300, 320 ka), one during stage 11 (undated), three during stage 13 (460, 490, 520 ka), and one each for stages 15 and 17/19 (590, 640 ka). Although the older Barbados reefs are not dated reliably enough to firmly establish correlations to insolation changes, they nonetheless demonstrate oscillating sea level over the past 700 ka.

In summary, by integrating reef ecology, radiometric dating, geomorphology, and regional tectonic history, researchers have amassed compelling evidence that Quaternary coral-reef tracts record glacioeustatic sea-level changes. No tectonic theories account for the cross correlation of New Guinea with Barbados reef ages located at two widely separated islands, although tectonic processes certainly alter the original elevation of reefs. For the most part, the ages of fossil coral reefs support orbital theory in that the ages of high sea-level periods correspond to periods of high insolation.

Oxygen Isotopes and Ice Volume

Dated emerged coral reef tracts compose only half the story linking global sea level to orbital forcing of climate change. The hypothesis that tropical reef tracts formed during periods of high insolation and low global ice volume was developed in tandem with evidence for ice-volume changes derived from the deep-sea oxygen isotope record of foraminifera. The oxygen isotope stage system that forms the basis of the SPECMAP chronology also provides a means to estimate first-order ice volume changes associated with orbital theory. Table 4-4 summarizes oxygen isotope stage terminology; by convention, odd-numbered stages represent interglacial periods and even-numbered stages glacial periods.

The principle of using the oxygen isotope curve as an ice-volume monitor stems from the fact that changes in ratios of $^{18}O/^{16}O$ of the

TABLE 4-4. Evolution of Marine Oxygen Isotope Stages

Author (year)	*Stages**	*Source***
Emiliani (1955)	1–5	Caribbean
Shackleton (1969)	5	Deep sea
Shackleton and Opdyke (1973, 1976)	6–23	Pacific Ocean sites V28–238,239
Ruddiman et al. (1986)	24–63	North Atlantic DSDP sites 607, 609
Raymo et al. (1989)	64–116	North Atlantic DSDP site 607
Sarnthein and Tiedemann (1989)	117–136	Northwest Africa ODP site 658
Shackleton et al. (1995a)	Revised system, kept stages 1–100; pre-2.6 Ma stages labeled by paleomagnetic chronology	

Note: Minor modifications to the definition of certain stages are made in more recent studies.

*Some stages are divided into substages denoted by letters (e.g., substages 5a through 5e). Another scheme divides them into numbered substages (e.g., 5.1 through 5.5)

**Abbreviations: DSDP = Deep Sea Drilling Program; ODP = Ocean Drilling Program.

calcium carbonate shells of tropical planktonic and deep-sea benthic foraminifers are a function of ice volume, water temperature, salinity, and vital effects among species (Emiliani 1955; Shackleton 1967). The total glacial-interglacial $^{18}O/^{16}O$ isotopic range is about 1.5‰ for mid-to-late Quaternary records, varying mainly by location. Shackleton and Opdyke (1973, 1976) argued the equatorial Pacific foraminiferal $^{18}O/^{16}O$ record exhibited mainly an ice-volume signal, with light ratios corresponding to low ice volume and heavy values to high ice volume (see Berger 1979; Mix 1992).

Radiocarbon and TIMS dating of submerged reefs formed during glacial times provide a way to date the glacial sea level and calibrate the deep-sea isotope curve to ice volume. Sea level dropped between 100 and 130 m when continental ice volume was at its maximum; probably about 121 m below present sea level (Fairbanks 1989). From this datum, Fairbanks and Matthews (1978) established that a 0.1‰ $\delta^{18}O$ change was equivalent to about 10 m of sea-level change. Today, it is generally accepted that of the total 1.5‰ glacial-interglacial change in deep-sea foraminiferal $\delta^{18}O$, about 1.2‰ is due to ice-volume changes and about 0.3‰ to temperature changes (Mix 1992; Kennett and Hodell 1993)

Through this line of reasoning, the deep-sea oxygen isotope curve was cross-checked with the radiometrically dated coral reef record, forming a single integrated model of first order orbitally driven glacio-eustatic sea-level changes. Figure 4-7 shows a typical oxygen isotope curve for the past few hundred thousand years that is believed by most researchers to represent orbitally driven sea-level oscillations.

Recent Developments in Dating Coral Reefs and Sea Level

Rapid progress in dating coral reefs in the last decade, largely through the application of TIMS dating and new stratigraphic analyses, has renewed debate about orbital theory and sea-level change. New data sets include corals from Barbados (Edwards et al. 1987, 1991; Bard et al. 1990b, Gallup et al. 1994); North America (Stein et al. 1991; Muhs 1992; Muhs et al. 1994); Houtman Abrolhos Islands off western Australia (Zhu et al. 1993), New Guinea (Stein et al. 1993); eolianites, or beach deposits and notches in the Bahamas (Neumann and Hearty 1996); and mollusks from the Balearic Islands, Mediterranean (Hillaire-Marcel et al. 1996).

Some researchers continue to find support for orbital theory. In a restudy of the classical Barbados reef tracts, for example, Gallup et al. (1994) found the correspondence between isotope stages, reef ages, and

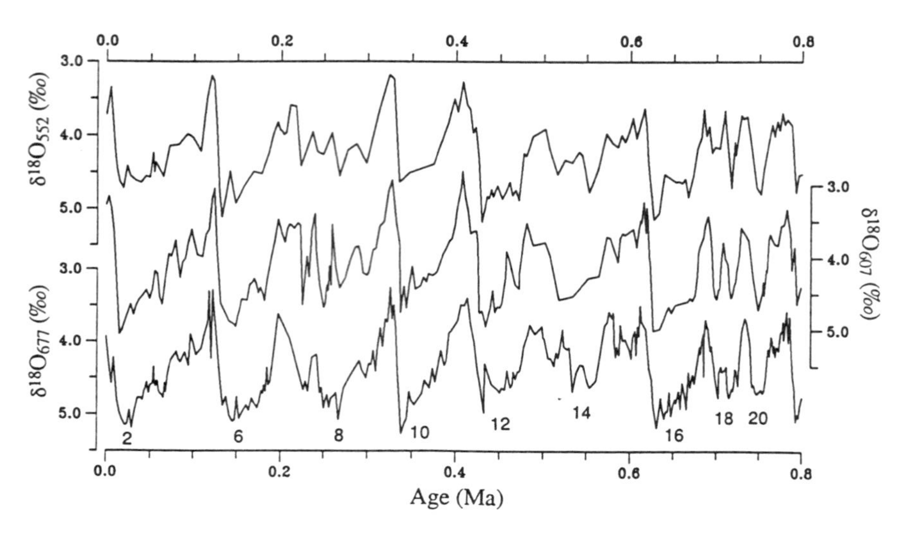

FIGURE 4-7 Oxygen and carbon isotopic variability in deep-sea cores 552 and 607 (North Atlantic) and 677 (Pacific) for the past 800 ka showing predominant 100-ka climate cycles. Courtesy of M. E. Raymo, with permission from Elsevier Science. From Raymo et al. (1990).

insolation maxima listed in table 4-5. They argued that this correlation of reef development to insolation changes was good enough to support orbital theory, suggesting the fit was especially strong during glacial terminations. They also postulated that there was a lag of 5000 ± 2000 yr between the time that insolation decreased and the subsequent sea-level drop. Gallup et al. (1994:796) concluded that "for the last three interglacial and two intervening interstadial periods, sea level peaked at or after peaks in summer insolation in the Northern Hemisphere. This overall pattern supports the idea that glacial-interglacial cycles are caused by changes in earth's orbital geometry." The correspondence of reef-tract ages to the insolation changes of the past 200 ka is believed by some to support orbital theory (Stirling et al. 1995; Edwards et al. 1997).

Some researchers, however, find disagreement between the sea-level record based on coral reefs and the basic tenets of orbital theory. Hawaiian coral reef ages from 131 to 117 ka (Szabo et al. 1994), for example, cast some doubt on orbital theory because they show that sea level rose too early and remained high for too long during the last interglacial period. If insolation was the driving mechanism controlling ice volume fluctuations, evidence that sea rose *before* insolation

TABLE 4.5. Oxygen Isotope Correlations in Barbados Coral Reefs

Oxygen isotope stage	*Reef age (ka)*	*Insolation maximum (ka)*
4/5	79.7	78
5a	83.3	86
5c	103	103
5d/5e	117	122
5e	120–130	128
5/6	> 130	134.5
6/7	190	194

began rising would be discordant with the basic tenets of orbital influence on global ice volume. The Hawaiian reefs showed that the timing of the inception of the penultimate deglaciation may have preceded the insolation maximum at 127 ka. If we assume that sea level took about 6–8 ka to rise from the glacial minimum to the interglacial maximum, as it did during the last deglaciation, called Termination 1 (Broecker and van Donk 1970; Fairbanks 1989), then sea level during Termination 2 must have begun to rise by about 140 ka, reaching its peak at least by the 131-ka date, long before insolation could have forced polar ice retreat. In fact, geomorphically distinct reefs older than about 127 ka have long been known on the Huon Peninsula (i.e., Chappell 1974) and have recently been dated in the Bahamas (Chen et al. 1991) and Barbados (Gallup et al. 1994) (see Stirling et al. 1995; Winograd et al. 1997).

For the LIG sea-level high stand, oxygen isotope stage 5e has traditionally been viewed as a short-lived climatic extreme of about 10 ka. The Hawaii U-series dates require LIG sea level to remain at or near the present level for more than 17 ka. Winograd et al. (1997) argued persuasively that the stage 5e interglacial lasted 5–10 ka longer than expected based on the ice-volume record in the tuned SPECMAP chronology.

Hillaire-Marcel et al. (1996) also found evidence for an extended interglacial high sea level. They examined the stratigraphy and U-series dating of mollusks from the Tyrrhenian ecostratigraphic stage in the Balearic Islands of the Mediterranean. Mollusks have generally been considered unreliable for U-series dating because they take up uranium after death, unlike coral uranium uptake during skeletal growth, and possible migration of ^{230}Th, which can confuse the dates (Kaufman et al. 1971). Still, Hillaire-Marcel et al. (1996) demonstrated only a brief interval, perhaps a few thousand years, of uranium uptake occurred before diagenetic processes closed the chemical system. Under these conditions, Hillaire-Marcel and colleagues postulated, the U-series ages derived from Tyrrhenian mollusks could produce more reliable ages than previously thought. They argued for an extended period of high sea level, lasting about 17 ka, punctuated by a brief sea-level fall, when a reddish conglomeratic beach rock was deposited. Their results demonstrate a relatively long but not completely stable period of sea level during isotope stage 5e. Neumann and Hearty (1996) also postulated a period of extended high sea level from 132 to 118 ka based on the sea-level record from the Bahamas.

Inconsistencies between reef records of sea level and isotopically derived ice volume estimates may reflect the influence of temperature on the oxygen isotope record. These inconsistencies revolve around the elevations of reefs and sea level during isotope substages 5a and 5c, dated at about 105 and 85 ka, respectively. Early workers suggested that sea level during substages 5c and 5a was as low as 45–48 m (Fairbanks and Matthews 1978), corresponding to heavy oxygen isotope values. Sea-level estimates from certain coastlines in temperate (Cronin et al. 1981; Szabo 1985) and tropical areas (see Muhs 1992; Muhs et al. 1994) contradicted these inferences and suggested that sea level during isotope stages 5c and 5a was closer to present levels. Consequently, the estimates for these substages have been revised considerably upward to 10–15 m below present levels (e.g., Hearty and Kindler 1995; Muhs et al. 1994). One explanation of the anomalously heavy oxygen isotope ratios is that deep-sea temperatures cooled during these late interglacial intervals (Chappell and Shackleton 1986; Mix 1992).

How is one to interpret the confusing array of late Quaternary coral reefs and sea-level history in terms of orbital theory? Hillaire-Marcel et al. (1996) summarized three competing models to explain the critical sea-level history during the interval 140 ka to 115 ka and its relationship to orbital theory. One model proposes a double high stand at about 135 and 120 ka with an intervening sea-level drop (Mercer 1981). A second possibility is an extended gradual sea-level rise beginning > 134 ka until a peak near 118 ka (Stein et al. 1993). A third explanation is an extended single period of high sea level between 134 ka and 118 ka (Szabo et al. 1994; Chen et al. 1991; Hillaire-Marcel et al. 1996). In all three cases it is difficult to account for the sea-level pattern if insolation is the sole factor controlling ice volume.

In summary, considerable evidence from fossil coral reefs indicates that their age and elevation signify glacio-eustatic sea-level changes, that the deep-sea oxygen isotope record is primarily an ice-volume monitor, and that both coral and isotopic records show first-order patterns due to orbitally driven sea-level oscillations. Although I have focused here on late Quaternary records, long-term deep-sea isotopic records extending back more than 5 Ma provide additional evidence for orbitally driven ice-volume fluctuations (Shackleton et al. 1995a) based on evidence from ocean-margin stratigraphic records (Cronin et al. 1994; Naish 1997). However, close inspection of sea-level trends during the penultimate deglaciation and the last interglacial period indicate that other factors, perhaps related to unknown aspects of ice sheet dynamics or to complex atmospheric and biogeochemical feedbacks, can complicate the ice-volume response to orbital forcing. A perfect match between coral-reef sea-level history and isotopic ice-volume records is inherently difficult to achieve, and the orbital theory–sea level link will continue to be debated as additional coral reefs are studied.

North Atlantic Sea-Surface Temperatures and Pelagic Ecosystems

There are clear linkages between variability in physical aspects of oceanic systems and variability in biological aspects of pelagic marine ecosystems as measured by biomass and other properties at various temporal and spatial scales (Mann and Lazier 1996). McGowen (1990) pointed out that the complex structure and function of pelagic marine ecosystems made mathematical modeling of fluxes of energy and matter difficult. Rather, McGowen (1990) suggested a phenomenological approach as an alternative way to understanding what drives pelagic ecosystems. This approach involves the comparison of time series of biotic systems and of associated physical changes such as those related to climate. In McGowen's scale-dependent scheme, for example, short-term, annual cycles of oceanic temperature affect biomass variability over spatial scales of 10–1000 km. In contrast, orbital climatic cycles represent a physical-biotic linkage that affects oceanic fronts (time scales of > 5 ka) and biogeographic provinces (time scales of > 1 ka). Consequently these cycles influence a much greater variability of pelagic biomass.

A phenomenological approach to reconstructing long-term pelagic marine ecosystem variability and its relationship to orbital cycles is a hallmark of paleoceanography. In this and the following sections, I describe evidence for orbital-scale cycles in several pelagic and deep-sea marine ecosystems. This section focuses on the impact of sea-surface temperature (SST) changes on foraminiferal populations in the vicinity of the North Atlantic Polar Front during Plio-Pleistocene climatic cycles. Planktonic foraminiferal assemblages are the most commonly used microfossil tool in North Atlantic paleoceanography; the value of foraminiferal ecology for SST reconstruction in particular, and orbital theory in general, cannot be overstated. Building on early work using foraminiferal assemblages for SST estimation by Phleger et al. (1953), Oba (1969), and others, John Imbrie and Nilva Kipp (1971) had the greatest influence on the use of pelagic microfossils for quantitative SST estimation and for testing orbital theory using the deep-sea paleoceanographic record. Imbrie and Kipp's goal was simple: "writing equations relating the portions of the biological side of this ecosystem to selected physical parameters of the oceans—and then using those equations on samples from cores to make fully quantitative estimates of past marine climates" (p. 71).

Imbrie and Kipp (1971) studied foraminiferal species census data from 61 core-top samples mostly from the Atlantic Ocean and developed the first widely used foraminiferal transfer function for estimating winter and summer SSTs. One aspect of Imbrie and Kipp's study that distinguishes it from many other studies of faunal and floral indicators is their careful discussion of underlying assumptions about foraminiferal species' ecology. One example is the assumption that the pelagic species and/or species assemblages have not changed their ecological responses to changes in oceanic conditions during the Pleistocene. Imbrie and Kipp also recognized the idea that each species might in theory have different optimal ecological conditions for growth and reproduction. We saw in the previous chapter that the concept of the species niche is prevalent in paleoclimatology, though not always explicitly so. Seldom has it been so carefully applied as by Imbrie and Kipp. Improved versions of the original foraminiferal transfer function as well as equations developed for other microfossil groups are now used on a regular basis in paleoclimatology.

William Ruddiman, Andrew McIntyre, Maureen Raymo, and their colleagues have applied foraminiferal ecology and transfer functions to evaluate orbital influence on subpolar North Atlantic ecosystems. Ruddiman et al. (1977) used the proportion of the polar species *Neogloboquadrina pachyderma* in the total foraminiferal assemblage to estimate North Atlantic SST variability during two glacial-interglacial transitions (isotope stages 2/1 and 6/5e). The abundance of this species in current subpolar regions would increase from 0% in the interglacial period to near 100% during full glacial conditions as the Polar Front migrated southward, then would decrease during the succeeding period of warmth. Using the foraminiferal transfer function to estimate SST during these climatic extremes, Ruddiman's group found that during glacial terminations in the mid- to high-latitude North Atlantic, SSTs changed 7–11°C. During periods of the most rapid oceanic change, when the Polar Front migrated northward, the SST rose at a rate equal to or greater than 1–5°C/ka.

Ruddiman and McIntyre (1981) also presented evidence that the northern North Atlantic Ocean is an important region for amplifying the relatively small insolation changes due to orbital factors into the large-amplitude climatic cycles evident in late Quaternary ice ages (figure 4-8). The North Atlantic sits adjacent to the large Laurentide Ice Sheet and the Fennoscandinavian Ice Sheet that 20 ka covered parts of northeastern North America and western Europe, respectively. Thus it serves as a major source of heat and moisture for ice sheet growth and decay. Ruddiman and McIntyre examined foraminiferal assemblages from deep-sea cores in the subpolar North Atlantic and discovered a 23-ka period in North Atlantic SST history probably linked to precession. They proposed that during periods of ice growth, the North Atlantic Ocean provides a mechanism to foster the growth of large ice sheets because subpolar and northern subtropical surface waters were still relatively warm during the ice-growth phase of late Quaternary climate change. They argued that the 3–5 ka lag of oceanic cooling after ice growth initiation provided a mechanism whereby the relatively warm surface waters of the North Atlantic served as a moisture source to nourish the growing Laurentide Ice Sheet.

Likewise Ruddiman and McIntyre showed that intervals nearly barren in planktonic microfossils (foraminifers, coccoliths) coincided with periods of high summer insolation at 65°N when catastrophic ice sheet disintegration occurred. Low surface-ocean productivity and extensive winter sea ice characterized the North Atlantic during glacial terminations. Sea ice would have cooled ocean temperatures, reduced heat storage, reduced salinities, and contributed to a greatly reduced moisture source for Northern Hemisphere ice sheets. More generally, Ruddiman and McIntyre argued, on the basis of the frequencies of SST, ice-volume oscillations, and astronomically calculated insolation changes, that precession, amplified by ocean mechanisms, accounted for abrupt terminations of Quaternary glacial climates.

Ruddiman and Raymo (1988; see Ruddiman et al. 1986, 1989) combined foraminiferal transfer function data with other indicators such as ice-rafted debris, percent carbonate, and oxygen isotopes, to extend

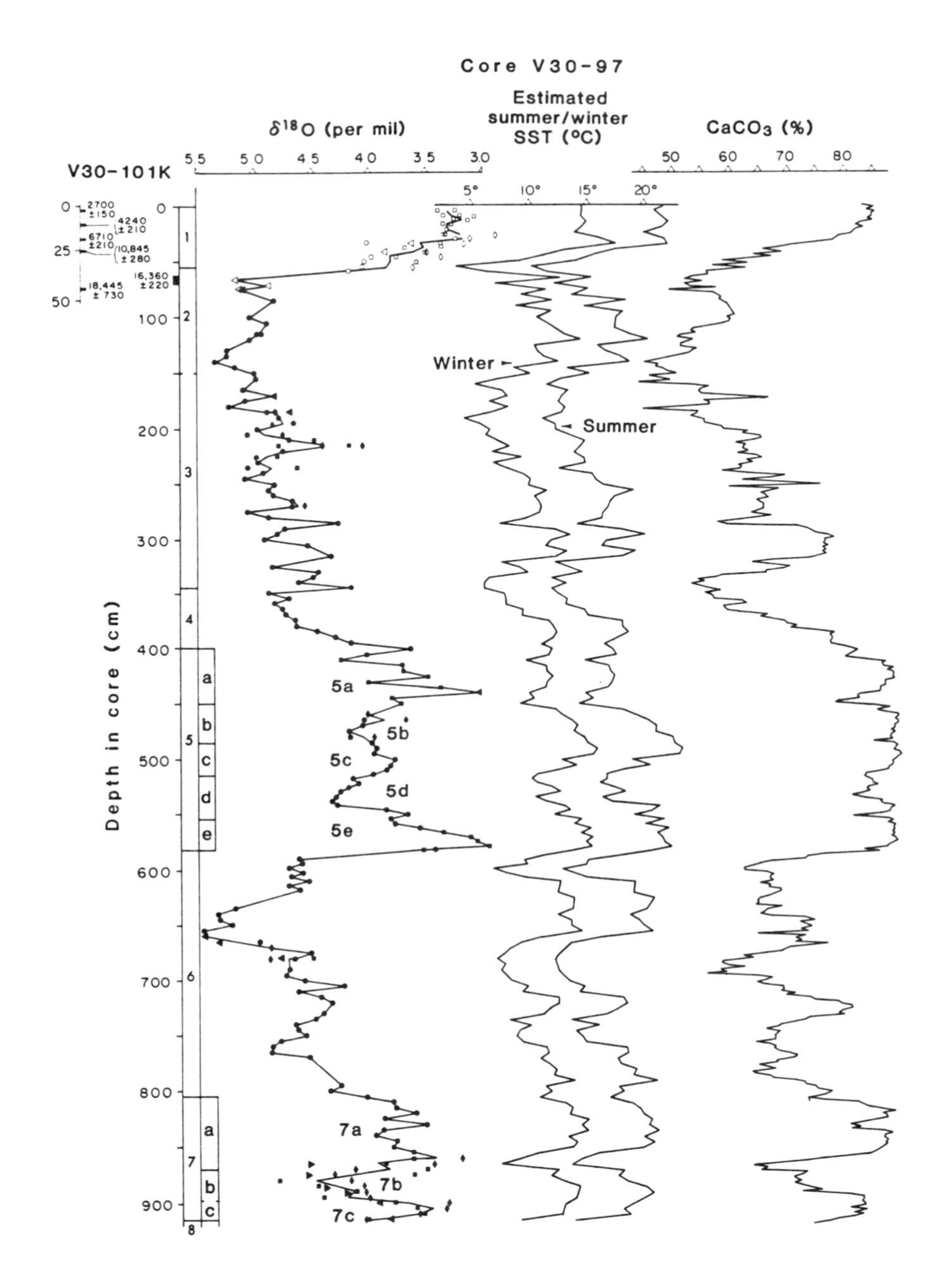
Core V30-97
V30-101K
δ18O (per mil)
Estimated summer/winter SST (°C)
CaCO3 (%)
Winter
Summer
Depth in core (cm)
5a
5b
5c
5d
5e
7a
7b
7c

North Atlantic climatic records back to 2.5 Ma. They confirmed that a predominant 100-ka period characterized long-term SST variability during the past 735 ka in the subpolar North Atlantic. However, the 100-ka oscillation contrasts strongly with climatic patterns during the prior 2 Ma. Ruddiman et al. (1986, 1989) and Raymo et al. (1986) found that foraminiferal assemblage, foraminiferal $\delta^{13}C$ and $\delta^{18}O$, and $\%CaCO_3$ variability at Deep Sea Drilling Program (DSDP) sites 607 and 609 oscillated mainly with the 41-ka period between 735 ka and 2.47 Ma. High-latitude SST and associated deep-ocean circulation oscillations were thus strongly and directly influenced by changes in earth's obliquity.

These studies confirmed that the climate system did not respond in a simple linear way to forcing from the 41-ka obliquity-driven insolation signal after about 700 ka. The important shift from a 41-ka obliquity period to a 100-ka eccentricity period has been an important and widely discussed puzzle in Quaternary paleoclimatology (Hays et al. 1976; Imbrie et al. 1984; Pisias et al. 1984). Internal climatic mechanisms such as ice-bedrock interaction, atmospheric CO_2 levels, and nonlinear response to precession were among the hypothesized causes. Ruddiman and colleagues offered a new explanation: The disappearance of the obliquity signal between 900 and 600 ka was the culmination of the progressive shift in the global climate system during the past 40 Ma caused in part by tectonic processes. They specifically suggested that late Neogene uplift of the Tibetan Plateau and the western North American plateau was a factor in the progressive increase in the amplitude of Plio-Pleistocene glacial-interglacial oscillations because of complex impacts on atmospheric circulation and global biogeochemical cycling (Ruddiman et al. 1986, 1989; Raymo 1988, 1994). Neogene plateau uplift remains a plausible mechanism to account for changes in orbitally induced insolation changes.

In sum, the North Atlantic pelagic ecosystem has been an important source of information on orbital forcing of climate and potential internal mechanisms that modify the impact of insolation changes. Ideas about oceanic moisture and heat feedbacks related to ice sheet growth and decay and the impact of tectonic processes on long-term

FIGURE 4-8 Oxygen isotope, sea-surface temperature (SST), and carbonate from deep-sea core V30-97 from the North Atlantic Ocean showing paleoceanic changes during the past two major glacial-interglacial cycles. Courtesy of W. F. Ruddiman, with permission from American Association for the Advancement of Science. From Ruddiman and McIntyre (1981).

climate evolution were developed largely on the basis of the deep-sea paleoceanographic record. Imbrie et al. (1993a:699) summed up one view of the impact of internal processes related to ice sheets on orbital forcing during the past 600 ka:

> When these ice sheets, forced by precession and obliquity, exceed a critical size, they cease responding as linear Milankovitch slaves and drive atmospheric and oceanic responses that mimic externally forced responses. In our model, the coupled system acts as a nonlinear amplifier that is particularly sensitive to eccentricity-driven modulations in the 23,000-yr sea-level cycle.

We will return to the nonlinear response to orbital forcing later in this chapter.

Deep-Ocean Circulation and Benthic Environments

Efforts to understand deep (> 2500 m) and mid-depth (1000–2500 m) oceanic circulation over orbital time scales complements studies of surface oceanic conditions. Deep-ocean circulation changes over orbital time scales are important elements in orbital theory because the deep sea provides a way to translate regional insolation changes affecting high latitudes of the Northern Hemisphere into a global climate signal via the ocean's thermohaline circulation. Recall from chapter 2 that deep-water formation occurs at high latitudes in the Nordic and Labrador Seas, which form North Atlantic deep water (NADW), and in the southern oceans, where Antarctic bottom water (AABW) drives deep circulation. In the north, Norwegian Sea deep water and Greenland Sea deep water spill over the Iceland Faroes and Denmark straight sill, where they form NADW. Evidence is emerging that deep-sea benthic environments are closely linked over orbital time scales to changes in surface conditions, especially SST, salinity, and pelagic phytoplankton productivity (Versteeg 1996).

The paleoceanographic data collected to quantify and explain NADW and AABW formation include carbon isotope ratios, which indicate circulation changes; %$CaCO_3$ (and other indices of carbonate dissolution) — the proportion of calcium carbonate to ice-rafted debris (IRD) — which shows the amount of sea ice and glacial activity; cadmium-calcium (Cd:Ca) ratios, a nutrient index; ostracode magnesium-calcium (Mg:Ca) ratios, an indicator of bottom-water temperature; and benthic faunal assemblages, which yield information on food and circulation. NADW has certain attributes relative to AABW that can be identified in deep-sea core sequences: relatively high foraminiferal $\delta^{13}C$ (NADW originates near the source of carbon), low nutrients (due to rapid convective sinking), high temperature (low-latitude origin of surface water), and high salinity (NADW high density causes it to sink above its freezing point). AABW southern ocean deep water by contrast has lower $\delta^{13}C$ values, is colder, is more nutrient-rich, and is less saline than NADW. NADW benthic communities generally have higher species diversity than those in AABW (Cronin and Raymo 1997)

Understanding the relationship between deep-sea ecosystems and Plio-Quaternary climate change requires a firm understanding of ocean conditions during extreme glacial versus interglacial climatic states. There is good evidence for altered glacial deep and mid-depth ocean circulation from measurements of $\delta^{13}C$ and Cd:Ca ratios in deep-sea benthic foraminifers (Curry et al. 1988; Boyle and Keigwin 1987; Boyle 1988a; Oppo and Lehman 1995). During glacial periods, foraminiferal $\delta^{13}C$ values for deep-sea sites are decreased throughout the North Atlantic region. Similarly, glacial deep water in the North Atlantic Basin is relatively enriched in nutrients as measured by the ratios of Cd:Ca in benthic foraminifers (Boyle 1992). This means that there was a greater influence of AABW in the North Atlantic Basin during the last glacial, very likely because of reduction in the formation of NADW.

Mid-depth paleoceanographic records (1000–2500 m) show that reduced influence of NADW in the deepest North Atlantic coincided with a greater influence in the mid-depth ocean due to weaker thermohaline circulation. Relatively enhanced $\delta^{13}C$ values contrast with those of the Holocene interglacial period (e.g., Oppo and Fairbanks 1987; Oppo and Lehman 1995).

Glacial periods are also characterized by major changes in North Atlantic deep-sea benthic communities. For example, distinct benthic foraminiferal assemblages from the past glacial period, once attributed to changes in oceanic circulation (Schnitker 1979), have recently been explained by changes in the supply of phytodetritus (food) settling in the water column (e.g., Gooday 1988; Thomas et al. 1995). Lower species diversity in ostracode assemblages from the North Atlantic Ocean also appears to be related to changes in food supply and other factors during glacial intervals (Cronin and Raymo 1997). Although these microfaunal groups represent only a subset of the deep-sea biotic community, they nonetheless suggest that the entire benthic ecosystem during glacial periods was diametrically different from those inhabiting the interglacial ocean.

Using these proxy indicators and the contrast between glacial and interglacial extremes, paleoceanographers have uncovered evidence for long-term geochemical and faunal cycles related to fluctuations in deep-ocean circulation and surface-ocean productivity over orbital time scales. Much of the evidence comes from the high-resolution sedimentary sequences obtained from Deep Sea Drilling Program (DSDP) and Ocean Drilling Program (ODP) core sites 607, 609, 552A, 704, 658, and 659 from the Atlantic Ocean (Ruddiman et al. 1986, Raymo et al. 1989, 1992; Hodell and Venz 1992; Sarnthein and Tiedemann 1990).

Well-defined cyclic fluctuations in NADW characterized the North Atlantic Ocean during the last 3 Ma at both 41-ka and 23-ka frequencies. Between 2.7 and 700 ka, when 41-ka cycles in SST, $\delta^{18}O$, and $\%CaCO_3$ were prominent (Ruddiman et al. 1986; Raymo et al. 1989, 1990) at DSDP site 607, periodic reductions in the strength of NADW formation occurred, as indicated by oscillations in benthic foraminiferal $\delta^{13}C$.

Diminished NADW and enhanced AABW during Plio-Pleistocene glacial periods are also manifested as decreased bottom water temperatures. The 41-ka period in the $\delta^{18}O$ record at site 607 is probably both an ice-volume and a deep-water temperature signal. Dwyer et al. (1995) confirmed this inference of decreased bottom water temperatures by measuring the Mg:Ca ratios in the benthic ostracode *Krithe*. They discovered that bottom water temperature (BWT) at DSDP site 607 oscillated at a 41-ka periodicity, with temperature change preceding ice volume changes by a few thousand years. BWT decreased during late Pliocene glacial periods and increased during succeeding interglacials, a pattern predicted from the stable isotopic data indicating greater AABW during the glacial periods.

A 23-ka cycle in $\delta^{13}C$ of benthic foraminifers was reported from site 607 (Raymo et al. 1989). This precessional cycle may be caused by an unknown low latitude influence on NADW formation or by changes in continental biomass that affect the carbon isotope signal. Because $\delta^{13}C$ changes may be due to remineralization of organic carbon or mixing between water masses of different origin, global changes in the amount of carbon sequestered on continents and shelves may be a factor altering the benthic foraminiferal signal (Duplessy et al. 1988).

Another measure of deep-sea cyclic change in the North Atlantic comes from benthic species diversity. The combined effects of reduced or more variable phytodetritus, and perhaps decreased bottom temperatures during glacial episodes (Thomas et al. 1995) apparently led to reduced benthic ostracode species diversity (Cronin and Raymo

1997). Because these crustacea play an important role in deep-sea benthic ecosystems, the diversity changes probably signify recurring large-scale restructuring of the benthic community during climate cycles. Versteeg (1996) provided independent evidence from Pliocene dinoflagellate assemblages at site 607 that orbitally induced (41-ka) changes in surface-ocean productivity also characterize the North Atlantic during this time. Periodic perturbations to the phytodetrital food supply falling to benthic communities may partly explain the observed deep benthic assemblage changes, although additional study of the surface-benthic coupling is needed.

In summary, Pliocene 41-ka cycles in %$CaCO_3$, $\delta^{18}O$, Cd:Ca and Mg:Ca ratios, and species diversity show convincing evidence that deep-sea environments in the North Atlantic changed cyclically in phase with orbital forcing. Glacial periods of the Pliocene, although only one third to one half the magnitude of those of the past 700 ka in terms of ice volume, were nevertheless characterized by nutrient enrichments, enriched carbon isotopic ratios, increased ice-rafted detritus, reduced carbonate production, reduced NADW influence, lower bottom-water temperatures, decreased benthic species diversity, and very likely a totally different benthic community structure.

Low-Latitude Climatic Cycles in the Indian Ocean Monsoon, Equatorial Pacific, and Atlantic Upwelling

Tropical and subtropical oceanic pelagic ecosystem variability is also characterized by orbital frequencies during the past few million years. Moreover, tropical and subtropical oceanic biomes probably play a reciprocal role in long-term climate change by regulating atmospheric-ocean carbon fluxes during large-scale climatic oscillations. For example, near-surface productivity changes, which vary seasonally due to wind-induced upwelling, also vary over glacial-interglacial time scales. Glacial periods are characterized by productivity that is higher than that during interglacial periods (Arrhenius 1952; see Berger et al. 1994; Zahn et al. 1994). Regions of oceanic upwelling where colder, nutrient-rich waters rise toward the ocean surface play an especially key role in circulation of the ocean's nutrients and carbon. This section describes biological records of climate change in two distinct types of oceanic upwelling systems—coastal upwelling off the Arabian Peninsula, Indian Ocean, associated with monsoonal variation, and open-ocean equatorial upwelling in the eastern equatorial Pacific Ocean and the Atlantic Ocean. In each region, oceanic processes that control ecosystem functioning and the composition of phyto- and zooplankton assemblages fluctuate over orbital time scales.

Summer Indian Ocean Monsoon

The southwest winds that blow off the Indian Ocean during summer drive the enormous rainfall associated with the summer monsoon of the Asian continent. Differential heating between the continent and the Indian Ocean cause southeast monsoonal winds and strong oceanic upwelling off the Arabian Peninsula. Upwelling in turn results in cooler water temperatures and increased nutrient concentrations in the surface waters during the monsoonal season. During the summer monsoon, SSTs are usually below 24°C, about 4°C lower than Arabian Sea SSTs for the rest of the year. The Arabian Sea upwelling system is the site of research on orbital climate cycles on subtropical ocean ecosystems.

Monsoon-induced SST changes influence temperature-sensitive planktonic foraminiferal species in a manner similar to latitudinal temperature gradients at convergence zones like those near the Polar Front in the North Atlantic Ocean described earlier. A widely used indicator is the species *Globigerina bulloides,* whose stable oxygen isotope ratios reflect SST variability (Prell 1984). The oxygen isotope ratios of modern *G. bulloides* are negatively correlated with surface temperatures; SST is in turn inversely correlated with wind-induced summer monsoonal upwelling (Prell and Curry 1981). Typically a subpolar species, *G. bulloides* lives in the southern oceans between the subtropical convergence and the Antarctic convergence zones. Its occurrence in subtropical latitudes of the Indian Ocean is directly attributable to the presence of cooler waters upwelling in coastal zones. Biogenic opal in the Arabian Sea is also controlled by surface-ocean productivity, which reflects nutrient conditions and upwelling (Leinen et al. 1986).

In addition to foraminiferal isotopes and opal, sediment mineralogy is also a useful proxy in the subtropical Arabian Sea for orbital glacial-interglacial climatic change. High percentages of quartz and clay reflect increased soil deflation during times of continental aridity (Street-Perrott and Harrison 1985) leading to increases in the amount of eolian (wind-blown) sediments carried to oceans. Eolian sediments are used as an indicator of atmospheric wind intensity and direction in continental sedimentary sequences (Kukla et al. 1988; Ding et al. 1994) and deep-sea ocean sediments (Janacek and Rea 1985; Rea 1994).

Hovan et al. (1989) demonstrated that the eolian material preserved in North Pacific Ocean sediments off eastern Japan represents dust blown 3500 km from its source in central China during five 100-ka orbital cycles.

In the Arabian Sea, sediment variability is expressed primarily in the relative percentages of calcium carbonate, eolian quartz and clay, and siliceous microfossils. The mass accumulation rate (MAR) of sediments and the lithic grain sizes are used as measures of eolian transport from the African continent. As such, they are good proxy indicators for continental aridity and the strength of the summer monsoon, respectively. Together with foraminiferal and opal data, they provide a means to link continental and oceanic paleoclimate records.

Clemens and Prell (1990) investigated orbital scale paleoclimate trends over the past few hundred thousand years preserved in sediments from Owens Ridge, northwest Arabian Sea. They reached four conclusions about the Arabian Sea record of eolian sediment: (1) Wind strength can be measured by grain size and MAR; (2) wind strength responds both to external insolation forcing and to internal changes in global ice volume (recorded by oxygen isotope ratios); (3) variability in MAR and grain size is corroborated by other proxy indicators (i.e., lithic grain size peaks at 42, 213, 254, 284 ka coincided with peaks in *G. bulloides*; only the peak at 319 ka did not); and (4) downcore patterns of eolian dust do not always record a global climate signal but must also be interpreted in a regional climatic context.

Clemens and Prell (1990, 1991) analyzed the frequency domain of isotopic, opal, and eolian records to examine the phase relationships between the Arabian Sea productivity, African eolian dust, and solar insolation. As shown in earlier studies (Prell 1984; Prell and Kutzbach 1987), the precessional frequency dominated the monsoon system. Precipitation increases as much as 38% and wind strength up to 75% at times when insolation increases by 19%. Clemens and Prell (1991) also found that MAR fluctuates in phase with oxygen isotopes, but it leads precessional insolation by 6 ka. Furthermore, they discovered that maximum lithologic grain size, an indicator of monsoonal strength, lagged behind orbital precession by 9 ka. Monsoonal strength therefore was linked to global or local ice volume and the availability of Indian Ocean latent heat.

In summary, monsoonal climate variability over the Indian Ocean and Arabian Sea is directly linked to precessional forcing of surface ocean productivity and continental aridity over the past few hundred thousand years.

Pacific Ocean Subtropical Convergence Zone

Earth's atmospheric concentration of CO_2 during Quaternary glacial periods is about 30% lower than it is during interglacial periods (about 190 versus 280 parts per million; see chapter 9). Several hypotheses have been advanced to explain the drawdown of atmospheric CO_2; most hold that enhanced oceanic biological productivity, due to ocean circulation changes, atmospheric or oceanic nutrient fluxes, increased upwelling, or other factors, led to greater glacial-age carbon sequestration by the oceans. In other words, there was an enhanced oceanic "biological pump" during glacial periods (Berger and Wefer 1991).

Equatorial upwelling ecosystems may play a role in causing the large glacial-interglacial contrast in atmospheric CO_2 because the high productivity oceanic regions are a likely sink for carbon transferred from the atmosphere during glacial periods. The eastern equatorial Pacific (EEP) near-surface pelagic ecosystem is one such important region of high biological productivity. EEP currents are dominated by the westward-flowing North and South Equatorial Currents (NEC, SEC), which are fed by the cool waters of the California and Peru Currents, respectively. The North Equatorial Countercurrent (NECC) flows eastward, separating the North and South Equatorial Currents in a zone characterized by equatorial divergence. The Equatorial under current (EUC) also flows eastward, near a depth of 200 m in the western and about 50–100 m in the eastern Pacific. The EUC originates as cool water in the subantarctic region; by the time it reaches the eastern Pacific, it has warmed and is defined by the 20°C isotherm.

Trade wind position and strength influence these equatorial currents and the degree of regional upwelling. Briefly, the process of surface equatorial divergence involves the replacement of surface water with upwelling water from depths of about 40–100 m. During interglacial periods, the surface layer is separated from the cooler deeper water by a permanent thermocline located at a depth of 10–40 m. Upwelling of colder nutrient-rich water in the eastern Pacific Ocean is driven by northeast trade winds, which blow at maximum strength in January to April, when they are in their most southerly position. In September to November, the northeast trade winds are at a northerly position and the North Equatorial Current and North Equatorial Countercurrent are at their strongest. South of the equator, the period September to November has the strongest trade winds and the EUC is strongest. Seasonal changes in the EEP are also influential during El Niño–Southern Oscillation (ENSO) events, as discussed in chapter 7. Their importance for the present discussion is that they create strong gradients in physical oceanographic properties that influence nutrient upwelling, biological productivity, and the near-surface pelagic ecosystem.

The EEP exhibits cyclic changes over the last 3 Ma that appear to be related to orbitally induced 100-ka cycles. Ocean Drilling Program Leg 138 was designed to examine long-term oceanographic variability related to the evolution of late Neogene climate (Pisias et al. 1995), using the exceptional oxygen isotope stratigraphy and gamma ray attenuation porosity evaluator (GRAPE) to analyze high-resolution sedimentary patterns (Shackleton et al. 1995b). At Leg 138 sites 846 and 847, a series of studies captured the response of near–surface ocean biological processes to long-term Quaternary glacial and interglacial cycles (Murray et al. 1995; Farrell et al. 1995; McKenna et al. 1995). To unravel how this complex oceanic system functioned over orbital time scales, Farrell et al. (1995) exploited the ecology of the planktonic foraminifers *Globigerina sacculifer,* which lives near the ocean surface layer, and *Neogloboquadrina dutertrei,* which calcifies in the shallow permanent thermocline at a depth of approximately 10–40 m (Curry and Matthews 1981). This apparent niche partitioning of the uppermost surface ocean allowed Farrell et al. (1995) to contrast the oxygen and carbon isotopic records from species living in two layers of the water column for the past 1.15 Ma.

Farrell and colleagues discovered a number of features about EEP ocean history. First, cycles of $CaCO_3$ preservation for the last 900 ka were primarily a record of oceanic productivity. Selective carbonate dissolution was relatively minor and was not of sufficient magnitude to overprint the ecological changes evident from the foraminiferal assemblages. Second, Quaternary glacial conditions were characterized by stronger equatorial upwelling, increased nutrients, reduced thermocline, cooler (by 1–3°C) surface-ocean temperatures, and higher surface-ocean productivity. Glacial intervals had a smaller vertical contrast in SSTs compared with that of interglacial periods—about 3°C versus 5°C—a greater nutrient injection into the mixed surface layer than did interglacial periods, and a greater accumulation of *N. dutertrei* and $CaCO_3$. Moreover, sea-surface temperatures may have been 1–3°C cooler than those of interglacial periods. Third, planktonic foraminiferal ecological data indicated that changing species abundances over glacial-interglacial time scales reflected changes in three specific factors: mixed layer depth, the source of EUC waters, and SST. The origin of EUC changes was linked to the subantarctic water, which affected SST, mixed layer depth, and oxygen and nutrient levels. Fourth, equatorial upwelling was generally strong during both glacial and interglacial periods but was somewhat reduced during glacial-interglacial transitions. Fifth, opal and carbonate patterns at site 847 showed that biological productivity was more important than carbonate dissolution in explanations of the long-term and short-term cyclic sedimentation at the site over the past 3 Ma.

They also reached important conclusions regarding orbital influence on tropical oceanic conditions. Fluctuations in $CaCO_3$ and opal-producing planktonic organisms exhibited a 100-ka frequency; the 41-ka and 23-ka obliquity and precessional frequencies are not readily apparent. Ravelo and Shackleton (1995) found evidence for 100-ka cycles at ODP site 851 to the west of site 847. Furthermore, the major contrast between glacial and interglacial climatic conditions near the equator began about 1.0 Ma. During glacial periods, the temperature difference between the upper ocean and the thermocline was small; during interglacial periods, the temperature differences were much larger. Between 800 and 700 ka, SSTs and temporal variations were mostly low. About 750 ka, the distinct interglacial-glacial contrast in upper surface temperatures reappeared until 130 ka, when the contrast again became large and invariant. In sum, there was a distinct time-evolving nature to the impact of orbital changes on the eastern equatorial Pacific ecosystem.

Other equatorial regions provide evidence that orbital-scale variability is a general feature of pelagic ecosystems, although these regions do not always behave like the EEP. For example, in the central equatorial Pacific, Pisias and Rea (1988) discovered nonlinearities in the periodicity of radiolarian assemblages over the past 850 ka. Two key species, *Pterocorys minithorax* and *Botryostrobus autitus-australis,* record the strength of the equatorial divergence zone, where they occur in high abundances in the eastern Pacific owing to cold SSTs and a shallow thermocline. This relationship to the thermocline depth, which varies across the Pacific, is well-known from studies of ENSO variability. Like the eastern Equatorial foraminiferal indices, these radiolarians are indicators of wind-driven sea-surface circulation. Pisias and Rea (1988) found evidence for a 31-ka periodicity that does not match the common orbital frequencies. Because the equatorial divergence indicators and the eolian grain-size measures of wind velocity were coherent and in phase, they concluded that the 31-ka cycles confirmed a link between atmospheric circulation (wind) and an oceanic response. They inferred a nonlinear response to orbital forcing as an explanation for the radiolarian signal, an idea also offered to explain patterns of coccolith variability in the equatorial Atlantic Ocean (McIntyre and Molfino 1996), suborbital scale variability in the EEP

(Hagelberg et al. 1994), and a 10.3-ka period to monsoon-related processes (Pestiaux et al. 1988).

Equatorial Atlantic

In the equatorial Atlantic, McIntyre et al. (1989) studied an equatorial transect of piston cores (V25-59, V30-40, RC-24-16) covering the last 250 ka of ocean history. Their goal was to relate orbital changes to monsoonal trade-wind strength, heat advection from the south, and ocean productivity in the equatorial divergence zone. They used 41 foraminiferal groups and the transfer function of Molfino et al. (1982) to estimate SSTs. Their four main foraminiferal assemblages—tropical, transitional, divergence, and subpolar—were similar to those of Imbrie and Kipp (1971). The divergence assemblage differed from the classic gyre assemblage in that it was dominated by three species (*Neogloboquadrina dutertrei, Globorotalia menardii,* and *Pulleniatina obliquiloculata*) indicative of the equatorial divergence zone (Molfino et al. 1982).

McIntyre et al. discovered that assemblages in the eastern equatorial Atlantic Ocean showed a much greater variability compared with the relatively stable assemblages from the western equatorial Atlantic. In Eastern Atlantic core RC-24-16, they found cold-season SSTs varying from 16 to 23°C. Spectral analyses showed that cold-season SST oscillated at a precessional 23-ka frequency; a weaker 41-ka signal was present in the warm-season SST record. The precessional power accounted for 26% and 40% of total variance in the central and eastern regions, possibly signifying a wintertime signal of seasonality changes. McIntyre et al. (1989) posited that, because the eastern equatorial Atlantic record is in phase with or slightly lags behind the Southern Hemisphere SST record, and is ahead of high-latitude SST and ice volume records, Southern Hemisphere insolation exerted a stronger influence on the equatorial Atlantic during late Pleistocene orbital changes than did Northern Hemisphere insolation. They attributed the SST variations to mechanisms related to trade-wind velocity and heat advection from the southern Atlantic Ocean. In McIntyre's model, when the aphelion is aligned with austral winter, trade-wind zonality, equatorial divergence, and mixed ocean productivity are at their maxima. Conversely, when the perihelion is aligned with boreal summer, North African monsoon dominates the record from the south to the northwest part of the African continent, and divergence and northward advection of minimal heat occurs via the Benguela Current.

Although the mechanisms by which solar insolation imparts its influence upon regional ocean ecosystems are not totally understood, equatorial pelagic ecosystems from the Indian, Pacific, and Atlantic Oceans all seem to be influenced by orbital insolation changes. Precession is the most common, but not the sole, frequency encountered in tropical oceanic records of the past few million years.

African and South American Vegetation and Orbital Cycles

Orbital climate variability is also reflected in the cyclic patterns of terrestrial vegetation and atmospheric circulation now reconstructed from the palynological and loess records from several continents (figure 4-9). The evolution of African terrestrial habitats over the past 3 Ma provides an excellent example of vegetation response to orbital forcing (Hooghiemstra and Sarmiento 1991; Hooghiemstra et al. 1993; Dupont et al. 1989; Leroy and Dupont 1994). Between the eastern Mediterranean and the equator lie eight west African vegetation zones: (1) Mediterranean pine forests, (2) Mediterranean desert-transition steppelike environments, (3) Sahara Desert, (4) the Sahel, (5) open dry grasslands with abundant *Acacia,* (6) Guinean and Sudanian wooded grasslands and open forest, (7) the Guinean-Congolian tropical rain forest, and (8) coastal mangroves. Each vegetation zone produces a signature pollen assemblage used in palynological analyses.

Pollen from African vegetation is transported westward by atmospheric circulation until eventually it is deposited in oceanic sediments off northwest Africa. Atmospheric circulation in this region of Africa is dominated by the African Easterly Jet (AEJ, also known as the Sahara Air Layer). At its maximum strength in July and August, the AEJ sweeps due east off the continent, carrying dust from the Sahel and southern Sahara Desert. In addition, at maximum strength in March, near-surface northeast trade winds flow just north of the equator. Dust is generated at 10–15°N, where it is injected into the mid troposphere, which is controlled by northern monsoon moisture, squall lines, and the summer position of the Intertropical Convergence Zone (ITCZ). Dust-flux maxima correspond to weak summer monsoons, when diminished vegetation characterizes the Sahel and Sahara.

The AEJ and the northeast trade winds carry large amounts of pollen, terrigenous material, phytoliths (cuticles from savanna grass), and freshwater diatoms (*Melosira*) blown from dried African lake beds (Pokras and Mix 1987; deMenocal et al. 1993) and deposit them in the eastern Atlantic Ocean off west Africa, where they quickly settle to become part of the oceanic sedimentary record. These two major atmospheric patterns carry distinct pollen assemblages in their dust load, which are used to track African vegetation and atmospheric history (Hooghiemstra et al. 1986). The AEJ is characterized by Chenopodiacea and Amaranthaceae (types of pigweed) and the northeast trade winds are characterized by grass pollen from Sahara and Sahel environments. AEJ dust flux therefore records fluctuations in continental aridity in the southern Sahara Desert and the Sahel.

Shifts occurred in both vegetation zones and in atmospheric circulation during orbital cycles. Northeast trade winds are generally stronger during glacial and weaker during interglacial periods, and the AEJ glacial dust load contains more Chenopodiacea and Amaranthaceae than does the interglacial AEJ. Extending the glacial-modern vegetation-pollen-atmospheric relationships back over orbital time scales, Dupont et al. (1989) demonstrated convincingly that west African vegetation oscillates at orbital frequencies that reflect variability in continental aridity and wind patterns. At ODP site 658, located directly in the path of the African Easterly Jet, several coherent pollen assemblages recur in sediment deposited during the past 670 ka. Moreover, Dupont's group detected a distinct relationship between vegetation and ice-volume changes measured by the oxygen-isotope curve. For example, the groups Poacea (grass) and Cyperacea (sedges) show a 40-ka periodicity, with Poacea leading ice volume by 5 ka and Cyperacea lagging behind it by about 4 ka. These particular plant groups appear to respond to high-latitude obliquity forcing of west African climate.

In contrast, other pollen types—*Ephedra,* Chenopodiacea, and Amaranthaceae—varied at periodicities of 20 ka and 23 ka, respectively. The presence of these groups probably reflects a precessional dominance of these components of the terrestrial vegetation of the Sahel and Sahara during the past 670 ka.

Parallel studies of dust flux during the past 5 Ma at ODP site 659 (Ruddiman et al. 1989; Tiedemann et al. 1989, 1994) and dust, phytoliths, and *Melosira* at sites 661 and 663 (Bloemendal and deMenocal 1989; deMenocal et al. 1993; deMenocal 1995) provide additional evidence for orbital cycles in African climate. Pokras and Mix (1987) and deMenocal et al. (1993) found 23- and 19-ka cycles in the abundance of *Melosira,* which is used to monitor African lake levels and monsoon-related summer precipitation. DeMenocal et al. (1993) argued that before about 2.4 Ma, African climate oscillated at 23- and 19-ka frequencies

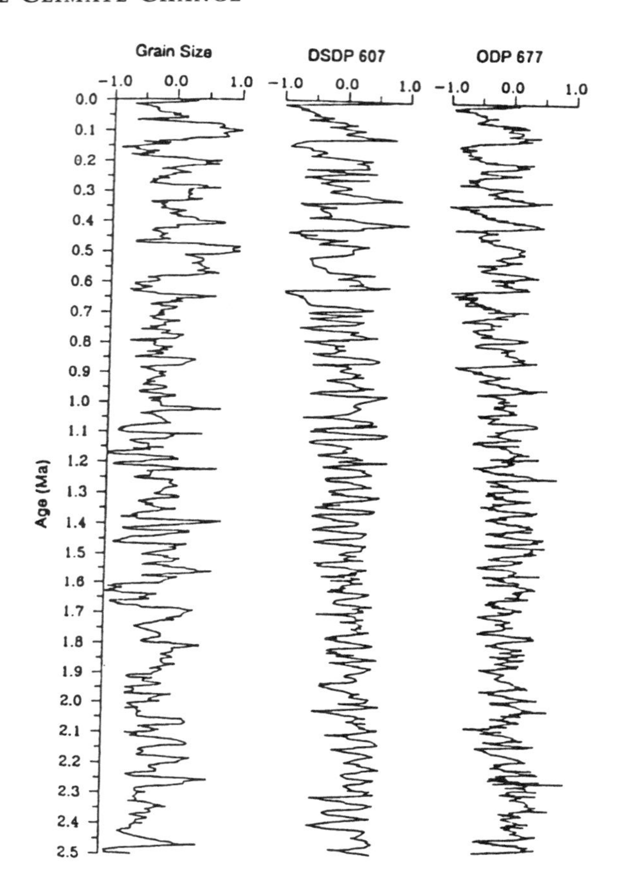

FIGURE 4-9 Terrestrial record of orbital climate change. Grain size of Chinese loess deposits is compared with oxygen-isotope records from deep-sea cores from Deep Sea Drilling Program (DSDP) site 607 and Ocean Drilling Program (ODP) site 677 for the past 2.5 Ma. Evolution of climatic variability from predominant 41-ka to 100-ka climate cycles is shown in both terrestrial and oceanic records. Courtesy of N. Rutter, with permission of Elsevier Science. From Ding et al. (1994).

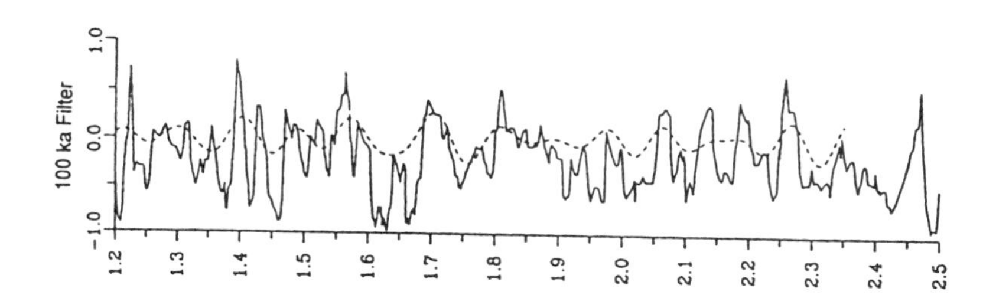
100 ka Filter
1.0
0.0
−1.0
1.2
1.3
1.4
1.5
1.6
1.7
1.8
1.9
2.0
2.1
2.2
2.3
2.4
2.5

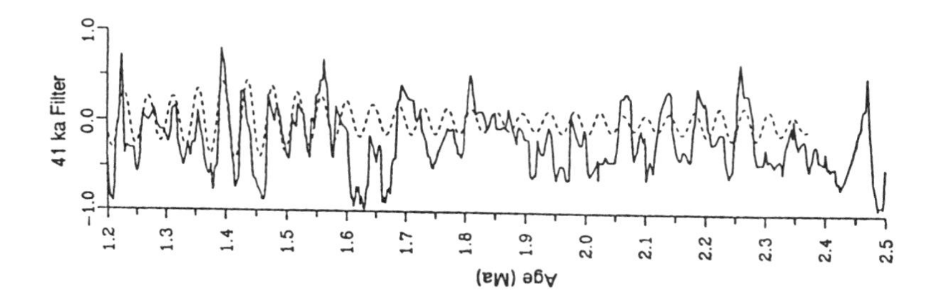
41 ka Filter
1.0
0.0
−1.0
1.2
1.3
1.4
1.5
1.6
1.7
1.8
1.9
2.0
2.1
2.2
2.3
2.4
2.5
Age (Ma)

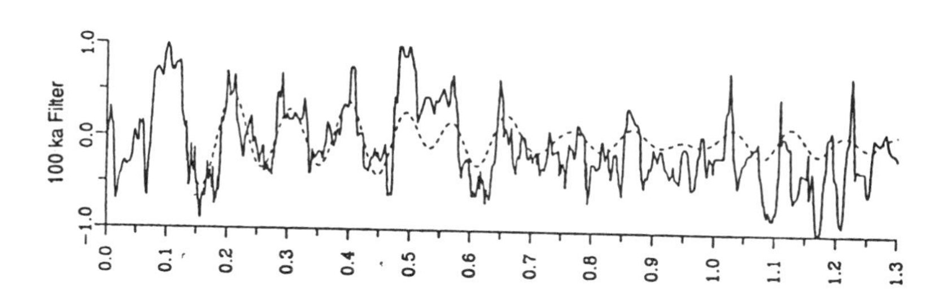
100 ka Filter
1.0
0.0
−1.0
0.0
0.1
0.2
0.3
0.4
0.5
0.6
0.7
0.8
0.9
1.0
1.1
1.2
1.3

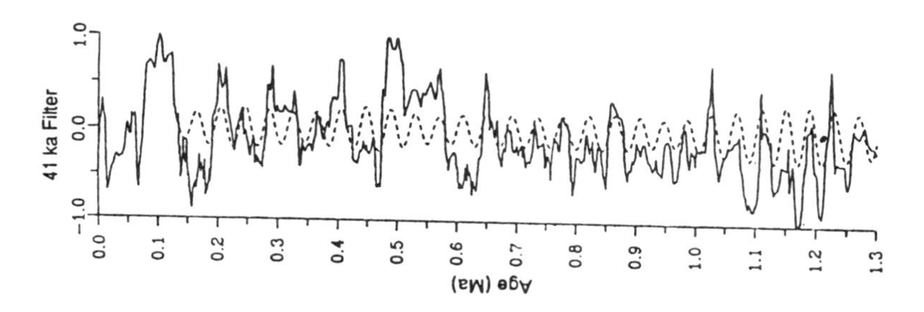
41 ka Filter
1.0
0.0
−1.0
0.0
0.1
0.2
0.3
0.4
0.5
0.6
0.7
0.8
0.9
1.0
1.1
1.2
1.3
Age (Ma)

because the Northern Hemisphere ice sheets were minimal and low-latitude precessional insolation changes dominated. They called this the "pre-ice" mode of climate change. Conversely, after 2.4 Ma, dust and phytoliths vary at 100- and 41-ka periodicities and indicate that African climate became linked closely to North Atlantic oceanic SST, which in turn varies due to high-latitude obliquity-driven insolation changes. DeMenocal referred to this as the "syn-ice" phase in reference to the distinct African climatic patterns following early development of large Northern Hemisphere continental ice sheets.

A fundamental shift in African climate during the mid-Pliocene also shows up in dust flux records at ODP site 659, where Tiedemann et al. (1994) inferred that low-latitude precessional forcing controlled African climate between about 5 and 3 Ma through land-mass heating and modulation of the intensity of the summer monsoon. Between 3 and 1.5 Ma, a period of transitional climate occurred when both precession and obliquity influenced the African dust record. During the past million years, a predominant 100-ka periodicity in dust flux is associated with high-latitude climate forcing.

Two other continental sites containing exceptional records of orbital-scale palynological cycles are the eastern Cordillera in the high plain of Bogota, Colombia (figure 4-10), and the Tenagi Philippon pollen record from Greece. The Bogota area is the locus of a high-elevation former lake bottom about 2550 m above sea level. This region has a long history of stratigraphic and palynological studies (e.g., van der Hammen 1973). Recent high-resolution pollen sequences from two sediment cores 357 m and 586 m long (Funza I and Funza II) studied by Henry Hooghiemstra and colleagues (Hooghiemstra and Sarmiento 1991; Hooghiemstra et al. 1993; Hooghiemstra and Ran 1994; Helmans and van der Hammen 1994; Hooghiemstra 1995; Hooghiemstra and Cleef 1995) reveal latitudinal shifts in vegetation during climate cycles of the past 3.2 Ma. Both eccentricity and precessional cycles were recognized in the Funza pollen record, as was a major shift in climate near 800 ka that is prevalent in paleoclimate records from many other regions.

Mommersteeg et al. (1995) documented a 975-ka record of pollen at Tenagi Philippon, in Macedonia, Greece. Based on analyses in the frequency domain of arboreal versus nonarboreal pollen and assemblages characteristic of four distinct climatic zones (warm, wet forest; warm climate with dry summers; open forest with cool wet climate; and cool, dry steppes), they detected orbital cycles at 95–99, 40–45, 24.0–25.5 and 19–21 ka as well as several other periods.

In summary, long-term continental records at orbital frequencies

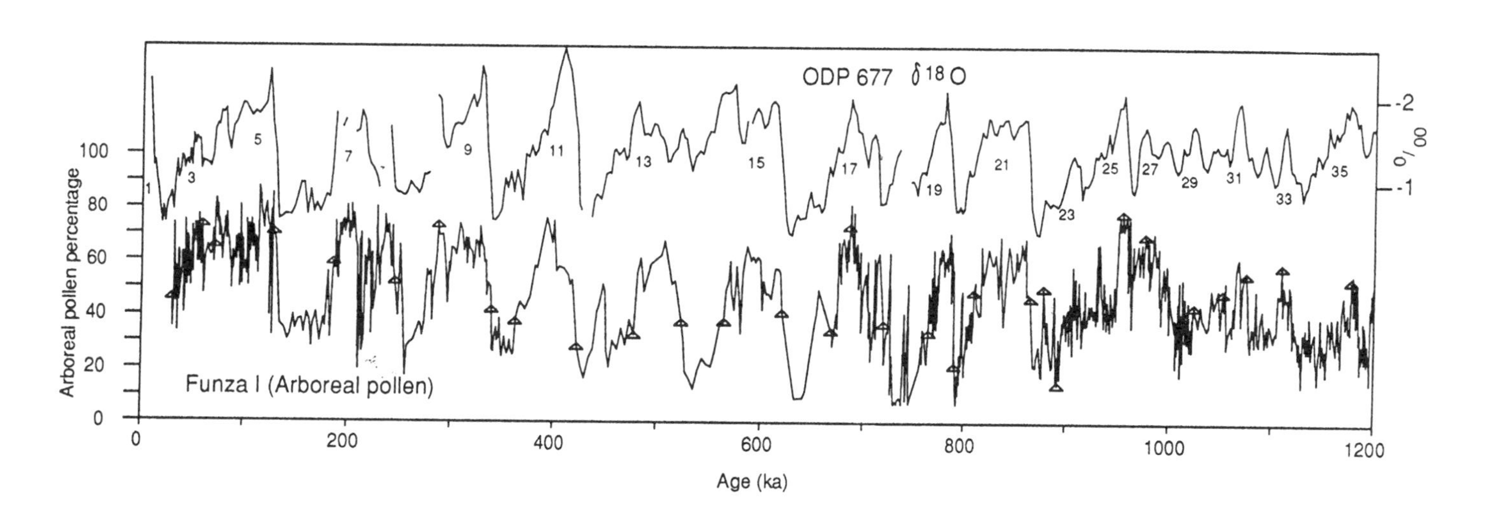

FIGURE 4-10 Correlation between vegetation history in Colombian Funza core pollen sequence and deep-sea oxygen-isotope record from Ocean Drilling Program (ODP) 677. Courtesy of Elsevier Science and H. Hooghiemstra. From Hooghiemstra and Ran (1994).

are less common than their oceanic counterparts, but more frequent use of indicators such as dust and pollen preserved in marine and lake sediments has led to advances in this subdiscipline over the past decade. Evidence suggests that low-latitude terrestrial ecosystems respond in complex ways to orbital insolation forcing, and more regional records are necessary to establish how other terrestrial ecosystems behave.

Mediterranean Oceanic Ecosystem Cycles and Tuning the Pliocene Climate Record

The striking layers of gray, white, and beige sediments exposed along cliffs of Calabria in the southern peninsular Italy and at Punta de Maiata in Sicily provide a vivid image of the remarkable regularity produced by orbitally driven biotic and sedimentary processes (figure 4-1B). The age and origin of alternating beds of high and low calcareous sediments of the Trubi marls and other Pliocene formations were examined in a series of studies on their stratigraphy and paleomagnetic record (Hilgen 1987; 1991a,b; Zijderveld et al. 1991; Hilgen and Langereis 1989, 1993), lithology (De Visser et al. 1989), marine micropaleontology and paleotemperature history (Zachariasse et al. 1989, 1990), and palynology and surface-ocean productivity (Versteeg 1994). The nearly uninterrupted paleoceanographic record from the Italian Pliocene sequence provided both a means to tune the paleomagnetic and biostratigraphic time scales to astronomical cycles and an exceptional record of Mediterranean oceanic ecosystems responding to precipitation, riverine nutrient input, and SST (see Lourens et al. 1996).

How do the Italian Pliocene sediments relate to the Mediterranean pelagic ecosystem in response to orbital cycles? The origin of the Trubi bedding reflects orbitally related changes in the surface oceanic ecosystem. At the Rossella section of Sicily, for example, the sediments have a quadripartite division into gray-white-beige-white layers. The Calabrian section, in contrast, lacks the intercalated beige layer and forms a simple bipartite white-gray alternation. Alternating rhythmic layers of carbonate-rich and poor layers can result from several processes, among them postdepositional diagenesis or carbonate dissolution in gray layers, changes in carbonate production, or dilution of carbonate by terrigenous influx. De Visser et al. (1989) and Zachariasse et al. (1989) analyzed the palynology, planktonic and benthic foraminiferal assemblages, isotope geochemistry, and mineralogy of the Trubi rhythmites for the interval 3.9–2.5 Ma. They concluded that the gray layers signified periods of high biological productivity, increased precipitation, riverine nutrient influx, and greater seasonality. The carbonate-poor beige layers were deposited under drier and cooler climatic conditions, perhaps when there was greater eolian dust influx and lower surface ocean productivity. The white layers represented an intermediate climate between warm, humid conditions and cool, dry ones.

Applying principles of astronomical tuning to the Trubi rhythmites also led to an important revision in standard Pliocene chronology. In Hilgen and Langereis's initial analyses, the quasi-cyclic nature of Italian Pliocene carbonates was found to vary at both short-term (15–18 ka) and long-term (290 and 360 ka) frequencies. At first glance, the short-term cycles seemed to be discordant with the 19 and 23-ka precessional periods. However, Hilgen and Langereis (1989, 1993) discovered a good reason for the discrepancy: the standard age estimates used for key Pliocene paleomagnetic polarity reversals (Kaena, Mammoth, Cochiti, Nunivak, Sidufjall, and Thvera Subchrons) were incorrect by a small but significant amount. These polarity boundaries had been dated by radiometric methods, which have inherent age uncertainties.

By revising subchron boundary age estimates—that is, by tuning the Pliocene paleomagnetic time scale between 5 and 3 Ma to the orbital time scale of Berger (1984)—and by keeping the Gauss-Gilbert Chron boundary at 3.4 Ma, the periodicity of the short-term Italian sedimentary cycles suddenly changed to 19 and 23 ka. The quasicyclic pattern of carbonate deposition was now clearly linked to precessional cycles, and the Italian sections have become an integral part of the new astrochronologically tuned standard Neogene time scale (Berggren et al. 1995, Lourens et al. 1996).

CHALLENGES TO ORBITAL THEORY

FROM A HISTORICAL perspective, the most persistent problem surrounding orbital theory has been age control. Support for the theory has waxed and waned like the ice sheets themselves, depending on the development and interpretation of new dating methods and revised time scales. Geologists continually seek to improve their ability to date the paleoclimatic record with a resolution that rivals that of celestial mechanics. The advent of radiocarbon dating initially posed problems for orbital theory (Imbrie and Imbrie 1979). More recently, the TIMS U-series method has raised a new challenge: high global sea level during insolation minima—an obvious contradiction to orbital theory. Here I touch upon another widely discussed paleoclimate record considered to be at odds with orbital theory.

Groundwater Precipitation of Vein Calcite in Devils Hole, Nevada

A shallow fissure in the Great Basin of central Nevada seems an unlikely challenger to the paradigm of orbital theory. Yet a remarkable climate record from Devils Hole (DH), an extensional fault zone in south central Nevada, poses just such a challenge. In a series of studies of the chronology and isotope chemistry of vein calcite at Devils Hole, Isaac Winograd and colleagues have raised important issues about the timing of mid-to-late Quaternary climate changes in relation to insolation changes (Winograd et al. 1988, 1992, 1997; Winograd and Landwehr 1993; Coplen et al. 1994).

Devils Hole has yielded two main calcite cores that provide an exceptionally well-dated climate record. DH-2 goes back 250 ka and DH-11 500 ka. DH-11 is a 36-cm long section of continuously precipitated Devils Hole calcite recovered in the subsurface from 15 m below the water table. The DH calcite originally formed from supersaturated groundwater emanating from the Ash Meadows discharge-recharge area, water that ultimately was derived from regional precipitation. Fourteen dates from alpha spectrometry and 21 dates from mass spectrometry U-series analyses provide probably the best directly dated continuous stratigraphic record covering the past few hundred thousand years yet available.

The key attributes to the Devils Hole vein calcite record are as follows: (1) The calcite is pure, as revealed by petrographic examination; (2) it was continuously deposited with no discontinuities; (3) the calcite is a closed geochemical system with no diagenesis; (4) the oxygen isotopic variation measured in the calcite constitutes a paleotemperature history of mean winter and summer precipitation in the recharge area, very likely a function of air temperature; (5) the calcite can be directly dated by the U-series method using mass spectrometry; and (6) the carbon isotopic record signifies either regional or global climate factors (Coplen et al. 1994).

The main paleoclimate proxies at Devils Hole are oxygen and carbon isotopes of the vein calcite. Sampled at minutely spaced intervals of 1.26 mm for 285 total samples, $\delta^{18}O$ and $\delta^{13}C$ show substantial variability almost certainly related to glacial-interglacial climate changes.

The oxygen isotope record is generally accepted to be a record of temperature history in the atmosphere of the recharge area. Indeed, the similarity between the DH oxygen-isotope curve and both the SPECMAP deep-sea oxygen isotope curve and the Vostok Antarctica ice core record of CO_2, CH_4, and atmospheric temperature is startling. Especially noteworthy is how the rapid glacial terminations in the DH oxygen-isotope curve have the same shape as those seen in deep-sea isotope curves.

The meaning of the carbon isotope record is not so clear. Coplen et al. (1994) found four $\delta^{13}C$ minima corresponding to Terminations 5 through 2. These minima lead the oxygen-isotope shifts by 7 ka. Because there is extensive buffering in groundwater that should keep the $\delta^{13}C$ values constant, Coplen et al. rejected a groundwater mechanism as the cause of the observed fluctuations and argued that carbon isotope variability must occur before the carbon enters the groundwater. Therefore, the fluctuations may represent changes in inorganic carbon in the recharge water due to either global variations in atmospheric $\delta^{13}C$, or variation in the extent and density of southwest Great Basin vegetation. The vegetation hypothesis says that the carbon-isotope minimum corresponds to the occurrence of the densest vegetation 5 ka before the insolation peak, when continental ice sheets had already receded. The carbon isotopes show a negative correlation with $\delta^{18}O$ at obliquity and precession cycles.

Despite the apparent correspondence of the DH record to deep-sea isotope curves, several important differences between the two paleoclimate records led Winograd et al. (1992) to claim the DH record is inconsistent with orbital theory's tenet that solar insolation triggered decay and growth of large ice sheets. Three discrepancies between the DH and most marine records are at issue. The first is the timing of climate trends. The beginning of the penultimate deglaciation—the isotope stage 6/5 transition—is too early to be forced by insolation changes. Deglaciation began at least 140 ka (probably a few thousand years earlier given the time it takes for groundwater to flow from the recharge area to the fissure), and full interglacial conditions were reached by 132 ka. This age is about 4 ka earlier than the insolation peak and the SPECMAP peak interglacial age. As described earlier in this chapter, this early deglaciation is supported by U-series dates on coral reefs from several coasts (see Winograd et al. 1997)

Second, the DH record shows a tendency for interglacials to become progressively warmer between 420 and 120 ka and for glacial cycles to increase in duration from 80 to 130 ka. Neither pattern is predicted by the insolation curve. Third, the past few interglacials lasted about 20 ka in duration, about 10 ka longer than postulated by the SPECMAP chronology. Winograd and colleagues argue that the extended duration of the past interglacial period at DH is supported by the Vostok ice core and U-series–dated coral reefs.

The major conclusion that Winograd and colleagues drew from Devils Hole is that insolation does not directly force late Quaternary deglaciations; rather, internal nonlinear feedbacks within the climate system better account for the age and paleoclimate pattern at Devils Hole. This notion has generated considerable discussion and differences of interpretation among prominent paleoclimate theorists. Are the U-series dates correct? Is the DH isotopic signal compromised, or is it a local signal unrelated to any global glacial-deglacial cycles?

Grootes (1993) accepts the DH dating. He argues that ice dynamics are a factor and that Southern Hemisphere insolation changes recorded in the Vostok ice-core climate record may also play a role. Grootes builds a conceptual model that links the Southern and Northern Hemispheres via a deep water teleconnection, suggesting that the Southern Hemisphere ice-core climate record reflects interhemispheric coupling during the past glacial-interglacial cycle.

Emiliani (1993) and Imbrie et al. (1993b) argue that the Devils Hole record is in fact an extremely compelling *confirmation* of Milankovitch theory. Emiliani, for example, says the DH record confirms his notion of Milankovitch theory with specific reference to cooling of the deep oceans during the coldest glacial periods. However, Landwehr et al. (1994) responded to Emiliani's comments by showing that there is in fact a poor match between insolation minima and Devils Hole oxygen isotope minima (two "predicted" minima do not occur; four other minima occur where insolation is not low).

Shackleton (1993) and Edwards and Gallup (1993) argue that the age dating may be compromised, possibly because of the adsorption of young ^{230}Th produced in the groundwater from ^{234}U onto the calcite wall during growth. Such a process can make the ages appear too old and explain the discrepancy with the tuned SPECMAP curve. They suggest this may be why the Termination 2 age at DH is as much as 5 ka too old.

Ludwig et al. (1993) counters this point by showing that (1) laminar flow occurs along the walls of Devils Hole cave, whereas turbulent mixing would be necessary to cause the adsorption of ^{230}Th; (2) particulate scavenging of ^{230}Th followed by gravitational settling would not bias the age; and (3) new measurements of U-series decay products from the walls of the calcite show negligible (< 1%) excess ^{230}Th, much too little to contaminate the age determinations. The DH petrology and aqueous environment support the U-series dating as valid.

Crowley (1994) accepts the dating and makes the counterintuitive but plausible climatic argument that the early Termination 2 warmth is due to the strength of the penultimate glacial episode (isotope stage 6). He argues that, during stage 6, the Polar Front was farther south by 5° latitude than during the LGM (stage 2) and that continental ice was more extensive, albeit thinner, in North America. The retreat of the ice front at about 145 ka corresponds to a precessional peak that may have led to full interglacial conditions. He cites evidence from lake levels from the western United States that isotope stage 6 lake levels were exceptionally high. So Crowley concluded that both the SPECMAP and DH chronologies are acceptable but that the Devils Hole signal is a regional climate response.

Whereas the exact relationship between the Devils Hole calcite and deep-sea marine paleoclimate records is unresolved, the idea that nonlinear dynamics influenced mid-to-late Pleistocene climate bears serious consideration (see Yiou et al. 1994).

MODELING ORBITAL CLIMATE CHANGE

"CAN MILANKOVITCH ORBITAL variations initiate the growth of ice sheets in a GCM [general circulation model]?" (Rind et al. 1989). Such a question is an ideal way to begin a brief, highly selective commentary on modeling the influence of orbitally induced solar variability on climate. To test orbital theory, many scientists have tried to model the response of the earth's climate to radiation changes due to orbital changes at 23-, 41-, and 100-ka frequencies. Several excellent reviews of modeling efforts are available (Berger et al. 1984; Imbrie and Imbrie, 1980; Imbrie et al. 1993a).

Modeling of orbital climate change has taken many forms. Some researchers have tried simply to establish statistical correlations between a climate proxy record, such as the oxygen isotope curve or the percentage of a key indicator species, with calculated insolation values at a particular latitude and season (Ruddiman and McIntyre 1981). The literature has many examples of this type of work. Other more complex studies have applied energy balance and general circulation models to the problem of orbital theory.

According to Imbrie and Imbrie (1980), two major obstacles had to be overcome in order to successfully test the major tenets of the theory. The first obstacle was how to test astronomical theory in the frequency domain. That is, do the frequencies of climate cycles match those calculated by celestial mechanics? This task required continuous time series of accurate climatic indicators from the geosciences. Much of this chapter has been devoted to reviewing a number of time series of climate proxies developed over the past few decades from the world's major ecosystems.

The second obstacle cited by Imbrie and Imbrie (1980) is that climatologists had to improve their understanding of the earth's radiation budget to quantify the earth's response to calculated seasonal and geographic changes. Successful attempts to use radiation models (Weertman 1976; Pollard 1978) and, more recently, model improvements described by Rind et al. (1989) generally seem to minimize these obstacles to the satisfaction of most paleoclimatologists.

As mentioned in the previous section, the shift from a 41-ka to a dominant 100-ka cycle has been a major puzzle in terms of modeling the intermittent waxing and waning of large-amplitude mid-to-late Quaternary ice sheets. How can small changes, due to eccentricity cycles, in total solar irradiance reaching earth's upper atmosphere be translated into such high-amplitude climate signals? Crowley and North (1991) and Imbrie et al. (1993a) describe many modeling studies that have attempted to model 100-ka cycles, and Imbrie et al. (1989, 1993a) narrow the list to seven groups of models attempting to solve this mystery. These models come in two classes. Members of the first have a free oscillation, such as those of Le Treut and Ghil (1983) and Saltzman and Sutera (1987), and model the interaction between orbitally forced oscillations and free oscillations of the atmosphere-ocean-cryosphere and lithosphere over a certain period. Free oscillations are best illustrated by the effects of the large-temperate ice sheets on the climate system. When ice thickness exceeds a critical value, free oscillations with periods ranging from 70 to 130 ka appear (Oerlemans 1982, Ghil and Childress 1987). The second class consists of models without a free oscillation, such as those of Imbrie and Imbrie (1980) and Pisias and Shackleton (1984). Each model has its own intrinsically valuable features, which are too varied to cover here. It is useful, however, to try to encapsulate the overall earth-system response to orbital forcing into a cohesive conceptual framework of climate change. Such a process model has been proposed by John Imbrie and a host of colleagues, each of whom individually has made major contributions to the documentation of orbital patterns in the geological record.

A Process Model of Quaternary Climate Change

The process model, described by Imbrie et al. (1992, 1993a) in "On the Structure and Origin of Major Glaciation Cycles: Parts 1 and 2," presents a unified model of earth's climate over the past few hundred thousand years. Part 1 describes the origin of the 23- and 41-ka cycles, and part 2 the more perplexing 100-ka cycle. Based heavily on oceanic data and tied to the oxygen isotope curve, the process model is the modern embodiment of Milankovitch's ideas and an authoritative synthesis of the causes of the late Quaternary ice ages.

Three fundamental aspects of their approach deserve special note. First, they focused on the causes of glaciation, which is a mid- to high-latitude northern hemisphere problem—a polar and subpolar climatic response to radiation. The model therefore does not explicitly treat those aspects of low-latitude climate change occurring at the 23-ka precessional frequency, such as discussed earlier with reference to monsoons, tropical dust, and African vegetation, nor some aspects of climate change related to CH_4 and water vapor found in ice-core records (chapter 9). All these processes clearly influence earth's climate.

Second, the cornerstone of the model is the amplification of the small insolation changes into major ice-age cycles through the action of the ice sheets themselves as measured by deep-sea oxygen isotopes as proxies of ice volume. All other proxies are pinned to the isotope/ice volume curve. Third, this version of the orbital theory of the ice ages recognizes the critical role of oceanic circulation changes (e.g., Broecker et al. 1985; Rind and Chandler 1991) in global climate change, specifically for translating insolation variability into local responses. Both sites of NADW formation, the polar Nordic Seas and the boreal Labrador Sea, are referred to as "heat pumps" because they pull low-latitude heat to higher latitudes. The switching on or off of these heat pumps controls to a large degree the climate response to insolation.

The model relies on determining the phase relationships between 16 paleoclimate indicators to depict the sequence of events within an idealized cycle (Imbrie et al. 1989). The indicators include Southern Hemisphere deep-sea circulation, SSTs, terrestrial aridity (dust), ocean dissolution, and nutrients, among others, all tied together by the marine oxygen-isotope curve. Thirteen proxy-record indicators extend back 300 ka or earlier. The lack of long-term time series for several key records (e.g., high Northern Hemisphere climate records) imposes the need to substitute well-dated ^{14}C events of the past deglacial termination (< 20 ka) in the Nordic Seas and the Arctic Ocean in lieu of a long time series. High-latitude climate trends of the past termination are assumed to be typical of prior terminations.

Three major conclusions emerge from Imbrie's review of orbital forcing:

1. The process model presents a spatial sequence of events in which the geographic response to forcing occurs in a single distinct sequence that is common to all 23-ka, 41-ka, and 100-ka periods for 11 of the proxies. The 23- and 41-ka insolation forcings have dominated the climate system, at least over the past 500 ka (as we saw earlier in the chapter, they influence climate at least 4–5 Ma back in certain records). The conclusion that the same sequence of events occurs for all periodicities was met with some surprise. Nonetheless, it is supported by both observation and theory, and it suggests that the same internal mechanisms operate during each type of orbital cycle.

2. The climate-system response to 23- and 41-ka cycles represents a linear, continuous response to insolation changes altering earth's radiation budget; the response to the 100-ka cycle is a nonlinear forcing in which large mid-latitude ice sheets amplify the initial insolation signal though various feedback loops.

3. Each ice-age cycle consists of at least four end-member states—glacial, deglacial, interglacial, and preglacial—that explain the evolution of the climate system. The initial triggering input is from Northern Hemisphere (65°N) June insolation. Each state is characterized by a different mode of deep-ocean circulation. The glacial period is characterized by a shutdown or weakening of the North Atlantic deep water formation; both the polar Nordic Sea and boreal Labrador Sea heat pumps are inoperative. In addition, during full glacial conditions, both northern hemisphere deep-water convective regions have reduced deep-water formation. During deglaciation, the Nordic heat pump is turned on. Under full interglacial conditions, the Labrador Sea boreal heat pump is also operative. Finally, during the preglacial state, insolation decreases in high latitudes; fields of ice and snow expand from their minimal conditions; and the Nordic pump is shut down or diminished.

CLOSING COMMENTS

MY DESCRIPTION OF Imbrie's process model is obviously a much condensed synthesis of an all-encompassing treatment of orbital theory and its evidence from the past few hundred thousand years. Perhaps the most important generality to emerge from a reading of Imbrie and colleagues' model and related literature on orbital forcing is that powerful internal processes modify or even overwhelm the insolation signal. These processes are specific to the regions and time scales involved. This conclusion may explain why acceptance of orbital theory is not universal (Broecker and Denton 1989; Winograd et al. 1992), and it reminds us of how prudent it is to frequently reevaluate predominant theories about climate change and consider the plurality of factors that cause climate variability. The next chapter, on millennial-scale climate variability, illustrates this point.

In closing, the patterns of climate change described in this chapter show that the study of glacial-interglacial cycles goes deeper than simply an effort to understand the growth and decay of ice sheets. In the quest to understand the astronomical-climate link, paleoclimatology also furnishes evidence about the ecological and evolutionary concepts of community structure, adaptation, and long-term intraspecific stability described in the previous chapter. While principles of energy fluxes, atmospheric and ocean circulation, and the physics of ice sheet growth and decay dominate certain aspects of orbital theory, one might argue that equally compelling theoretical reasons stem from biology as to why climatic cycles produce the cyclic faunal and floral patterns commonly preserved in the geological record. There seem to be fundamental biological characteristics that unite marine and forest communities so that they reassemble periodically despite repeated perturbations to the ecosystem. Fundamental characteristics of species also allow them to maintain evolutionary stability over millions of years of cyclic climatic change. Evolutionary and ecological processes, though not always in the forefront, are a basic component of orbital climate research.

V

Millennial-Scale Climate Change

So great the climate of Europe changed, that in Northern Italy, gigantic moraines, left by old glaciers, are now clothed by vine and maize.
CHARLES DARWIN, 1859 (P. 353)

DRYAS OCTOPETALA

THE ARCTIC FLOWER *Dryas octopetala* now lives in cold climate zones of high Northern Hemisphere latitudes. During frigid climates of the past glacial period about 20 ka, when ice sheets covered Scandinavia, northern Germany and much of the British Isles, *D. octopetala* lived far south of its modern geographic range. Like other plants whose pollen settles in sediments in bogs and lakes, *D. octopetala* is a telltale relict of glacial-age climates in temperate latitudes. As solar insolation rose and climate warmed about 15 ka, *D. octopetala*'s habitat shifted northward, and it disappeared from Quaternary European palynological records, replaced by pollen of taxa indicating warmer climates. Jensen (1935) named this climatic warming the Allerød pollen zone from his study of pollen from Allerød, Denmark.

A sudden, inexplicable change then occurred in the vegetation record of Denmark. Jensen (1935) found a surprising reappearance of *D. octopetala*, at approximately 12.5 ka, during a brief period sandwiched between the preceding Allerød and the succeeding Preboreal periods. The return of *D. octopetala* to Europe is a widespread regional event that has come to be called the Younger Dryas stadial (Mangerud et al. 1974), a climatostratigraphic term that signifies a partial return toward glacial conditions. For decades, this abrupt climatic reversal engaged the attention of only a relatively small group of glacial geologists and palynologists trying to unravel the glacial-interglacial transformation. Recently, though, the Younger Dryas had occupied center stage for a new generation of paleoclimatologists who seek to understand its cause. The Younger Dryas is perhaps the single most researched paleoclimate event of the past decade, certainly for the Quaternary, and a symbol of the dynamic nature of earth's climate.

How did the Younger Dryas emerge from relative obscurity to impart a new identity to paleoclimatology and lead to a paradigm shift in how many scientists view climate change? One major impetus was the paper by Wallace Broecker and colleagues (1985) proposing that the Younger Dryas represents more than a recolonization of Europe by an Arctic flower but a global climate event triggered by changes in North Atlantic ocean temperature and salinity that altered global oceanic circulation with climatic repercussions around the world. Seizing upon advances in modern and glacial-age ocean circulation and chemistry (e.g., Broecker 1982), Broecker proposed his "conveyor belt" theory to explain millennial-scale climatic events during glacial periods. Today the warm North Atlantic Drift provides about 30% as much heat to the North Atlantic and Europe as does direct solar insolation, because wintertime cooling of surface waters causes them to sink and form deepwater. This convective overturn drives deep-ocean circulation, which like a conveyor belt carries heat around the world. Broecker argued that a full or partial shutdown of North Atlantic thermohaline circulation and deep-water formation could lead to the Younger Dryas stadial by depriving Europe's atmosphere of its heat source.

The cornerstone of Broecker's conveyor belt theory (a version also referred to as the "salt oscillator" hypothesis, developed in later papers [e.g., Broecker et al. 1988a,b, 1989, 1990]) was that North Atlantic deepwater (NADW) was vulnerable to the influence of glacial meltwater from the adjacent North American ice sheet. North Atlantic thermohaline circulation may have been affected when Laurentide Ice Sheet meltwater, which drained via the Mississippi River into the Gulf of Mexico about 15–12.5 ka, was suddenly diverted into the North Atlantic via the Great Lakes and St. Lawrence River, a route previously blocked by the retreating Laurentide Ice Sheet (Dyke and Prest 1987; LaSalle and Elson 1975; Cronin 1977; Hillaire-Marcel and Occhietti 1980; LaSalle and Shilts 1993; Rodrigues and Vilks 1994). The resulting decrease in surface salinity made North Atlantic surface water less dense and slowed down NADW formation in the Nordic Seas. Without deep-water formation pulling warm water from the North Atlantic Drift–Gulf Stream system northward from low latitudes, northern Europe would be deprived of the oceanic current that was its heat-source. *D. octopetala* and other cold-climate indicators could thus return to northern Europe.

In the wake of Broecker's Younger Dryas hypothesis, paleoclimate orthodoxy was pushed aside as paleoclimatologists began to investigate ocean-atmosphere-ice links during climate changes that were too rapid to be driven by orbital insolation cycles (figure 5-1): Did thermohaline circulation shut off the transport of heat to northern Europe, triggered by meltwater export via the St. Lawrence River, catalyzing a series of global events linked through a deep-sea circulation conveyor belt (Teller 1990)? Was the Younger Dryas global in extent and if so, how strong was the signal outside the North Atlantic? Were climate oscillations during deglaciation synchronous in the Northern and Southern Hemispheres? What was the role of ice sheets and icebergs in thermohaline variability?

Interest in the Younger Dryas was also spurred by advances in dating (i.e., AMS radiocarbon dating, chapter 2) and high resolution paleo-

FIGURE 5-1 Calving icebergs off Antarctica depict the role of ice in climate changes of the past 20,000 yr. Courtesy of J. Bernhard.

climate records of deglaciation from Greenland ice cores (chapter 9). Research rapidly expanded into climatic instability during the glacial period preceding the Younger Dryas and during the last interglacial period, which has traditionally been viewed as a relatively stable, quiescent interval of global warmth. Together, this class of climate change is referred to as millennial-scale, or "suborbital," climate change. Millennial climate events are climate oscillations occurring over 1–10 ka that are sometimes quasicyclic in nature. They can begin and/or end abruptly, within centuries or less, and include Heinrich events, Bond cycles, and Dansgaard-Oeschger Events, Younger Dryas and other oscillations. In this chapter I describe evidence for millennial-scale climate variability during three distinct climate states of the late Quaternary: (1) the late glacial-Holocene transition (between 20–10 ka, also called Termination 1 and marine oxygen isotope stage 2/1 transition), (2) the past glacial period (about 115–18 ka, including the past glacial stage known in North America as the Wisconsinan and in Europe as the Wurm or Weischelian glaciation), and (3) the past interglacial period (about 140–115 ka, the Sangamon of North America, the Eemian interglacial of Europe) and briefly the Holocene interglacial.

Each period—deglaciation, glacial, and interglacial—signifies a fundamentally different set of climate boundary conditions (land- and sea-ice distribution, atmospheric circulation and chemistry, oceanic circulation, insolation) within which short-term climate change occurs. Thus the causes of millennial-scale climate change may differ during deglaciation, glacial, and interglacial climates. Each period harbors physical and biotic evidence for rapid climate change, such as that which inspired Charles Darwin's comments on the modern climate of formerly glaciated areas of northern Italy. Darwin's comments are taken from his perceptive discourse on Quaternary climate change and the dispersal capability of plant species in response to postglacial warming. They are among the earliest discussions of climatically induced biogeographic shifts in species distributions, such as that of *D. octopetala,* that we now use routinely in paleoclimatology. Before describing evidence for short-term climate variability within each climatic state, I wish to briefly accredit some early studies that form the foundation of contemporary research on rapid climate events.

EARLY EVIDENCE FOR RAPID CLIMATE CHANGE

UNTIL THE PAST decade, millennial-scale climate change was overshadowed by discoveries of long-term climate cycles at the frequencies of orbital variations. In the introductory chapter to their excellent volume entitled *Abrupt Climatic Change,* Berger and Labeyrie (1987) noted how little paleoclimatologists appeared to know about climate change within the Holocene interglacial and that there was no satisfactory explanation for the Younger Dryas climate reversal. Most chapters in their book leave the reader with a sense that just a decade ago, the paleoclimatology of rapid events was in its formative stages.

Nonetheless, early studies show that the geologic record held clues about brief, intense climate events. The seminal study of Quaternary oxygen isotope ratios from the Camp Century Greenland ice core by Dansgaard et al. (1971) found isotopic evidence for wide swings in atmospheric temperature over Greenland—as much as 5–10°C occurring in just a century. Quaternary beetles from England also provided strong evidence for rapid climate change. Russell Coope's (1977, see Elias 1994) classic investigations showed that beetles have ideal attributes for documentation of rapid climate change: evolutionary stability, species-specific morphological complexity, abundance in Quaternary organic silts, physiological constancy, and rapid response to temperature change. Coope combined beetle ecology and taxonomy with glacial geology to demonstrate rapid climate transitions during the late Quaternary, including a return of Arctic beetle species to the British Isles during the Loch Lomond stadial, the age equivalent of the Younger Dryas stadial. Coope's beetles showed summer temperatures dropped several degrees at the inception of the Loch Lomond and then rose again by as much as 8°C, apparently *within a few centuries.* Coope (1977) also found beetle assemblages reflected climatic oscillations in the older Weischelian glacial period of the British Isles in the Chelford, Upton, Warren, and Windermere interstadials.

Rapid climate change also punctuated the North American Wisconsinan and the European Weischelian glacial stages. For example, glacial stratigraphy in the eastern Great Lakes and St. Lawrence Valley of the United States and Canada reflects a dynamic, oscillating Laurentide Ice Sheet margin, represented by alternating glacial and nonglacial deposits like loess, lake sediments or peat (Dreimanis and Karrow 1972). The Port Talbot, Plum Point, Erie, Mackinaw, and North Bay Interstadials all testify to dynamic change within the glacial period caused by either climate change, nonclimatic ice sheet behavior, or both (Clark 1994). Outside North America and Europe, especially in the Southern Hemisphere, important precursors of recent studies laid the groundwork for the global search for the Younger Dryas and other millennial-scale events (see Suggate 1990).

ICE AND MILLENNIAL-SCALE CLIMATE CHANGE

IN THE PREVIOUS chapter, we saw that large-scale oscillations between two end members of global climate—glacial and interglacial modes—have characterized the earth for the past few million years. Before discussing the possible causes of millennial-scale change within a glacial period, a glacial-interglacial transition, or an interglacial period, I must first describe the boundary conditions that distinguish the glacial state from our current interglacial period. Glacial periods had colder global mean annual temperatures (about 5°C average). In some areas such as the North Atlantic Ocean and adjacent continents, sea-surface and atmospheric temperatures decreased a full 10°C. Large, mid-latitude lakes appeared owing to higher regional precipitation. Equatorward shifts in the geographic distribution of cold-loving species of terrestrial plants, vertebrates (mainly mammals), invertebrates (beetles, mollusks), marine plankton, and benthos occurred. Glacial intervals experienced a different mode of deep-ocean circulation in which there were altered deep-sea nutrients, temperature, and chemistry. Glacial periods also had increased aridity in subpolar regions, higher atmospheric dust content, lower concentrations of atmospheric CO_2, CH_4, and other gases, and different quantities of various acidic and alkaline chemical species. Finally, glacial periods are naturally also characterized by large mid-latitude ice sheets, a point upon which we need to elaborate in order to understand millennial climate change.

Geologists sometimes use the term *glacial periods* somewhat informally to refer to extended periods in the geological record when there was abundant continental ice in at least some regions. Long-term periods of glaciation include the late Ordovician, the Carboniferous-Permian, and the post-Eocene part of the Cenozoic (Crowley and North 1991; Frakes et al. 1992). The climate during these periods was generally colder that that of the intervening periods. Pre- to Late-Cenozoic continental ice sheet history is beyond the scope here; henceforth, I will limit my discussion to late Cenozoic glacials. In the Cenozoic, a substantial amount of Antarctic polar ice has existed since at least the Oligocene (about 35 Ma [e.g., Matthews and Poore 1980]); and in Greenland and Arctic regions since at least the late Miocene (Jansen and Sjøholm 1991). The late Cenozoic is thus widely viewed as a glacial period compared with the relatively warm Eocene or mid-Cretaceous.

The presence of large mid-latitude Northern Hemispheric ice

sheets in North America and Europe during the Quaternary distinguishes the glacial world from the modern interglacial one, in which there are the two high-latitude polar ice sheets in Greenland and Antarctica. The extent of Quaternary North American and European ice sheets (figure 5-2) has been reconstructed by mapping glacial deposits and landforms (Flint 1971; Denton and Hughes 1981; Grosswald 1993; Ruddiman and Wright 1987; Wright et al. 1993; Teller and Kehew 1994). The North American Laurentide Ice Sheet extended

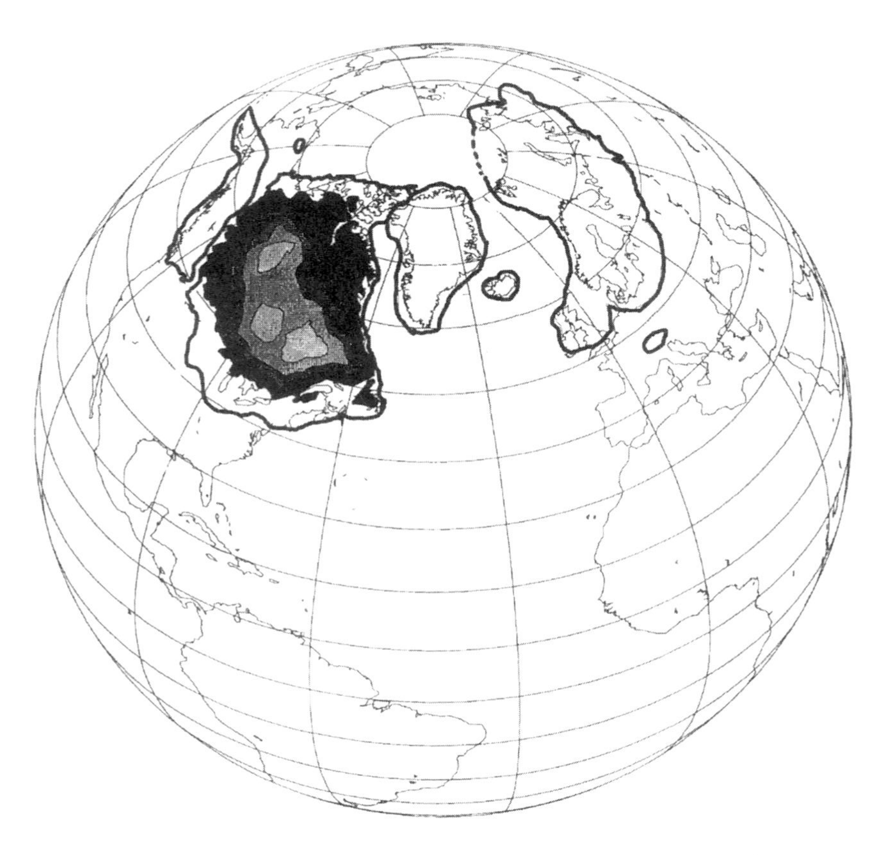

FIGURE 5-2 Major late Pleistocene ice sheets in the North Atlantic region. Dark line shows approximate extent of ice during last glacial maximum (LGM) about 20,000 yr ago; dark area shows approximate Laurentide Ice Sheet thickness during deglaciation about 16 ka (14 ^{14}C ka). Courtesy of P. Clark and American Geophysical Union. Modified from Clark et al. (1996a).

south and northward from its center of growth near Hudson Bay; its southern margin stretched from the northern Great Plains of the mid continent to the continental shelf off New England and the Canadian maritimes (figure 5-2). The Canadian Arctic Islands were covered by the Innuitian (or Franklin) Ice Sheet, Greenland by a slightly larger ice sheet that the modern one. The Cordilleran Ice Sheet stretched along the western backbone of North America from the Pacific northwest to southeastern Alaska. In Europe, the Fennoscandinavian Ice Sheet centered on Scandinavia and mainland Europe, the British Isles Ice Sheet on part parts of the British Isles, and the marine-based Barents Ice Sheet along the northern border of the Fennoscandinavian Ice Sheet.

Ice sheets are important in climate change for many reasons. Imbrie et al. (1992, 1993a) and Alley (1995), among others, have postulated that large ice sheets play a key role in orbital-scale climate change by amplifying the climatic effects of solar insolation variability. Large ice sheets are also so thick that they can change regional atmospheric circulation simply by blocking the large-scale circulation of atmospheric air masses (Manabe and Broccoli 1985; Kutzbach and Guetter 1986; COHMAP 1988). Ice sheets also increase the total reflected radiation from earth's surface, and these ice-albedo effects are thought to cause a positive climatic feedback.

On shorter time scales, large continental ice sheets can amplify or even trigger millennial-scale climate events, especially through iceberg discharge from marine-based ice sheet margins or through meltwater discharge. Northern Hemisphere ice sheets ringed the North Atlantic and Nordic Seas where they could potentially influence oceanic conditions in this manner. Ice-sheet meltwater discharge is a basic tenet of Broecker's salt oscillator hypothesis to explain the Younger Dryas reversal during the past deglaciation.

During full glacial conditions, the interplay between the Laurentide Ice Sheet margin and the ocean in the Hudson Strait region of eastern Canada may have led to a process of ice-sheet drawdown through ice streams that calved numerous giant icebergs that drifted eastward across the North Atlantic, cooling and freshening ocean surface waters. Iceberg calving and melting from a dynamic but relatively small part of the Laurentide Ice Sheet margin are the predominant feature of Heinrich events, the rapid deposition of ice-rafted debris (IRD) across much of the subpolar Atlantic (Bond et al. 1992, see below).

Sea ice is also a critical part of the climate system with regard to rapid climate Quaternary change (figure 5-1) (Peltier 1993). Sea ice influences climate by reflecting solar radiation, altering deep-oceanic convection in high latitudes, decreasing heat exchange between the ocean and atmosphere, cooling the adjacent continent, altering surface-ocean productivity by photosynthetic algae, and changing atmosphere-ocean carbon fluxes. Sea ice that originates from fresh water continental ice-sheet sources, such as Northern Hemisphere ice sheets, forms icebergs that upon melting decrease surface salinity and slow thermohaline circulation. Sea ice formed from frozen seawater in much of the Arctic (Aagaard et al. 1985) and the periphery of Antarctica affects ocean circulation through a process called brine rejection. Brine rejection may occur when sea ice forms and upper layers of the ocean become saltier and heavier, leading to deep-water formation. This process is seen an important mechanism in the Northern Hemisphere deep-water circulation budget (Aagaard and Carmack 1989; Rudels and Quadfasel 1991). During the last glacial maximum (LGM) and deglaciation, dynamic changes in sea-ice distribution in the Nordic Seas (Jansen and Bjorklund 1985; Sarnthein et al. 1994; Koc and Jansen 1994) and the Arctic (e.g., Stein et al. 1994; Cronin et al. 1995, 1996) occurred.

This brief summary should give a glimpse of glacial-age boundary conditions and an idea of the importance of continental and sea-ice relevant to the following discussion of millennial climate change.

THE YOUNGER DRYAS AND OTHER RAPID CLIMATE EVENTS DURING DEGLACIATION

THE YOUNGER DRYAS cold snap occurred between about 12.5 and 11.5 ka throughout Europe, the North Atlantic Ocean, and eastern North America (figure 5-3) (Kennett 1990). Whether this event was also represented globally is a question that has persisted for more than three decades. Recent paleoclimate evidence, derived largely from high-resolution sampling efforts, shows that the Younger Dryas—and other rapid climate events—punctuated the deglacial period of most continents and oceanic regions. Although they are seldom of equal magnitude or precise chronology to Younger Dryas–like events recorded in the North Atlantic theater, rapid climate shifts during glacial-interglacial transitions are a characteristic feature of earth's climate. A brief review of the evidence from land and oceans follows.

Younger Dryas on Continents

Europe

In Europe the terrestrial climate record from pollen zones and lacustrine sediments shows the following post-LGM deglaciation sequence:

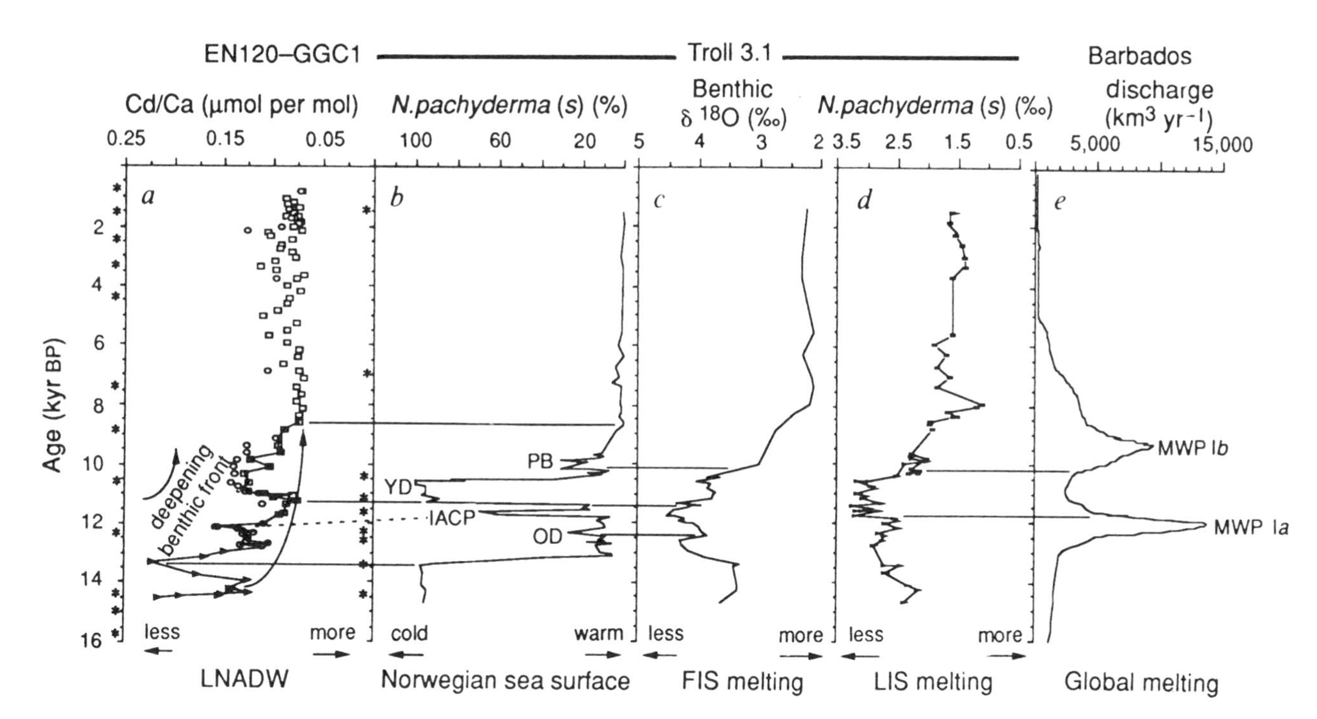

FIGURE 5-3 Rapid climate and sea-level changes during the past deglaciation showing relationship between the Younger Dryas (YD) record of deep-sea Lower North Atlantic deep water (LNADW), Norwegian Sea surface water (proxy: *Neogloboquadrina pachyderma*), ice-sheet melting ($\delta^{18}O$ and *N. pachyderma*), and Barbados coral isotopic record of meltwater pulses (MWP1a, 1b). Cd:Ca ratios reflect changes in lower North Atlantic deep water. LIS = Laurentide Ice Sheet; FIS = Fennoscandanavian Ice Sheet; PB = Preboreal; IACP = Intra-Allerød Cold Period; OD = Older Dryas. Courtesy of S. Lehman with permission from Macmillan Magazines. From Lehman and Keigwin (1992).

Oldest Dryas (cool), Bølling (warm), Older Dryas (cool), Allerød (warm), Younger Dryas (cold), and Preboreal (early Holocene, warm) (figure 5-4). The Bølling and Allerød are often lumped as a single interstadial event when the stratigraphic resolution precludes recognition of the Older Dryas.

Early evidence for a climate reversal during deglaciation came from the continental record of fossil pollen (e.g., Jensen 1935; Iversen 1954). Mangerud et al. (1974) and Larsen et al. (1984) reviewed the Younger Dryas in Europe and referred to it as a climatic event—a stadial—based on evidence from stratigraphy, glacial geology, and pollen biozones. The vegetation changes during the Bølling–Allerød–Younger Dryas and the Younger Dryas–Preboreal transitions were generally considered rapid and synchronous events across a wide area of Europe.

Denton and Karlen (1973) studied alpine glaciers in the Sarek and Kebnekaise Mountains of Swedish Lapland, which are approximately

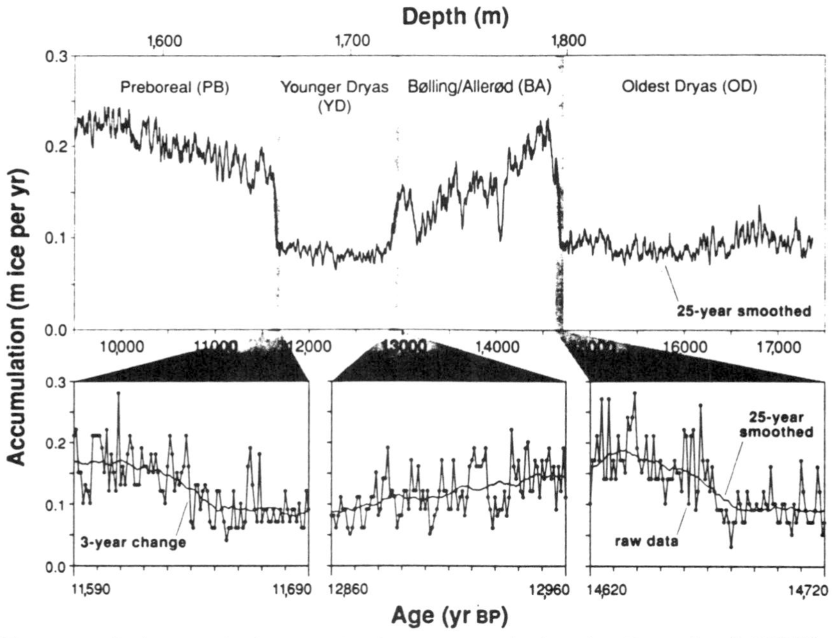

FIGURE 5-4 Rapid changes in ice accumulation in Greenland GISP2 core during deglaciation. Low accumulation characterizes the cold Older and Younger Dryas periods. Courtesy of R. B. Alley with permission from Macmillan Magazines. From Alley et al. (1993).

1500–2289 m above sea level, and in a companion study (Denton and Karlen 1976) glaciers in the St. Elias Mountains of the Yukon Territory, Canada, and Alaska. They mapped multiple late glacial and Holocene moraines, including glacial expansions during the Little Ice Age of the fifteenth through nineteenth centuries (see next chapter). They postulated that the regional climate during this time was in fact analogous to that of the Younger Dryas and that glacial readvances occurred every 2.5 ka back to 12 ka, perhaps related to solar variations. Denton and Karlen clearly recognized the rapidity of these events: they argued that glacial expansions took about 900 yr, contractions about 1750 yr.

The intensity of Younger Dryas cooling in Europe is underscored by Mangerud (1987), who summarized evidence for the rate of cooling in Scandinavia based on cirque glaciers in western Norway. He found that summer temperatures dropped 5–6°C in less than 200 yr, probably within a few decades. Fossil ice wedges suggest that mean annual temperature may have fallen to as much as 13°C colder than modern temperatures in the region. In Scandinavia, the Younger Dryas stadial signified essentially a return to near glacial conditions of only a few thousand years before. Mangerud (1987) postulated that the Younger Dryas episode was related to the changing position of the warm North Atlantic Current off the western coast of Norway. The ocean there has two modes, one in which the warm current bathes Norway in warmth as it does currently, and another, like the Younger Dryas stadial, in which the current is deflected slightly southward, toward Portugal, depriving Norway of its oceanic heat source and leaving Norwegian coasts in glaciallike climates.

The case for synchroneity of Allerød–Younger Dryas–Preboreal climatic chronology throughout the North Atlantic region continues to gain support. Investigators have obtained near-calendar-year chronologies from German (Hajdas et al. 1995) and Polish (Goslar et al. 1995) lake rhythmitic sediments, German tree-ring dendrochronology (Becker 1993), Swedish glacial varved sediments (Stromberg 1994), and Greenland ice cores (Alley et al. 1993; Dansgaard et al. 1993; Johnsen et al. 1992). As discussed in chapter 2, the calibration of accelerator mass spectrometry radiocarbon dates to high-resolution chronologies with annual resolution extends back to 11.5 ka, that is, to the Younger Dryas termination (Wohlfarth 1996). Before 11.5 ka, slightly less reliable calibration is achieved by uranium-series (U-series) dating of corals (Bard et al. 1990a,b, 1993).

Building on the classic work of De Geer (1912), Stromberg analyzed varve sequences from central and southern Sweden and showed that the Younger Dryas–Preboreal transition was slow at first and then suddenly extremely rapid for a 200-yr period. Stromberg calculated that the rate of ice margin retreat was at first 50–75 m/yr, then 100–200 m/yr for a period of 1500 yr. Björck et al. (1996a) proposed a "synchronized chronology" of European Younger Dryas events in which the dendrochronological age of the Younger Dryas–Preboreal transition is precisely dated at 11,450–11,390 ±80) yr. At this time German pine tree ring widths increased, and tree-wood isotope chemistry and $\delta^{14}C$ changed. In addition to the European continental evidence, Björck and colleagues point to isotopic evidence from the Greenland ice-core project (GRIP) indicating that the Younger Dryas cooling began and terminated abruptly, as well as to ^{14}C isotopic data from trees (Stuiver and Braziunas 1993) that Younger Dryas cooling was accompanied by reduced deep-water formation as the North Atlantic conveyor belt slowed down. They argue that, in addition to fresh water influx from North American glacial lakes and ice from Hudson Strait, the draining of the ice-dammed Baltic Ice Lake in Europe may have contributed to the shut-down of the conveyor belt.

Establishing a calendar-year chronology also enabled Björck and colleagues to find evidence for another late glacial cooling event in Europe—the Preboreal oscillation (PBO)—that occurred about 300 yr after the end of the Younger Dryas. The PBO lasted only about a century and half, between 11,200 and 11,050 yr, but it is clearly evident in data from tree-ring width and isotope chemistry, in salinity indicators in the Baltic ice lake, and in European lake pollen and lithologic records. After a brief period of strong oceanic ventilation at the inception of the Preboreal warming, a brief but significant decrease in atmospheric temperatures and ocean ventilation occurred. The PBO is also evident in the GRIP ice core $\delta^{18}O$ record (e.g., Johnsen et al. 1995). The upshot of these and other studies is that climate oscillated during the recovery from the Younger Dryas stadial in several other proxy records from the Nordic Seas (Lehman and Keigwin 1992), Europe (Haflidason et al. 1995), and Greenland ice cores (Taylor et al. 1993a).

North America

In North America, palynologists and glacial geologists have long debated the evidence for climate events coeval with Europe's Bølling–Allerød–Younger Dryas events. Pollen records from North America gave a confusing signal. Some early studies were considered at odds with the concept of a North American Younger Dryas climatic equivalent (e.g., Davis 1967, 1981; Bernabo and Webb 1977). Although temporal variability in late glacial pollen assemblages was evident, a preferred explanation was that ecological and/or taphonomic factors were sufficient to explain the paleovegetation record.

In the North American midcontinent, stratigraphic and glacial geological evidence suggested a dynamic ice margin during deglaciation. The Two Creeks Forest Bed in Wisconsin (Broecker and Farrand 1963) and the Valders readvance of the Michigan Lobe of the Laurentide Ice Sheet are examples. But it was not clear if these events were the age equivalent of the Younger Dryas. Kaiser (1994), for example, recently redated the Two Creeks Forest Beds using accelerator mass spectrometry (AMS) radiocarbon dates and spruce tree dendrochronology and concluded that it represented a brief 252-yr-long interstadial at 12,050–11,730 yr BP, perhaps the equivalent of Europe's Older Dryas.

Gosse et al. (1995) uncovered some of the best evidence available that a Younger Dryas-cooling event occurred in Western North America. They dated the Inner Titcomb Lakes moraine in the Wind River Mountains using the cosmogenic isotope beryllium-10 at 11.5–13.8 ka. The ages for the Titcomb Lakes glacial readvance overlap the ages of the Younger Dryas in the North Atlantic.

In the St. Lawrence Valley of Quebec, the St. Narcisse Moraine was built when the Laurentide Ice Sheet readvanced into the Champlain Sea, an arm of the Atlantic Ocean occupying the isostatically depressed St. Lawrence Valley in eastern Quebec, Canada (Elson 1969; LaSalle and Elson 1975; LaSalle and Shilts 1993). Although the late glacial period was a time of rapid environmental change in the region (Cronin 1977; Corliss et al. 1982), the relationship of regional glacial events like the St. Narcisse to the European deglacial sequence was unclear. Rodrigues and Vilks (1994) argued on the basis of radiocarbon dating of glacial Lake Agassiz drainage into the Champlain Sea that freshwater runoff from large glacial lakes occurred too late to trigger the Younger Dryas episode, although it may have sustained the Younger Dryas cold climate. Recently de Vernal et al. (1996) found dinoflagellate and isotopic evidence from off the mouth of the St. Lawrence, the outlet for Laurentide ice to discharge into the North Atlantic, that suggested the St. Narcisse readvance did in fact coincide with decreased freshwater discharge during the Younger Dryas interval. They found colder Younger Dryas oceanic surface temperatures in Cabot Strait and Laurentian Channel and evidence for reduced salinities during deglaciation in surface waters of the Gulf of St. Lawrence. However, they argued that meltwater influence was too weak to be detected farther offshore or in deeper water. Thus, although de Vernal et al. (1996) found support for a Younger Dryas event in eastern Canada, their evidence contradicts the idea that oceanic dilution by St. Lawrence River meltwater discharge was sufficient to affect large-scale North Atlantic circulation.

Peteet (1987) reviewed the eastern North American pollen record and made the case that the sudden increase in spruce and fir tree pollen near 11–10 ^{14}C ka marked a climate event comparable to those recorded in Europe. More recently, detailed studies provide compelling evidence that Younger Dryas–age cooling was characterized by climatically induced changes in vegetation and fauna in maritime Canada (New Brunswick and Nova Scotia) and Maine (Mott and Stea 1993; Mott et al. 1986, 1993; Stea and Mott 1989; Mayle et al. 1993; Levesque et al. 1993, 1994, 1997; Cwynar and Levesque 1995). Pollen and chironomid midge larva from late glacial lake sediments indicate Younger Dryas–age regional cooling of 12°C. At this time, a forest-tundra ecosystem returned to eastern Canada for a prolonged period. The addition of the chironomid midge proxy data to augment palynological data dispelled concern that deglacial vegetation changes might be explained in terms of ecological factors. The mean radiocarbon ages for the beginning and end of the Younger Dryas event in eastern North America were about 10.7 and 10.0 ^{14}C ka, respectively—ages generally consistent with European calibrated-year ages of about 12.5 and 11.5 ka (see chapter 2 and Wohlfarth 1996).

Intensive study of deglacial paleoclimate records unveiled evidence for other cooling episodes. Levesque et al. (1993) discovered palynological evidence for a pre–Younger Dryas cooling, which they named the Killarney oscillation, dated at about 11,290–10,960 ^{14}C yr. Although this was not as intense a cold snap as the Younger Dryas, atmospheric temperatures still dropped about 8°C.

Other Continents

Any review of the global nature of the Younger Dryas event must begin with the comprehensive assessment by Rind et al. (1986), who reviewed 80 European and 78 North American palynological and lake-level studies for evidence of a millennial scale cooling of Younger Dryas age. They found definitive evidence for cooling throughout Europe. Less conclusive evidence for cooling came from eastern parts of North America (now confirmed), the North Atlantic Ocean foraminiferal assemblages, and isotopic data from the Camp Century and Dye 3 Greenland ice cores. Evidence for Younger Dryas–age cooling was also found in eastern and southeastern Africa, in northern

South American, and in parts of Asia, but not in the midcontinent of North America or in Australia. Rind and colleagues estimated that the extent of Younger Dryas atmospheric cooling based on pollen, isotopes, glacial geology, and beetles ranged from -2 to -10°C in various regions of Europe, Chile, and New Zealand today. Overall, the Younger Dryas was more than just a regional event, but little or no data supporting it were available from many areas of the world.

In the succeeding decade, numerous studies of the continental record of the Younger Dryas interval have been published, including those focusing on Europe, North and South America (Peteet 1993), Central America, southern and eastern Africa, southeast Asia, and New Zealand (Peteet 1995). These investigations show that many terrestrial pollen records from around the world contain evidence for climate reversal, although only the European records give good data about the amplitude of cooling and precise chronologies.

Antarctic ice cores have yielded considerable information indicating a two-step deglaciation punctuated by a cooling event (Jouzel et al. 1989, 1995 and references therein). The ice core record for rapid climate change will be taken up in chapter 9. Suffice to say here that Jouzel and colleagues make a convincing case for a cold snap less intense than that which occurred in the North Atlantic Younger Dryas and possibly preceding the North Atlantic event by about 1000 yr. They call this the Antarctic cold reversal (ACR), dated at about 13.5–12.5 ka (Jouzel et al. 1995), and it provides firm evidence the Southern Hemisphere experienced a climate reversal during deglaciation prior to the Younger Dryas.

One dissenting voice in the debate about the global extent of the Younger Dryas has been Markgraf (1991, 1993), who argued that midlatitude pollen records from Chilean lakes and peat from Tierra del Fuego and southern Patagonia reveal complex vegetation shifts that do not support the idea of a simple climatic reversal toward glacial conditions in mid and high latitudes of South America. In higher latitudes, there is marked short-term variability in vegetation, but the timing of changes in specific plant taxa do not match each other. This asynchroneity among taxa led Markgraf to suggest that the changes signified vegetative responses to local events. Furthermore, the presence of intermittent charcoal in the deglacial sediments suggests that fire was a major type of disturbance to which vegetation changes could be attributed (Markgraf 1993). Singer et al. (1998) reached a similar conclusion that there was no Younger Dryas cooling in New Zealand on the basis of pollen evidence.

Although Markgraf's caution about ecological factors controlling vegetation shifts is well stated, abundant paleoclimate evidence (e.g., Denton and Hendy 1994; Lowell et al. 1995) supports Younger Dryas–age ice advances and climate-related vegetation shifts in mid latitudes of the Southern Hemisphere. Disputes about the Younger Dryas in the Southern Hemisphere also reflect the probability that certain proxies are not sensitive to Younger Dryas cooling and that the climatic signal of millennial-scale climate change can be spatially variable.

Younger Dryas in the Oceans

North Atlantic

The Younger Dryas cold snap is widely recognized in the world's oceans on the basis of planktonic foraminiferal data, although paleoceanographers have sometimes used a different terminology with reference to Younger Dryas–age cooling. The deglaciation is called *Termination 1* (Broecker and van Donk 1970), and it has a characteristic two-step shape in many isotope and faunal curves from deep-sea cores. Duplessy et al. (1981) found oceanic oxygen isotopic evidence for Younger Dryas cooling and ice-volume changes that interrupted the deglaciation, and they believed the climate reversal was important enough to justify splitting of the deglacial episode into two parts: Termination 1a, 15–13 ka, when about 50% of continental ice had melted, and Termination 1b, when the remainder melted. Termination 1a and 1b were separated by a period (13–12 ka) when ice volume seemed to remain constant.

Fairbanks (1989) later found that Termination 1a and 1b could also be distinguished in the sea-level record of Barbados corals and discovered a period when sea level rose at a much slower rate than it did during Termination 1a and 1b, when climatic warming was faster. Fairbanks (1989) referred to these two warming periods as meltwater pulses (MWP) 1a and 1b to signify two distinct periods of Laurentide Ice Sheet discharge and the break between them, the oceanic equivalent of the Younger Dryas stadial.

The surface ocean record of the Younger Dryas episode revolves around two key parameters, sea-surface temperature (SST) and sea-surface salinity (SSS). Both can be reconstructed using faunal and isotopic proxy methods. One species of planktonic foraminifer, *N. pachyderma,* is dominant in high-latitude assemblages and is particularly critical for recording a Younger Dryas shift in SST in North Atlantic and Nordic Seas (figure 5-5). *N. pachyderma* has two forms (ecophenotypes or morphotypes), whose ecological preferences are known from living populations (e.g., Be 1977; Carstens et al. 1997). The left-coiling or sinistral (*N. pachyderma* sin) morphotype lives in oceanic water colder than 10°C and dominates (> 95%) foraminiferal faunas living in water temperatures colder than 5°C. This cold-loving morphotype usually reaches 100% of the total foraminiferal assemblage in the Arctic Ocean. The right coiling or dextral form (*N. pachyderma* dext) lives in warmer waters.

Ruddiman and McIntyre (1973) and Ruddiman et al. (1977) documented reduced SSTs in North Atlantic oceanic surface waters when they found planktonic foraminiferal assemblages with *N. pachyderma* (sin) reaching as much as 60%. Such dominance signified a tongue of cold surface water that extended into midlatitudes about 10.5 ^{14}C ka between 40°N and 65°N latitude. Ruddiman and McIntyre recognized that this SST cooling might correlate with the Younger Dryas stadial in Europe. Ruddiman et al. (1977) estimated that winter and summer SSTs cooled from 8.2°C to 1.8°C and from 14.3°C to 7.4°C, respectively. This cooling equates to 11°C per 1000 yr, an astonishing rate of oceanic SST change. During the Younger Dryas-Preboreal warming, winter and summer SSTs rose from 1.8 to 10.1°C and from 7.4 to 15.2°C, respectively.

In the Norwegian Sea, the Younger Dryas cooling has also been firmly established (Jansen and Bjorklund 1985; Jansen 1987; Lehman and Keigwin 1992; Koc and Jansen 1994). Jansen (1987) used three separate paleoclimate proxy methods—foraminifera, radiolaria and stable isotopes—to show that SST in the southeastern Norwegian Sea dropped from +7°C during the warm Allerød to near glacial temperatures during the Younger Dryas. Mangerud (1987) made a strong case that the Norwegian Sea oceanic record of the Younger Dryas matched the continental and shallow marine record of Norway.

Lehman and Keigwin (1992) conducted one of the first deep-sea studies to document ocean SST changes over short time scales in the Younger Dryas. They examined the high sedimentation rate (5 m/1 ka) Troll core from the Norwegian Sea trench (figure 5-3). Lehman and Keigwin (1992) found that the percentages of *N. pachyderma* (sin) in the Norwegian Sea foraminiferal assemblage declined from near 100% at the glacial maximum to 0% after approximately 1000 yr, near the end of deglaciation. But the glacial-deglacial decrease in *N. pachyderma* (sin) was punctuated by a sharp spike between 11.2–10.5 ^{14}C yr, when *N. pachyderma* reached more than 90% of the total foraminiferal assemblage. Here was the oceanic equivalent to the pollen, ice core, and terrestrial isotopic evidence indicating a return to glacial conditions and dividing the deglaciation into two distinct periods of climate warming. Lehman and Keigwin also discovered evidence from *N. pachyderma* for a brief pre–Younger Dryas cooling episode.

Changes in SSS correlative with climate cooling on land would be another oceanic litmus test of a millennium-long climatic reversal. The diminished influence of meltwater should be detected in surface faunas and isotopes. Indeed, early ideas that meltwater would impact on ocean SSS were supported by modeling and stable-isotope studies (Berger 1978), some suggesting most of the world's ocean was covered by a low-salinity meltwater lid during deglaciation. Although this extreme scenario was later discounted by studies of sediment mixing and dissolution (i.e., Jones and Ruddiman 1982), meltwater influence on ocean circulation remains a cornerstone of oceanic mechanisms to explain rapid climate. Indirect evidence for reduced salinities comes from the existence of stratigraphic zones in the North Atlantic in which deep-sea sediments barren of planktonic foraminifers alternate with more typical foraminiferal-rich sediments. Isotopic evidence from planktonic foraminifers can also reflect salinity changes, though the complexities of decoupling salinity from temperature in both isotope and faunal proxies has remained a difficult problem since the earliest isotope studies of planktonic foraminifers.

Several new methods have been developed to measure the SSS response during the Younger Dryas stadial. Sikes and Keigwin (1994) separated SST and SSS signals by comparing temperature estimates based on long-chain alkenones of the nannofossil *Emiliana huxleyi* to those derived from foraminifera. The UK-37 alkenone methodology yields SST estimates about 1–3°C warmer than those derived from faunal estimates. The alkenone record of deglaciation shows that Termination 1b was characterized by a major temperature rise, whereas the foraminiferal data from the same region show Termination 1a exhibited the greater increase. The difference may arise in part because *E. huxleyi*'s habitat is the ocean surface but some foraminiferal species occupy more than one depth zones. Sikes and Keigwin concluded that the Younger Dryas event in this region of the Atlantic was characterized by a 3–6°C temperature drop, and the alkenone data suggest that surface salinities decreased in parts of the Atlantic during the deglaciation, supporting other lines of evidence for meltwater influence in the northeastern North Atlantic Ocean during Termination 1a.

A second approach to SSS estimation was mentioned above in the work of de Vernal et al. (1996) who used dinoflagellates to track SST history, as it related to Laurentide Ice Sheet decay, in the northwestern Atlantic off the mouth of the St. Lawrence River.

A third group of microfossils are diatoms, which provide an additional tool for reconstructing SSS. Diatoms are photosynthetic algae, which like many dinoflagellates, require sunlight and live in the photic zone where climate-driven SST and SSS changes will be most apparent. Diatoms are preferable to the foraminifer *N. pachyderma* for paleosalinity reconstruction because *N. pachyderma* lives below the oceanic surface layer from 50 m to several hundred meters (Carstens et al. 1997). Koç and Jansen (1992, 1994) used a new diatom transfer function to evaluate SST and sea-ice history in cores from a transect across the Norwegian Sea. They showed that the Younger Dryas cooling in the Norwegian Sea was essentially synchronous with the Younger Dryas in European continental lacustrine and Greenland ice core records. Koç-Karpuz and Jansen (1992) estimated that SST rose 9°C at the Younger Dryas–Preboreal transition in < 50 yr, and the polar front and maximum sea-ice margin rapidly migrated northwestward to near Greenland.

Gulf of Mexico

The Gulf of Mexico was another place to look for evidence for a Younger Dryas meltwater SST and SSS response to a reversal of deglaciation because early deglacial Laurentide Ice Sheet meltwater drained southward through the Mississippi River valley. Kennett and Shackleton (1975) found isotopic and faunal evidence for meltwater discharge during early deglaciation. Ecological data on temperature preferences of four cool water foraminiferal species—*Globorotalia inflata, Globigerina falconensis, Globigerinella aequilateralis,* and *Globigerina bulloides*—reveal SST cooling in the Orca Basin between 11,400 and 9800 ^{14}C yr, roughly equivalent to the Younger Dryas chronozone. The Younger Dryas assemblage resembled that living in this part of the Gulf of Mexico during the peak of the past glacial period before 14 ka (Kennett et al. 1985).

Flower and Kennett (1990) restudied the Orca Basin paleoceanographic record using oxygen isotopes and planktonic foraminifers and found evidence that Laurentide Ice Sheet glacial meltwater was first discharged via the Mississippi and then later may have been diverted eastward through the Great Lakes system and the St. Lawrence Valley. The life histories of planktonic foraminiferal species were central to interpreting the isotopic record and testing the hypothesis for freshwater discharge into the Gulf of Mexico. Flower and Kennett (1990) used experimental and field evidence on the ecology of *Globigerinoides ruber* showing that this species is tolerant of salinities as low as 22 parts per thousand (ppt). In contrast to other species, which migrate into deeper layers when fresh water influx from the Amazon River lowers surface salinities, *G. ruber* remains near the surface. This trait made *G. ruber* an ideal proxy for analyzing the isotopic spike marking the influx of Laurentide ice meltwater into the Gulf of Mexico. Flower and Kennett contrasted the oxygen-isotope record of *G. ruber* with that of *Neogloboquadrina dutertrei,* a species that seems to prefer higher salinities. The former species shows a Younger Dryas–age $\delta^{18}O$ isotopic excursion reaching –4‰, followed by a rise to +0.8‰. In contrast, the isotopic record of *N. dutertrei* during the same interval shows no isotopic excursion suggestive of a meltwater spike. The distinct isotopic responses of the two species reflect their ecology: One species stayed near the surface in the low-salinity lid, while the other dove perhaps 100 meters to find higher-salinity water.

Flower and Kennett (1990) also used distinct white and pink morphotypes of *G. ruber* to determine how fast the Younger Dryas event ended in the Gulf of Mexico. The white morphotype lives in the winter, the pink in the summer. The isotope record of the white variety shows a rapid decrease in isotopic values during the Younger Dryas, whereas the isotope curve for the pink variety showed no such excursion. This result may indicate meltwater discharge continued during the summer, even during the cool Younger Dryas. The pink-white *G. ruber* isotopic contrast is evident over only 10 centimeters of sediment, which led Flower and Kennett (1990) to conclude that the Younger Dryas meltwater spike event ended in < 130 yr, a remarkably abrupt cessation of Mississippi River freshwater discharge.

Tropical Ocean

Improved paleotemperature techniques have led to the discovery of Younger Dryas–age cooling in equatorial oceans. For example, in the tropical Indian Ocean between 20°N and 20°S, Bard et al. (1997) used alkenone chemistry and found that deglaciation began about 15.1 ka with an abrupt 1.5°C warming followed by a slight Younger Dryas cooling 12.2–11.5 ka, followed by another 1°C warming. Bard's group showed that Younger Dryas cooling was synchronous in the tropical Indian Ocean and the North Atlantic cooling but that it lagged behind the climatic cooling recorded in the Southern Hemisphere. Using strontium-calcium (Sr:Ca) ratios in coral skeletons, Beck et al. (1997)

found evidence from Vanuatu in the equatorial Pacific for early Holocene oceanic temperatures that were as much as 6.5°C cooler about 10.3 ka. However, the coral record was discontinuous, and they could not determine the chronological sequence relative to the North Atlantic.

Kennett and Ingram (1995) found increased $\delta^{18}O$ of benthic and planktonic foraminifers; greater proportions of cool-water *N. pachyderma;* more massive, nonlaminated sedimentation; and more diverse benthic foraminiferal assemblages in Younger Dryas–age sediments (13.0—11.2 ka) in the Santa Barbara Basin off southern California. They suggested that during the Younger Dryas, increased ventilation of North Pacific intermediate water caused the normally oxygen-poor silled basin to become well ventilated and support a diverse benthic community. Kennett and Ingram explained the apparent synchroneity between the Younger Dryas Santa Barbara Basin event with North Atlantic oceanic changes as a strong coupling of North Atlantic and eastern Pacific Ocean climate systems due either to atmospheric or oceanic processes.

Decadal to centennial-scale oscillations in surface ocean biological productivity were determined back to 12.6 ka from sedimentology and diatom floras from the finely laminated sediments of the Cariaco Basin off Venezuela by Hughen et al. (1996). Light-colored planktonic foraminifera–rich layers that form during the dry upwelling seasons alternate with darker terrigenous layers formed during wet, nonupwelling periods. The control of biological productivity in the Cariaco Basin was attributed to variations in winds and nutrients. Hughen et al. found clear evidence for a Younger Dryas episode coeval with that in the North Atlantic and concluded that trade-wind strength was the immediate cause of the surface-ocean changes and the ultimate cause may have been linked to thermohaline circulation.

Deep Sea

Evidence that NADW formation diminished during the Younger Dryas episode would provide additional support for Broecker's conveyor belt hypothesis of altered thermohaline circulation. Boyle and Keigwin's (1987) evidence from carbon isotopic and cadmium-calcium ratios in deep-sea benthic foraminifera indicating a brief shutdown, or at least a diminished flow, of NADW during deglaciation, supported this tenet of the conveyor belt theory.

Moreover, if surface ocean cooling, NADW reduction, and ice-sheet discharge could be linked, then the connection between the oceans, ice, and climate during rapid climate change might become better established. Keigwin et al. (1991) studied high sedimentation–rate cores from the Bermuda Rise and Blake Outer Ridge of the Western North Atlantic to demonstrate a close correspondence between surface and bottom processes over millennial time scales. They documented four distinct periods of NADW reduction during the last glacial termination—14.5, 13.5, 12.0, and 10.5 ^{14}C ka, with the youngest event corresponding to the Younger Dryas. At the same time, a period of meltwater discharge to the surface ocean is recorded in planktonic foraminiferal oxygen isotopes and the glacial geology from the continental record. Keigwin and colleagues also tied each NADW reduction to meltwater from a particular late glacial ice sheet. For example, the 14.5 ^{14}C ka event matched the timing of the Barents Ice Sheet decay, and the 10.5 ka event matched the Laurentide Ice Sheet decay via the St. Lawrence during the Younger Dryas event.

The complex relationships between ice sheet decay, deglacial meltwater discharge, and deep-sea sedimentation and foraminiferal isotopic excursions such as those described by Keigwin have been revised by Clark et al. (1996a) and Björck et al. (1996a). Because these events are intimately tied to sea level records of deglaciation, they are taken up again in chapter 8.

In summary, deep-sea circulation changes occurring on millennial time scales provide an important line of evidence supporting the hypothesis that the Younger Dryas and other climatic reversals involve deep oceanic mechanisms.

Brief Stadials in Older Terminations

Younger Dryas–like events occur in older glacial terminations, although few studies have explicitly searched for them. For example, Sarnthein and Tiedemann (1990) discovered that, during several of the past six glacial terminations, apparent reversals toward glacial-like climate characterized the deep-sea isotopic record of foraminifera from Deep Sea Drilling Program (DSDP) site 658 off west Africa. Haake and Pflaumann (1989) and Sarnthein and Altenbach (1995) discovered a reversal in the δ^{18}O of *N. pachyderma* from the Nordic Seas during Termination 2 that has a Younger Dryas look to it. Seidenkrantz et al. (1996) reviewed evidence for a Younger Dryas–type event during Termination 2, which they named the Zeifen-Kattegat oscillation. They found evidence for this rapid climatic reversal, based on stable isotopes in speleothems, biostratigraphy, and pedostratigraphy in soil sequences, as well as ice-core climate records, at no fewer than 24 lacustrine and marine sites.

Some evidence indicates that older millennial-scale climate reversals may differ from the Younger Dryas event in the North Atlantic Ocean. Oppo et al. (1997) studied Termination 2 at the isotope stage 6/5 transition in faunal and isotopic records from high sedimentation–rate cores from the Bjorn Drift in the subpolar North Atlantic Ocean. They found evidence for only a minor pause in the rise in SSTs and a small cooling in the middle to late stages of deglaciation. Sea-surface temperatures remained cool for much of Termination 2, rising 8–10°C relatively late in deglaciation, near the inception of full interglacial conditions. Oppo et al. point out that the lack of an evident climatic reversal during Termination 2 in many parts of the subpolar North Atlantic may result from the persistence of cold SSTs during deglaciation and delayed warming.

The Cause of the Younger Dryas

Several causes have been invoked to explain the Younger Dryas cooling event: ice-sheet surging, a solar mechanism (Denton and Hughes 1981), European ice-sheet decay leading to an increase in icebergs (Mercer 1969), Laurentide Ice Sheet meltwater influence on thermohaline circulation (Broecker et al. 1985, 1989), and atmospheric mechanisms related to water vapor (Kennett and Ingram 1995; Lowell et al. 1995; Bard et al. 1997). In almost all cases, the Younger Dryas is regarded as a complex event somehow linking the ice, oceans, and atmosphere during a period of increasing solar insolation.

John Mercer showed remarkable foresight with regard to the interplay of North Atlantic ocean, ice, and atmosphere and European climate when he discussed the Allerød–Younger Dryas oscillation in a paper published in 1969. Mercer's conclusions almost 30 yr ago now seem prescient: "The Allerød warm interval ended abruptly in Europe, over centuries" (Mercer 1969:232) and the succeeding Younger Dryas cool interval lasted about 650 yr. Mercer was among the first to directly address the question of whether European glacial-climatic events such as the Allerød–Younger Dryas climate transition were local climate events or a sign of a global climatic cooling. Accepting the consensus of that time that a Younger Dryas–like event did not occur in North America (Davis 1967), Mercer postulated that this Eurocentric event was related to the export of sea ice from the Arctic Ocean and Norwegian Sea (including sea ice derived from the disintegrating

Barents Ice Shelf) into the North Atlantic Ocean. He did not believe that the disintegration of continent-based ice was responsible for the Younger Dryas, as stipulated in the conveyor belt model, but he nevertheless linked ice, ocean, and atmosphere in arguing that the export of huge amounts of ice from the Arctic to the North Atlantic would deprive Europe of the warmth of the North Atlantic Drift. Mercer even went so far as to quantify the export as 6000 km^3/yr, resulting in a 30-cm-thick fresh-water layer annually over much of the North Atlantic, which explained early oxygen isotope evidence. Although Mercer recognized that surface ocean events in the North Atlantic could have worldwide repercussions via ocean-atmosphere thermodynamics, he believed their impact would diminish with distance from the North Atlantic region.

Broecker's conveyor belt theory revised early ideas about the extent to which the North Atlantic oceanic circulation in general, and NADW reduction due to ice sheet meltwater in particular, might trigger rapid climate change and affect global climate during Younger Dryas–type events. One plausible ultimate cause of the Younger Dryas is the diversion of Laurentide Ice Sheet meltwater from the Mississippi to the St. Lawrence River system (Broecker et al. 1989). Although many studies cited earlier support the idea that glacial meltwater triggered the Younger Dryas, several discoveries pose problems for certain tenets of the conveyor belt hypothesis. One challenge involves the degree to which freshwater Laurentide Ice Sheet discharge via the St. Lawrence River affected North Atlantic Ocean.

The argument that meltwater from the St. Lawrence did not trigger Younger Dryas cooling has been made on two grounds: The regional paleosalinity history from of the St. Lawrence River mouth was not strong enough to affect the North Atlantic (de Vernal et al. 1996), and the timing and magnitude of meltwater pulses inferred from the Barbados sea-level record does not fit the Younger Dryas record (Fairbanks 1989). Although abundant evidence from postglacial stratigraphy (Hillaire-Marcel and Occhietti 1980), lake levels (Teller and Kehew 1994), and paleoenvironments (Cronin 1977, 1988; Rodrigues and Vilks 1994) show a shift in Laurentide Ice Sheet discharge from south to eastward at approximately 11 ka, de Vernal et al. argued on the basis of foraminiferal stable-isotope and dinoflagellate data that surface salinities in the Cabot Strait area off the mouth of the St. Lawrence decreased continually during the entire Champlain Sea episode 11.7–10 ka, while salinities farther offshore remained near their present value 32‰ for the entire deglaciation. De Vernal concluded that the effect of freshwater runoff from glacial lakes on oceanic salinity vanished in the nearshore neretic zone and was not strong enough to influence the open ocean salinity, which remained constant or even rose during the Younger Dryas interval, unless thermohaline circulation is more sensitive to small salinity fluxes than is generally believed.

Fairbanks (1989) found indirect evidence that meltwater did not trigger the Younger Dryas cooling. Although he found a clear two-step pattern to sea-level rise and meltwater discharge based on the oxygen isotope record from Barbados corals, he also found that sea level rose only a few meters during the Younger Dryas. Thus the rate of sea-level rise decreased, but the overall trend did not reverse. Fairbanks (1989) estimated the volume of freshwater discharge during MWP 1a and MWP 1b to be 14,000 km^3/yr and 9500 km^3/yr, respectively. During the Younger Dryas, the discharge rate was reduced fivefold to 2700 km^3/yr, a large amount compared with modern Mississippi (560 km^3/yr) and St. Lawrence (330 km^3/yr) River discharge rates. Fairbanks (1989) thus argued that the salinity change in the Gulf of Mexico was caused by a shift in the net rate of discharge, not the locus of discharge, as called for in Broecker's theory.

The link between SSS and meltwater as a driver of late glacial climate events was also discussed by Clark et al. (1996), who reviewed the timing of meltwater events and deep-sea isotopic excursions. They argued that MWP 1a does not have a clear signature from any Northern Hemisphere ice sheet and that meltwater diversion through the St. Lawrence River is a likely mechanism to explain Younger Dryas cooling in the North Atlantic region.

Today active discussion continues about the processes that account for rapid climate change during glacial-deglacial transitions continues. As new data are obtained, the consensus grows that abrupt climate reversals are evident from changes in relative abundance and isotopic composition in ecologically sensitive organisms. Several conclusions can be drawn from this evidence:

1. Continental and marine climate indicators show the Younger Dryas cold snap was synchronous and severe throughout northern Europe and the adjacent surface and deep North Atlantic Ocean and Greenland Ice Sheet.
2. Marine faunal and floral paleoceanographic evidence indicates a reduction in North Atlantic deep-ocean ventilation during the Younger Dryas. For example, increased Younger Dryas

atmospheric ^{14}C concentrations in tree rings reflect slower ocean uptake of this cosmogenic isotope (Stuiver and Braziunas 1993; Björck et al. 1996a). In addition, Antarctic bottom water (AABW) production may have increased during the mid to late Younger Dryas episode after the initial decrease in NADW formation (Broecker 1998).

3. Pollen, insect, foraminiferal, and glacial evidence in deglacial records from North America and Europe points to other abrupt, often low-amplitude climate reversals before and after the Younger Dryas.
4. Extensive evidence from ice-core, terrestrial, glacial, deep oceanic, and surface oceanic records outside the North Atlantic region leaves no doubt that abrupt 1000-yr climate reversals are a characteristic feature during transitions from glacial to interglacial climate throughout most of the world and very likely during older Quaternary glacial terminations.
5. The synchroneity between Northern and Southern Hemisphere and equatorial abrupt climate events is still not established (Mayewski et al. 1996). Some evidence indicates tropical regions cool in phase with North Atlantic cooling (e.g., Bard et al. 1997), but the Antarctic cold reversal indicates that the South Polar region cooled about 1000 yr before the Younger Dryas episode in the Northern Hemisphere (Jouzel et al. 1995). The Antarctic cold reversal may be due to another mechanism. Also, the tropics and Southern Hemisphere exhibit a much lower amplitude signal than that in the North Atlantic region (Jouzel et al. 1995; Lowell et al. 1995).
6. Emerging evidence from the Southern Hemisphere (e.g., Charles et al. 1996), the tropics (Bard et al. 1997), and Greenland ice cores (Severinghaus et al. 1998) indicates that atmospheric processes also play a role in abrupt climate events (Kennett and Ingram 1995; Broecker 1995, 1998; Lowell et al. 1995; Overpeck et al. 1996; Beck et al. 1997).

The triggering mechanism that caused the Younger Dryas will become evident when improved chronology is obtained that permits precise correlation between the two poles and the tropics. The oceans' thermohaline circulation might take 500–1000 yr to transfer the impact of a North Atlantic meltwater lid and reduced NADW formation throughout the world, thereby producing an interhemispheric asymmetric climate pattern. Abrupt cooling in one hemisphere would be followed by cooling in the other about 1000 yr later. Conversely, when polar and tropical regions experience synchronous climate change, atmospheric as well as oceanic mechanisms probably together play the important roles in abrupt climate oscillations. A change in water-vapor content is one plausible explanation for interhemispheric synchroneity, because this gas is radiatively active and traps heat. Atmospheric carbon dioxide (CO_2) is not likely to cause rapid high-amplitude climate change, because concentrations of CO_2 change too slowly, over thousands of years of deglaciation (see chapter 9). Although methane (CH_4) concentrations change more abruptly than CO_2 during the Younger Dryas, which probably reflects climate-related changes in wetland CH_4 production, the radiative forcing capacity of CH_4 is not as strong as that of water vapor or CO_2, and CH_4 probably was not a major forcing of the Younger Dryas (see chapter 9).

HEINRICH EVENTS AND DANSGAARD-OESCHGER CYCLES: MILLENNIAL-SCALE CLIMATE CHANGE DURING GLACIAL PERIODS

SUBORBITAL CLIMATE events known as Dansgaard-Oeschger (D/O) cycles and Heinrich events punctuate the glacial interval of about 90–18 ka. These two phenomena were originally discovered from very separate lines of evidence: D/O cycles from the Greenland ice-core isotope record and Heinrich events from deep-sea North Atlantic faunas and ice-rafted debris (IRD) sediments. In general, glacial-age millennial events are viewed as manifestations of ice-ocean-atmosphere processes similar to those that produced the Younger Dryas in that (1) they are short-term cooling, producing IRD in North Atlantic sediments; (2) they occur at times when substantial mid-latitude ice exists on the continents around the North Atlantic, (3) they begin and end abruptly. They differ, however, from the Younger Dryas in that solar insolation was generally lower and the earth was experiencing an extended glacial period. Furthermore, the amplitude of D/O cycles were not as great as the Younger Dryas event, except perhaps during marine isotope stage 3. Dansgaard-Oeschger cycles also differ in magnitude from low-amplitude millennial-scale oscillations characteristic of interglacial periods (see later).

Figure 5-5 from Bond and Lotti (1995) depicts the complex record of glacial-age millennial-scale climatic events recorded in the Summit GRIP ice core, and the surface North Atlantic Ocean that form the basis of much of the following discussion. The figure graphically illus-

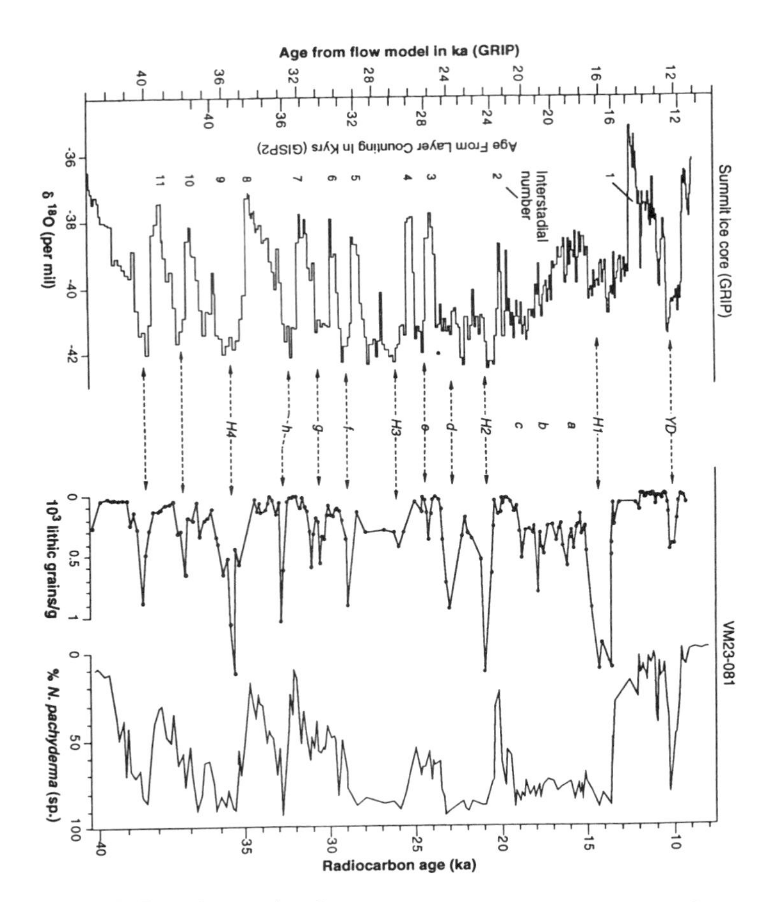

FIGURE 5-5 Relationship between Summit GRIP ice core, Greenland, millennial climate events, iceberg discharges (lithic grains), and sea-surface temperatures (SST) (% *N. pachyderma*) in North Atlantic deep-sea core VM23-081. H1–H4 are Heinrich events numbers 1–4; YD is Younger Dryas cool interval. Courtesy of G. Bond with permission from the American Association for the Advancement of Science. From Bond and Lotti (1995).

trates the relationship between the latest 11 D/O interstadials, numbered 1–11, and the Younger Dryas preserved in the oxygen-isotope record from the GRIP ice core. It also shows the past four Heinrich events (H1–H4) evident in the lithic and foraminiferal record from North Atlantic core VM23-081 for the past 40 ka.

Dansgaard-Oeschger Events

Studies by Dansgaard et al. (1982, 1993), Dansgaard and Oeschger (1989), and Oeschger et al. (1985) revealed 23 interstadial events between about 115 and 10 ka, characterized by rapid swings in oxygen-isotope ratios of ice recovered from Greenland ice cores. In the literature, the interstadial warming is often called a Dansgaard-Oeschger event; the full climatic cycle including the subsequent return to glacial conditions is referred to as a Dansgaard-Oeschger cycle. Most of the interval between 115 and 20 ka is characterized by low to moderate variability in high-latitude solar insolation, yet D/O interstadial isotopic excursions reveal atmospheric temperature increases of >5°C, a 30–60% increase over background glacial temperature in Greenland (figure 5-5). The oldest two D/O events (number 23 and 21) correspond to marine isotope substages 5c and 5a at about 100 ka and 80 ka, and the youngest interstadial occurs during the Termination 1 deglaciation (figure 5-5). The duration of each D/O event is about 2—3 ka; most seem to begin gradually or in a step-wise fashion and end abruptly—in only decades to centuries. In addition to the rapid temperature increases, interstadial climates during D/O events are characterized by snow-accumulation rates double those of intervening cold periods and by greater atmospheric turbulence.

Given the growing evidence that suborbital glacial-age millennial scale climate reversals occur in paleo-atmospheric records, Broecker et al. (1988a:1) posed the question: "Can the Greenland climatic jumps be recognized in records from ocean and land?" In other words, do D/O events signify global climate changes, and if so, what causes high-frequency climate oscillations at times when extensive Northern Hemispheric temperate ice sheets existed? We return to this question after first examining the nature of Heinrich events.

Heinrich Events

The North Atlantic Ocean during the past glacial period also records rapid millennial scale climate variability preserved in faunal, stable isotopic and lithologic deep-sea records. Helmut Heinrich (1988) discovered a cyclic pattern of ice-rafted dropstones occurring during the past 130 ka in a transect of deep-sea sediment cores from the Dreizback Seamount in the northeastern Atlantic Ocean. Heinrich proposed that the observed 11-ka periodicity in dropstone layers was linked to summer and winter insolation minima caused by precessional orbital cycles.

The name *Heinrich event* was applied by Broecker et al. (1992) and Bond et al. (1992) to the lithologic IRD layers discussed by Heinrich (see Brocker 1994). Broecker, Bond, and colleagues used two well-established paleoceanographic tools—detrital carbonate lithic grains that comprise the IRD and planktonic foraminiferal assemblages—to establish the nature and extent of Heinrich layers in the subpolar North Atlantic. They named six distinct Heinrich layers (H6–H1) between 70 and 14 ka (dated by volcanic ashes, oxygen isotopes, and ^{14}C dating) and spaced approximately 5–10 ka apart. Some researchers refer to the Younger Dryas as Heinrich event H0. The oldest event, H6, is dated at about 65–60 ka; H5, H4, H3, and H2 are dated at 44.0, 33.2–35.1, 26.0, and 22.0 ka; H1 (15–13 ka) precedes the Younger Dryas by approximately 2–3 ka. On the basis of foraminiferal and lithological evidence, Broecker and colleagues determined that Heinrich events, in their simplest form, are a two-step process. First, a drop in SST is accompanied by a decrease in the net flux of planktonic foraminiferal shells to the ocean bottom. The planktonic foraminiferal species *N. pachyderma* was pivotal in documenting rapid oceanic changes; it reached 80% to >90% of foraminiferal assemblages at DSDP site 609 during Heinrich events. Second, there is a short-lived but massive discharge of icebergs from the eastern Canadian portion of the Laurentide Ice Sheet into the North Atlantic Ocean. Strictly speaking, the sediment deposited by these icebergs comprises the deep-sea Heinrich layer. The debris-laden iceberg path could be traced 3000 km in a west-to-east series of cores as a swatch of coarse-grained sediment covering the sea floor from Canada to near Portugal.

Ice-rafted debris has long been recognized as evidence for iceberg activity in the subpolar North Atlantic. Ruddiman (1977) pioneered study of IRD by examining the > 62-μm fraction of sediment in 32 subpolar North Atlantic cores to establish patterns of IRD deposition during the past interglacial-deglacial cycle. Ruddiman found that IRD deposition during the past interglacial and the Holocene were similar and that there were two incremental increases in IRD deposition corresponding to climatic cooling at the isotope substage 5e/5d (115 ka) and 5a/4 (75 ka). The shift at 75 ka was accompanied by a shift in the axis of IRD deposition to an east-west zonal position characteristic of what are now known as Heinrich layers.

Ruddiman estimated the total volume of sand that blanketed the North Atlantic during distinct climate modes. In the central and peripheral subpolar Atlantic during the past interglacial period 125–115 ka, he estimated that about 60×10^{11} grams of sand were deposited; in contrast, during the subsequent glacial interval, this figure increased to about 220 × 10^{11} grams. In one pod of high IRD deposition located under the path of icebergs exiting the Greenland sea near the Denmark Strait, interglacial IRD levels were 200 mg/cm^2, rose to 800 mg/cm^2 during glacial isotope stage 4, and rose even higher to > 1200 mg/cm^2 during the glacial maximum. Parallel changes in IRD deposition occurred in the narrow stretch of ocean floor between the Labrador Sea and the Northern Iberian Peninsula. Maximum glacial IRD deposition occurred near 50°N latitude, where the confluence of southeastward-flowing icebergs from Laurentide ice met with those from Scandinavian ice in the subpolar North Atlantic cyclonic gyre, where warm North Atlantic Drift water caused them to melt.

Recent studies of IRD layers by Heinrich (1988), Bond et al. (1992, 1993), Andrews and Tedesco (1992), Andrews et al. (1993b), Alley and MacAyeal (1994), Dowdeswell et al. (1995), and Grousset et al. (1993) provide the vital statistics of a Heinrich event. Heinrich layers were deposited about every 7–12 ka in a quasicyclic pattern; enough ice was discharged during each event to thin continental ice sheets by as much as 1200–1460 m, though these are only rough estimates of ice-sheet thinning. The deposition of the lithic layer itself took as little as > 250 yr, on average 750 yr. Each event deposits about 5 × 10^{15} kg of sediment, blanketing an area of 5 × 10^{12} m^2, in a layer of lithic grains reaching a maximum thickness of 1 m near the glacial outflow and a minimum thickness of 1 cm. The average layer is about 10 cm thick. The flotilla of icebergs reduced oceanic SST and salinity and affected the composition of pelagic oceanic ecosystems.

As shown in figure 5-5, these intense pulses of ice rafting in the North Atlantic are part of a more complex series of oceanic events linked to interstadial D/O events. By examining the North Atlantic record of Heinrich layers and associated faunal events at higher temporal and spatial scales, Bond et al. (1993) and Bond and Lotti (1995) discovered that between 80 and 20 ka, Heinrich events occurred at the end of cooling cycles, lasting 10–15 ka, and each cycle ending in a period of major iceberg discharge. Heinrich events thus occur at the culmination of progressive cooling of oceanic temperatures. A large and rapid increase in SST followed the IRD event as subpolar SST reached nearly interglacial warmth. This saw-tooth, step-wise cooling pattern in which a succession of brief (2—3 ka) interstadial warm SST events, each one slightly cooler than the previous one, were bundled together over a 10- to 15-ka period (called *Bond cycles* by Lehman [1993]). The brief SST warmings were the oceanic equivalent of D/O ice core events, and Heinrich events are actually superimposed on the more dominant, frequent, and rapid short-term D/O cycles.

The amplitude of oceanic response to Heinrich events may be somewhat lower than that recorded in Greenland ice-core air-temperature records. For example, Cortijo et al. (1997) dissected regional variation in surface hydrology during Heinrich event 4, dated at 33.2–35.1 ka, in the North Atlantic Ocean between 40–60°N using foraminiferal isotopic records. They documented a 1–2°C SST decrease and a 1.5–3.5‰ surface salinity decrease between 40°N and 50°N, but little salinity change north of 50°N. The amplitude of SST change was clearly less than the 10–15°C drop recorded in Greenland ice cores and suggests that the North Atlantic surface ocean had a dampened response to this D/O event or that the $\delta^{18}O$-temperature relationship is not well constrained.

If the source of Heinrich event icebergs could be determined, then it might be possible to establish the mechanism causing these sudden glacial discharges. Geochemical analyses of IRD grains showed that not all Heinrich layers are lithologically the same (e.g., Grousset et al. 1993). For example, efforts to distinguish event H3 (about 26 ka) from events H1, H2, H4, and H5 revealed several characteristics that set H3 apart (Bond et al. 1992; Gwiazda et al. 1996). Event H3 began gradually, in contrast to the abrupt beginning of other events. H3 also had a lower detrital content compared with that of other Heinrich layers and was characterized by a depletion in foraminiferal abundance reflecting a low-productivity surface ocean. Strontium and neodymium analyses of the silicate fraction show H3 had a different source of origin than other layers. Gwiadza's paper also presented data on lead isotopes of individual feldspar grains and showed a Greenland-Scandinavian source region for the lithic grains, as opposed to a Canadian Shield source for other Heinrich layers coming from the Hudson Strait area. Finally, in most Heinrich events they found that the radiometric age of the fine fraction carbonates based on K/Ar dating is much older than the age of material in background sediment; in H3, the K/Ar ages of Heinrich layer and background sediment are about the same.

Bond and Lotti (1995) also fingerprinted IRD sediments by analyzing 15 different types of lithic grains from cores of North Atlantic sediments off western Ireland deposited during cool events. They identified three main source areas: Hudson Strait (Andrews and Tedesco, 1992; Andrews et al. 1993b; Dowdeswell et al. 1995); the Gulf of St. Lawrence, draining portions of the Laurentide Ice margin; and the Icelandic Sea, carrying ice into the North Atlantic from eastern Greenland, Svalbard, and the Arctic Ocean via the Denmark Strait. Hematite-coated grains came from the Gulf of St. Lawrence discharge, and basaltic glass could be traced to the predominantly volcanic region of Iceland. Regarding the timing of IRD, they suggested that generally, within the age uncertainty, there may have been nearly *synchronous* iceberg discharges from *separate* ice sheet sources. If the Bond-Lotti chronology is correct, however, discharges from Hudson Strait may lag slightly behind those icebergs originating from the other two sources. One possible explanation of asynchroneity is that sea-level rise associated with initial massive iceberg discharge from northern Europe's ice sheet could draw down the ice stream feeding Hudson Strait.

Are Millennial-Scale Dansgaard-Oeschger Cycles Global Climate Events?

An ongoing search for suborbital-scale climate change in other glacial-age proxy records indicates that D/O cycles were very likely global climate events that affected deep-oceanic circulation, Southern Hemisphere glaciers and climate, western North American climates and alpine glaciers, and Pacific Ocean environments.

Deep Atlantic Ocean

If rapid climate changes in the North Atlantic region involve massive discharge of icebergs, then their impact on sea-surface salinity and temperatures should likewise affect thermohaline circulation and deep-sea benthic environments. Keigwin and Jones (1994) explored carbonate cycles, planktonic foraminiferal oxygen isotopes, and benthic foraminiferal carbon isotopes in the GPC-5 and GPC-9 cores from the Bermuda Rise and Blake Outer Ridge, where high sedimentation rates make them ideal sites to investigate millennial-scale deep-sea paleoceanography (figure 5-6). They found that deep-ocean cycles occur at a frequency of about 4 ka. They attributed periods of decreased carbonate percent to three possible climate-related factors: increased flux of terrigenous detrital material and lower carbonate percent during enhanced iceberg flow, resuspension of sediment due to shifting bottom water currents associated with deep kinetic eddies, and/or increased carbonate dissolution during cold climate due to diminished NADW and/or enhanced Antarctic bottom water. Isotopic oscillations in these cores reflect both SST and deep-water circulation changes that Keigwin and Jones (1994) argued supported the salt oscillator (conveyor belt) concept to explain millennial-scale events in the deep sea.

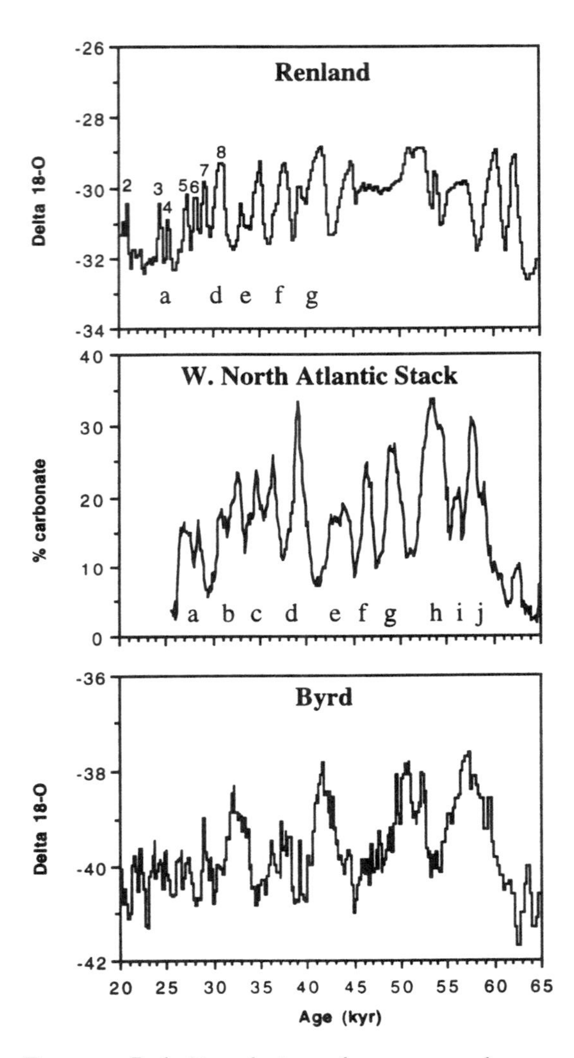

FIGURE 5-6 Correlation of percent carbonate in deep-sea cores from the Bermuda Rise and Bahama Outer Ridge between 20 and 65 ka. Positive excursions in North Atlantic oceanic carbonate content (a through j) signify relatively warm interstadial events corresponding to Dansgaard-Oeschger (D/O) events in Renland, Greenland, and Byrd, Antarctica, ice cores. Courtesy of L. D. Keigwin and American Geophysical Union. From Keigwin and Jones (1994).

Oppo and Lehman (1995) confirmed that suborbital changes in deep water formation punctuate the interval spanning isotope stages 4 through 2 (75–13 ka) on the basis of carbon and oxygen isotopic foraminiferal evidence from the northeastern North Atlantic Ocean. They found that surface-water cooling and reduced deep-water formation were general features of Heinrich events, although not all events had the same intensity or geographical pattern of cooling (e.g., H5 had relatively warm surface ocean conditions in the far eastern parts of the North Atlantic). NADW production gradually weakened during the prolonged cooling that preceded most Heinrich events, and the rapid climatic warming that followed Heinrich events manifested itself in rapid shifts in benthic foraminiferal carbon isotopes, signifying increased strength of NADW formation. Oppo and Lehman found isotopic evidence that stadials during early oxygen-isotope stage 3 were less intense glacial episodes in terms of NADW suppression than later events, a pattern also evident in Greenland ice cores.

Deep-sea foraminiferal species are sensitive to changes in food resources and other parameters, and their populations react quickly to benthic habitat changes associated with D/O climate events. Rasmussen et al. (1996) studied D/O events since 58 ka in the Faeroe-Shetland Channel, between Iceland and Norway. This is a critical ocean gateway between the North Atlantic Ocean and the Norwegian Sea, lying directly in the path of warm surface water entering polar seas and newly formed NADW exiting across the shallow sill into the North Atlantic. Using multiproxy data (benthic and planktonic foraminifers, magnetic susceptibility, isotopes, IRD), they discovered that all 15 D/O events and Heinrich events H1–H5 were expressed in the surface and deep-water record of the Faeroes-Shetland Channel. Three distinct modes of oceanic conditions characterized the glacial period: a warm interstadial mode with strong NADW formation and warm surface temperatures, a transitional period of oceanic cooling when NADW production diminished, and a cold stadial mode characterized by peak abundances of the glacial species *N. pachyderma* (sin) and minimal NADW influence.

Rasmussen et al. (1996) used ecological data on benthic foraminifers from the North Atlantic and Polar Seas (e.g., Vilks 1989; Scott and Vilks 1991; Gooday 1993) to infer benthic environmental change. During interstadials the region was inhabited by *Nonion zaandamae* and *Epistominella vitrea.* The former species requires a high influx of food such as that available during the modern interglacial period; the latter is an opportunistic species that thrives on seasonal phytodetritus influx. A gradual decrease in *Stainforthia fusiformis* during the period 58–30 ka indicates progressively weaker deep-water formation. *Elphidium excavatum,* an adaptable species tolerant of widely varying salinity, temperature, and food resources predominated during the relatively unstable transitional climate regimes between the stadial and interstadial modes. The coldest climate periods, the stadials, were inhabited by nearly monospecific assemblages of *Cassidulina teretis,* a typical Arctic Ocean species that thrives in intermediate water depths under conditions of low and unstable food supply.

The equatorial Atlantic provides a record of tropical surface and deep-sea response to millennial-scale climate events between 38 and 3 ka. In their study of planktonic and benthic foraminiferal isotopes from a core from 4000 m water depth, Curry and Oppo (1997) made startling new discoveries about D/O events. First, they found evidence from $\delta^{18}O$ of *Globigerinoides sacculifer* for low-latitude surface water cooling of 2–3°C during D/O events. This cooling was about one third to one fourth the magnitude of glacial-age cooling. Second, deep-bottom water production was reduced during stadials, confirming evidence from other North Atlantic regions (Keigwin and Jones 1994; Oppo and Lehman 1995). Third, periods of short-term tropical SST cooling coincided with lower air temperatures, and shown by ice-core records, during cold periods between D/O interstadials. Curry and Oppo's major conclusion was that observed SST changes cannot be accounted for simply by changes in meridional heat transport associated with the conveyor belt's thermohaline circulation. Changes in atmospheric greenhouse gas concentrations of CH_4, CO_2, and water vapor may help explain the synchroneity of stadial-interstadial events over > 2–3 ka time scales, but the precise phasing of short-term events remains clouded by lack of decadal and centennial-scale resolution.

Santa Barbara Basin, Pacific Ocean

The paleoceanographic history of the eastern Pacific Ocean provides stark confirmation that the impact of D/O events was felt throughout the world's oceans. The Santa Barbara Basin off the California coast is a silled basin, about 600 m deep, with sill depths of 475 m to the Pacific Ocean and 230 m to the Santa Monica Basin to the south. At depths below these sill depths, oxygen-depleted waters are derived from the upper Pacific intermediate waters. In the oxygen minimum zone, dissolved oxygen is only about 0.4–0.7 ml/L. Anoxia or hypoxia leads to little or no benthic activity and the formation of laminated sediments. Moreover, the sedimentation rate is extremely high in the

Santa Barbara Basin, about 120 cm/ka. The combination of intermittent anoxia and high sedimentation rate allows the preservation of high-resolution centennial- to millennial-scale events.

Behl and Kennett (1996) documented 19 of 20 D/O events since 75 ka in the sedimentary and micropaleontological record of the Santa Barbara Basin from Ocean Drilling Program (ODP) site 893. Between 60 and 13 ka, 17 high-frequency alternations occurred. Laminated sediments signifying anoxic bottom conditions were dominated by the benthic foraminifers *Bolivina, Suggrunda,* and *Rutherfordoides.* They alternated with massive (nonlaminated) sediments, signifying oxic conditions dominated by the genera *Epistominella, Nonionellina, Nonionella,* and *Uvigerina* dominated. Near-anoxic conditions (bottom oxygen levels reduced to < 0.1 ml/L) correlate well with D/O interstadial periods in Greenland Ice cores. Under these extremely stressful conditions only a few foraminiferal species can survive. Restricted circulation and high biological surface productivity in the Santa Barbara Basin apparently amplified the impact of large-scale climate events on intermediate ocean waters in the Santa Barbara Basin, leading to greater decreases in bottom oxygen levels than is typical for other regions of the Pacific Ocean Margin.

In summary, in just a few years, D/O climate events have been found to affect surface and deep oceanic conditions in the Atlantic Ocean, Nordic Seas, and Pacific Ocean. Surface- and bottom-dwelling foraminiferal and algal species with distinct ecological adaptations to specific environmental conditions were highly sensitive to altered physical conditions (temperature and salinity), dissolved oxygen, and food resources during millennial-scale climate events.

Western North American Continental Climate Record

Evidence has emerged that climate change associated with Heinrich events is reflected in glacial and paleolimnological records of western North America. Mid-latitude glaciers might, in theory, be sensitive to suborbital climate change, and although the chronology is inadequate, dynamic Wisconsin ice advance and retreat has been known for some time for parts of the Rocky Mountains (e.g., Colman and Pierce 1981) and Pacific northwest (Porter et al. 1983).

Phillips et al. (1990) used an improved dating technique, cosmogenic ^{36}Cl, to augment ^{14}C dates and obtain ages for large boulders on Sierran moraines and other landforms. These early studies revealed a first-order correspondence between the Tioga, Tahoe, and other glacial episodes and the deep-sea oxygen isotope record. Phillips et al. (1996)

later found that some Sierran glacial advances very likely corresponded with Heinrich event discharges from Atlantic ice-sheet margins, notably, events H5, H3, H2, and H1. The earliest advance was dated at about 49 ka, a few thousand years older than the ages for H5 in marine sediments. Between 40 and 28 ka, a series of minor, undated glacial advances may have occurred, but a correlation with event H4 is inconclusive. Tioga stages 1, 2, and 3—dated at 28, 22–23, and 16 ka, respectively—immediately preceded H3, H2, and H1 events, respectively. Phillips et al. (1996) pointed out that the Sierra Nevada glacial record involves local hydrological factors as well as climate, but they nonetheless support the theory that the Heinrich events had a global signature.

Clark and Bartlein (1995) reviewed the glacial history of the Yellowstone and Colorado Front Range areas of the Rocky Mountains, the Puget Lobe of the Cordilleran Ice Sheet, and the Cascade Range of the Pacific Northwest for evidence for short-term glacial advances and retreats that might correlate with Heinrich and D/O events. Using ^{14}C, ^{36}Cl, and U-series dating of travertines in the Yellowstone area, they found that at least four major glacial advances occurred, bracketed with dates of 47–34, 30.0–22.5, and 19.5–15.5 ka; another event was estimated at 54–47 ka. Clark and Bartlein argued that Pacific coast glacial advances correlate with North Atlantic stadial climate events expressed as Heinrich events and glacial retreats correspond to interstadials in the North Atlantic region. On the basis of the better-dated glacial sequences in northern Yellowstone outlet glaciers and the Puget lobe, Clark and Bartlein (1995) argued that glaciers advanced to near their maximum terminus up to several thousand years before the corresponding Heinrich event in the Atlantic. There was no equivalent to event H3, and event H5 is relatively poorly expressed, perhaps owing to high summer insolation values at these times.

Millennial-scale climate events are also expressed in the sedimentary and desiccation history of Owens Lake in the Great Basin of California (Benson et al. 1996). Owens Lake is sensitive to hydrological variations stemming from precipitation and springtime runoff. Using analyses of total inorganic and organic carbon, oxygen isotopes, and magnetic susceptibility of sediments, Benson and colleagues discovered 19 distinct glacial advances in the Sierra Nevada (labeled A in figure 5-7) between 52.5 and 23.5 ka, most of which are represented by decreased discharge to Owens Lake. Intervening glacial recessions are labeled R events (figure 5-7). Periods of Owens Lake closure labeled C in figure 5-7 were identified by oxygen isotopic maxima and represent periods when the lake level receded below the core depth.

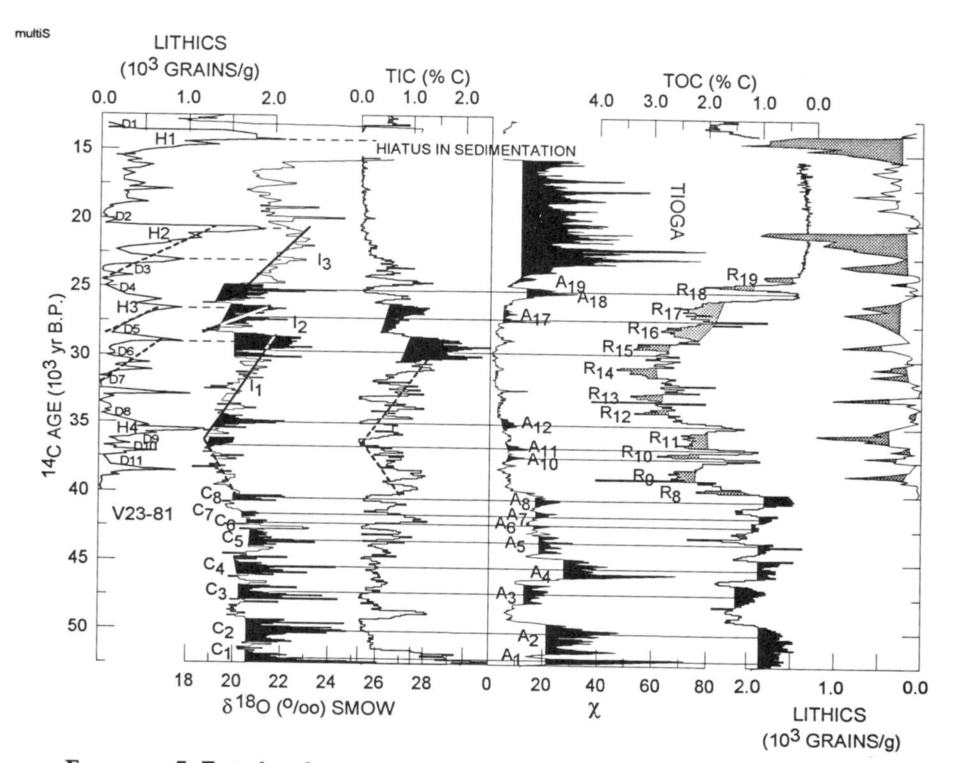

FIGURE 5-7 Paleoclimate record of millennial climate events from Owens Lake and Sierra Nevada mountains, California, compared to deep-sea record from core V23-81 from 50–15 ka. Lithic grain record from deep-sea records of Heinrich (H) and D/O interstadial events (D1–D11), Owens Lake hydrological closure (C1–C8), Sierra Nevada glacial advances (A1–A19) and recessions (R6–R19), and periods of decreasing wetness (I1–I3) are indicators of climate changes in the western United States. Courtesy of L. V. Benson with permission from American Association for the Advancement of Science. From Benson et al. (1996).

Millennial-scale events in Owens Lake appear to correspond to the rapid D/O cycles and Heinrich events recorded in the North Atlantic region. For example, Benson's group postulate that at least 9 of 11 major Sierran ice recessions correspond to lithic events recorded in Atlantic core V23-81 raising the possibility that Sierran and eastern Laurentide Ice Sheet melting was coeval. Although dating limitations preclude precise correlation between short-term eastern and western North American glacial events, the number of millennial-scale climatic events in both regions are similar. Benson et al. suggest that changing air temperatures over North America during the past glacial period might be responsible for this pattern. Finally, the frequency of the 19 glacial cycles, one every 1500 yr or so, is very close to that recorded in Greenland ice cores and North Atlantic climatic records, strongly suggesting a common climatic forcing.

Florida Vegetation and Equatorial Ocean

Until recently, the tropics were generally considered less sensitive than high-latitude areas to major climate changes like those accompanying Quaternary glacial episodes. In chapter 2, we saw how paleoclimatologists have remolded their thinking about tropical sensitivity to climate change as evidence accumulates for cooling and precipitation changes in low latitudes during the LGM.

Were the low latitudes affected by D/O rapid climate events? In a study of late Quaternary Florida vegetation history, Grimm et al. (1993) discovered that multiple vegetation changes occurring over the course of the past glacial period showed a strong correspondence to ice-core and North Atlantic records. Palynological records from an 18.5-m-long sediment core from Lake Tulare contain an exceptional pattern of pollen spectra, dominated by alternating *Pinus* (pine), *Quercus* (oak), and *Ambrosia* (ragweed), representing the last 50 ka. Pine episodes (up to 60% of the total assemblage) correlate with Heinrich events, signifying periods of wet regional climate. The *Ambrosia* phases include pollen from common herbs (ragweed, marsh elder, sunflowers), indicating much drier climates and an open grassland habitat. The oak pollen assemblage indicates open oak woodland or savanna habitats. Spectral analyses showed a dominant 5.7-ka cycle in the Florida pollen record. Grimm et al. concluded that the pollen record might be explained by rapid shifts in atmospheric conditions, due more likely to decreased temperatures than increased precipitation, or perhaps oceanographic events in the Gulf of Mexico related to meltwater discharge.

Chinese Loess and Paleosol Record of Millennial Climate

Porter and Zhisheng (1995) and Zhisheng and Porter (1997) recently discovered six dust events dated between 110 and 70 ka in the central Chinese loess record using grain-size analysis and magnetic susceptibility as climate indicators. Dust flux in central China is influenced by the strength of northwesterly winter monsoon winds; magnetic susceptibility is an indirect measure of the summer monsoon and its control of pedogenesis. Cold events marked by these proxies matched cold peaks C19–C24 known from North Atlantic sediment cores (McManus et al. 1994). Zhisheng and Porter also found two dust events that may signify brief cooling events dated about 121 ka and 115 ka, within the Eemian interglacial period.

Southern Hemisphere Chilean and New Zealand Glacial Chronology

The puzzle of millennial-scale climate change will remain unsolved unless well-dated paleoclimate records of the Southern Hemisphere's glacial and climatic events can be correlated to climatic records from mid-to-high latitudes of the Northern Hemisphere and from tropical regions. Recently, Denton and Hendy (1994) and Lowell et al. (1995) addressed interhemispheric climatic events through extensive field mapping of glacial geology in the Andes of Chile and the Southern Alps of New Zealand. In Chile, a series ridges 3–20 m high called the Llanguihue moraines formed during the past glacial period about 40–13 ka, when alpine snowlines were 1000 m lower than they are now. Lowell et al. (1995) found that at least six late Quaternary piedmont glacier advances reached the outer moraine belt dated by radiocarbon at 14,890, 21,000, 23,060, 26,940, 29,600, and > 33,500 ^{14}C yr. These glaciers alternatively advanced and retreated with a regularity and frequency usually reserved for more continuous paleoclimate records obtained from marine and lacustrine sedimentary sequences. The ages of the Llanguihue moraines indeed matched, to a first approximation, the timing of suborbital-scale stadial intervals in the North Atlantic region.

In addition to the extensive glacial evidence, pollen and beetle data were also an integral part of documenting the late Quaternary Chilean climate record. In the Chilean Lake District and on Isla de Chile, Lowell et al. (1995) discovered that changes in the Chilean flora that paralleled those in evidence from the glacial advance-retreat chronology. Each stadial episode was marked by 4–5°C cooling, which was nearly as great as the 6°C cooling that occurred during the LGM (Ashworth and Hoganson 1993). Paleovegetation was primarily comprised of Subantarctic Parkland ecosystems dominated by the southern beech tree *Nothofagus* in open landscapes of grasses and composites. These ecosystems are quite distinct from those of the North Patagonian rainforest vegetation that entered the region at about 14 ^{14}C ka, when final deglaciation accelerated.

Combining the Chilean climate record with that from New

Zealand (from Denton and Hendy 1994 and other studies), Lowell et al. (1995) proposed that because both regions are under the influence of Southern Hemisphere westerlies, the composite record provides an adequate test of interhemispheric climatic symmetry during Heinrich events. They concluded (p. 1541): "These glacial maxima in mid-latitude mountains rimming the South Pacific were coeval with ice-rafting pulses in the North Atlantic Ocean." They also proposed that mid-latitude Southern Hemisphere glacial and vegetation history reflect globally synchronous climate changes that cannot be explained solely by changes in North Atlantic thermohaline circulation or by Laurentide Ice Sheet surges. Instead they might be linked to atmospheric changes, most probably related to changes in atmospheric water-vapor content. The Southern Hemisphere continental record, while providing convincing evidence that millennial-scale climate events dominate the late Quaternary record, still requires improved temporal continuity to establish the precise phasing between North Atlantic and Southern Hemisphere mid-latitude glacial events.

Causes of Millennial Scale Events During Glacial Periods

As we have seen in the preceding sections, a number of causes have been proposed for rapid climatic events during glacial periods. These include internal dynamics of the Laurentide Ice Sheet whereby basal lubrication causes periodic surges about every 7 ka (MacAyeal 1993; Alley and MacAyeal 1994); ice-ocean-atmosphere interactions in the North Atlantic circulation (Bond et al. 1993; Bond and Lotti 1995); continental ice-sheet modification of atmospheric circulation (Clark and Bartlein 1995); changes in low-latitude hydrological cycles (Chappellaz et al. 1993); tropical atmospheric changes in trade winds, probably ultimately linked to thermohaline circulation (Hughen et al. 1996); and atmospheric changes in water vapor (Lowell et al. 1995) and perhaps other greenhouse gases linked to nonlinear responses to orbital precession (Curry and Oppo 1997).

What is the preferred mechanism? General circulation modeling of global climate indicates that climate change due to greenhouse gas forcing will show hemispherically synchronous and symmetric patterns of climate change about the equator. Also, polar regions are expected to cool more than low-latitude regions. Conversely, in the case of climate change forced by changes in oceanic thermohaline circulation, changes in polar climate will be asymmetric about the equator—the subpolar North Atlantic Ocean and adjacent continents will warm as thermohaline circulation carries heat to high latitudes and releases it to the atmosphere; the southern oceans will cool correspondingly (Manabe and Stouffer 1988; Rind and Chandler 1991; Rahmstorf 1994). Thus, in brief, if interhemispheric synchroneity of rapid climate events can be documented, that is, if both polar regions and equatorial regions cool and warm simultaneously, then atmospheric circulation and chemistry must be considered a leading cause. If asynchronous climate events in both hemispheres are found, regional events such as thermohaline circulation switches that propagate from the North Atlantic Ocean to other regions might be more plausible explanations.

The Laurentide Ice sheet may have been a trigger for millennial-scale climatic changes such as Heinrich events. MacAyeal (1993) modeled the Laurentide Ice sheet and proposed internal dynamics could account for Heinrich events. This mechanism holds that when an ice sheet is thin, the temperature at its base is < 0°C (a cold-based ice sheet). As the ice sheet thickens, the basal temperature of the ice sheet increases until it reaches 0°C and basal melting occurs. Basal melting can lubricate the bed, causing ice to surge, eventually thinning and refreezing. Ice-sheet surging can lead to the discharge of icebergs into the North Atlantic surface layer, lowering the salinity and reducing deep-water formation from the conveyor belt. As thermohaline circulation diminishes, less heat is drawn to high northern latitudes, which acts as a negative feedback leading to increased ice growth in the Laurentide area. A self-sustaining cycle is created, which has come to be known as the *binge-purge model* (MacAyeal 1993).

Bond and Lotti (1995) postulated that both Heinrich and D/O events were driven by the same forcing, and their fingerprinting of IRD in the North Atlantic raises the problem of asynchronous iceberg discharge from several different ice sheets. If icebergs drove the North Atlantic oceanic system, then massive ice-sheet discharge from Greenland, the Barents Sea, or Scandinavia must have been extremely dynamic, recurring every 2–3 ka.

Clark and Bartlein (1995) also favored a role for the Laurentide Ice Sheet, but suggested its primary impact was on atmospheric patterns. They discussed four possible mechanisms to explain correlative climate events from both the North Atlantic region and western North America. They dismiss the atmosphere and the ocean as the primary drivers in part because general circulation models do not predict that western North America should cool to the degree indicated by the glacial stratigraphy. Nor do climate models suggest that climatic forcing by rising levels of atmospheric CO_2 would cool the west substantially.

Instead Clark and Bartlein favor the hypothesis that the Laurentide Ice Sheet itself exerted an influence on temperature and atmospheric circulation. Drawing on the analogous situation of the LGM, when the ice sheet depressed western North American temperatures 4–10°C and shifted the jet stream southward (Imbrie et al. 1992), they propose a similar situation may have occurred during millennial-scale climate change, when a second-order climatic variability was superimposed on predominant long-term orbitally driven cycles. In essence, this theory holds that short-term glacial expansion before a Heinrich event would increase the glaciation potential in the west, and decreased ice sheet size would cause the reverse. These ice sheet effects acted in concert with changing levels of solar insolation during the past glacial period about 50–20 ka.

Lowell et al. (1995) concluded that the Northern Hemisphere ice rafting and isotopic climate events were synchronous with Southern Hemisphere glacial advances. Thus they discount thermohaline circulation, internal ice dynamics, and orbital factors as the main triggers of D/O events during the past glacial period and the rapid climate shifts during final deglaciation. Instead, they favor an explanation for interhemispheric synchroneity of global climate changes that involves atmospheric forcing, perhaps from water vapor changes that would produce globally synchronous climate change.

Direct evidence for low-latitude changes in temperature and moisture balance (Grimm et al. 1993), wind strength and oceanic biological productivity (Hughen et al. 1996; Curry and Oppo 1997), and terrestrial CH_4 production (Chappellaz et al. 1993) support the hypothesis that atmospheric changes are intimately involved with millennial-scale events. Water vapor changes originating in the tropical oceans were also a favored explanation by Curry and Oppo (1997), who argue that meridional thermohaline circulation cannot account for the degree of tropical cooling and NADW reduction nor for the synchroneity of tropical and high-latitude climate change during D/O climate events.

One is thus drawn to several conclusions about glacial age millennial-scale climate change. First, high frequency, high-amplitude climatic events, expressed in the North Atlantic region as D/O events, are a fundamental feature of the past 100 ka. Enough strategically located evidence has accumulated from low latitudes and the high latitudes of the Southern Hemisphere, however, to indicate that D/O events are probably global millennial-scale climate events occurring throughout much of the past glacial period from 100 to 20 ka. Their ultimate causes are still under discussion, and they may be related to orbital precessional cycles that through nonlinear feedbacks cause biotic changes in low latitudes, which in turn produce changes in atmosphere content of water vapor and CH_4. Globally synchronous and symmetrical climate change over the past 50 ka, and probably over the past 100 ka, supports this idea. However, ice sheets and oceanic circulation changes certainly play major roles in amplifying and altering the regional impacts of these effects. Determining the cause of millennial-scale climate changes will most likely remain a major challenge until centennial- and decadal-level chronologies become available in tropical and extratropical Southern Hemisphere regions.

THE EEMIAN: CLIMATIC VARIABILITY DURING THE LAST INTERGLACIAL PERIOD

THE CONCEPT THAT two distinct climate states existed during the Quaternary—one with and one without large temperate ice sheets—is an oft-cited theme in paleoclimatology (Broecker et al. 1985) and climate modeling (Manabe and Stouffer 1988). Many paleoclimatologists believe that Quaternary interglacial climate variability may be of a fundamentally distinct nature from that of glacial periods. The traditional view of interglacial climate has held that it is less sensitive to perturbation by high-amplitude, high-frequency climate oscillations of glacial periods (e.g., D/O events), in part because glacial events involve climatic modulation of low-frequency orbital events by low-latitude ice sheets (see Boulton and Payne 1994). Minimal continental and sea ice in critical regions, especially along continental margins of the North Atlantic, may render interglacial climate less subject to meltwater discharges and oceanic and atmospheric feedback loops.

In recent years, interglacial climate variability and sensitivity has been thrust into the spotlight for several reasons. First, humans live in the now 10,000-yr-old Holocene interglacial period, during which atmospheric CO_2 levels had remained relatively stable until the past century of emissions resulting from human activity (chapter 9). Paleoclimatologists must establish whether the twentieth-century atmospheric temperature rise evident from instrumental records reflects elevated CO_2 and CH_4 concentrations or whether it reflects natural climatic variability typical of an interglacial. A corollary question is will future climate remain stable or will it change, at least in terms of a Quaternary interglacial period, in an unprecedented way?

A second catalyst for interglacial climate research was the surprising discovery by Dansgaard et al. (1993; GRIP 1993) of evidence that the past interglacial atmosphere over Greenland was punctuated by a series of rapid (70–750 yr) temperature oscillations recorded in the GRIP ice-core isotopic record. These hypothesized temperature swings were almost as high in amplitude—perhaps 10°C—as glacial-interglacial swings, but they occurred over just a few centuries during the Eemian interglacial period. Other GRIP proxies also indicated substantial short-term atmospheric variability during the Eemian.

Since the original report by the GRIP researchers, several studies have challenged the idea that extreme temperature instability punctuated Eemian climate, instead proposing that the stratigraphy of the deep Greenland ice might be compromised by ice flow due to pressure (e.g., Grootes et al. 1993, see below). This issue is still unresolved (see Larsen et al. 1995 and later), but most researchers accept that the integrity of the deeper Greenland ice core record is compromised. Nonetheless, the results from the GRIP ice core raised the specter of interglacial climate instability. To give a full understanding of this problem, I must first clarify what an interglacial period signifies.

The last period of global climate warmth before the Holocene is known as the last interglacial (LIG) period. Paleoclimate evidence shows the LIG period is characterized in many regions, especially in North America, Europe, and high northern latitudes, by climatic conditions slightly warmer than current ones. Warmer climate is often reflected in a more northerly geographic range of temperature-sensitive species of plants, coastal mollusks, shallow water ostracodes, foraminifers, and other organisms; in ice-core records of atmospheric temperatures, soil development, and other continental indicators; in minimal Northern Hemisphere ice sheet volume (except perhaps Greenland); and in a global sea level about 4–6 m higher than today (see chapter 8). On a global scale, global paleoclimate reconstruction of the LIG climate carried out by CLIMAP (1984) showed that mean annual temperature was only about 1°C warmer than it is currently. The CLIMAP study and others have led to the general perception of the LIG period as a close cousin to the Holocene, a period of relative stability that followed the step-wise cooling and climate instability of the last glacial period.

Choosing an appropriate term to identify the last period of global warmth is complicated by the many local and regional lithostratigraphic, geomorphic, and climatostratigraphic terms used to identify the LIG period. Here I follow a cadre of paleoclimatologists who use the term Eemian to refer to this time. In Europe, the Eemian interglacial stage was first named for shallow water marine deposits near the Eem River in the eastern Netherlands (see Zagwijn 1961; Mangerud et al. 1979). Eemian sediments overlie deposits of the

Warthe glacial (in part Saalian) and underlie Wurm deposits from the past glacial period (Bowen 1978). Eemian marine faunas, like many other high latitude marine faunas, contain relatively warm water species, known in Europe as Lusitanian elements, for their modern distribution in the Lusitanian faunal province. Lusitanian faunal elements occur in Eemian-age deposits along European coastlines. Eemian pollen assemblages are also distinct from those found in earlier interglacial periods and are indicative of regional European climates warmer than current climates (Selle 1962; van der Hammen et al. 1971).

Currently, the Eemian and its age equivalents throughout northern Europe, Greenland, and the North Atlantic Ocean are probably the single most-studied group of Quaternary interglacial deposits. Eemian shallow marine and continental deposits are also confidently correlated with marine oxygen isotope substage 5e (Mangerud et al. 1974, 1979) and are the age equivalent of the Ipswichian stage in England (Mitchell et al. 1973; Bowen 1978) and part of the Sangamon interglacial stage in the midcontinent of the United States (Follmer 1983; Curry and Follmer 1992), which is also called the Sangamon Geosol (Curry and Pavich 1996). Whereas the Eemian was originally described on the basis of marine deposits containing warm-water faunas, the Ipswichian was defined on stratigraphic and palynological criteria and the Sangamonian stage on soil development criteria. Bowen (1978) and Flint (1971) provide useful reviews of the historical development of European and North American interglacial stratigraphy.

The exact age and duration of the Eemian interglacial is another disputed topic. In the absence of precise geochronologic data, most deep-sea marine and continental paleoclimate studies use the SPECMAP time scale of Martinson et al. (1987), which dates the marine isotope curve by tuning it to astronomically dated insolation cycles. The SPECMAP time scale model holds that the LIG period, marine isotope substage 5e, was only about 10 ka long (130–120 ka), which is similar to Shackleton's (1969) early estimate of LIG duration.

Slowey et al. (1996) recently defended the SPECMAP age of the LIG period by directly dating deep-sea sediments using uranium-thorium dating. They circumvented the problem of low uranium content in foraminifera by dating uranium-rich algal and inorganic aragonitic sediments from the Bahama Bank slope deposited during the LIG period. They argued that the Bahama slope has little terrigenous input, so detrital uranium is minor, and that the shallow water depths mean thorium scavenged from the water column is low. Seven uncorrected dates ranged in age from about 128.0 to 136.5 ka. Applying corrections for detrital-hosted uranium and thorium scavenged from seawater, Slowey et al. (1996) gave an age range of 127–120 ka for isotope stage 5e, essentially in line with the SPECMAP chronology. Additional dating will be required and the radiometric corrections confirmed before the age of the Eemian interglacial period in deep-sea sediments is resolved (Winograd et al. 1997).

Efforts to directly date and correlate continental and shallow marine interglacial deposits using a host of age-dating techniques (amino acid racemization, thermoluminescence, electron spin resonance) have been successful to various degrees in correlating continental deposits to the marine isotope record, but rarely do they provide ages of sufficient accuracy to determine the exact beginning and termination of the interglacial. Likewise, as we saw in chapter 4, the duration of the Eemian high sea-level stand that was responsible for coastal interglacial deposits and coral terraces several meters above present sea level has been disputed.

Nevertheless, several lines of evidence suggest the Eemian interglacial was longer than 10 ka. Among them is the revisionist dating of emerged Eemian-age tropical coral reefs described in the past chapter. New dating of coral reefs indicate that the duration of the last interglacial high sea level may have been at least twice as long as the SPECMAP age model, as much as 17 to 20 ka. In addition, Winograd et al. (1988, 1992) dated the Devils Hole, Nevada, paleoclimate record of the LIG period using multiple U-series dates, which suggested that the penultimate glacial-interglacial transition began as early as 140 ka and the LIG period ended about 20 ka later. Thus, the warm Eemian climatic state lasted up to twice as long as estimated from the astronomical SPECMAP isotope chronology (Winograd et al. 1997).

European Eemian pollen records derived from annually laminated diatomite sediments at Bispingen (Müller 1974) yields a floating chronology based on simple laminae counting that supports a LIG duration of at least 11 ka. Although only part of the Bispingen Lake sedimentary record is laminated, Field et al. (1994) estimated that the interglacial period lasted about 11.2 ka at Bispingen. The Eemian interglacial climate persisted for at least this long (perhaps 15 ka) at La Grande Pile according to Guiot et al. (1989).

Finally, Kukla et al. (1997) conducted a detailed comparison between the climate records of marine isotope stage 5 in the North Atlantic and the Eemian at La Grande Pile, France. They concluded that, contrary to prior opinion, the first Weischelian (post-Eemian) interstadial in Europe does not correspond to marine isotope substage 5d and the onset of glacial conditions in Europe does not coincide with the isotope stage 5e/5d transition. Kukla and coworkers argue instead that the Eemian interglacial period in Europe is coeval with marine isotope substage 5e *and* part of substage 5d. In this case, the Eemian interglacial period, which ended abruptly in Europe with the demise of the *Picea-Abies-Carpinus* deciduous forest, would be as much as twice as long as previously believed, lasting 19 ka during the interval 130–111 ka. Thus the Eemian was twice as long as the current Holocene interglacial.

The distinction between a 10-ka and a 20-ka interglacial period is obviously critical to the interpretation of high-resolution paleoclimate records of the LIG period. Evidence appears to be growing for a "long" chronology in which the Eemian interglacial lasted at least 15 ka.

European Vegetation During the Eemian

There are four major Eemian vegetation profiles from northern Europe: the type Eemian in the Netherlands (Zagwijn 1961; van der Hammen 1971), La Grande Pile bog (Woillard 1978; Woillard and Mook 1982; de Beaulieu and Reille 1992; Soret et al. 1992; Guiot et al. 1989, 1992, 1993), Les Echets in France (de Beaulieu and Reille 1984), and Bispingen Lake in northwest Germany (Müller 1974; Field et al. 1994). Table 5-1 summarizes the stratigraphy of La Grande Pile and Bispingen compared with deep-sea isotope chronology.

TABLE 5-1. Simplified LIG Stratigraphy and Climatic Zones of Europe

Stratigraphy		*Pollen zones*		
Isotope substage	*La Grande Pile*	*Bispingen*	*Amsfoort*	*Summer temperature (°C)*
5a	St. -Germain II			
5b	Melisey II			
5c	St. -Germain I			
5d	Melisey I			
5e	Eemian	RW-V	E-6	10–15
		RW-IV	E-5	17–22
		RW-III	E-4	17–22
		RW-II	E-2	16
		RW-I	E-2	10

Note: Integrated by Larsen et al. (1995).

Here is a brief summary of the salient aspects of Eemian climate variability derived from European lake records, keeping in mind that each region has been studied using different paleotemperature and precipitation methods (e.g., modern analog, response surface, Field et al. 1994), and the same pollen sequence has been examined using different techniques (Guiot et al. 1992, 1993). Still, the basic structure of the Eemian interglacial vegetation is apparent. The paleovegetation data generally indicate that the early part (about 3 ka) of the Eemian was the warmest part of the interglacial period. An early Eemian summer temperature maximum is evident in the Netherlands (Larsen et al. 1995) and at La Grande Pile (Guiot et al. 1993). The genus *Taxus* (yew) is an especially good indicator of early Eemian warmth at La Grande Pile (de Beaulieu et al. 1992). At Bispingen, Germany, the earliest 2.9 ka of the Eemian had reduced seasonality and increased moisture compared with the preceding glacial period.

A step-wise drop in winter temperature occurred at Bispingen about 7.6 ka after the interglacial began, which was followed by a gradual temperature decline, with a major oscillation in winter temperature about 6.1 ka into the interglacial period. This exceptional event was marked by the abrupt disappearance of warm floral elements such as *Corylus* (hazelnut), *Fraxinus* (ash) *Tilia* (basswood) *Ulmus* (elm), and *Taxus* (yew). The cold-tolerant genus *Betula* (birch) increased. The early temperature maximum in the Netherlands is followed by a drop in temperatures of about 4°C, then by a steady cooling for the remainder of the interglacial period. At La Grande Pile, later Eemian pollen assemblages include greater percentages of cold-tolerant taxa like *Pinus* (pine), *Abies* (fir), and *Picea* (spruce).

These vegetation and temperature oscillations are small when compared with those of other Quaternary climate transitions, but they are still quite significant. For perspective, the glacial-interglacial winter temperature contrast at La Grande Pile is 28°C for Termination 1 and 20°C for Termination 2. Comparable shifts of 17–18°C occurred in England during Termination 1. At Bispingen, the transition from glacial to interglacial in Termination 2 was abrupt and large—a rise in mean cold month temperatures was from – 13°C to +5°C over only 700 yr.

Several tentative conclusions can be made regarding the Eemian pollen record of Europe. First, the Eemian is not a mirror image of the Holocene; there are significant vegetation changes and especially noteworthy millennial-scale changes in seasonality. This variability takes the form of changes in minimum coldest month temperatures. Second, the Eemian climate in northern Europe as recorded by these records shows a strong influence of the North Atlantic Ocean, particularly related to changing sea-ice conditions and its influence on atmospheric circulation patterns. Finally, the biological life histories of individual tree types suggest that each species responded distinctly to interglacial climatic variability.

North Atlantic Interglacial Variability

The deep-sea North Atlantic record of Eemian climate suggests the oceanic realm is not subject to the temperature oscillations evident in European pollen records. For example, Ruddiman and McIntyre (1981) documented the role of the mid-latitude North Atlantic Ocean surface ocean in amplifying the 23-ka precessional climate cycles as a source of moisture for ocean and adjacent continental ice growth and heat transfer and storage. Their planktonic foraminiferal SST record for the LIG period clearly showed oceanic evidence that isotope substage 5e had very minor fluctuations in winter and summer temperatures relative to the glacial-interglacial transition and within the past glacial.

Recent studies of interglacial North Atlantic subpolar oceanic variability in areas with high sediment rates have confirmed the idea of interglacial marine stability. McManus et al. (1994) provided firm evidence from two climate proxies—IRD and planktonic foraminiferal assemblages—that climate variability in oceanic surface conditions was "extremely muted" (figure 5-8). Patterns of variation in the proportions of *N. pachyderma* (sinistral) at DSDP site 609 and core V29-191 were indicative of this stability: *N. pachyderma* remained near 0% in sediments deposited during marine substage 5e. Only upon the inception of marine isotope substage 5d, coincident with the appearance of IRD and common *N. pachyderma* (20–40% of the total foraminiferal assemblage), can one identify the end of peak interglacial conditions and the onset of cooler climate. Recall from the previous discussion of Heinrich events that frequent repositioning of the polar front (the boundary between waters originating in the cold Arctic and the subtropical Gulf Stream) over millennial time scales during the glacial was marked by changes in *N. pachyderma.* No such variability was found for the marine equivalent of substage 5e. The likely sources of glacial-age variability—ice discharges and reduced SSTs—were not prominent features of the Eemian interglacial climate regime in this region.

McManus et al. (1994) considered several possibilities to explain the conflicting data for Eemian instability in the GRIP ice core and the contrasting stable climates in the adjacent North Atlantic Ocean:

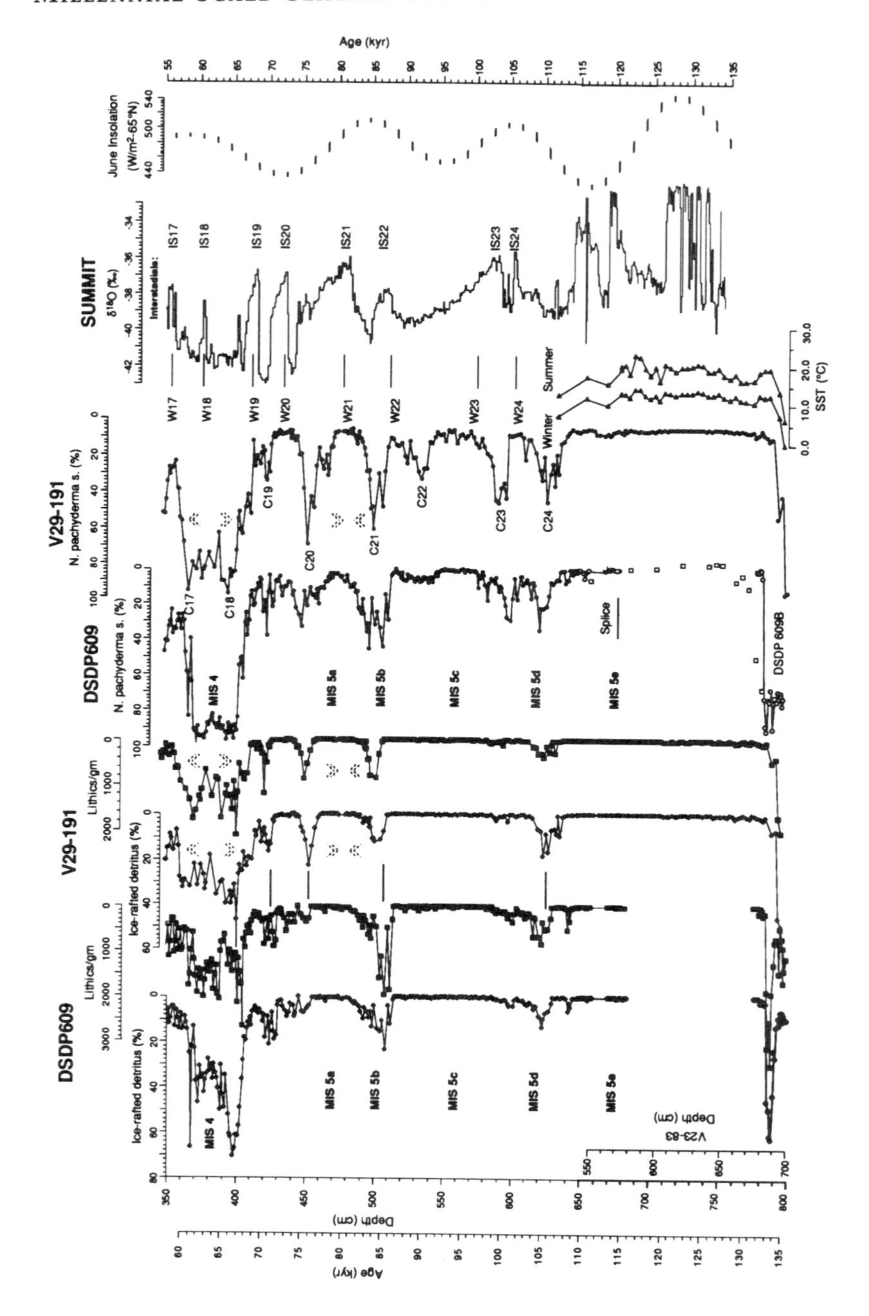

Age (kyr)
June Insolation (W/m2-65°N)
SUMMIT
δ18O (‰)
Interstadials:
IS17
IS18
IS19
IS20
IS21
IS22
IS23
IS24
W17
W18
W19
W20
W21
W22
W23
W24
Summer
Winter
SST (°C)
V29-191
N. pachyderma s. (%)
C17
C18
C19
C20
C21
C22
C23
C24
DSDP609
N. pachyderma s. (%)
Splice
DSDP 609B
MIS 4
MIS 5a
MIS 5b
MIS 5c
MIS 5d
MIS 5e
V29-191
Lithics/gm
Ice-rafted detritus (%)
DSDP609
Lithics/gm
Ice-rafted detritus (%)
MIS 4
MIS 5a
MIS 5b
MIS 5c
MIS 5d
MIS 5e
V23-83
Depth (cm)
Depth (cm)
Age (kyr)

The difference was due to a lower sensitivity of the polar fauna to changing temperatures; the North Atlantic climate was decoupled from the Greenland air temperature; the Greenland isotope instability signifies moisture-source changes far from the North Atlantic–Greenland area rather than temperature over Greenland; or the GRIP ice core record before about 110 ka was disturbed by ice flow at depth.

Additional evidence for Eemian North Atlantic climatic stability was provided by Oppo and Lehman (1995) and Oppo et al. (1997), who showed that SST and NADW production varied at very low amplitudes during the Eemian compared with the variation occurring on glacial-interglacial time scales and within the past glacial period. The study by Oppo et al. (1997) was particularly important because they examined cores taken along a bathymetric transect from 1450 to 2600 m water depth, producing a paleobathymetric profile of mid-depth variability during isotope substage 5e. A convincing case was made that strong NADW formation occurred during the entire 5e interval, indicating vigorous thermohaline circulation during most of the Eemian. Isotopic evidence for minor deep-water and corresponding SST changes within substage 5e merit additional study.

Interglacial Climate in the Nordic Seas

The Nordic Seas (Greenland, Norwegian, and Icelandic Seas) comprise an extremely sensitive area with strong east-west SST gradients resulting from the warm inflowing Norwegian Current; the southward flowing East Greenland Current, which exports cold Arctic Ocean outflow; and the counterclockwise northward branch of the Irminger Current. Fronval and Jansen (1996) found that planktonic foraminifers in a series of cores taken across the modern oceanographic gradient

FIGURE 5-8 Interglacial climatic variability in deep-sea cores from Deep Sea Drilling Program (DSDP) site 609 and core V29-191 compared with Summit GRIP ice-core oxygen-isotope record and June insolation between 135 and 60 ka. Interglacial marine isotope stage 5e (MIS 5e) is generally characterized by little or no ice-rafted lithic grains or cold SST-dwelling *N. pachyderma,* showing oceanic climate stability during the peak interglacial period. During MIS 5d–5a, periodic oscillations of IRD and *N. pachyderma* signify alternating warm interstadial events (W24–W17) that correspond to Summit interstadials (IS24–IS17). The Summit ice-core record older than about 110 ka may be compromised by ice flow. Courtesy of J. F. McManus with permission from Macmillan Magazines. From McManus et al. (1994).

revealed three cool snaps within the Eemian. They dated these cool events at roughly 127–126 ka, 122–121 ka, and 117 ka. During each event, they found that the east-west temperature gradient weakened relative to its current situation. Fronval and Jansen (1996) postulated that these marine oscillations may correspond to cold events recorded in other marine sediment cores from off Greenland, in the Eemian continental record from Europe, and in the GRIP ice core.

Evidence from temperature-sensitive benthic foraminiferal species from sediment cores from shallow marine deposits off northern Denmark indicates at least two major cold events occurred during the Eemian interglacial period (Seidenkrantz et al. 1995). After an initial period of maximum interglacial oceanic temperatures, the arctic foraminiferal species *Cassidulina reniforme, Stainforthia loeblichi,* and *Islandiella* spp. and the subarctic species *E. excavatum* punctuate the Eemian benthic assemblages, reaching a combined 20–50% of the total assemblage. Simultaneous declines in warm temperature species are recorded. The pattern of early Eemian warming followed by large-scale cooling is reminiscent of the pollen records from Bispingen and La Grande Pile described above. Seidenkrantz et al. correlated two of the Danish foraminiferal events to Eemian-age Greenland ice-core isotope events corresponding to isotope stages 5e–2 and 5e–4 (Dansgaard et al. 1993).

Interglacial Deep-Sea Circulation

The isotopic and sedimentological record of Eemian deep-oceanic circulation was studied by Keigwin et al. (1994) who found little evidence for high-amplitude Eemian climate instability related to changes in NADW production. Deep-sea core GPC-9 from the west flank of the Bahama Outer Ridge is an extremely sensitive region of deep sea circulation because it is located in the Western Boundary Current of the North Atlantic at 4758 m water depth and records changes in NADW and Antarctic bottom water. Variability in $\delta^{13}C$ of benthic foraminifers and in weight percent carbonate record deep-sea oceanographic change at the GPC-9 site (figure 5-9). Periods of high carbonate correspond to relatively warmer climates. Keigwin et al. (1994) matched previously unrecognized changes in deep oceanic circulation to interstadial-stadial fluctuations in both the Greenland ice cores (D/O events #19–24, between 65 and 110 ka) and paleovegetation zones at Grande Pile, France; Tenagi Philippon, Greece; and Clear Lake, California, for substage 5e through early stage 4 intervals. In contrast to the instability they found between 100–65 ka, the Bahama

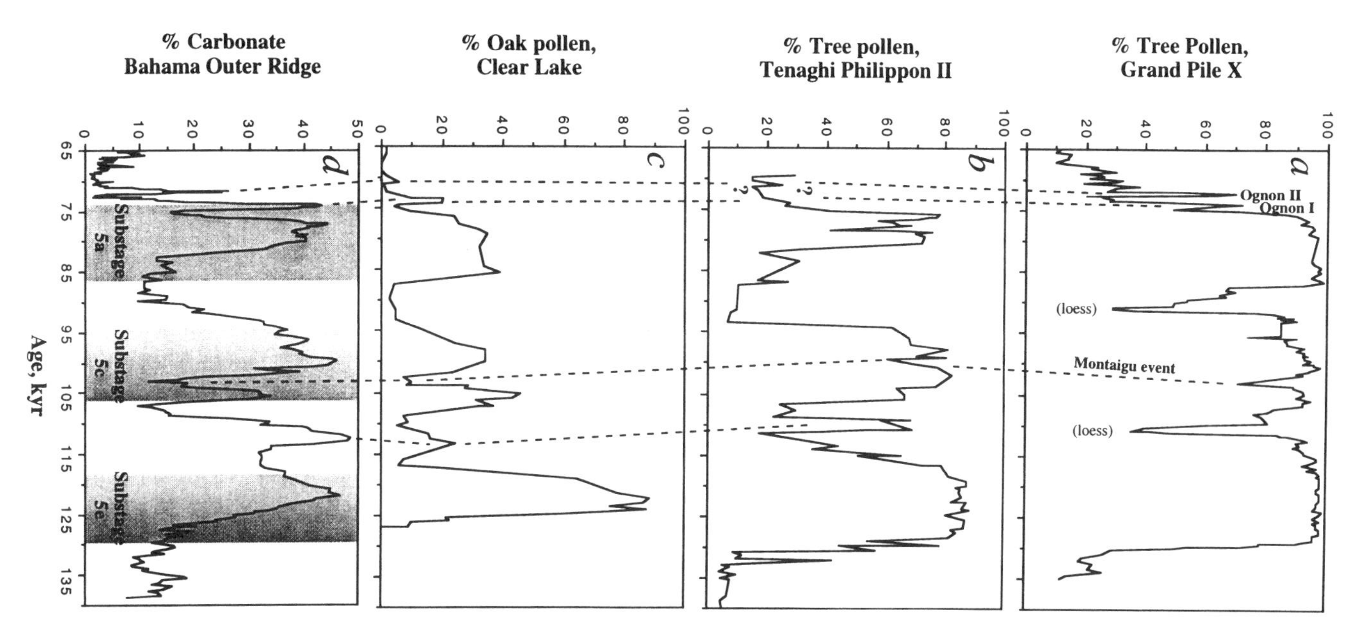

FIGURE 5-9 Interglacial climate variability recorded in percent carbonate at the Bahama Outer Ridge; oak pollen at Clear Lake, California; tree pollen at Tenagi Philippon II, Macedonia; and Grande Pile X bog, France. Montaigu and Ognon I and II events are brief climate reversals that occurred during marine isotope substages 5c and 5a. Courtesy of L. D. Keigwin and with permission from Macmillan Magazines. From Keigwin et al. (1994).

Outer Ridge record of Eemian deep-sea circulation showed no evidence of diminished NADW production before 110 ka such as that indicated in the GRIP Eemian isotopic record. Thus, these results contradicted the Eemian record from GRIP ice core; the GPC-9 Eemian isotopic and carbonate patterns resembled more the record of surface North Atlantic stability shown by McManus et al. (1994). Keigwin's group concluded that the lack of large mid-latitude ice sheets around the North Atlantic during peak Eemian interglacial warmth played a role in minimizing the amplitude of climate changes during the Eemian in the North Atlantic Ocean, at least in terms of the strength of the conveyor belt.

In summary, Eemian interglacial Greenland ice core oscillations are more likely an artifact of ice disturbance than a true climate signal. Nonetheless, considerable evidence from European marine and continental records suggests at least a moderate degree of ocean- and air-temperature variability within the Eemian interglacial. Deep-sea subpolar North Atlantic surface and deep-water climate records, on the other hand, suggest a much higher degree of interglacial climate stability in terms of SST and thermohaline circulation compared with the European continent and the Nordic Seas.

Holocene Interglacial Instability

Recently, Bond et al. (1997) discovered millennial-scale climatic cycles in the Holocene interglacial record of the North Atlantic. Using essentially the same proxy tools used to document glacial-age Heinrich events (lithic grains and petrologic tracers, planktonic foraminifers, stable isotopes), Bond's group studied North Atlantic ice-rafting events in the Denmark Strait, off Labrador, and off western Ireland, three regions that might be sensitive to minor fluctuations in sea-ice conditions. They came to five main conclusions. First, abrupt shifts in North Atlantic SST of 2°C occurred during the Holocene, coincident with the occurrence of IRD. Second, these abrupt climatic coolings last about 1–2 ka, the same duration as glacial-age D/O events. Third, Holocene IRD–temperature cooling events were an order of magnitude smaller than D/O events of the preceding glacial period. Fourth, there was a periodicity of 1470 {+/}- 500 yr to these Holocene events, a period that matched cycles in IRD for the preceding glacial interval. Fifth, Holocene climatic events coincide with atmospheric changes discovered in the nearby GISP2 ice-core records (e.g., O'Brien et al. 1995). Bond and colleagues made the important statement that the Holocene interglacial events "appear to be the most recent manifestation of a pervasive millennial-scale climatic cycle operating independently of the glacial-interglacial climate state." (1997:1257). They suggest that Atlantic surface circulation has two modes: a warm mode with minimal IRD and warm surface waters like current conditions and a cooler mode when Greenland-Iceland Sea ice was advected farther southward.

SUMMARY

A MOUNTAIN FLOWER, an Arctic-loving protist, Canadian midge larvae, salinity-intolerant foraminifers, and other ecologically sensitive species, used in conjunction with isotopes, IRD, moraines, and other paleoclimate indicators, paint a picture of contrasting amplitudes of climate variability at millennial time scales during deglacial, glacial, and interglacial climate states. Both deglacial and glacial-age climates are extremely sensitive to minor changes in hydrological balance, oceanic temperature and salinity, ice sheet dynamics, atmospheric dust, and other factors. The level of variability within the Eemian interglacial period, although not nearly as great as that during deglacial and glacial intervals and still largely unknown for most regions, nonetheless shows hints of millennial-scale changes. Recent compelling evidence from ocean sediments and ice cores shows the Holocene interglacial period underwent, at least regionally, millennial-scale cycles seemingly independent of long-term glacial-interglacial states.

The immediate causes of some Quaternary millennial-scale climate change certainly reside in internal processes related to the ocean-ice-atmosphere system. What the ultimate causes are remains unknown. External forcing by insolation changes related to orbital cycles does not provide a likely explanation for short-term climatic events, although during deglacial transitions it is probably a factor. Changes in atmospheric dust, water vapor, and other trace gases may be suspect in their role in amplifying or dampening the climate signal, but atmospheric CO_2 levels change too slowly and CH_4 responds to climate change rather than leads it (see chapter 9). Internal dynamics of ice sheets probably sometimes help amplify the signal through effects on thermohaline circulation, but evaporation and precipitation changes can have similar effects. Finally, solar variability, hypothesized by Denton and Karlen (1973, 1976) among others to explain suborbital climate changes, must be further investigated.

The pervasive evidence for millennial-scale climate change discovered over the past decade was almost unthinkable a few years ago. Its discovery has challenged theorists and empirical paleoclimatologists alike to find the causes of rapid climate change. Paleoclimatologists have a wealth of new climate records; almost everywhere they look closely enough, they have discovered millennial events, albeit some of low amplitude. Unfortunately, interregional and interhemispheric records are still difficult to correlate and thus interpret in terms of initial causes and feedback mechanisms. While the link between earth's orbit and climate change is easily traced to the work of Croll, Milankovitch, and other early astronomers and geologists, discoveries in suborbital climate variability have only recently fostered closer looks at causal mechanisms. At the risk of oversimplification, the 1960s and 1970s belonged to Milankovitch orbital cycles, whereas the 1980s and 1990s are dedicated to suborbital time scales. If insolation changes due to orbital cycles are the pacemaker of the ice ages, then the conveyor belt, icebergs, atmospheric circulation, and solar variability are among the many agents of climate that keep the pacemaker ticking.

VI

Holocene Centennial and Decadal Climatic Variability

. . . for close upon 70 years, the ordinary progress of the solar cycle, as we have been accustomed to it, was in abeyance — in abeyance to such a degree that the entire records of those 70 years combined together would scarcely supply sufficient observations of sunspots to equal one average year of an ordinary minimum such as we have been accustomed to during the past century.

E. Walder Maunder, 1922.

POLAR BEARS AND POTATOES: HUMAN HISTORY AND HOLOCENE CLIMATE

In the Middle Ages, according to Hubert H. Lamb in his 1995 book *Climate History and the Modern World,* Icelandic tradition called for the carpeting of church floors with polar bear skins. Such skins were plentiful in Iceland during the twelfth through fourteenth centuries, a sign of abundant sea ice in the surrounding waters. But their availability mysteriously declined in the fifteenth century, and eventually because of the scarcity of skins, the Danish monarchy was forced to restrict trade in them. Lamb suggests that the demise of polar bears (*Ursus arctus*) might have coincided with reports of decreasing sea ice around Iceland at about the same time that Europe was entering into the period known as the "Little Ice Age."

Climate change, of course, is believed by some scholars to render far worse consequences to human history than the rising price of medieval floor decor. In early nineteenth-century Ireland, the humble potato was a dominant staple crop, grew on thousands of small, subsistence farms, and fed 40% of Ireland's populace. Ireland's agriculture — indeed its entire course of history — was forever changed, however, in 1842 when the aptly named fungus *Phytophthora infestans* was carried to Ireland from America, leading to the Irish Potato Famine, known also as The Great Hunger. This famine, which lasted from 1846–1852, was most acute in the years 1846 and 1847. Just six years after introduction of the fungus, the blight had resulted in more than 1 million human deaths, more than 1 million emigrants, and a reduction by 25% of Ireland's population of 8.5 million.

The Irish Potato Famine conjures images of coffin ships and potato people. It still stirs up emotion, myth, and misconceptions about the Irish people and their culture, as well as debates about nineteenth-century British policy toward Ireland. To some students of recent climatic history, it may also represent a manifestation of the impact of climate on human history. Lamb (1995) has pointed out that climatic conditions in Ireland during the fatal years 1846–1847 should also be part of the equation in the famine. The immediate culprit in the potato debacle, *P. infestans,* is a fungus that multiplies quickly, causing potatoes to rot and depriving people of their primary food source. The fungus, however, requires certain environmental conditions in which to thrive. Warm (>10°C), humid (> 90%) weather, those very conditions recorded in diaries from the most deadly year, 1846, are ideal for the spread of *P. infestans.* Indeed, *P. infestans* has made a resurgence in the United States and Canada in the 1990s (Fry and Goodwin 1997).

The histories of Icelandic polar bear skins and the Irish Potato Famine raise the same question: What role has climate change played in affecting human society and history? More particularly, was climate and its impact on sea ice a primary agent in Iceland's decline in polar bears? Or were the ecology, the behavior of *U. arctus,* and/or changing Norse hunting patterns also factors in diminished polar bear numbers? Was Ireland's potato famine caused by an oscillating climate that produced a few sultry summers that escalated the scale of the famine to disastrous proportions? Or was it the inevitable outcome of "a monolithic economy trapped in a downward spiral of poverty..[an] inevitable outcome of years of improvidence" (Kinealy 1994:11), the manifestation and vindication of Malthusian principles of political economy, as perceived by some historians? Or, alternatively, as argued cogently by Kinealy (1994), might the image of a misguided Irish people itself be misguided and the famine a result of political sentiment and policy in Britain leading to limited export of disaster relief to a starving Irish nation?

This chapter deals with the complex and challenging subject of climate changes at decadal and centennial time scales during the Holocene interglacial period covering the past 10 ka (figure 6-1).

FIGURE 6-1 Commonwealth Glacier, Antarctica. Decadal and centennial variability in the Holocene is recorded in both fluctuations of glacier margins and in the paleoclimate record in annual layers of glacial ice. Courtesy of J. Bernhard.

Holocene climate change poses several unique and perplexing problems for paleoclimatology. First, Holocene climate change is inextricably linked to human cultural evolution. The Irish Potato Famine is one of countless, equally complex historical events linked to various degrees to climate history. Were the tenth-to-twelfth century cultural zenith and the subsequent late twelfth-to-thirteenth century demise of the American Anastasi in some way related to climate change during the Medieval Warm Period and Little Ice Age? Or were cultural factors more important (Petersen 1994; Dean 1994)? Did fifteenth-century Norse colonies disappear from western Greenland because of increasingly frigid climates and isolation from Iceland and Europe by sea ice (Dansgaard et al. 1975)? Or were competitive clashes with native American Eskimo populations or the cultural problems of small isolated colonies to blame (Lamb 1984)?

These climate-history links are enigmatic because establishing causality between climatic and historical events is a difficult task. Nonetheless, the study of the impact of climate change on human cultures should be anchored by the application of the principles of modern paleoclimatology, judiciously integrated with archaeology and history. Each cultural period deserves its own careful evaluation involving an assessment of regional climate and environmental changes, as well as historical events. Such an approach can lead to revisionist explanations of cultural trends such as the recent conclusions of Hodell et al. (1995) on Mayan civilization and late Holocene climate.

A second unique aspect of the Holocene is that it is also central to understanding humanity's impact on earth's present and future climate. The past few millennia constitute the benchmark, the natural baseline, from which climatologists must seek to differentiate between twentieth-century human-induced climate change and natural climate variability. This imperative exists in part because the buildup of radiatively active atmospheric trace gases (carbon dioxide [CO_2], methane [CH_4], and nitrous oxide [N_2O]) continues and because model projections of the climatic response to these gases, albeit controversial, portend large-scale effects (Houghton et al. 1996). Indeed, late Holocene climate variability, which for years was relegated to paleoclimatology's back burner behind research of high-amplitude glacial-interglacial and deglacial climate changes, is now actively studied. This key aspect of the Holocene can be summed up in the following question: Is the observed twentieth-century warming unusual for the Holocene or is it part of a natural climatic cycle, a recovery from the Little Ice Age?

A third factor is that paleoclimatologists have at their disposal sources of climatic data unavailable for older, "geological" intervals. These sources come from diaries; flood records; official documents; farm records; and, since the sixteenth century in Europe (later in other areas), instrumental data such as temperature, precipitation, and tide gauge records. Climate change has been chronicled in unlikely indexes such as the famous relationship between the price of a bushel of wheat and sunspot numbers proposed by Sir William Herschel in 1801. Climate change has also been linked to the demise of English vineyards and oscillations in the wine harvests of France, as well as more conventional measures such as glacier advance and retreat upon villages in the Alps. To this day, chronicled climatic events provide useful sources of information captured in classic texts by Ladurie (1971), Lamb (1977, 1995), Grove (1988), Eddy (1983), Bradley and Jones (1993), and Hoyt and Schatten (1997), among others.

Instrumental and historical sources have the advantage of accurate age control but they are often spatially and temporally spotty and sometimes suspect. Paleoclimatologists often use historical climate records to calibrate paleoclimatological proxies and sometimes to splice reconstructed records of the past millennium with the more reliable historical data available. But paleoclimatic records of the past few centuries have improved greatly over the past few years. They now often yield superior climate data about decadal- and centennial-scale trends (e.g. Overpeck et al. 1997).

A fourth unique aspect to Holocene climatic variability involves the factors most likely to cause climate change over decades to centuries — volcanism, solar irradiance (figure 6-2), oceanic circulation, or atmospheric trace gases. Climate models indicate that these forcing factors effect a relatively small amplitude signal on global and hemispheric patterns of temperature and precipitation. Consequently, the signal-to-noise ratio is extremely low compared with that seen over a glacial-interglacial cycle (Rind 1996). For example, global mean temperature has oscillated by only 0.5–1°C during the past few millennia. Rind suggests that the most plausible mechanisms that force climate to change over short time scales within an interglacial period generate variability of about 0.2–0.5°C. This low Holocene signal-to-noise ratio was one reason that Covey (1995) advocated that paleoclimatological research be carried out on past periods of global warmth when the climatic signal was much larger than the Holocene signal. In short, climate change on hemispheric and global scales within the Holocene has traditionally been difficult to detect relative to that occurring over longer time scales.

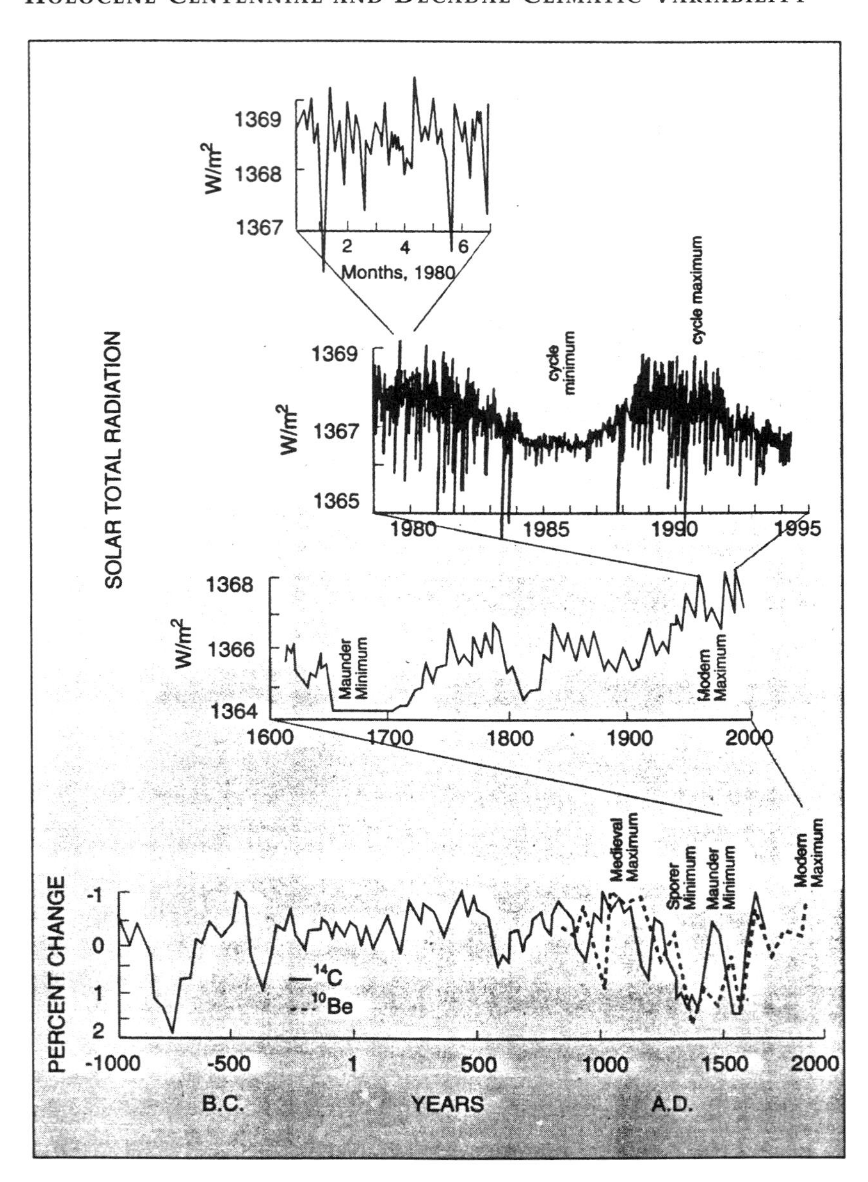

FIGURE 6-2 Variability in total solar radiation during the past three millennia recorded from cosmogenic isotopes (^{10}Be and ^{14}C), during the past 400 years showing the Maunder sunspot minimum and maximum, during the past 20 years showing a modern solar cycle, and during the year 1980. Courtesy of J. Lean and J. Eddy.

A fifth aspect of the study of Holocene decadal- to centennial-scale climate change is that it is a fledgling field. Although new volumes appear almost annually (Bradley and Jones 1992; Diaz and Markgraf 1992; Hughes and Diaz 1994a; Martinson 1995; Jones et al. 1996) and Holocene ice core records are making great advances (e.g., Meese et al. 1994; O'Brien et al. 1995; Alley et al. 1997) no clear consensus has emerged about interhemispheric climate correlations nor about the dominant forcing factors. Moreover, large gaps exist in the paleoclimate database for the Holocene, especially certain oceanic regions.

European climate change over the last five or six centuries spawned a chronology of climate change based mainly on unorthodox, though often accurate, human observations of their surroundings. The next section describes this classical Holocene climate terminology derived from historical events in Europe and elsewhere within the context of modern Holocene paleoclimatology.

CLASSICAL HOLOCENE CLIMATE CHRONOLOGY

THE HOLOCENE STAGE encompasses the past 10 ka of earth history. Originally named by the 1885 International Geological Congress to designate the "wholly recent" period (Bowen 1978), the Holocene (also called the Flandrian interglacial in Europe) is a period distinguished from the preceding Pleistocene glacial epoch as a time of relatively warm interglacial climate. Traditionally viewed as a time of stable climate relative to that of the glacial and deglacial periods, the Holocene has in fact experienced significant climate changes over various time scales.

The definition of the base of the Holocene has a long history (Morrison 1969; Bowen 1978; Roberts 1989). Now it is almost universally placed at 10 ka, roughly at the inception of the current interglacial period. The round number 10,000 yr is not merely one of convenience; Northern Hemisphere summer insolation peaked 8% above present and winter insolation troughed 8% below present between 12–9 ka (COHMAP 1988), and pre- and post-10-ka climates differed substantially in many paleoclimate records (figure 6-3).

As a time stratigraphic unit, the Holocene stage is usually not subdivided into formal substages (Bowen 1978), but it can be informally divided into an early Holocene (10–8 ka), a middle Holocene (8–4 ka), and a late Holocene (4 ka to present). The early-middle and middle-late Holocene boundaries correspond to major changes in temperature, precipitation, and other parameters in many regions and paleoclimate records (e.g., Dean 1997; Stager and Mayewski 1997).

Although climatic changes 8 and 4 ka are not necessarily manifested as single global and/or synchronous events, they are convenient boundaries for discussing Holocene climate variability at various time scales.

Before describing Holocene climatic variability, a brief digression is warranted to explain several climatic terms that permeate the literature on Holocene climates. These "classical" Holocene paleoclimatological terms, originally defined to describe regional climatic intervals of cold or warmth, are listed with references in table 6-1.

Altithermal, Hypsithermal, and Little Climatic Optimum

The early-middle Holocene has been recognized as a period of climate warmth and altered precipitation in many regions. Often this period is referred to as the "altithermal," the Atlantic warm period, the "little climatic optimum" in Europe, the postglacial "hypsithermal" (Deevey and Flint 1957), and (in some archaeological literature) a cul-

TABLE 6-1. Classical Holocene Climate Chronology

Period	*Approximate age*	*Reference*
Holocene	10 ka to present	International Geological Congress, Portugal 1885 (see Roberts 1989)
Altithermal or hypsithermal (also called little climatic optimum)	9–5 ka	Deevey and Flint 1957
Neoglaciation	4 ka to present	Porter and Denton 1967
Medieval Warm Period (also called Medieval solar maximum)	9th–4th centuries	Lamb 1965
Little Ice Age	15th–19th centuries	Grove 1988; Matthes 1939
Maunder sunspot minimum*	1645–1715	Eddy 1976; Maunder 1922
Spoerer sunspot minimum*	1450–1540	Spoerer 1889
Wolf sunspot minima*	1290–1350	Wolf 1868
Dalton sunspot minimum*	1795–1823	Hoyt and Schatten 1997
Schwabe 11-yr sunspot cycle*		Schwabe 1844
Gleissberg 80- to 90-yr sunspot cycle*		Gleissberg 1966

*These solar cycles are discussed in Schove (1983), Anderson (1992a, 1993), and Hoyt and Schatten (1997).

tural "golden age." Early-mid Holocene warmth is variously dated between 9 and 5 ka, depending on the region. Climatic warmth with temperatures exceeding those of the late Holocene was inferred from early palynological, shallow marine, tree-line, and sea-level records (Fairbridge 1961) and from glaciological evidence. Many early studies carried out at mid to high latitudes of Europe and North America suggested that local and regional climatic conditions were much warmer than those of the preceding glacial period and slightly warmer than those of the late Holocene. In low latitudes, hypsithermal Saharan lake levels were also known to be relatively high due to enhanced early Holocene precipitation.

Neoglaciation

Neoglaciation is a term that has been applied to the late Holocene because early research indicated unequivocal glaciological evidence that the late Holocene was a period of overall advancing alpine glaciers. Matthes (1939) reported on late Holocene glacier activity and moraines in the Sierra Nevada and coined the term "Little Ice Age" to refer to "an epoch of renewed but moderate glaciation which followed the warmest period of the Holocene" (p. 518). Porter and Denton (1967) renamed the Holocene interval of North American glacial readvances the "Neoglaciation" and the term "Little Ice Age" has since become restricted to the period encompassing the fifteenth through the late nineteenth centuries (Grove 1988). In addition to Northern Hemispheric alpine glaciological evidence, late Holocene cooling is clearly seen in east Antarctic ice core paleo-atmospheric temperature trends (Mosley-Thompson 1996), some Greenland ice-core isotopic records (Fisher et al. 1996), decreasing frequency of summer melt layers in the Greenland GISP2 ice-core (Alley and Anandakrishnan 1995), Southern Hemisphere glacial history (Clapperton 1990), and deep-sea isotopic (Keigwin 1996) and bottom-water temperature (Dwyer et al. 1995) records. However, the past few thousand years of late Holocene climate is not characterized by a simple unidirectional trend toward cooling climate but instead is punctuated by the complex oscillations discussed next.

Medieval Warm Period

The ninth through the fourteenth centuries are often referred to as the Medieval Warm Period (Lamb 1965). A more narrowly defined period,

1100–1250 A.D., is sometimes defined as the Medieval Warm Epoch, which coincided with a period referred to as the Medieval solar maximum (Jirikowic and Damon 1994).

Medieval warmth is inextricably linked to the Little Ice Age, which followed it in Europe; they are often contrasted as extended periods of warm and succeeding cold climate, respectively. Although the historical literature on the Little Ice Age is larger, the Medieval Warm Period, as with other periods suspected of global warmth (Budyko 1982; Dowsett et al. 1994; Webb et al. 1993), has lately received considerable attention because of projected future global warming. Jirikowic and Damon (1994:314) offered the following opinion: "If greenhouse gas emissions are a geophysical experiment on a grand scale (albeit poorly designed), the Medieval Solar Maximum does provide a control run."

Given the importance of understanding the causes of climatic warmth, Hughes and Diaz (1994b) posed the question in the title of their paper "Was there a Medieval Warm Period, and if so, where and when?" They reviewed the paleoclimate evidence from a number of sources and concluded that, although there was no sustained Medieval Warm Period–Little Ice Age sequence, there is good evidence for warmth during the mid twelfth and early thirteenth centuries and the early fourteenth century. Elevated summer temperatures occurred in parts of Scandinavia, China, the Sierra Nevada Mountains of California, the Canadian Rocky Mountains, and Tasmania. Regions such as the southeastern United States, southern Europe, and the Mediterranean show few climatic differences from those current conditions. We discuss climate variability in medieval times in a later section.

Little Ice Age

The concept of a "Little Ice Age" — a centuries-long period of cool climate—developed mainly in Europe after the original introduction of the term by Matthes (1939) to describe late Holocene glacial advances in the western United States. What evidence led to the concept that climate was cold during the fifteenth through nineteenth centuries? Most came from glacial readvances historically documented in alpine glaciers in northern Europe. Ladurie (1971) in his book *Times of Feast, Times of Famine* (a translation of Ladurie's *Histoire du climat depuis l'an mil*) gives one of the more extensive and informative reviews. Although a short summary of this vast documentation of Little

Ice Age Europe is beyond the scope of this chapter, we will take a brief look at the behavior of Iceland's largest ice cap, Vatnajøkull, which epitomizes the scale of Little Ice Age glacial readvances and their dramatic impact on local inhabitants and early naturalists. Vatnajøkull is the world's largest temperate ice cap, about 8500 km^2. Since the earliest Norse colonization of Iceland during the tenth century, Iceland's glacial (see Grove 1988) and sea-ice (Bergthorsson 1969) history have been well-documented. Historical accounts described by Grove (1988) indicate that during the late nineteenth century, surging by the Bruarjokull front of Vatnajøkull could be heard by Icelanders living 50–60 km away from the glacier. During one major surge, it advanced to cover an additional 1400 km^2. Similar accounts of glacial advances in the Swiss Alps and Scandinavia, augmented by dramatic late-nineteenth-century photographs and engravings, attest to the impact that alpine glacier movement had on nineteenth century naturalists, among them Louis Agassiz as he developed his glacial theory. Although glacial advances can be caused by internal ice mechanics and not necessarily climatic factors, there was ample evidence from regional glacial activity to develop the concept of a Little Ice Age, although the means to correlate the timing of glacial advances was still limited.

In addition to glaciological evidence for a colder climate, the history of solar activity during the fourteenth through the nineteenth centuries, especially trends in sunspot numbers, led to the hypothesis that there may be an association of diminished solar activity and cool intervals during the Little Ice Age. Major periods of sunspot minima called the Wolf (1290–1350), Spoerer (1400–1510), and Maunder (1645–1715) minima, coincided with cold European winters (Eddy 1976).

The Little Ice Age has recently been the subject of intensive research (Grove 1988; Bradley and Jones 1992, 1993; Overpeck et al. 1997; Kreutz et al. 1997). To a large degree, a keen interest in Little Ice Age climate history stems from the need to establish whether the twentieth century rise in surface temperatures is part of a natural climatic oscillation following Little Ice Age cooling or an anomalous warming caused by human activity. Consequently, it is imperative to avoid confusion and misinterpretation about the concept of the Little Ice Age. Three aspects of global climate between 1500 and 1900 A.D. are now generally accepted. First, the regional cooling and glacial advances experienced in some regions during the fifteenth through nineteenth centuries must be viewed in the broader context of a longer-term Holocene climatic evolution. Greenland ice-core records show that the Little Ice Age is the most recent episode in a series of Little Ice Age–like climatic events identified in ice accumulation rates (Meese et al. 1994) and glaciochemical records of dust and sea salt (O'Brien et al. 1995) dated at > 11,000, 8800–7800, 6100–5000, 3100–2400, and 600–0 years ago.

Second, the ages for the inception of Little Ice Age glaciological events vary with author in part because glacial readvances are generally asynchronous in different regions. Grove (1988), for example, concluded that the Little Ice Age maximum occurred at different times in Scandinavia, the Alps, and other regions of Europe. Regional asynchroneity should not be surprising in light of the internal effects that glacial dynamics and regional climate (i.e., summer temperatures and winter precipitation [see Gillespie and Molnar 1995]) exert on the advance and retreat of alpine glaciers. Recently, however, Kreutz et al (1997) determined from records of atmospheric circulation that the inception of the Little Ice Age was synchronous in Siple Dome, Antarctica, and in central Greenland. They identified the beginning of the Little Ice Age as a shift to enhanced meridional atmospheric circulation in both polar regions that occurred about 1400 A.D. The Little Ice Age began within 20 yr in Greenland and 28 yr at Siple, Antarctica. The combined dating error of the two annually dated cores is 12–20 yr. Kreutz et al. thus concluded that the Little Ice Age began about 100 yr after the period of low solar activity (Wolf, Spoerer, Maunder minima), and thus solar activity was not solely responsible for its inception.

A third aspect of the Little Ice Age is that it was not a sustained multicentury period of climatic cooling. Hughes and Diaz (1994a) found some support for early seventeenth-century cooling at least partially consistent with the concept of a Little Ice Age but did not conclude there was a multicentury period of cooling. Landsberg (1985) and Bradley and Jones (1992) also objected to the notion of the Little Ice Age as a sustained period of global cooling. Luckman (1996:85) summed up the opinion of many researchers about the Medieval Warm Period and Little Ice Age as follows: "The concept of a 'Medieval Warm Period' and 'Little Ice Age' as distinctive climatic intervals of several centuries duration is clearly a gross oversimplification of glacier advances and intervening warmer periods that occurred throughout the last millennium." Paleoclimatic evidence from more continuous records such as ice and sediment cores increasingly indicates that complex decadal-scale temperature, precipitation, and other atmospheric changes characterized the fifteenth through nineteenth centuries (e.g., Overpeck et al. 1997).

Solar Activity and Climate Change

The link between solar activity and climate change is traced by Douglas V. Hoyt and Kenneth H. Schatten in their book *The Role of the Sun in Climate Change* (1997) to the fourth century B.C. After a lapse, active solar observation resumed during the European Renaissance. The sun experiences sunspots, faculae, solar flares, coronal mass ejections, and other processes that influence its total irradiance. Sunspots are dark, cool regions that form on the face of the sun through magnetic processes. Faculae are bright regions surrounding sunspots that emit a greater amount of energy than the sunspot regions. Both sunspots and faculae affect the *solar constant,* the sun's total solar irradiance. The solar constant varies over short and long time scales around a mean value of approximately 1367 W/m^2. In addition to the total irradiance emitted, solar output also varies in the wavelengths of irradiance emitted by solar activity such that short (ultraviolet [UV]) wavelengths vary more than longer wavelengths. Hoyt and Schatten (1997) present a case that UV emissions may be a critical aspect of the long-disputed relationship between the sun and observed climatic patterns.

Several important solar cycles (e.g., the 11-yr Schwabe cycle, 80- to 90-yr Gleissberg cycle) and four periods of reduced sunspots (the Wolf, Spoerer, Maunder, and Dalton sunspot minima, table 6-1) have been important to paleoclimatology. Each is named after its discoverer who through meticulous observation established secular trends in sunspots and other phenomena. We will see later that many paleoclimate records appear to preserve evidence that solar variability affects earth climate, although the mechanisms translating minor solar irradiance changes into climate changes are poorly understood (figure 6-2).

EARLY AND MIDDLE HOLOCENE CLIMATIC HISTORY

Let us now examine some of the more important paleoclimate records of Holocene climate variability at centennial and decadal time scales. Recent studies have shown the early to mid Holocene was not a uniform period of global warmth but rather a complex interval of spatially and temporally variable temperature and precipitation. The Cooperative Holocene Mapping Project (COHMAP 1988) is the most extensive single investigation into early- to mid-Holocene climate.

COHMAP researchers found that orbitally induced solar insolation may have been an important forcing in early Holocene climates. They determined that summer insolation in highlatitudes of the Northern Hemisphere was 5–8% higher than it is now; winter insolation was lower,; mean annual polar insolation was greater; and low-latitude insolation was slightly lower. COHMAP paleoclimate reconstructions generally indicated on a global scale that high-latitude regions were relatively warm during parts of the early Holocene. COHMAP also discovered that enhanced monsoonal conditions also characterized Eurasia and parts of Africa where there was greater annual precipitation.

Continental climate during the Early Holocene has been documented in numerous other regional studies. In central Europe, Starkel (1991) describes multiproxy evidence for three phases of early Holocene climate: early warming (10.3 ka), Preboreal-Boreal transformation (10–8.5 ka), and the change to oceanic climate (8.7–7.7 ka). The first two phases he calls the Eoholocene because they represent a period of cold winters, hot summers, and minimal precipitation. The shift to oceanic climates occurred approximately 8.5 ka. Starkel's continental climate reconstruction is generally in accord with that obtained from the paleo-atmospheric record from the central Greenland GISP2 ice core (Meese et al. 1994). In western North America, Anderson and Smith (1994) studied stratigraphic records from the California Sierra Nevada Mountains, documenting dry microclimates about 6 ka, shifting abruptly about 4.5 ka to the deposition of peat and the appearance of plant taxa indicative of moister climate conditions. In eastern North America, lake levels were also dynamic (Dwyer et al. 1996).

Recent paleo-atmospheric and paleoceanographical reconstructions also reveal high-latitude early Holocene warmth in the Northern Hemisphere. Webb and Wigley (1985) estimated some Northern Hemisphere regions were 1–2°C warmer than current temperatures on the basis of palynological data. Arctic micropaleontological and isotopic data indicate strong early Holocene North Atlantic inflow of warm water, perhaps related to reduced sea ice relative to present conditions (Gard 1993; Stein et al. 1994; Cronin et al. 1995). Isotopic evidence from some Greenland ice cores (Renland and Agassiz sites, see chapter 9) indicates Holocene cooling followed an early Holocene climate maximum (Fisher et al. 1996), but surprisingly, in other cores (GRIP, GISP2) a Holocene cooling trend is not so apparent. Meese et al. (1994) found that GISP2 ice accumulation increased rapidly in a stepwise fashion from the Younger Dryas until about 9 ka, after which it oscillated about a narrow mean. O'Brien et al. (1995) likewise found oscillating Holocene climate in the GISP2 glaciochemical record.

Low-latitude climate records of the early Holocene also show a distinct contrast with glacial and late Holocene conditions. For example, Thompson et al. (1979, 1985) reconstructed tropical climate conditions from the Quelccaya ice core in the Andes of Peru on the basis of dust, isotopic, and glaciochemical evidence. Their evidence clearly shows a strong contrast between colder, wetter, and dustier atmospheric conditions during the late glacial interval and warmer, drier, and diminished aerosol dust content during the early to mid Holocene.

Early Holocene sea level rose rapidly as the deglaciation ended and may have reached a level slightly higher, by a meter or so, than the current level. Chappell and Polach (1991) and Eisenhauer et al. (1993) dated Holocene Pacific reef complexes and postglacial sea-level rise; Eisenhauer's study also estimated a peak sea level near 6.3 ka. Scott and Collins (1996) review the topic of the mid-Holocene sea-level high stand and argue that about 5.0–3.5 ka sea level was as much as 2 m higher than it is now. They suggest that high sea level resulted from climatic factors other than a surging Antarctic Ice Sheet or an isostatic crustal adjustment due to water loading of ocean basins. However, estimates of the precise eustatic sea level during the early to mid Holocene warm period is still complicated by isostatic, geoidal, and sedimentological and compaction processes (e.g., van de Plassche 1986; Pirazzoli 1991; see chapter 8).

In summary, considerable evidence supports relative warmth, altered precipitation, altered oceanography, and perhaps slightly higher sea level during the early Holocene. However, high-resolution interregional comparison of various paleoclimate proxies have yet to be carried out, and additional research is needed on this important time interval (Webb et al. 1993).

LATE HOLOCENE CLIMATE HISTORY

THE MAJOR SOURCES of paleoclimate history of the past few thousand years are tree rings, tree cellulose, carbon isotopes, ice cores, tropical corals, lake sediments, and marine sediments. Each source has its own strengths and weaknesses; each has contributed greatly to documenting decadal- and centennial-scale variability in different components of the climate system. The following section discusses a selected group of studies that use these methods and that have been instrumental in fostering a reevaluation of the classical interpretation of late Holocene climate history.

Tree Cellulose $^{13}C/^{12}C$ and ^{14}C Production

The chemical signal from tree cellulose carbon isotopes yields two types of important climate information about paleo-atmospheric conditions and past solar activity. Variation in $^{13}C/^{12}C$ ratios provides a history of relative humidity and perhaps also atmospheric temperature. Stuiver and Braziunas (1987) studied 19 North American conifer trees to reconstruct 2000 years of climate history. Arguing that humidity is a primary and temperature a secondary control on the $^{13}C/^{12}C$ ratios, and assuming that $^{13}C/^{12}C$ ratios lagged behind atmospheric change by 70–90 yr owing to biological factors, they obtained an excellent ($r = 0.8$) correlation between $^{13}C/^{12}C$ and climate records from ice-core acidity and temperature for the interval 1100–1850 A.D.

The $\delta^{14}C$ variability due to solar variability also provides a record of climate history. Stuiver and Braziunas (1993) studied GISP2 ice core ^{14}C and $\delta^{18}O$ records and reached the important conclusion that 11-yr cylces of atmospheric and tree ^{14}C are caused by solar modulation of cosmic ray fluxes and also that interdecadal-scale variability is due to solar modulation, thermohaline circulation changes affecting ^{14}C storage in the world's oceans, or both (figure 6-3). Stuiver et al. (1997) recently reached more conclusive evidence linking Greenland climate to solar activity. They state (p. 259): "The timing, estimated order of temperature change, and phase lag of several maxima in ^{14}C and minima in $\delta^{18}O$ are suggestive of a solar component to the forcing of Greenland climate over the past millennium. The fractional climate response of the cold interval associated with the Maunder sunspot minimum (and ^{14}C maximum), as well as the Medieval Warm Period and Little Ice Age temperature trend of the past millennium, are compatible with solar climate forcing, within an order of magnitude of solar constant change of ~0.3%."

Tree Rings, Precipitation, and Temperature

Dendroclimatology—the study of climate change using tree rings—provides more evidence for climate change over the past few millennia than any other single method. Most tree-ring paleoclimatology is carried out in either high latitudes of the Northern Hemisphere or alpine high-elevation sites, and the majority of the data pertain to summer-temperature history. In simplistic terms, cooler temperatures produce narrower rings, but many other factors discussed briefly here also influence tree rings. The basic dendroclimatological methodology, statis-

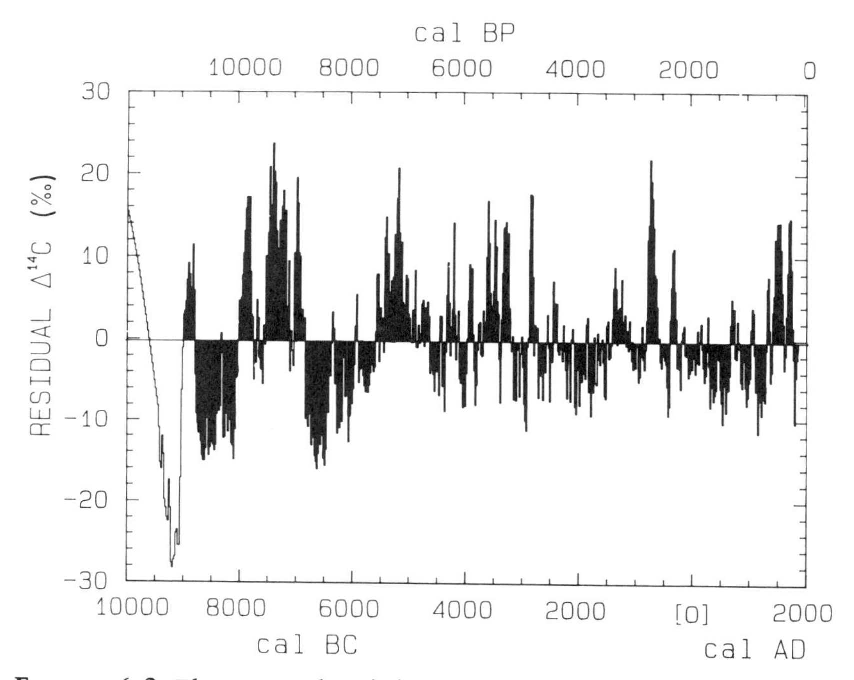

FIGURE 6-3 The past 12 ka of climate recorded in residual "$\Delta^{14}C$ from tree-ring records and corals (before 9500). The Younger Dryas event is recorded at 9190 B.C. (= 11,140 calibrated years). The eighteenth-century positive excursion is the Maunder minimum, when there was high cosmogenic ^{14}C production. Courtesy of M. Stuiver with permission from Edward Arnold. From Stuiver and Braziunas (1993).

tical standardization procedures, and discussion of ecological and physiological factors influencing tree growth can be found in the comprehensive texts by Fritts (1976), Schweingruber (1988), and Stokes and Smiley (1996); statistical approaches and pitfalls in dendrochronology are treated by Cook (1992, 1995).

There are several principal advantages to using tree rings for paleoclimatology. First, they yield extremely high-quality chronology. Tree-ring width and maximum density are the two staple measurements made in dendroclimatological studies. Counting rings back from the present and/or cross-checking unique ring sequences between living and subfossil trees, or wood obtained from building material, often yields exceptional, annual temporal resolution. As discussed in chapter 2, tree-ring chronologies are a primary means of calibrating the radiocarbon time scale for the period 13–8 ka.

A second advantage is the abundance and accessibility of trees for sampling. Schweingruber and Briffa (1996) summarized data from living tree-ring records from more than 100 sites from a broad swath across Siberia and northern Europe, alpine areas of southern and central Europe, Canada, and the western United States that constitute the Northern Hemispheric Densitometric Network. Although only a handful of millennial-length records are available, Schweingruber and Briffa (1996) point out that work is underway to establish additional long-term records in many regions.

Hughes and Graumlich (1996) state that the conterminous western United States has the most spatially complete regional dendroclimatology record anywhere, with no fewer than 80 tree-ring chronologies of at least 1000 years duration and 23 chronologies that extend 2000 years. Such coverage is important because the guiding maxim in paleoclimatology—that one should not extend paleoclimatologic inferences too far beyond the region of study—holds even more strongly in dendroclimatology (Cook 1995) owing to the many complex local environmental conditions (soil, microclimate, etc.) and biological processes (age, interspecific, interpopulation variability, etc.) that influence tree growth and sometimes obscure the climate signal.

One encounters several potential problems when interpreting climate history from dendroclimatological data. The most obvious is that different tree species living in different climatic zones and at different elevations have differing responses to climate factors. Temperature controls tree growth most strongly in regions characterized by cool, moist summers (Schweingruber and Briffa 1996); consequently, most of the highest quality records come from high-latitude or high-elevation regions. Spruce and larch tree-ring records may be a better indicator of temperature in northern Eurasia, whereas giant sequoia is suitable in the Pacific coast of the United States (Hughes and Brown 1992) and bald cypress in the southeastern U.S. coastal plain (Stahle et al. 1988). In general, tropical and subtropical regions, with little seasonality in temperature, have yielded relatively little paleoclimate data from dendroclimatology because warm-climate trees often have multiple-growth intervals during a single year. Semiarid climatic zones, in contrast, have a much greater potential for tree-ring climate histories.

Another possible complication in dendroclimatology is that the calibration of living tree growth during the past century to climatological parameters (temperature and precipitation) may be compromised by the effects of anthropogenic factors on tree growth. The most notable influence may stem from elevated atmospheric CO_2 concentrations and enhanced nitrogen fertilization. It is generally accepted that increased CO_2 levels enhance the growth of many tree types and can mask the influence of other environmental factors.

Briffa et al. (1996) point out two additional considerations confronting the use of tree-ring records for century-scale paleoclimate reconstruction. First, they point out that many early tree-ring studies concentrated on interannual and decadal trends at the expense of centennial-scale climate change. In doing so, however, statistical standardization techniques tended to filter out any low-frequency (century-scale) climate signals that may have been present. This is in contrast to many ice-core and sedimentary climate records where interdecadal and century scale climate variability were often the primary focus and high-frequency interannual time scales were either overlooked or impossible to evaluate because of the temporal resolution. Future tree-ring studies of low-frequency patterns are sure to remedy this bias.

Another still unexplored factor considered by Briffa and colleagues is the question of species and population adaptation to gradually changing environmental pressures. Evolutionary aspects of tree growth in response to environmental change may receive more attention as genetic and physiological research on tree response to climate change matures.

Among the most important ecological aspects of tree growth is the aging problem. Ontogenetic variability in tree-ring development can obscure the climatic controls, especially during early growth years. The aging process has led to statistical standardization methods designed to model tree-ring growth and decouple ecophysiological factors from the climate signal (Fritts 1976). Other biological factors that must be considered are the impact of fires and pests on tree growth. Also, site conditions (soil chemistry and moisture, local hydrology) can change during the lifetime of a tree, potentially altering its growth pattern. These problems can usually be overcome by using multiple tree records from a small area (e.g., Jacoby et al. 1996a).

With this background, we turn to a representative subset of the considerable tree-ring evidence for regional-scale climate oscillation over the past few millennia, much of it focused on the Medieval Warm Period and the Little Ice Age.

Western North America

As mentioned above, the western conterminous United States has more tree-ring records than anywhere else, many extending back a millennium or more. Comprehensive discussions of these records can be accessed through the papers by Graumlich (1993) and Hughes and Graumlich (1996), as well as the book edited by Diaz and Markgraf

(1992). Tree-ring data are supplemented by many pollen records from lake sediments (see Petersen 1994). Firm evidence from some regions supports relative warmth about the time of the European Medieval Warm Period. In fact, Graumlich (1993) found that summer temperatures between 1100 and 1375 A.D. exceeded modern temperatures in the Sierra Nevada region. The precipitation record, however, was dominated by decadal-scale variability.

High-Latitude Regions

Numerous tree-ring records come from high Northern Hemisphere latitudes (see Cook 1995; D'Arrigo and Jacoby 1993; Jacoby et al. 1996b), some of the more important of which extend back at least 1000 years (figure 6-4). It should be emphasized that like any tree

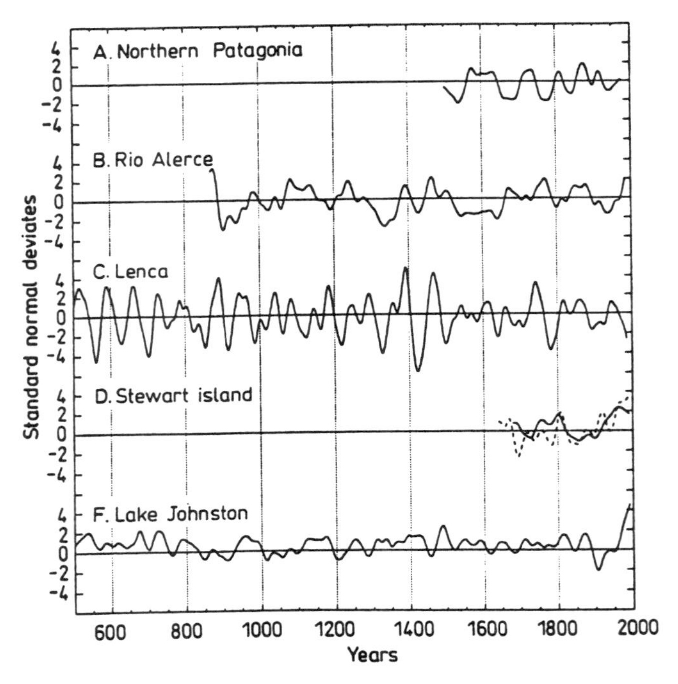

FIGURE 6-4 Southern (A) and Northern (B) Hemisphere temperature records obtained from tree rings for the period 600 A.D. to present. Trends are expressed as 50-year smoothing splines scaled as standard normal deviates to compare to the reference interval 1822–1957. Courtesy of E. Cook with permission from Springer-Verlag. From Cook (1995).

ring–derived temperature curve, the standardization and statistical methods used can substantially influence reconstructed temperature history (see Cook 1995). Some high-latitude records show evidence that summer temperatures during the Medieval Warm Period exceeded those during the twentieth century. The Tornetrask, Sweden, tree-ring record (Briffa et al. 1992) shows both this medieval warmth as well as relatively cool conditions during the Little Ice Age.

In other regions, the record is not so clear. The tree-ring record from the Polar Ural Mountains (Graybill and Shiyatov 1992) shows slightly less medieval warming than the Swedish record but exhibits a clear late-fifteenth century cooling. In Alaska, there are cool intervals in the sixteenth and seventeenth centuries and unprecedented warming in the twentieth century, a trend also found in the Mongolian tree-ring record (Jacoby et al. 1996b). Jacoby et al. (1996a) summarized the

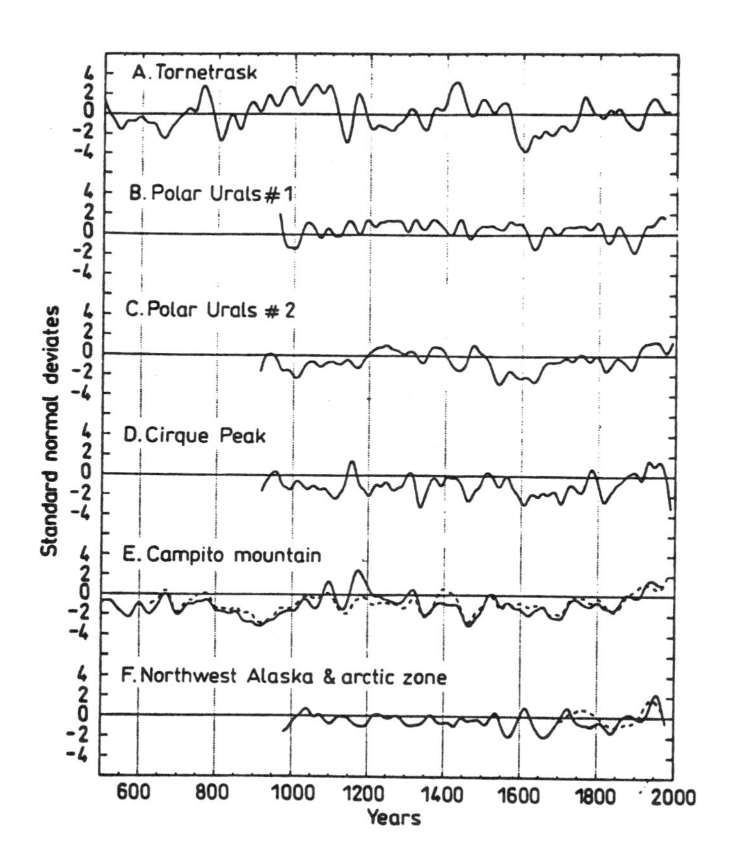

spatially and temporally complex tree-ring record from Canada and Alaska and generally found that for certain periods of the Little Ice Age, cooler climates prevailed in several parts of North America. They note that this pattern of continental cooling is inconsistent with climate model simulation results for Little Ice Age climate forcing due to both North Atlantic Ocean circulation changes and decreased solar irradiance (Rind and Overpeck 1993).

Southern Hemisphere

Most of the limited tree-ring climate data from the Southern Hemisphere come from South America in studies by Villalba (1990, 1994) and Lara and Villalba (1993), as well as others summarized by Villalba et al. (1996). In addition, Cook et al. discuss Tasmanian data (1991, 1996). Villalba (1994) reconstructed climatic conditions from the tree-ring record from northern Patagonia and recognized relative warmth from 1080 to 1250 A.D. and a long cold and moist interval from 1270 to 1660 A.D. with peaks at 1340 and 1640 A.D. A generally good correspondence exists between the Patagonian tree-ring and glacier records (Rothlisberger 1986) and the tree-ring record from central Chile. Villalba comments that if interannual climate variability during the Medieval Warm Period was like that of the current time, then it was characterized by a predominance of warm El Niño events, which now bring above-average precipitation to Chile and higher temperatures to Patagonia.

The major conclusion reached by Cook et al. (1996) from the long-term Tasmanian record is that significant interdecadal warm-season temperature variability in Tasmania has occurred for at least the past 3000 yr concentrated at frequency bands of 31, 57, 77, and 200 years. In addition, Cook et al. warn that these interdecadal variations may obscure future greenhouse warming trends due to elevated CO_2 levels.

Eastern United States

Early tree-ring studies of the Hudson River region (Cook and Jacoby 1979) and the Potomac River watershed (Cook and Jacoby 1983) established the use of annual ring width to study drought history (a paleo–Palmer Drought Severity Index [PDSI]). In the Hudson Valley, Cook and Jacoby (1979) examined eastern hemlock, pine, and chestnut trees spanning the interval 1694–1972. They discovered quasiperiodic oscillations with statistically significant periodicities of 11.4 and

26 yr, the former near the frequency band of 11-yr sunspot cycles, the latter near a soli-lunar cycle that may have influenced western zonal atmospheric flow. In the Potomac River watershed, Cook and Jacoby (1983) extended the PDSI index to reconstruct paleoriver discharge back to 1730. Eighteenth-century streamflow oscillated only weakly between 1730 and 1820, when coincident with Arctic and North Atlantic climate changes, the pattern shifted to higher amplitude variability with a pronounced low-flow interval of 1850–1873.

Cook and Mayes (1987) analyzed 74 tree-ring climate records from eastern North America for the period 1700–1972. They found anomalously abrupt changes in tree growth at 1810 A.D. and 1880 A.D., with the period 1827–1835 A.D. showing the most persistent high growth. Records of drought and cyclones from 1902–1908, an interval of similar climate and enhanced tree growth, indicated that the early nineteenth century tree rings reconstructed faithfully represented regional patterns of precipitation. With an abrupt shift at about 1837–1847 A.D., the region entered a distinct climatic phase that ended in 1880, coincident with other proxy records for the end of the Little Ice Age. This interval was characterized by cool, dry, stable air masses over much of the eastern United States.

Stahle et al. (1988) examined a 1614-yr record of rainfall from bald cypress from North Carolina and recognized alternating wet and dry periods of about 30-yr duration during the Medieval Warm Period. They also found interdecadal variation during the Little Ice Age. In general, however, the Little Ice Age between about 1300 and 1600 A.D. was relatively wet and ended in the time 1650–1750 A.D., when drier summer conditions prevailed. Stahle and Cleveland (1994, 1996) summarized work on a growing network of southeastern U.S. bald cypress tree-ring records extending from Virginia to Georgia and west to Arkansas and Louisiana. Eleven records extend back to about 1200 A.D. The spring rainfall record confirmed earlier conclusions that interannual variability changed significantly during the Medieval Warm Period and Little Ice Age but that there were no sustained century-long periods of climate wetness or dryness. Instead, the evidence points to extremely wet years in the late sixteenth and early seventeenth centuries that corresponded to widespread cold conditions elsewhere during the Little Ice Age. Perhaps more telling, Stahle and Cleveland (1996) concluded that interdecadal spring rainfall variability in the southeast is related, via tropospheric processes to decadal-scale climate anomalies in the Pacific Ocean measured in indices such as the Pacific–North American (PNA) index.

Other Regions

Serre-Bachet and Guiot (1987) studied tree-ring records from the Alps and the Mediterranean for the period 1172–1972 and reviewed the literature for the Mediterranean and North Africa. They found clear evidence in both Alpine and Mediterranean records for widely varying summer temperatures during both the Medieval Warm Period and Little Ice Age, but the low-frequency signal seemed to indicate relatively warm temperatures before 1335 A.D., a period of cooler climate until 1456 A.D., and another cooling through the duration of the Little Ice Age until about 1860 A.D.

Fire Scars

Paleoclimatic interpretation about precipitation and temperature can be deduced from analysis of fire scars in some trees because forest fires produce lesions in tree rings and short-term growth surges are often associated with fire scars. Swetnam (1993) discussed a 2000-yr fire-scar record of giant sequoia trees from California. By matching fire scars with nineteenth-century historical records of forest fires and correlating scars with tree-ring widths, Swetnam found that precipitation was the most important factor over time scales of years but that temperature was more important over decades to centuries. He found that the relative size of the fire increased exponentially as the time interval between fires lengthened, presumably because fuel accumulated on the forest floor. In addition to interdecadal variation in fire scars, he identified three broad intervals of fire-scar climate history: 500–1000 A.D. and after 1300 A.D. were intervals of few widespread fires of high intensity, and the intervening period 1000–1300 A.D. saw many small, low-intensity fires. Swetnam's results also led him to revise ideas on sequoia ecology, pointing out that climate-related fire history is at least as critical as soil, topography, and other local factors in controlling the evolution of sequoia-forest dynamics in California.

Ice Cores

Polar (Greenland, Devon Island, Canada, and Antarctica) and low-latitude ice cores (Andes Mountains, Peru, and Tibetan Plateau, China) have provided a wealth of information on paleo-atmospheric evolution. Chapter 9 covers the methodology and results of the flourishing field of ice-core research; here the discussion is limited to ice-core studies addressing climate change during the past few thousand years.

The advantages of ice cores are their annually resolved chronology obtained from distinct seasonal layering and the multiproxy data sets that can be extracted from ice and trapped air within the ice, yielding information on temperature, precipitation, atmospheric circulation, trace gas chemistry, and other parameters. As with any paleoclimate proxy method, many complex processes together produce the final signal in ice cores. The major disadvantage of ice cores is that there are relatively few areas of the world where large ice sheets and ice caps occur, limiting the potential spatial coverage for global paleo-atmospheric reconstructions.

Low-Latitude Ice Cores

In a series of studies of high elevation, low-latitude ice cores from Quelccaya ice cap, Peru (Thompson et al. 1979, 1985, 1986); the Huascarán ice cap of Peru (Thompson et al. 1995b); and the Dunde (Thompson et al. 1988, 1989) and Guliya ice caps, Qinghai-Tibetan Plateau in China (Thompson 1996; Thompson et al. 1995a, 1997) have provided a wealth of information on low-latitude climate change during the past few thousand years (figure 6-5). On the eastern flank of the Andes Mountains, in the western Amazon Basin at 5670 m above sea level, ice accumulates on the Quelccaya ice cap at a rate of about 1.5 m/yr. The 180 m of ice yielded a 1500-yr record of precipitation derived from the Atlantic Ocean to the east and atmospheric dust from the dry Altiplano to the west (Thompson et al. 1985, 1986).

Using microparticle content and electrical conductivity to measure soluble impurities in ice meltwater and $\delta^{18}O$ of the ice, Thompson's group showed that climate changed during the Little Ice Age, between 1500 and 1880 A.D. The oxygen isotope values were about 0.9‰ lower, and conductivity and microparticle content were both 30% higher during this time span than during the fourteenth, fifteenth, and twentieth centuries. An abrupt climatic termination about 1880 was characterized by sharply reduced wind and increased snow accumulation, signifying greater amounts of precipitation. Dry conditions were also recorded in the isotope record during the periods 1928–1947, 1864–1905, and 1452–1550 A.D.

Thompson (1996) compared net-accumulation-rate trends for the Guliya and Quelccaya ice caps for the past 2000 yr and found three periods of high accumulation and wet climate: before 1000 A.D., 1400–1775 A.D., and 1900 to the present. Extended dry periods occurred at 1075–1375 A.D. and 1775–1900 A.D. They noted that the striking parallels between these two ice-core records, located 20,000

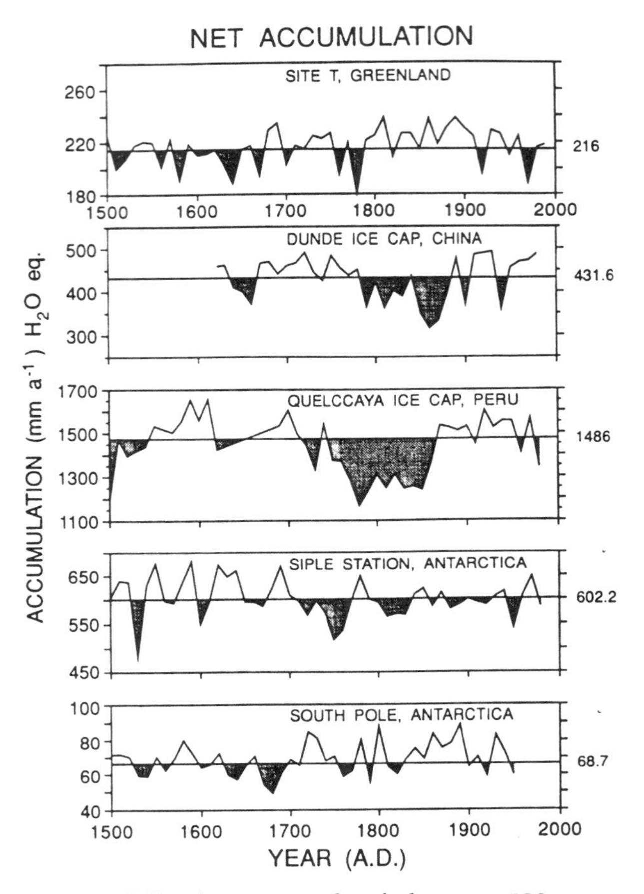

FIGURE 6-5 Climate trends of the past 500 years recorded in changes in net ice accumulation in Greenland, Tibetan Plateau (Dunde, China), Peruvian (Quelccaya), Antarctic Peninsula (Siple), and South Pole Antarctica ice core records. Periods of low accumulation characterize certain intervals of the Little Ice Age, but there is no simple uniform period of cooling between 1600 and 1900. Courtesy of E. Mosley-Thompson with permission from Elsevier Publishing. From Mosley-Thompson et al. (1993).

km apart, provides a convincing argument for an Asian–South American climate teleconnection to explain contemporaneous centennial-scale patterns over the past two millennia. Whereas Pacific-wide interannual climate teleconnections have been known for decades as the Southern Oscillation atmospheric part of the El Niño–Southern Oscillation (ENSO) phenomenon, the Peruvian and Chinese ice-core records provide the first firm evidence for the equivalent of a century-scale atmospheric linkage.

Polar Ice Cores

The earliest studies of Greenland ice cores already showed evidence for regional Little Ice Age cooling in the Northern Hemisphere (Dansgaard et al. 1971). More recently, climate records from ice sheets in Greenland (Meese et al. 1994) and Antarctica (Mosley-Thompson et al. 1993; Cole-Dai et al. 1995) have shed considerably more light on late-Holocene polar climate, indicating a much more complex climatological response to the Medieval Warm Period and Little Ice Age.

One of the most important results to emerge from an interhemispheric comparison of net accumulation rate (a good indicator of precipitation) of ice in Greenland, Antarctica, Peru, and China over the past 500 years, is that there is no obvious decadal-scale synchroneity. Mosley-Thompson et al. (1993) stress that precipitation is a climatic variable that would be expected to modulate widely over interhemispheric distances. Moreover, a second conclusion emphasized by Mosley-Thompson et al. (1993; Mosley-Thompson 1996) is the "antiphase" relationship between east and west Antarctica as exhibited in the oxygen isotopic, dust, and accumulation-rate trends from South Pole and Siple (Antarctic Peninsula) ice cores. At Siple, the interval 1600–1830 A.D. was especially warm and less dusty than today, the opposite signal that one might expect during the Little Ice Age. Conversely, at the South Pole, parts of the Little Ice Age interval were colder than today. Cole-Dai et al. (1995) studied anion glaciochemistry (chloride, nitrate, sulfate) from an additional core from the Dyer Plateau, Antarctic Peninsula, and confirmed that there may have been an increase in accumulation during the Little Ice Age, but there was no apparent long-term trend in anion concentrations between 1509 and 1989. A preliminary generality stemming from these results is that zonal climate patterns across the Pacific Ocean seem to change in phase, at least on centennial time scales but that north-to-south pole, pole-to-equator, and even intrapolar climate comparisons reveal complex phasing and asymmetries at shorter time scales, depending on the proxy record.

Peel et al. (1996) compared ice core stable isotopic records from the Antarctic Peninsula adjacent to the Weddell Sea with the glacial record from the South Orkney, South Shetland, and South Georgia

Islands and Patagonia. They found evidence for an isotopically cool period about 1750–1820 and a warm period in the early eighteenth century, most likely related to changes in atmospheric precipitation and shifting moisture sources. Although the exact nature of the ice-core isotopic fluctuations is complex and remains poorly understood, in general the variations correspond with well-dated glacial advances at 1750–1850 A.D. on Signy Island. Peel et al. also cautioned that ice core sites located in higher elevations in the Antarctic Peninsula seem to be better monitors of large-scale climate trends and that lower-elevation sites are influenced greatly by local factors.

The evolution of trace gases over the past millennium has provided a critical baseline against which to compare levels elevated owing to anthropogenic activity. Several papers provide important reviews of the reliability of ice cores for tracking natural levels of trace gases. For the past millennium, the best evidence so far indicates that CO_2 concentrations fluctuated approximately 10 ppmv about a mean of approximately 280 ppmv; methane showed changes of approximately 70 ppbv. Barnola et al. (1995) described the evidence for natural variability in CO_2 from Greenland and Antarctica and concluded that the Antarctic record of CO_2 is essentially an accurate record, unaltered by chemical processes within the ice. In contrast, the integrity of the Greenland CO_2 record may be compromised because of chemical impurities, altering the original concentration by an estimated 20 ppmv. The atmosphere in the Northern Hemisphere generally contains more particulate material than the Southern Hemisphere. When it is incorporated into Greenland ice, this dust can chemically alter the original air chemistry. Barnola's new data supported the preliminary results of Siegenthaler et al. (1988) that there may have been a medieval (end of the thirteenth century) increase in CO_2 of about 10 ppmv. It is as yet unclear how this oscillation is related to climatic conditions of the Medieval Warm Period, but this natural rise in CO_2 might represent a small imbalance in the global carbon cycle equivalent to about 0.3 gross tons carbon per year.

Blunier et al. (1993) carried out an interlaboratory comparison of methane concentrations in a 1000-yr record from an ice core from Greenland (figure 6-6). They established that the "natural" preindustrial baseline levels of methane vary around a global mean of 700 ppbv but that during the interval 1100–1200 A.D. methane concentrations were as much as 50 ppbv higher than during the succeeding period 1250–1450 A.D. Glacial-age methane concentrations were around 400 ppbv. Blunier and colleagues attributed at least part of the abnormally

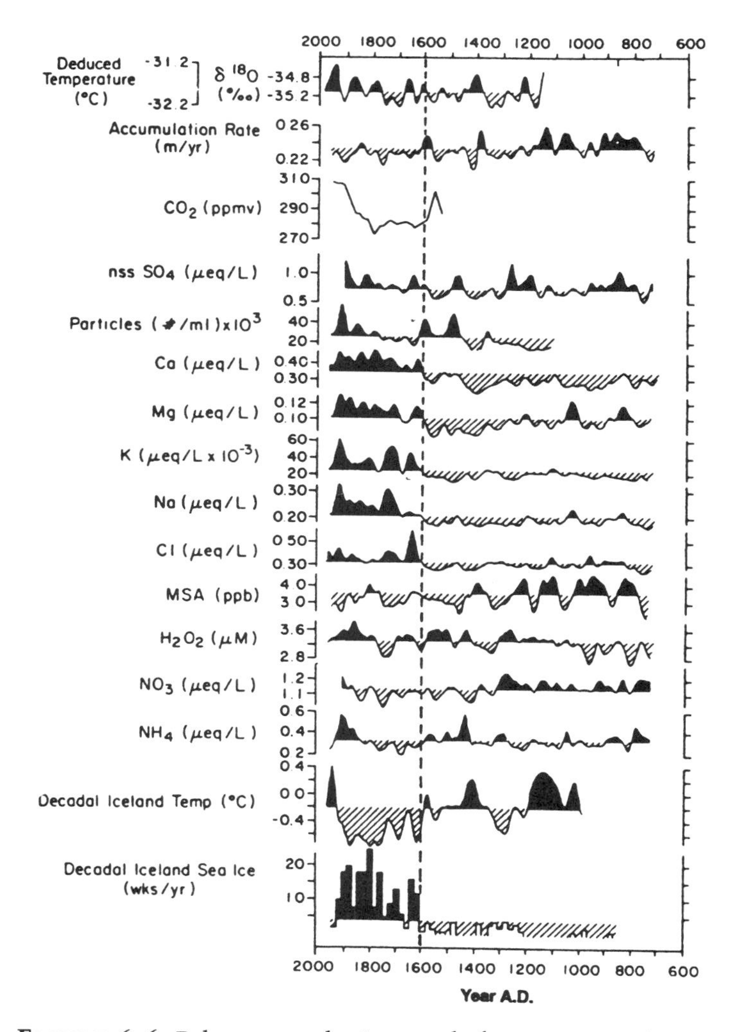

FIGURE 6-6 Paleo-atmospheric record of temperature, ice accumulation, and other glaciochemical indicators from the past 1400 yr from the GRIP ice core. Courtesy of P. M. Grootes and GISP Management Office. From Grootes (1995).

high medieval methane concentrations to increased population growth and rice cultivation (a major source of methane production) in China.

Raynaud et al. (1996) reviewed the record of radiatively active trace gases from the Holocene as measured from six ice cores, five in Antarctica, one in Greenland. They concluded that CO_2 concentrations dropped from 280 to 250 ppmv between 10 and 8 ka and rose again to preindustrial levels of 280 ppmv by 5 ka. Similarly, CH_4 concentrations decreased by 150 ppbv about 8.2 ka before rising to their preindustrial levels of 700–750 ppbv. Changes in the global carbon cycle after the later stages of deglaciation and in variable wetland methane production in Northern and Southern Hemispheres may explain these fluctuations.

Ice-core trace-gas records have been successfully spliced with instrumental records from the past few decades to produce a fairly continuous and internally consistent record of CO_2 and CH_4 concentrations for the past few centuries (Prather et al. 1995). In addition, trace-gas oscillations during the past millennium may in fact be *responses* to high-frequency climate changes (relative to low-frequency glacial-interglacial climatic cycles) or other factors that influence global and hemispheric CO_2 and CH_4 budgets. Trace gases may not necessarily be the causes of low-amplitude climate changes. Research aimed at sorting out the various forcing factors is still needed before conclusions about the cause of these climate changes can be reached.

In summary, both low-latitude and polar ice cores offer substantial data on centennial-scale climate oscillations during the past 3000 years. The greatest contributions of ice core paleoclimatology to decadal- and centennial-scale climate changes will come in the not-to-distant future.

Coral Scleroclimatology

Hermatypic (reef-building) coral skeletons contribute to late-Holocene paleoclimatology of shallow-water tropical regions in the subdiscipline called scleroclimatology. Although most coral paleoclimatology has been focused on ENSO interannual variability discussed in the next chapter, several long coral records extend to the seventeenth century.

The aragonitic skeletons of corals capture geochemical signals of climatatic significance. Three main proxies can be extracted from corals: cosmogenic ^{14}C production as a measure of solar activity; stable isotopes (oxygen and to a lesser extent carbon), which are a record of water temperature and, to a lesser extent, precipitation; and trace elements (e.g., strontium, magnesium, barium) which are geochemical tracers of ocean properties or temperature.

Druffel (1982) discovered an increase during the early seventeenth century in the $^{14}C/^{12}C$ ratios in corals from the Florida Straits that seemed to reflect an increased ratio in the atmospheric source of ^{14}C. The temporary ^{14}C increase supports evidence from tree-ring cellulose and observational records for a reduced level of solar activity during the Maunder sunspot minimum. Druffel also noted that oxygen isotopic data from the same corals suggested a slight sea-surface temperature (SST) drop in the Florida Straits and nearby Sargasso Sea. Evidence for elevated early eighteenth century $^{14}C/^{12}C$ from independent sources such as subtropical corals and trees supports the idea that at least some of this variability, about half according to Stuiver and Braziunas (1993), is due to solar modulation of ^{14}C flux.

Oxygen-isotope records have been used to document SST history back to about 1607 A.D. for corals from Isabela Island in the Galapagos (Dunbar et al. 1994) and to 1635 A.D. for a colony from Abraham Reef in the Great Barrier Reef. The Galapagos and Great Barrier Reef temperature records show contrasting patterns for the period 1700–1850: in the Galapagos, relatively warm SSTs characterized the 1700s, but the Great Barrier Reef temperatures were slightly cooler than average. Dunbar et al. found that the Galapagos oxygen isotope record indicates cooler conditions during 1600–1660 and 1800–1825 A.D. Cook (1995) also cites data from Great Barrier Reef coral growth banding records—also considered a temperature proxy—indicating cooler temperatures during much of the interval between 1600 and 1800 A.D.

In addition to SST records, Linsley et al. (1994) obtained a long record of oxygen isotopes from corals from the Pacific off Central America, a location where the oxygen-isotope signal is influenced less by temperature variability and more by sea-surface salinity (SSS) variability due to fluctuations in seasonal and interannual rainfall. They examined high- and low-frequency isotopic oscillations and found that several large-amplitude oxygen-isotope variations occurred during the Little Ice Age but that the pattern of high-frequency oscillations in precipitation did not change from the Little Ice Age through the twentieth century. One important conclusion was that in the Panamanian coastal region mean precipitation, SST, or both have increased since the early 1800s.

Two points about the coral record of the past few hundred years should be stressed. First, many more records will be required to develop a fuller understanding of centennial-scale Pacific Ocean temperature history. Second, these long-term trends are superimposed on shorter oscillations associated with the quasicyclic ENSO discussed in the next chapter.

Lacustrine Sediments

Paleolimnological records of Holocene climate obtained through geochemical and micropaleontological analyses of lake sediments can serve as rain gauges to indicate changes in lake levels due to precipitation variability regulating the inflow of fresh water. For instance, McKenzie and Eberlie (1987) used oxygen isotopic data to produce a Holocene climate record from the Great Salt Lake in the western United States. After a playa stage of extremely dry climate, the generalized record of the past 5.5 ka showed three periods of high lake levels and two intervening periods of low levels, which McKenzie and Eberli suggested may represent low-frequency (2.5 ka) quasicyclic climate variability.

The multidisciplinary investigations of the Holocene history of Elk Lake (Bradbury and Dean 1993) is a premier example of how to reconstruct regional-scale climatic variability over centennial time scales. Anderson (1993) studied the late-Holocene record of Elk Lake, Minnesota, where he uncovered approximately 200-yr cycles of variability in sedimentation and lacustrine environmental conditions. These cycles can be linked to solar cycles found in some tree-ring and marine climate records (see next section), with the usual caveat that the mechanism of climate change due to solar influence is not clear at this time. Dean (1997) found dominant periodicities of 400 and 84 yr in the Elk Lake record of bulk-sediment aluminum and other elements for the past 1500 yr. He concluded that Little Ice Age climates in Minnesota were dry and windy, similar to but not as intense as the mid-Holocene climates of central North America. Although the Elk Lake record suggested links between the mid- to late-Holocene North American climatic and solar activity, Dean considered the connection to be speculative.

Laird et al. (1996) studied the Holocene climate history through diatom analysis from Moon Lake in Missouri. During periods of dry climate, lake salinity rises and the lake harbors a characteristic high-salinity diatom flora; the converse occurs during periods of wet climate. Laird et al. recognized four hydrological periods in the Holocene record of Moon Lake: an early Holocene transition from an open freshwater to a closed saline system (by 7.3 ka), a mid-Holocene period of high salinity (7.3–4.7 ka), a transitional period with poor diatom preservation (4.7–2.2 ka), and the late-Holocene period of variable lower salinity. The lowest Moon Lake salinities since the last glacial-Holocene transition occurred during the Little Ice Age. The past 2.2 ka of climate at Moon Lake was characterized by fluctuating moisture conditions. Laird et al. compared late-Holocene records from other parts of central North America to the Moon Lake record. They concluded that there was significant temporal and spatial variability throughout this part of North America over the past millennium. Nonetheless, the Medieval Warm Period was generally a period of drought and, in contrast, the Little Ice Age was one of moist, cool climate.

Lakes also preserve important Holocene pollen records of vegetation changes during the past few millennia. Campbell and McAndrews (1993) studied a 1000-yr pollen record from sediments from Crawford Lake, southern Ontario, to compare actual vegetation history during the Little Ice Age with models that simulate forest growth after a disturbance. Both the Crawford Lake pollen record and model-simulated vegetation history reflected similar responses to a 2°C Little Ice Age cooling. Campbell and McAndrews discovered that, for 650 years between about 1200 and 1850 A.D., southern Ontario forests remained in ecological disequilibrium. This was a much longer time period for forest response to disturbance than previously modeled (e.g., Davis and Botkin 1985). A strong Little Ice Age signal is apparent in the vegetation record in Ontario for several reasons, including the ecotonal nature of the region, abundant pollen of key tree types, and minimal fire disturbance. Campbell and McAndrews's evidence for combined seral succession and climatic impact from rapid cooling emphasizes the importance of distinguishing ecological and climatic influences, especially over centennial time scales.

Marine Sediments

Marine sediments contain a potentially superb record of oceanographic history related to Holocene climate change. I discuss here selected examples of three separate types of marine records containing evidence for decadal- to centennial-scale Holocene climate change: laminated sediments from semi-enclosed silled basins, often preserving interannual records; deep-sea sediment drift regions, providing a record of open ocean variability; and the continental margin of Antarctica, yielding a record of Southern Hemispheric ice shelf–glacier–ocean interactions.

In chapter 5, high-resolution, late Quaternary oceanographic records obtained from semi-isolated silled marine basins revealed important rapid climatic changes in tropical regions (e.g., Hughen et al. 1996). An equally refined temporal resolution can be achieved from silled marine basins near continental margins. Silled basins often experience seasonal oceanographic changes, including bottom water anoxia or hypoxia, such that aerobic respiration is absent or reduced. Without bioturbation of sediments resulting from benthic activity, varves (annual sedimentary layers) often characterize the sedimentary record of silled basins, providing a high-resolution record of climatic and oceanic variability analogous to that obtained from annual layers in tree rings and ice cores. However, the paleoceanography of anoxic basins is extremely complex and requires an in-depth understanding of the biological, chemical, and physical processes that govern sedimentation in each basin.

The Santa Barbara and Santa Monica Basins off California, Guaymas Basin in the Gulf of California, and the Cariaco Basin off Venezuela are excellent examples of well-studied marine basins yielding detailed Holocene climate records. Hagadorn et al. (1995) studied the Santa Monica Basin record, in which they found a dominant decadal scale variability in bottom-water oxygen depletion. Using multiproxy indicators of surface and bottom water conditions ($\delta^{13}C$, $\delta^{18}O$, benthic foraminifers, total organic carbon, and X-ray density of light-dark banding of sediments), they proposed that organic flux to the bottom was due to oscillating surface-water productivity, coincident with SST changes. The formation of laminated sediments was influenced mainly by primary surface productivity and, to a lesser extent, by changing terrigenous influx. The two dominant frequency bands were 26 and 15 years. Interestingly, the decadal scale trends in the Santa Monica Basin were more prominent that the weak ENSO signal observed in historical records of the region. Hagadorn et al. suggest that a 3- to 6-yr ENSO signal might be more difficult to discern in California basins than previously believed (Kennedy and Brassell 1992). In any case, the Santa Monica Basin provides evidence for decadal-scale variability in oceanic conditions that potentially can be correlated with large-scale Pacific Ocean climate trends.

Is there evidence for anomalous Little Ice Age climate variability in marine sediments? Dunbar (1983) found foraminiferal isotopic evidence in varved sediments from the Santa Barbara Basin for sustained periods of relatively cool oceanic conditions (especially a 30-yr period before 1780, the period 1850–1880, and the interval 1895–1915). Cooler surface-ocean conditions can be attributed to an intensification and equatorward shift in trade winds resulting in enhanced upwelling of cooler waters. Noting that the Little Ice Age was not a single cool isotopic event but rather a series of periodic excursions, Dunbar found a pattern of high-frequency climate variability within the Little Ice Age interval, generally consistent with many recent terrestrial records that also indicate that this period was not one of sustained cool climate.

In the well-studied Guaymas Basin in the Gulf of California, Schrader and Baumgartner (1983) used silicoflagellate populations to identify six major stages of increased surface-ocean productivity over the past 500 years: 1490–1520, 1570–1580, 1660–1700, 1770–1810, 1840–1850, and 1920–1930. Silicoflagellates are reliable proxies of the varying influence of cool California and warmer Equatorial currents and atmospheric conditions that induce upwelling. The six periods of high productivity visible in four sediment cores corresponded to periods of dry climate, enhanced northerly winds, and oceanic upwelling. A comparison of the Guaymas Basin record with the continental tree-ring record from the nearby Sierra Madre occidental showed an inverse relationship between tree-ring width and productivity. That is, narrow tree rings, signifying dry regional climates, were coeval with strong upwelling and productivity in the Gulf of California; more humid continental climates matched periods of reduced productivity. For some intervals, sedimentological and benthic foraminiferal trends and inferred bottom dissolved oxygen levels also varied over the past few centuries. In general, the silicoflagellate record supports the notion that high-amplitude oceanic oscillations occurred during the Little Ice Age interval.

Even finer intra-annual records of oceanic variability can be obtained in some marine laminated sediments. Sediment-trap data showing seasonal sedimentation and biotic processes in the Gulf of California define a complex five-stage series of events that can be identified and applied downcore to produce interdecadal climate change history. Thunnell et al. (1993) showed that during stage 1, high lithogenic flux occurs during summer and autumn due to heavy rainfall. This is followed by a biogenic opal flux maximum during early winter, when a mixed diatom assemblage, dominated by the diatom genera *Stephanophyxis* and *Rhizosolenia,* is deposited. This stage is followed quickly by a *Coscinodiscus*-dominated layer. During winter a mixed diatom flora is deposited, followed by a nearly monospecific assemblage consisting of *Thalassiothyrix longissima* deposited during, or just after, coastal upwelling of the following spring.

The life history of *T. longissima* provides a remarkable tool for ultrarefined temporal resolution in paleoceanography. This diatom forms distinct diatomaceous mats that in some taxa are known to regulate their own buoyancy and migrate through the water column, apparently to take advantage of nutrient availability. Pike and Kemp (1996, 1997) studied these diatomaceous mats using back-scattered electron imagery (BSEI). By identifying 119 separate mats in a 50-cm-long section of core deposited over a 289-yr interval between about 9200-9500 yr, and with knowledge of the seasonal sedimentation and diatom biology, they could determine interannual oceanographic variation in the biogenic and lithogenic pulses. The minute layers of diatomaceous mats in the sediment core signify incursions of Pacific surface water into the Gulf of California. A different number of mats, ranging from zero to four, formed each year. Pike and Kemp (1997) inferred that the periodicity of the number of annual diatom mats could be used to infer climatic history of the early Holocene. Their results revealed 11-, 22-, and 50-yr periods in mat frequency. Pike and Kemp (1997) reviewed a large literature from lake sediments, other Guaymas Basin sedimentary records, lake records, and tree rings and historical data, which also suggested that 11- and 22-yr frequencies were common in Holocene paleoclimate records.

The obvious similarity of the Guaymas microstratigraphic and diatom record to 11-yr Schwabe and 22-yr Hale solar cycles raises the possibility that these biogenic pulses may ultimately be linked to climate-related solar activity. Pike and Kemp also point out that an additional 50-yr diatom mat periodicity seems to match the frequency found in California fish population cycles that have been linked to decadal-scale climate and oceanographic variability. The precise mechanism linking solar and ocean processes is still unknown.

Deep-sea sediments deposited in open-ocean settings are perhaps the most underexploited source of decadal- and centennial-scale climate variability, in part because sedimentation rates are often too low to obtain the requisite sampling resolution. Such is not the case, however, in sediment-drift regions where rates of sediment accumulation reach 100–200 cm/1000 yr, rivaling rates on continental margins. Keigwin (1996) examined carbonate percent and planktonic foraminiferal isotopes from Bermuda Rise sediments in the northern Sargasso Sea to obtain a record of North Atlantic Ocean centennial-scale oceanic change for the past three millennia. The Sargasso Sea region provides an oceanic record that may be representative of basin- or hemisphere-scale events. Having established that late glacial and Holocene oceanic variability is quasiperiodic with a 4000-yr cycle

(Keigwin and Jones 1994), Keigwin made two startling discoveries with broad implications for high-frequency paleoceanography. First, evidence exists for a distinct carbonate minimum centered about 400 years ago, in the middle of Little Ice Age, preceded by a carbonate maximum about 1000 years ago corresponding to the Medieval Warm Period. The Little Ice Age carbonate minimum, the largest in the past 10 ka, was attributed to enhanced terrigenous influx possibly related to North Atlantic climatological factors such as increased storminess that might have had an impact on Gulf Stream circulation. It is uncertain whether changes in North Atlantic deep water influenced the Bermuda Rise record of the Little Ice Age in a manner similar to that of past glacial intervals.

A second discovery was that the Little Ice Age and Medieval Warm Period were characterized by surface oceanic cooling of 1°C and warming of equal magnitude, respectively. To reconstruct SSTs, Keigwin calibrated the oxygen-isotope record of *Globigerinoides ruber* to historical SST records and found that $\delta^{18}O$ minima were centered about 500, 900, and 1100 yr B.P. His isotopic estimates of SSTs 1500–1700 yr B.P. were relatively cool, about 22°C, then rose to 23.5–24.0°C between 900 and 1100 yr, and fell again to 22.0–22.5°C between 300 and 500 yr. These temperature oscillations of 1.5–2.0°C are twice the amplitude of those recorded over the past few decades. Moreover, these trends indicate that the twentieth century regional oceanic warming of 0.5°C is not unprecedented in the past three millennia in the Sargasso Sea area.

What caused this large-amplitude, centennial-scale oceanic variability? Though much uncertainty remains, the Bermuda Rise pattern of Little Ice Age ocean cooling is consistent with climatological and oceanographic conditions that characterize periods when the North Atlantic Oscillation (NAO) index attains minimal values (Hurrell 1995). Low North Atlantic Oscillation indices, signifying shifting storm tracts, could represent climate conditions analogous to those that prevailed during the Little Ice Age. Because higher-frequency (decadal) variability is as yet not available from North Atlantic sediments, direct comparison with recent trends is difficult. Whatever the cause, Keigwin's study demonstrated that the Bermuda Rise and other high-sedimentation-rate regions hold clues to large-scale climate events over the past few millennia.

Whereas ice core records are usually cited as a primary source of century-scale Holocene atmospheric changes over Antarctica, there is also excellent sedimentological, geochemical, and micropaleontological evidence for correlative variability in oceanic, sea ice, and glacial conditions. On the western side of the Antarctic Peninsula, Domack et al. (1993), Leventer et al. (1996), and Domack et al. (1995) studied the late Holocene record from Andvord Bay, Palmer Deep, and the Muller Ice Shelf, respectively. In Andvord Bay, a fjord located off the northwest Antarctic Peninsula, 12 cycles of total organic carbon and biogenic silica, proxies of oceanic productivity related to sea-ice conditions, characterize the past 3000 yr. During the approximately 278-yr cycles, minima in total organic carbon and silica deposition characterized the relatively cool climate intervals. Domack et al. (1993) found the lowest minimum was dated at 330 yr B.P., during the Little Ice Age.

Leventer et al. (1996) also found a 200- to 300-yr periodicity in magnetic susceptibility and diatom assemblages in Palmer Deep basins near Andvord Bay. High magnetic susceptibility corresponds to *Chaetoceras*-dominated diatom assemblages; low magnetic susceptibility has a more variable assemblage. The preservation of distinct diatom assemblages results from a complex set of biological factors, including distinct life histories of the mat-forming diatom *Rhizosolenia,* production of resting spores versus vegetative cells in *Chaetoceras,* and the production of gametangia in *Corethron,* as well as oceanographic and atmospheric processes related to sea-ice conditions and winds. In general, past periods of high productivity were most likely driven by meltwater-induced stabilization of the ocean water column. Leventer et al. (1996) documented approximate 200-yr cycles in diatom assemblages and magnetic susceptibility, which they attributed to solar-induced regional climate changes, noting that similar fluctuations have been recorded in other laminated sediments such as the Cariaco Basin off Venezuela (Peterson et al. 1991).

A third late Holocene Antarctic climate record was derived from sedimentological, geochemical (total organic carbon), and benthic foraminiferal assemblages obtained from the Lallemond Fjord on the Antarctic Peninsula. Domack et al. (1995) discovered evidence that a large advance of the Muller Ice Shelf about 400 years ago was coincident with the inception of Little Ice Age conditions and coeval with other glacier advances in the Southern Hemisphere (Clapperton 1990). Domack proposed that, during this Little Ice Age event, the Muller Ice Shelf advanced into Lallemond Fjord despite relatively warmer atmospheric conditions in the region (evidence from the Siple ice core discussed earlier). They argue that the apparent conflict posed by an advancing ice shelf and warm oceanic surface conditions can be explained by changes in circumpolar deep water. In contrast to modern conditions, in which circumpolar deep water enters Lallemond Fjord, the Little Ice Age may have been a period when relatively warm circumpolar deep water was excluded from the Lallemond Fjord, leading to decreased basal melt rates of the ice shelf and an ice shelf advance. If their hypothesized climatic reconstruction is correct, then a complex situation arises in which an advancing ice shelf coincides with altered deep-oceanic circulation at a time of regional atmospheric warmth over the Antarctic Peninsula, but cooler temperatures over the central Antarctic continent as recorded by the South Pole ice core isotopic record.

Taken together, the Antarctic Peninsula oceanic and glaciological records and the continental ice core record provide multiproxy evidence of temporally and regionally variable climatic instability over the past few thousand years. With recent improvements in knowledge of the ecology of diatom and benthic foraminiferal species that are most commonly preserved in southern ocean sediments, future studies will undoubtedly improve the Holocene paleoclimate record from these key areas.

Global Correlations of Centennial- and Decadal-Scale Climate

Mann et al. (1996) recently made one of the first quantitative studies of global-scale climate variability at centennial to decadal time scales. They used both paleoclimate proxy data from several of the dendroclimatic, scleroclimatic, ice core, and varved sedimentary records described earlier and data from the historical temperature record. They specifically addressed the issue of whether decadal scale (15–35 yr) and centennial-scale (50–150 yr) oscillations, which on the basis of climate modeling are postulated to emanate from internal oscillations of the climate system, could also be identified in paleoclimate records. They concluded that consistent interdecadal-scale climate variability was most evident in the paleoclimate proxy records from tropical and subtropical regions. In contrast, high-amplitude centennial-scale climate oscillations were confined mainly to the Arctic and North Atlantic region. Additional hemispheric and global comparisons of high-frequency Holocene records will undoubtedly appear in the next few years as new records are obtained. In the meantime, the results of Mann et al. are very encouraging that the forcing factors of recent climate events can be unraveled.

FORCING MECHANISMS OF CENTENNIAL AND DECADAL HOLOCENE CLIMATE

WHAT ARE THE causes of centennial and decadal climatic variability during the Holocene interglacial? Five mechanisms are considered to be the most likely candidates (Rind and Overpeck 1993).

Explosive Volcanism

Volcanic eruptions can send large quantities of particulate aerosols into the stratosphere, having a cooling effect on the earth. Historically, one of the most famous was the 1815 eruption of Tambora, in the Pacific Ocean, which led to the famous year without a summer in Europe in 1816. The Tambora eruption can be identified by a sharp spike in sulfate and dust content in ice cores from Peru, Greenland, and Antarctica. The dramatic eruption of Mount Pinatubo in the Philippines in 1991 gave atmospheric scientists a chance to observe in detail the global effects of a giant eruption.

Whereas individual eruptions influence climate for a period of several years, extended periods of increased explosive volcanism have been proposed as a cause of late Holocene cooling (e.g., Porter 1986). The Crete, Greenland, ice-core record, for example, has clear spikes in electrical conductivity (Hammer 1977) that indicate that volcanic activity may have influenced climate of the past few thousand years. Bradley (1987) evaluated the impact of explosive volcanism since 1851 and found that volcanic particulates have a cooling influence on climate, but the temperature drop varies seasonally and latitudinally. The greatest impact (0.6-1.0°C) was felt in high northern latitudes during summer and fall.

Crowley et al. (1993) and Crowley and Kim (1993) argued, however, that other processes, such as the production of dimethyl sulfide by marine organisms, may play a more significant role in producing atmospheric particulates than previously thought. The atmospheric effects of volcanism in particular and other processes that produce sulfate aerosols in general are the subject of intense research (Delmas 1993). In brief, many researchers tend to discount a major role for volcanism in producing extended periods of climate cooling over decadal and centennial time scales during the Holocene, although individual volcanic events certainly can have a major short-term impact (Overpeck et al. 1997).

Solar Variability and Climate Change

The solar engine that drives earth's climate system is an obvious potential source of climate variability (figure 6-2). Indeed, in their chapter entitled "Cyclomania," Hoyt and Schatten list almost 40 separate solar-induced cycles that have been reported in the literature. Cycles ranging in frequency from 6.6 days to 334 years supposedly influence climate or climate-related processes like thunderstorm frequency and precipitation. Among the solar-related climate cycles that appear most commonly in paleoclimate records of the past few millennia are the 11-yr Schwabe sunspot cycle, the 22-yr Hale "double sunspot" or solar-magnetic cycle, and 80- to 90-yr Gleissberg sunspot envelope-modulation cycle, and the 180- to 206-yr Suess cycle (Stuiver and Braziunas 1993; Jirikowic and Damon 1994). The central question in the solar-climate debate is: Did solar activity influence earth's climate over the past few millennia? The answer is probably. Though one cannot do justice to this large literature, in this section, I will try to capture the essence of the solar-climate argument by citing sources of data, key studies, and representative opinions.

Several direct measurements of solar activity are used in climate studies: mean solar irradiance, aurorae, sunspot number and solar cycle length, sunspot structure and sunspot decay, and solar rotation (Hoyt and Schatten 1997). Data on these properties are derived from satellite observations over the past 20 years and historical records, which with few exceptions (e.g., Chinese records) cover at most a few centuries. Mean solar irradiance and solar cycle length are probably the most important indicators of solar activity. One way to estimate mean solar irradiance is by measuring two of the sun's surface features—bright faculae and dark sunspots—which rise and fall with an 11-yr Schwabe cycle (Lean et al. 1992). Changes in the number of faculae and sunspots are related to the sun's surface magnetism. During a "typical" recent 11-yr solar cycle, solar irradiance varied about 0.1% from maximum to minimum (Hoyt and Schatten 1993). Lean et al. (1992) estimated solar irradiance on the basis of combined reduced sunspots during the Maunder minimum equaling a 0.24% drop in solar irradiance. The value 0.24% is quite high, exceeding the modern 0.1% value for a solar cycle, but Crowley and Kim (1996) give several reasons why the Maunder minimum irradiance drop might have exceeded the amplitude of modern sunspot cycles.

Another measure of solar output is the length of the sunspot cycle. Hoyt and Schatten traced the link between solar-cycle length and climate back to turn-of-the-century observations. They reevaluated early work by Clough on variations in solar-cycle length compiled from aurora and Chinese records and recent work by Friis-Christensen and Lassen (1991), who found a strong positive correlation between that the length of sunspot cycles and climate. Hoyt and Schatten found that solar-cycle length seems to correlate moderately well with climatic reconstructions back to 1300 A.D.

Space-based solar monitoring and most historical ground-based measurements and observations are, however, not adequate to record the full range of solar variability that occurs over long time scales. Consequently, paleoclimate proxies such as ^{14}C and ^{10}Be have been used to estimate longer-term solar trends. Depending on the period of time under consideration, cosmogenic isotopic variability in geological and glacial records is believed to be caused not only by solar (heliomagnetic) processes but also by earth (geomagnetic) processes involved with changes in earth's magnetic dipole, oceanic circulation, carbon cycle, or a combination of these.

Modern paleoclimate research of the solar-climate link began with the classic papers by de Vries (1958), Suess (1965, 1968), and Stuiver (1965), who pioneered the study of paleosolar activity through the use of ^{14}C anomalies mainly in tree rings. Changes in cosmogenic isotopic concentrations, mainly ^{14}C derived from tree cellulose (Stuiver and Braziunas 1993; Stuiver 1994) and ^{10}Be from ice cores (Beer et al. 1988, 1994) have become the most reliable proxies of past solar variability. The cosmogenic isotope ^{14}C is produced in the atmosphere by cosmic ray bombardment and ultimately incorporated into organic material and ocean sediments through the following steps:

	Cosmic rays and neutron production	Oxidation			Photosynthesis	
^{14}N	$\Rightarrow$	^{14}C	$\Rightarrow$	$^{14}CO_2$	$\Rightarrow$	Terrestrial carbon
^{14}N	$\Rightarrow$	^{14}C	$\Rightarrow$	$^{14}CO_2$	$\Rightarrow$	Ocean carbon (dissolved, organic, carbonate)

It is estimated that 100 kg of ^{14}C is produced annually through cosmic-ray bombardment and neutron production. Carbon-14 decays at a rate of about 1% every 83 yr (Stuiver 1994). Changes in the earth's magnetic field can also influence ^{14}C production in the atmosphere.

Once ^{14}C is formed from nitrogen, it is oxidized into $^{14}CO_2$. Then, processes internal to the climate system, such as changes in ocean circulation and ocean photosynthetic productivity, influence the eventual ^{14}C concentrations found in earth's atmosphere, oceans, and marine (e.g., coral skeletons) and terrestrial (e.g., tree cellulose) organisms.

In contrast to ^{14}C, the most significant processes influencing the deposition of ^{10}Be in polar ice are short-term atmospheric processes such as wind transport, which can complicate the primary ^{10}Be signal resulting from solar and cosmic ray variation.

Eddy's (1976) proposal that the Maunder sunspot minimum may have caused Little Ice Age cooling sparked a resurgence of solar-climate research. Since Eddy's paper, many edited volumes have been published addressing the sun's role in climate change (McCormac 1983; Castagnoli 1988; Pecker and Runcorn 1990; Sonnett et al. 1991; Nesme-Ribes 1994). Eddy (1983) presents an entertaining historical essay on the solar-climate link, and Lean et al. (1995) and Hoyt and Schatten (1997) give recent reviews.

Some of the more influential papers on the late-Holocene solar-climate link are those by Foukal and Lean (1990), Lean et al. (1992, 1995), and Lean and Rind (1994) (figure 6-2). Lean and colleagues spliced 20-yr observational data from space-based radiometers with evidence for variable solar activity from sunspot number and cosmogenic isotopes. They concluded that during a period of historically cool climate like that of the Little Ice Age, global temperatures may have fallen up to 0.5°C, perhaps because solar insolation may have decreased by about 0.24% of the mean level for the period 1980–1986 (sunspot cycle 21).

Hoyt and Schatten (1993, 1997) and Crowley and Kim (1996) also support the hypothesis that solar variability might have influenced climate over the past few hundred years. Hoyt and Schatten (1997) build a case that solar ultraviolet variations are an important parameter that might influence earth's climate because (p. 100): "solar spectral irradiance fluctuations are proportionally larger at short wavelengths...and because they carry a significant fraction of the total solar energy (about 20% below 300 nm according to Lean in 1991)." Crowley and Kim (1996) suggest that 30–55% of the 0.5–0.6°C warming (omitting 0.2–0.3°C nineteenth century warming from human influence) that has occurred since the coolest part of the Little Ice Age interval, about 1700 A.D., can be explained by solar changes.

Using satellite-derived measurements from a method called active cavity radiometer irradiance monitor (ACRIM), Willson (1997) recently discovered an upward trend in the total solar irradiance of

0.036% per decade between the minima of solar cycle numbers 21 and 22 (the period 1980 to present). Willson concluded that such a rate of irradiance change occurring in the opposite direction over 200 years would be sufficient to account for the Little Ice Age cooling of about 0.4–1.5°C. Moreover, if the current trend is sustained, Willson surmises that solar irradiance could influence earth's future climate.

Paleoclimate research has shown that external (solar) and internal (oceanic) processes very likely act in concert to produce paleoclimate records of solar activity over different time scales. Stuiver and Braziunas (1993), building on earlier work (Stuiver and Quay 1980; Stuiver and Braziunas 1989; Stuiver 1993), quantitatively analyzed the past 10-ka ^{14}C record from five fir trees from the Pacific coast of the United States (figure 6-3). On the basis of the periodicity in ^{14}C, they proposed the following solar-ocean-atmosphere linkages: The 10- to 11-yr component (Schwabe cycle) was in part associated with external forcing by solar modulation of cosmic-ray flux, whereas multidecadal variability was more likely tied to changes in thermohaline circulation (see later), which caused changes in the rate of carbon storage in the oceanic carbon reservoir. The 206-yr cycle was attributed mainly to solar modulation and oceanic circulation. They also identified a 2- to 6-yr cycle related to ENSO ocean-atmosphere dynamics and a strong Younger Dryas (12.5–11.5 ka) increase in ^{14}C attributed to diminished North Atlantic deep water formation and global oceanic upwelling.

Examining oceanic records, Reid (1993) noted a positive correlation between ocean SSTs and long-term solar activity measured by sunspot number. For terrestrial records, Friis-Christensen and Lassen (1991) argued that instead of sunspot number, the length of the sunspot cycle provides a more valid means to examine the solar-climate linkage. High solar activity is associated with short sunspot-cycle length and vice versa. Friis-Christensen and Lassen demonstrated a statistical correspondence between sunspot cycle length and Northern Hemisphere surface land temperature for the past 130 yr and concluded that their results supported, but did not prove, the idea that solar activity influences earth's surface temperatures.

Anderson et al. (1992) has proposed that solar activity may be responsible for variability in the long-term centennial and decadal trends in the ENSO. As indicated by historical reconstructions of ENSO history, ENSO events have occurred in greater frequency in cycles of about 90, 50, 24, and 22 years since 622 A.D. Solar modulation of ENSO activity, according to Anderson, may have produced significant impacts on regional climate in the Pacific region.

Briffa (1994) takes a more guarded view on the role of solar activity and periods of cooler climates of the past millennium. In a paper entitled "Grasping at Shadows? A Selective Review of the Search for Sunspot Related Variability in Tree Rings," Briffa searched the tree-ring record from Scandinavia for the postulated 11-yr, 22-yr, and 80- to 90-yr cycles and the putative link between sunspots, solar variability, and Fennoscandinavian climate during the period 1700–1970. Briffa (1994:431) concludes: "These observations argue against any simple conclusions linking periods of low sunspot activity with generally cooler climate."

In sum, although secular variability in cosmogenic isotopes has been convincingly linked to solar activity, proving that this solar variability has caused climate to change is still difficult. Some scientists remain skeptical that mechanisms exists to translate the small (0.10–0.24%) magnitude of solar irradiance into a change in climate. Still, the emerging literature from paleoclimatology and astronomical research supports the idea that solar climatic forcing has occurred over the late Holocene. As concluded by Hoyt and Schatten (1997:202): "recent studies make a good case that the sun's radiant output varies over decades and longer time scales and that these variations are playing a significant role in climate change." All signs point to a continued debate about how short-term solar irradiance influences climate.

Oceanic Thermohaline Circulation

The possibility that oceanic thermohaline circulation may have influenced global or regional climate of the past few millennia is rapidly gaining impetus as oceanographic observational records, ocean modeling, and paleoceanographic research uncover secrets about ocean-atmosphere dynamics and their influence on climate. One atmospheric and oceanic event that epitomizes decadal-scale oceanic variability is the Great Salinity Anomaly (Dickson et al. 1975, 1988). In the 1950s and 1960s, a positive pressure anomaly developed over Greenland, leading to abnormal amounts of polar water exiting the Icelandic Sea region. Cooler and less salty waters did not mix with underlying water; as a result the temperature and salinity decreased in a layer about 200–300 m below the surface. Temperature and salinity minima were reached about 1967–1968. Oceanographers were then able to trace the circulation of this low-salinity layer for the next two decades as it circled around the North Atlantic subpolar gyre (Dickson et al.

1988). Mann and Lazier (1996) reviewed substantial evidence that this great salinity anomaly had a significant impact on the biology of the Nordic Seas and the North Atlantic.

Hurrell (1995) discussed 150 years of meteorological data on atmospheric oscillations in the North Atlantic region, referred to as the North Atlantic Oscillation, and its relationship to interdecadal variability in European weather. He suggested that decadal atmospheric variability might also be linked to rapid changes in paleo-atmospheric conditions such as those recorded in Greenland ice cores. Keigwin (1996) extended this idea to centennial-scale variability in North Atlantic deep-oceanic circulation and found isotopic evidence for reduced North Atlantic deep water during the Little Ice Age. Deep-ocean circulation changes in the North Atlantic have also been linked by Cook et al. (1996) to isotopic excursions in Greenland ice cores and other climate proxy records.

Decadal-scale climate variability has also been discussed in the Pacific Ocean. For example, Trenberth and Hurrell (1994) reviewed evidence for decadal climate changes associated in part with changes in ENSO frequency and intensity. They concluded that decadal-scale climate change resulting from Pacific atmosphere-ocean coupling has a smaller amplitude than interannual ENSO-related changes but can have climatic impacts in extratropical regions, including North America, via teleconnections. Several short-term paleoceanographic records will be discussed later, and interannual tropical climate is discussed in the next chapter.

Given the recent flurry of research activity into short-term late-Holocene climate, the link between decadal- and centennial-scale oceanic, atmospheric, and continental climate changes will improve as additional high-sedimentation rate deep-sea cores are examined.

Atmospheric Trace Gases

Large-scale changes in atmospheric concentrations of the trace gases carbon dioxide (CO_2) and methane (CH_4) play an important role in earth's climate over tens of thousands of years (chapters 4, 9). During the past millennium, the atmospheric concentration of carbon dioxide has varied about 10 ppmv around a mean value of about 280 ppmv as observed in Antarctic ice cores (Siegenthaler et al. 1988; Barnola et al. 1995). The analytical uncertainty in paleo-CO_2 from trapped air bubbles is about ± 3 to 5 ppmv. There was a small peak in atmospheric concentrations of CO_2 between about 1200 and 1400 A.D. Methane ice core records from Antarctica (Etheridge et al. 1992) and Greenland (Nakazawa et al. 1993; Blunier et al. 1993) show that concentrations varied around a global mean of 720–740 for the past 500–1000 years, but interhemispheric differences in methane concentrations reflect different methane-source production on continents. Preindustrial methane excursions as high as 70 ppbv may reflect climate (precipitation) or agricultural effects.

Are these small changes in radiative trace gases a major forcing of climate change over the past 3000 years? Answering this question is complicated by anthropogenic enhancement of atmospheric CO_2 and CH_4 levels, which may have contributed to post–Little Ice Age warming of the past century. In general, the small natural variability in these gases before the industrial revolution is probably not sufficient to cause major centennial-scale climate changes (Rind and Overpeck 1993), although higher-resolution records of paleo–trace gases might shed additional light on this subject.

Summary of High-Frequency Forcing

In a climate modeling study, Rind and Overpeck (1993) considered the questions: What should the characteristic signal be from these various centennial- and decadal-scale forcing mechanisms? Do solar variability, thermohaline circulation, or trace gases exert a strong enough influence to affect earth's climate? Using a general circulation model, they simulated about 50 years of climate change in response to a hypothesized reduction in solar irradiance during the Maunder minimum. They suggest that a reduction in global insolation like that occurring during the Maunder minimum could cause the cooling of 0.5°C observed in proxy records. Furthermore, the solar variability proposed for the Maunder minimum should be manifested by a global climate signal, with possible enhanced cooling in extratropical continental areas. In contrast, a more regional and hemispherically asymmetric imprint with the strongest signal near regions of deep-water formation would result from oceanic thermohaline-circulation changes. Likewise, a global signal, enhanced in high latitudes, would result from atmospheric trace gas forcing. Rind (1996) recently cautioned, however, that the signal-to-noise ratio in short-term late-Holocene climate change is so small that considerably more work is required to be more certain of the causal mechanisms of such change.

CONCLUDING REMARKS

SIX REMARKS ABOUT decadal- and centennial-scale paleoclimatology place this emerging field in the perspective of research on longer-term climate history. First, one overriding principle of paleoclimatology applies to centennial- and decadal-scale climate research: The amount of climate information extracted from any particular record will be directly proportional to how well-known are the processes that produced the record. This maxim is particularly true in the case of tree rings, scars, forest assemblages, corals, and marine organisms. Biotic processes including physiology and growth, dispersal and migration, as well as evolutionary adaptation and adaptability of species in response to environmental disturbance are paramount for paleoclimate interpretation.

Second, decadal-scale variability is prominent in many records. But in many cases, either the chronology is not adequate for interhemispheric or intercontinental correlation or the proxy method used is not precise enough for valid comparisons. Moreover, multiple climate indicators from the same record, such as temperature and precipitation, may not show variability in the same frequency bands. Given the small amplitude of some short-term climate signals, calibration and quantification of proxy methods need improvement.

Third, some recent summaries of late Holocene climate history adopt a somewhat pessimistic assessment of our understanding of climate change during past millennium, particularly with reference to the Medieval Warm Period and the Little Ice Age. Jirikowic and Damon (1994:309) quote Malcolm Hughes's opinions about whether we can define the Medieval Warm Period: "at least it wasn't a Little Ice Age." The Little Ice Age has also been criticized as a viable concept in paleoclimatology by Luckman (1996). These comments stem in part from researchers' inability to identify global, synchronous signals for climate variability over the past millennium.

The apparent propensity toward "globalism"—the view that climate change of the past few millennia must be all-or-nothing—is mysterious. Apparently, to some researchers, a climate oscillation such as Little Ice Age cooling would have to be globally synchronous and of similar magnitude everywhere to meet their criteria for significance. However, why should decadal-long periods of climatic warmth in the North Atlantic necessarily correspond to synchronous climate oscillations of equivalent amplitude in the South Pole or anywhere else? Rind (1996) made the important points, perhaps stemming from his experience with long-term climate changes, that the radiative forcing of short-term climate due to solar, volcanic, and trace-gas factors is of relatively small magnitude, at least compared with orbital forcing, and that the direct radiative signal might be overwhelmed by advective processes in the atmosphere. In effect, this complication would lead to hypotheses that these various processes should impart a regional heterogeneity to the paleoclimate signal, exactly what seems to be emerging from paleoclimate archives.

Cole (1996:341–342) also expressed this point about regionally heterogeneous interdecadal climate patterns, with specific reference to those reconstructed from corals from the Pacific region for the past few centuries: "All of these records indicate significant decadal variability, but in most cases this variability does not correlate among records. This incoherence should be expected, given that most of these records were generated independently and there is no obvious climatological framework that implies they should all correlate."

In this same vein, I would contend that the Medieval Warm Period and Little Ice Age, although spatially and temporally variable and still poorly reconstructed, remain useful concepts in paleoclimatology, much the way the Younger Dryas, the Eemian interglacial period, Heinrich events, and the El Niño phenomenon are valuable. Each is prototypical of a distinct genre of climatic variability. All these climate events were originally defined on the basis of local or regional events, which eventually stimulated active research into their global manifestation, timing, and broader climatic significance.

The historical development of paleoclimate research on larger-scale climate forcing, like orbital changes or ocean-basin configurations over millions of years, shows that the search for overly simplistic paleoclimate patterns can be futile because of complex feedbacks and processes operating in different components of the climate system. Taking the example of orbital climate change, insolation changes associated with 41-ka obliquity cycles induce changes in high-latitude regions (deep-water formation, sea-ice distribution, albedo effects, etc.) that are propagated through the climate system over the course of thousands of years. The phasing of climatic response to this regional triggering carries a heterogeneous signature. Even as obliquity-driven changes are occurring, precession-related climate impacts occur in low-latitude regions that themselves produce an even more complex pattern.

The Younger Dryas episode, discussed in chapter 5, is another example of a complex climate response manifested in various magnitudes in the Northern and Southern Hemispheres. Younger Dryas cooling is much greater in virtually all North Atlantic and northern

European proxy records; a dampened Younger Dryas response is characteristic of extra–North Atlantic regions. Moreover, the Younger Dryas is emerging as an asynchronous event in the Northern and Southern Hemispheres, even though it may have been forced by the same initial triggering mechanism. Likewise, decadal or centennial-scale climate changes forced by solar activity may not show the same amplitude response in Antarctica as occurs in the North Atlantic regions; asymmetric climate response may be expected in the case of oceanic circulation changes.

A fourth point concerns future directions in short-term climate research. Several studies have argued on statistical grounds that 10–20 paleoclimate records from judiciously placed sites would be adequate to decipher patterns of centennial-scale change over the past two millennia (e.g., Diaz 1996; Jones and Briffa 1996). I would stress an additional need to focus on oceanic records in key areas such as convergence zones and sites of upwelling and deep-ocean convection, especially in high-sedimentation-rate environments. There is enormous untapped potential for climate history preserved in lake and marine sediments, where knowledge of the ecology of species may surpass the level of understanding of physiological and ecological processes influencing tree growth and, to a lesser extent, coral growth. In any case, better understanding of biotic processes is a sine qua non for improved short-term climate studies.

Fifth, there has been an overwhelming emphasis on events occurring in the past two millennia (with few exceptions, such as Mosley-Thompson and Thompson's ice-core studies) at the expense of high-resolution reconstructions of early- to mid-Holocene climate. These earlier periods also deserve attention, especially because the proxy tools are already available or under development as a result of late-Holocene studies.

Finally, I wish to comment on perhaps the most ominous issue, that of current trends in climate. The significance of the twentieth-century rise in global mean temperatures is in need of a full text of its own. Many studies show that observed twentieth-century warming is anomalous, often equaling or exceeding even the regional warmth reconstructed for the intervals during the Medieval Warm Period. Indications of progressive warming have been found in Mongolian and Alaskan tree rings; Southern Hemisphere tree-ring, oceanic, and atmospheric records; lacustrine environments; glaciological history; and tropical and some polar ice cores.

This leads us back to Icelandic polar bears and Irish potatoes and the vexing problem of causality. The wealth of new quantitative paleoclimate proxies briefly outlined in this chapter offers compelling evidence that 10- to 100-yr oscillations in atmospheric temperature and precipitation took place in the North Atlantic region during the past few centuries. Decadal-scale periods of cooling within the Little Ice Age were of sufficient magnitude to reduce the density of sea ice around Iceland or to bring unusually warm and humid summers to Ireland. But correlation does not imply causality. Although the climatic mechanisms that might have initiated short-term climate change are complex and multifaceted, the human dimensions of overhunting, population growth, farming practices, and disaster assistance seem even more tangled and indecipherable. Indeed, it may never be possible to establish indisputable causality between important cultural events and climate change—that is, to completely decouple human and natural climatic factors in the demise of polar bears, the intensity of the potato famine, or other historical events. Short-term climate change of the Middle Ages through the twentieth century must be viewed through the lenses of human cultural activities such as hunting, farming, and polluting.

One should likewise keep in mind the potential for paleoclimate history to be misused, inadvertently or deliberately, by invoking past climatic events as the primary cause of cultural events such as Ireland's nineteenth-century famine. Climate can provide a convenient scapegoat for historical events that may have much more plausible explanations in political and cultural institutions. There are many examples of abuse of scientific evidence to support political or social agendas; the rekindling of the race-IQ controversy provoked by Hernstein and Murray's book *The Bell Curve* (see Jacoby and Glauberman 1995) is just one. Paleoclimatologists should proceed with caution as they integrate increasingly sophisticated reconstructions of paleoclimate histories of the past few millennia with efforts to understand the causal factors behind past human history and future climate change.

VII

Interannual Climate Change in the Tropics: ENSO

Because of the numerous species affected and its global impact, it is probably fair to include severe ENSO events among the greatest natural perturbations known on our planet.

PETER W. GLYNN, 1990

LA CORRIENTE DEL NIÑO

FOR CENTURIES, the coastal current off Peru and Ecuador brought a rich harvest of fish from the cold, nutrient-rich upwelling ocean to northwestern South American fishing villages. Around Christmas time, the waters became warm—fisherman named this annual event *El Niño,* which is Spanish for the Christ child, or *La Corriente del Niño* or *Las Dias del Niño.* Every few years, however, the El Niño wintertime ocean warming was more intense and sometimes earlier than usual. When this happened, the upwelling of nutrient-rich waters was suppressed, and warm, nutrient-depleted surface ocean waters entered the coastal regions, knocking the entire pelagic marine ecosystem out of balance. The outcome of this chain of events could be catastrophic for the local people as the staple fish harvest plummeted.

Now oceanographers use the local term, *El Niño,* to describe the intensified quasiperiodic (every 3–7 yr) occurrence of these anomalously strong, extended periods of oceanic warming. During El Niño events, a major, oceanwide event occurs across the entire tropical Pacific Ocean, characterized by warm sea-surface temperatures (SSTs) in the eastern and central regions and by changes in the depth of the thermocline and the strength of surface ocean currents. The opposite of El Niño—called *La Niña*—occurs in years when the December-March warming is unusually weak and relatively cool surface temperatures characterize the eastern Pacific Ocean.

The atmosphere's complement to the ocean's El Niño is called the Southern Oscillation (SO). It too has deep roots in observed weather anomalies that severely affect society. George Philander (1990) relates in his enlightening book on El Niño that the Southern Oscillation was discovered in the early twentieth century when Great Britain dispatched mathematician Gilbert Walker to India to investigate the periodic catastrophic breakdown of the Asian monsoon, an event that was particularly severe in 1899–1900 and led to widespread starvation. The Indian subcontinent depended upon the summer monsoon for its agricultural production and food supply, and droughts associated with the breakdown of the monsoon are well established (Rasmussen and Carpenter 1982; Quinn et al. 1987; Philander 1990). In typical years, the summer monsoon, when the western Pacific and Indian Oceans are characterized by low pressure systems, brings extensive rainfall to the Indian subcontinent. Conversely, high surface pressure typically dominates the summer atmosphere in the eastern equatorial Pacific and off the Pacific coast of South America.

Walker examined atmospheric data from around the world and discovered a periodic flip-flopping of barometric sea-level pressure in the tropics when the normal weather patterns reversed themselves (Walker 1924; Walker and Bliss 1932). High-pressure systems that typically dominate the eastern Pacific move westward, and low-pressure systems that characterize the African to Australian region, including the Indian subcontinent, shift eastward. During these unusual years, like the 1899–1900 event, the Indonesian-Asian monsoon low-pressure system is suppressed and drought results. The eastward migration of low pressure brings heavier-than-usual rains to coastal South America. These atmospheric anomalies correspond to times of depleted fish harvests off the South American coast and of intensified rainfall along coastal Peru and Ecuador, making semidesert landscapes green. Walker's landmark research on the periodic breakdown of the Asian monsoon led to his discovery of the Southern Oscillation.

Decades after Walker's studies, the ocean-atmosphere linkage of El Niño–Southern Oscillation (ENSO) was established by Jacob Bjerknes (1966, 1969) of the University of California at Los Angeles. Bjerknes is credited with first establishing the "coupling" between the east-west atmospheric circulation changes in the Pacific discovered by Walker and the oceanographic changes in the eastern Pacific. On the basis of oceanographic and meteorological observations, Bjerknes recognized that the coupling of the tropical ocean and atmosphere resulted from changes in wind stress, which in turn influenced equatorial upwelling and the thermal structure of the tropical ocean. Changes in upwelling led to anomalously high SSTs in the eastern Pacific. The normal zonal atmospheric cell circulation, which Bjerknes termed Walker circulation, was upset during ENSO events, and precipitation diminished in the western Pacific and increased in the eastern Pacific. During ENSO years, the western Pacific warm pool, a large region where annual SSTs exceed 28°C, shifted eastward, suppressing cool upwelling water and causing the thermocline to deepen. Bjerknes also expanded on Walker's ideas that these interannual climatic anomalies influenced weather events in extratropical latitudes, and he introduced the term "teleconnection" to describe their impact on extratropical weather patterns.

ENSO is now a widely recognized climatic phenomenon that originates as a tropical ocean-atmosphere disturbance and leads to global climate repercussions extending into temperate zones (see Philander 1990; Glynn 1990) (figure 7-1). Climatologists believe that during strong ENSO years, not only does the Indonesian-Asian monsoon system fail, but Africa and the western Pacific can experience severe drought; western North American summers can be hot and dry; and the southeastern United States can receive anomalously high winter rainfall. Bunkers et al. (1996) discovered anomalous rainfall in parts of the Great Plains during ENSO events. Rainfall patterns in other extratropical regions can be offset during many or most ENSO events (Ropelewski and Halpert 1986, 1987; Diaz and Kiladis 1992). The well-publicized 1997–1998 El Niño has caused large precipitation anomalies in California and the southeastern United States. During the most extreme modern-time ENSO events such as the historic 1982–83 ENSO, drought and flooding have caused billions of dollars in property damage and agricultural losses, as well as the loss of many lives (Glynn 1990).

ENSO is a climate phenomenon that has recurred for centuries, about every 3–7 yr, depending on the period under consideration and the measures used to define an ENSO year (Rasmussen et al. 1990; Halpert and Ropelewski 1992). Documentary evidence suggests 9 strong and 14 moderate El Niño events have occurred over the past century (Quinn et al. 1987), and a total of 82 ENSO events have occurred between 1607 and 1953 (Quinn 1992). That is an average rate one event every four years, with strong events occurring about every

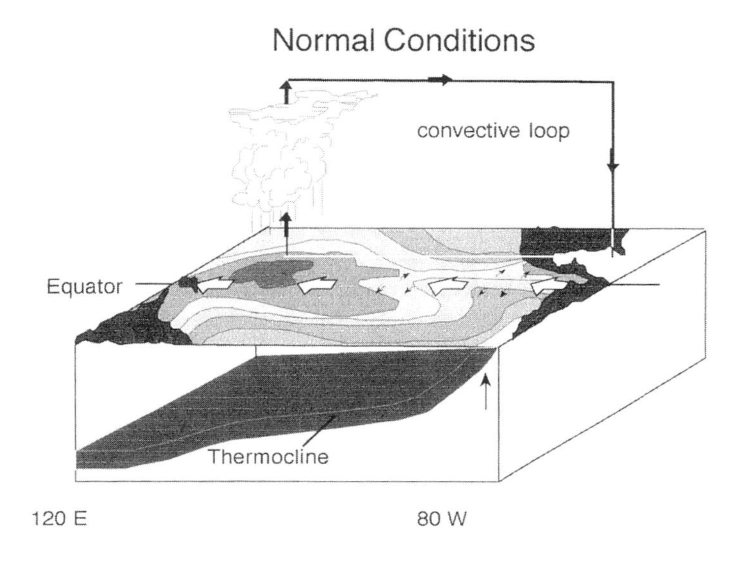

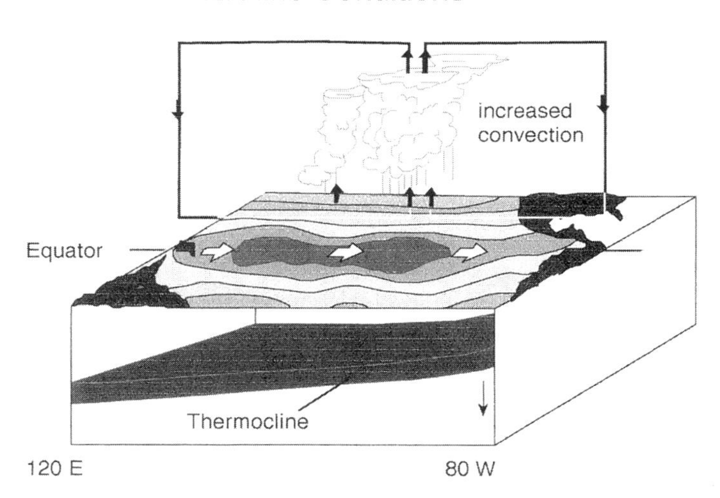

FIGURE 7-1 Schematic diagram of oceanic and atmospheric conditions in the Pacific Ocean under "normal" and El Niño conditions. El Niños are characterized by an eastward shift in tropical convection, warmer eastern Pacific SSTs, and a shallowing of the thermocline in the western Pacific Ocean. The convective loop represents Walker cell atmospheric circulation. Courtesy of M. MacPhaden and National Oceanic and Atmospheric Administration.

decade. Diaz and Kiladis (1992) and many others, however, stress that the occurrence of ENSO events can be quite irregular. For example, ENSO events as measured by instrumental records were absent or minor during most of the 1940s.

The causes of ENSO oscillations are related to internally generated dynamics between the ocean and the atmosphere and, as currently understood, are probably not due to external forcing mechanisms (Cane 1986; Trenberth and Shea 1987). El Niño–Southern Oscillation, however, remains a quasiregular phenomenon; there is a random element to it, due either to chaos associated with the deterministic aspects of ENSO or simply to noise in the climate system. In addition to its quasiregular recurrence, ENSO frequencies vary over decadal and centennial time scales as we shall see below, further complicating our understanding of it.

Despite its complexity, integrated ocean-atmosphere observational programs (Sarachik 1996) that model ENSO events months and years ahead are rapidly being developed (Zebiak and Cane 1987; Latif et al. 1994; Chen et al. 1995a). In fact, the massive 1997–1998 El Niño was predicted by several ENSO models. A more complete understanding of the long-term development of ENSO in the tropics and its extratropical teleconnections, however, will ultimately rest on a model that combines the ocean-atmosphere coupling and an understanding its long-term history and its relationship with other climate variables. This chapter describes paleoclimatology's fledgling role in this effort.

INTERANNUAL PALEOCLIMATOLOGY

THE BRANCH OF paleoclimatology devoted to seasonal, interannual, and interdecadal climate reconstruction is playing a pivotal role in establishing the history of ENSO and its impact on tropical and global climate. The paleoclimatology of ENSO over centuries and millennia is quickly emerging as the premier example of the ability of researchers to develop new means of measuring short-term tropical climate oscillations from coral scleroclimatology, ice core studies, and tree-ring dendroclimatology introduced in chapter 6. These three paleoclimate proxies of interannual climate variations form the foundation of the Annual Record of Tropical Systems (ARTS), an international program carried out under the auspices of the Past Global Changes (PAGES) and Climate Variability (CLIVAR) programs. Corals, ice cores, tree rings, and to a lesser extent microfossil evidence from laminated oceanic and lacustrine sediments are biological archives that enable paleoclimatologists to document centennial, decadal, interannual, and even seasonal variability in rainfall, SST, salinity, vertical ocean mixing, and water-mass circulation, as well as human influence on ocean environments. Several coral records now extend back four centuries, whereas ice-core and tree-ring records of temperature and precipitation cover the past few millennia.

Corals, low-latitude ice cores, and tree rings provide three major pieces to the puzzle of long-term ENSO evolution. First, they help answer the question of whether ENSO is stable over decadal and centennial time scales. Establishing natural long-term variability in ENSO, especially interdecadal cyclicity (Anderson et al. 1992), is particularly important because observational records suggest a major change in ENSO behavior occurred in the 1970s (Trenberth and Hurrell 1994; Wang 1995). For the past two decades the La Niña phase, as measured by the atmospheric Southern Oscillation index, has been extremely weak and the El Niño phase has become more frequent and enhanced. The development of long-term time series from key geographic regions of the tropics will help establish whether this shift is part of a longer-term natural cyclicity inherent to the climate system or a directional trend associated with anthropogenic perturbation of the atmosphere. The paleoclimate perspective gives researchers the ability to assess the tropical ocean-atmosphere system as it was before instrumental records became available and before anthropogenic perturbations.

Second, paleoclimatological records can potentially establish how far back in time the ENSO phenomenon has been a characteristic of earth's climate system. Was ENSO behavior different or even absent in the early or middle Holocene or during periods of warmer or colder climates, such as the Medieval Warm Period or the Little Ice Age? To answer these questions, the causes of long-term ENSO variability must ultimately be viewed from the standpoint of multiple climatic forcing factors that include atmospheric gas content, changes in solar and volcanic activity, and ENSO itself.

Third, paleoclimatology can provide information on the regional ecological impacts of seasonal and interannual tropical climate variation. Extreme ENSO events have devastating impacts on tropical ecosystems, especially coral-reef communities (Glynn 1990; D'Elia et al. 1991). Seasonal northward migration of the Intertropical Convergence Zone (ITCZ) brings spring rain to regions off Central America, affecting tropical terrestrial and marine ecosystems. The biotic response to ecological stress associated with departures in rainfall, oceanic salinity and temperature, and nutrients can be measured through coral-colony physiology and growth and biogeographical shifts of plant and animal populations. Extratropical ecological impacts also can be reconstructed through tree ring and other paleoclimate records.

Corals are uniquely qualified as an archive of interannual tropical paleoceanography and as a complement to the low-latitude ice-core and tree-ring records. Rapidly growing, long-lived hermatypic coral colonies provide the basis for an improved understanding of tropical climate during the past 500 yr, including the Little Ice Age and twentieth-century warming, as well as the more high-frequency interannual ENSO events. Later in this chapter, I describe the principles and application of coral growth (skeletogenesis) and physiology to document climate changes in the tropics. I also describe the interannual paleoclimate records available from the tropics and extratropical regions. Before proceeding, however, a brief review of the features of the coral record of tropical climate and the history of ENSO derived from documentary evidence is in order.

ASPECTS OF THE MODERN EL NIÑO–SOUTHERN OSCILLATION

The Tropical Ocean and Atmosphere

In contrast to meridional circulation, which flows in a north-south direction in Hadley, Ferrel, and Polar cells, tropical circulation in Walker cells flows east to west along low latitudes and across longitudes in the Pacific Ocean (figure 7-1). The rising limb of a Walker cell occurs in the western Pacific, where convective clouds bring heavy rainfall. The upper-level atmosphere flows eastward, and by the time it reaches the eastern Pacific coastal areas of Peru and Ecuador, it sinks; relatively moisture-poor, it causes arid climatic zones. Return low-level atmospheric flow is wind driven, flowing westward near the ocean surface, and is associated with the major equatorial surface currents. This lower limb of the Walker cell completes the cycle of zonal circulation.

The trade winds, flowing from the northeast in the northern tropics and the southeast in the southern tropics, are part of the meridional circulation and also play a major role in the ENSO phenomenon. The Intertropical Convergence Zone (ITCZ)—the boundary between northeast and southeast trade winds in the eastern Pacific—is a narrow latitudinal band of atmospheric convection where clouds develop as moisture-laden trade winds converge upon a warm ocean. About 40% of the world's total precipitation falls within 15° latitude of the equator, much of it associated with the ITCZ. Anomalous shifts in the annual migration of the ITCZ to southerly latitudes occurs early during ENSO events, and seasonal and interannual migration of ITCZ convection is visible in satellite images. ITCZ variability is also evident from the geochemical records obtained from tropical corals.

Tropical Pacific Ocean currents are dominated by the westward flowing, wind-driven North and South Equatorial Currents. Eastward-flowing countercurrents run in the opposite direction; the North and South Equatorial Countercurrents are strongest in the winter season of their respective hemispheres, when trade winds are strongest. Even more prominent than the surface countercurrents is the Equatorial Undercurrent, which flows across the Pacific at a core depth of 200 m and transports large volumes of water. The western and eastern Pacific Ocean waters contrast strongly with one another in terms of SST, nutrients, and biology. The western Pacific Ocean warm pool harbors the warmest SSTs in the world's oceans, exceeding 28°C on an annual basis, compared with parts of the eastern Pacific, where SSTs vary from 21°C to 26°C. The ITCZ lies near the boundary between the warm pool and the cooler eastern Pacific region, which is dominated by trade winds and equatorial upwelling.

Typically, the thermocline in the Pacific Ocean is relatively deep in the western Pacific and shallow in the east where equatorial upwelling leads to cooler SSTs. During ENSO warm phases, this west-east thermocline gradient diminishes, and the relative deepening of the eastern Pacific thermocline is reflected in warmer SSTs.

Development of a Typical ENSO Event

Although each ENSO event has its own unique personality, in the most general terms, several unifying characteristics are evident. How do the atmospheric and oceanic conditions evolve during a typical ENSO event? ENSO is not an on-off phenomenon. Most ENSO events owe their existence to longitudinal asymmetries in thermal circulation of the atmosphere and oceans. ENSOs develop seasonally, in a time-transgressive nature across the Pacific Ocean, ultimately affecting weather patterns downstream in mid latitudes. Generally, the normal seasonal atmospheric and oceanic trends become perturbed during ENSO events. In the tropical Pacific atmosphere, these seasonal changes mainly involve the location of the ITCZ. The ITCZ remains in the Northern Hemisphere all year, where it is associated with SSTs greater than 27.5°C and easterly winds. Small SST variations in the western Pacific warm pool associated with ITCZ migrations are also a prominent seasonal anomaly enhanced during ENSO events. An important oceanic anomaly during ENSO years is the much warmer December-March SST in the eastern Pacific recognized for centuries by indigenous South Americans.

The development of ENSO events during a 12-month period was divided by Philander (1983; see also the scheme of Rasmussen and Carpenter 1982) into three intervals: the precursors, which begin in summer and fall; the mature phase in November through January; and the decline, when normal conditions resume during the following spring.

The precursors of ENSO are in some respects seasonal trends in precipitation and SST carried to extremes. For example during the early phase of an ENSO event, the ITCZ, which normally migrates southward to the equator or even several degrees south of the equator, brings heavy rains to regions that are typically extremely arid. Elevated SSTs in the eastern Pacific in fall (sometimes earlier) signal an ENSO event.

During the mature phase of ENSO, there is rapid propagation of heavy rainfall into the central equatorial Pacific during the months of November, December, and January. The western Pacific warm pool migrates eastward, leading to warmer-than-normal SSTs off northeastern South America. The thermocline during ENSO loses its steep west-east gradient, and the eastern Pacific thermocline deepens. This thermocline shift reflects the eastward migration of the warm pool, much weakened trade winds from the east, and reduced upwelling. The western Pacific thermocline lies at a substantially shallower depth during the ENSO mature phase than it does during normal years (figure 7-1).

The sea level across the Pacific Ocean also changes by a few centimeters during ENSO events. The significance of sea-level variations as a manifestation of ENSO events was pioneered by Wyrtki (1973, 1975). Satellite measurements indicate that over short time scales sea level rises in the Pacific Ocean when the thermocline deepens; sea level declines when the thermocline is shallow. Such variations may result from low-frequency, low-amplitude oceanic waves—the eastward-propagating Kelvin and westward-propagating Rossby equatorial waves generated by ENSO-scale climatic events (see chapter 8). The pulse of warm water flowing eastward and the diminished trade winds and equatorial upwelling lead to an increase in sea level in the eastern equatorial Pacific visible in TOPEX-Poseidon satellite photographs.

A final characteristic of the mature phase of ENSO is the effect on climatic conditions in extratropical regions. One manifestation of ENSO in middle latitudes is intensified Hadley cell circulation, which can affect atmospheric conditions in the northern Pacific Ocean and ultimately weather in North America (Bjerknes 1966; Trenberth and Hurrell 1994).

ENSO events end and normal conditions resume at different times in the eastern and western Pacific Ocean. For example, in the east, cool SSTs and strong trade winds first resume in the southeastern Pacific and propagate westward. Not until about 12 to 18 months after the ENSO event began do normal conditions return to the entire Pacific Ocean. In the atmosphere west of the international dateline, winds shift from eastward to the more typical westward direction; the thermocline deepens again; and precipitation decreases in the vicinity of 160°W longitude.

In summary, a generalized ENSO event can be viewed as a time when the typical pattern of tropical atmospheric and oceanic warming due to high equatorial solar radiation is perturbed. Both the location of maximum Pacific SSTs and the region of maximum convection and rainfall shift eastward from near Indonesia to about the International Date Line in the central Pacific Ocean. Currents and wind patterns also shift eastward. Pacific trade winds weaken, and the thermocline develops at greater depths in the eastern Pacific. Meteorologically, severe drought results in Australia, Indonesia, and parts of Asia, excessive rainfall characterizes the normally dry eastern Pacific regions.

Indices to Measure ENSO

We can quantify atmospheric and oceanographic anomalies that accompany the development of El Niño and La Niña events in several ways. Among the most commonly used is the Southern Oscillation Index (SOI), a standardized measure of the atmospheric pressure difference between Darwin, Australia, and the island of Tahiti in the eastern Pacific. The lower the SOI value, the stronger the intensity of the ENSO; the reverse is true for La Niña.

Sea surface–temperature variability can be used instead of atmospheric pressure to quantify ENSO conditions. Philander (1983) tracked seasonal ENSO development using SST values taken 100 km off South American coast between 3°S and 12°S latitude. The Japan Meteorological Office developed an SST index using the extensive Pacific Ocean database (O'Brien et al. 1996). Consistent patterns of sea-level change at Truk Atoll (152°E, 7°N) also faithfully record El Niño events (Philander 1983).

Rasmussen and Carpenter (1982) focused on SST anomalies and changes in wind fields to characterize ENSO events. Rasmussen and

Carpenter (1982) and Ropelewski and Halpert (1986) plotted 24 consecutive months of SST measurement to identify the evolution of El Niño events. They refer to the period from the July in the year before ENSO with a minus sign (–1 yr), the July of the beginning of ENSO as the 0 (zero) year, and the following 12 months with a plus sign (+1 year) (see Sarachik 1996 for a review).

The choice of El Niño index depends on one's purpose and, because observational data are spotty prior to the past few decades, how far back in time one needs to examine trends. Trenberth and Shea (1987), for example, suggested that sometimes the SOI and SST measurements do not correspond. The use of a particular index or method of reference, however, can be critical because it may influence whether or not a certain year is characterized as El Niño, La Niña, or neutral. The exclusion or inclusion of a particular year can obviously lead to mistaken correlations between tropical and extratropical climatic trends.

One final complication in the study of ENSO paleoclimatology is how one refers to the calendar year in which an El Niño or La Niña climate anomaly occurred, because ENSO climate anomalies cover an interval spanning two calendar years. This issue can be especially problematic with reference to long-term records such as those described later. Some authors have adopted the convention of referring to the El Niño (or La Niña) by the first calendar year in which ENSO conditions began to develop in the summer or fall. Thus the 1982–1983 El Niño would be referred to as the 1982 event. Bunkers et al. (1996) found labeling of El Niño years (especially before 1935) problematic in their study of springtime precipitation anomalies associated with ENSOs in the U.S. Great Plains. Some years that Bunkers et al. (1996) considered to be El Niño years were not exactly the same as those defined by other authors. To remedy this problem, it is preferable to define an ENSO event by the years during which the complete ENSO cycle develops and terminates—for example, the 1982–1983 ENSO event.

Extratropical Anomalies Associated with ENSO

El Niño effects dominate the tropical Pacific Ocean climate but they also have climatological impacts far from the Pacific region. Walker and Bjerknes noted early on that the Hadley cell would be intensified in the winter hemisphere during periods of weakened Walker cell circulation. A large literature now describes anomalous global weather patterns associated with ENSO events, sometimes finding significant teleconnections. We will mention just a few that are relevant to our discussion of ENSO paleoclimatology.

Rainfall anomalies are among the strongest teleconnections. Whetton and Rutherford (1994) studied 500 years of ENSO-related rainfall anomalies in the Eastern Hemisphere by comparing the South American ENSO documentary record of Quinn and colleagues (Quinn et al. 1987; Quinn 1992) with long-term historical records of Nile floods, Indian droughts, and North China rainfall, as well as Java tree-ring widths. They discovered a statistical significance between South American ENSOs and Eastern Hemisphere climate proxies for the period 1750 to the present; historical data prior to 1750 A.D. were insufficient to establish definitive relationships before that time.

Kousky et al. (1984) studied rainfall patterns in the Southern Hemisphere during ENSO events and found persistent evidence for El Niño-related drought over Australia, Indonesia, India, Western Africa and northeastern Brazil, and enhanced rainfall in the eastern Pacific Ocean, Peru, Ecuador, and southern Brazil. They attributed the anomalous precipitation patterns to shifts in the upper-troposphere subtropical jet streams and to lower-level atmospheric changes in the strength of the trade winds in both the Pacific and Atlantic Oceans.

In many studies, rainfall and drought patterns in North America have been deemed associated with ENSO on the basis of long-term instrumental records from numerous weather stations. Examples of anomalies occurring during most, but not all, ENSO years include wet winters in Florida and the Gulf of Mexico (Douglas and Englehart 1981; Rasmussen and Wallace 1983; Ropelewski and Halpert 1986), high precipitation in the Northern Plains (Bunkers et al. 1996) and the Great Basin (Ropelewski and Halpert 1986), high temperatures in northwestern North America (Ropelewski and Halpert 1986), and heavy winter precipitation and stream flow in western North America (Cayan and Webb 1992).

Major weather events have also been linked to El Niño activity. For example, Trenberth et al. (1988) attributed the 1988 North American drought that severely affected the north central area of North America to conditions originating in the tropical Pacific during the end of the 1986–1987 ENSO event. Diminished rainfall was due in part to anticyclonic (high-pressure system) conditions linked to northward displacement of the ITCZ in the eastern Pacific Ocean. According to Trenberth et al. (1988), these anomalous wave-train patterns across North America occurred when the jet stream and its associated rainfall were displaced northward of their typical position. Negative SST anomalies during the

April through July, 1988, period in the eastern Pacific—the opposite of ENSO positive anomalies—reached extremes, leading to northward displacement of the ITCZ precipitation. The 1988 North American weather patterns may have been part of global atmospheric anomalies stemming from the prior years' anomalies.

Relation of ENSO to Other Climate Phenomena

ENSO is the dominant mode of interannual climate variability in the Pacific Ocean, but other interannual and interdecadal weather and climatic patterns are often superimposed upon ENSO activity. Although not as well known as ENSO, they merit attention because more and more researchers are seeking evidence for their existence in paleoclimate records.

The Quasi-Biennial Oscillation (QBO) is a climate fluctuation occurring every 24–30 months and developing in the lower stratosphere, 25–35 km altitude, in tropical latitudes 0–15° on either side of the equator. Rasmussen et al. (1990) found a biennial oscillation in lower-level wind fields in the Pacific, which they associated with ENSO activity. They suggested that the 2-yr and 4- to 5-yr ENSO cycles were in phase in 1982–1983 and contributed to producing the massive 1982–1983 ENSO event. Conversely, the development of the 1974 event was aborted because the two cycles were out of phase.

The Pacific–North American (PNA) pattern is a weather pattern that develops in the northern Pacific Ocean (Wallace and Gutzler 1981) that may be linked to ENSO activity. Halpert and Smith (1994) showed that during March-May of 1993, over the course of several years of persistent ENSO conditions in the Pacific, anomalous weather patterns related to the Pacific–North American pattern continued unabated over the North Pacific and North America. This teleconnection between the Pacific–North American and tropical Pacific climate is characterized by above-normal temperatures in Alaska and western North America and below-normal temperatures in the southeastern United States. Halpert and Smith (1994) noted that weather patterns in the Southern Hemisphere were consistent with typical ENSO patterns during a period they referred to as a continuing warm episode. Another mode of climate variability occurring in the Pacific is the Pacific (inter) Decadal Oscillation (PDO) (Mantua et al. 1997).

An additional climate pattern is the North Atlantic Oscillation (NAO), which develops in the North Atlantic and influences weather in Europe and the eastern portion of North America (Hurrell 1995). The NAO is measured by the pressure difference between Iceland and the Azores and is responsible for producing mild winters in Europe. The periodicity of the NAO is about 7–10 yr; the linkages between the NAO and ENSO are not yet well established.

Long-term trends in the Quasi-Biennial, Pacific Decadal, and North Atlantic Oscillations, as well as the Pacific–North American pattern, and the relationship of each climate pattern to ENSO will be an important field for future paleoclimatological research (see Charles et al. 1997).

Ecological Impacts of ENSO

The global climatic impacts of ENSO events can be particularly devastating to biological systems. Populations of species with sensitive physiologies can be decimated throughout their ranges. Entire ecosystems, such as pelagic marine and coral reef ecosystems, can be knocked out of equilibrium through perturbations to SST or salinity or through nutrient depletion. The ecological impacts of ENSO are described in the book on the 1982–1983 ENSO edited by Glynn (1990) and in the book on ENSO history edited by Diaz and Markgraf (1992). Research has been carried out on a variety of taxonomic groups. Coral bleaching (the loss of a colony's symbiotic zooxantheallate algae) and high coral mortality rates are other major ecological consequences of extreme El Niño events (Glynn 1990; D'Elia et al. 1991). Glynn estimated that up to 70–90% of the coral reefs in Costa Rica, Panama, and Colombia and as much as 95% of the corals in parts of the Galapagos Islands were killed as a result of SST warming during the 1982–1983 El Niño event. Decade-long warming before the 1982–1983 event may have rendered Eastern Pacific corals more vulnerable than they otherwise would have been. Considering the many additional negative effects on coral-reef ecosystems, Glynn (1990) estimates that reefs will take decades or even centuries to recover from this large ENSO. The paleoclimate record described later indicates that tropical ocean-atmosphere disturbances of ENSO have severe effects on the growth and skeletal geochemistry of colonial corals throughout the Pacific Ocean as well as on entire coral ecosystems. The studies of Carrasco and Santander (1987) of zooplankton (especially copepods) off Peru and Ochoa and Gomez (1987) of dinoflagellate algae present other examples of ENSO impacts on marine ecosystems.

ENSO events can also indirectly disrupt terrestrial ecosystems far from the tropical origins of El Niño by altering precipitation patterns. Swetnam and Betancourt (1992), for example, documented extreme variability in fire-related tree-growth perturbations in the southwestern

United States during ENSO events. They inferred that fires and their impacts were caused by ENSO-related precipitation anomalies.

One final point should be reiterated before examining the paleoclimatology of ENSOs. No two ENSO events are identical—each has its own character. The 1982–1983 and the 1997–1998 ENSOs are good examples. The 1997–1998 event was the most severe on record as measured by several oceanographic and ecological measures. Like the 1982–1983 event, it did not develop like ENSO events of the prior 25 years (Philander 1983). Similarly, the recent protracted 1990–1994 ENSO conditions as measured by the SOI were not as intense as the 1982–1983 single event, and oceanographically did not have the same SST anomalies as other recent ENSO events. There have been, however, similar periods of extended low SOI index values, such as the period 1940–1942, suggesting the 1990s pattern may not be unique. Variations in the historical record of ENSO events is born out by changes in the relative frequency and amplitude over decadal and centennial time scales evident in the paleoclimatological record of corals.

ENSO History From Documentary Records

The recognition that ENSOs are critical events for global climatology has led to not only extensive satellite and oceanic observation programs over the past few decades, but also efforts to reconstruct long-term history of ENSO over the past few centuries from historical records. Historical records of ENSO overlap and complement the paleoclimate records; the two must be used in tandem in efforts to determine long-term trends in interannual climate history of the past millennium.

The best-known reconstructions of El Niño are probably the compilations of W. H. Quinn and colleagues (Quinn et al. 1978, 1987; Quinn and Neal 1992; Quinn 1992), who reconstructed ENSO for the past four and a half centuries from historical records around the world. Quinn's original studies were based mostly on records from the eastern Pacific–northwestern South American region. Quinn expanded his historical research geographically to include Nile flooding and temporally back to 622 A.D. (Quinn 1992). Although the confidence in the record for specific years declines rapidly before the late 1700s, Quinn's records remain the standard against which paleoclimate records can be compared (Whetton et al. 1996). Another long-term reconstruction is that for the eastern Pacific and South America compiled by Hamilton and Garcia (1986), who reconstructed the history of Peruvian precipitation from 1531 to 1841 A.D.

Several studies have incorporated both historical and paleoclimate data into studies of long-term patterns of ENSO. Diaz and Markgraf (1992) edited a major volume synthesizing the history and paleoclimate record of ENSO events in the Pacific and throughout the world. The general conclusion one reaches upon reading the papers in the Diaz and Markgraf volume is that ENSO events do indeed have global repercussions outside tropical areas. For example, Diaz and Pulwarty (1994) reviewed seven indices, five from historical and two from paleoclimatic sources. The paleoclimate indices included records from tree rings and the Quelccaya ice cap (Thompson et al. 1984). Diaz and Pulwarty (1994) found statistically significant 2- to 6-yr cycles in most of the intervals covered by all seven records. They reached the important conclusion that the El Niño phenomenon has been a significant part of climate variability in the Pacific region for the past millennium. Furthermore, longer cycles of variability also showed up in several records.

Whetton and Rutherford (1994) and Whetton et al. (1996) used an eclectic group of indices—Javan tree rings, Nile flooding, extreme droughts and famine in India, and rainfall in northern China—to investigate Eastern Hemisphere teleconnections. Some of these records went back to 1525 A.D., but the most reliable records of ENSO phenomena extended back only to 1750. When they compared these records to the documentary ENSO record of Quinn, they discovered strong positive correlations between ENSO events and Chinese rainfall patterns back to 1525 A.D. and ENSO and Javan tree-ring widths back to 1670 A.D.

When Did El Niños Begin?

Very little information pertains to the origins of ENSO as a dominant mode of interannual climate variability in the tropics. Sandweiss et al. (1996) found fossil tropical water faunas as far south as 10°S latitude that are radiocarbon dated between 8–5 ka. On the basis of this paleontological evidence, as well as faunal evidence from archaeological sites in northern Peru, they suggest ENSOs may have begun in the middle Holocene about 5 ka. They further argue that the warmer-than-modern SSTs off Peru in the early- to mid-Holocene represent a very different oceanographic situation from the present one in which the Humboldt Current keeps this region relatively cool and inhabited by temperate faunas. Currently, the major source of oceanic and climatic variability off Peru is the quasiperiodic ENSO phenomenon.

Low-resolution archaeological and paleontological data are very indirect means by which to infer a mid-Holocene onset of ENSO. DeVries et al. (1997) for example, have criticized the hypothesis of Sandweiss et al., suggesting instead that the thermally anomalous molluscan assemblages reflect changes in coastal morphology and not climatic factors. High-resolution paleoceanographic studies should in the future provide a better means of detecting the origin of ENSO-like paleoclimate history during the Holocene.

PALEOCLIMATE RECORDS OF TROPICAL SEASONAL AND INTERANNUAL CLIMATE

CLIMATOLOGICAL OBSERVATIONS from satellites, oceanographic buoys, and other devices, combined with historical and documentary analyses of ENSO have revealed much about patterns and global consequences of interannual tropical climate history. Why don't we rely solely on these sources to study ENSO? What role do paleoclimate data play?

Paleoclimatological records derived from corals, ice cores, and tree rings have three distinct advantages over observational and historical data for understanding interannual, decadal, and seasonal climate history. First, observational data, although more abundant, owing to station density, than data from coral colonies, are available only from the past few decades. This period is too short examine ENSO phenomena within the context of decadal and centennial-scale climate change. Moreover, it only allows us to study ENSO during the latter part of the twentieth century, a period when human activity is suspected to have altered natural climatic patterns.

Historical records, as valuable as they are, are also less reliable and less complete than paleoclimatology for the investigation of ENSO trends, and they are not immune to periodic revision and modification. Corals, ice cores, and tree rings provide access to periods of time overlapping with and predating historical records and to regions where no early observational records were kept. At present coral paleoclimatology for ENSO study extends back about 500 yr. Coral records are especially important in remote areas of the Pacific Ocean where El Niño originates but where oceanic and atmospheric measurements are sparse.

A second advantage is that coral research can be carried out with analytical consistency in the laboratory. Unlike historical information, which is undoubtedly uneven in quality and usually qualitative, laboratory methods can ensure the consistency of sample preparation and instrumentation called for in paleoclimate research. Gaps due to lost historical records, changing methods of record keeping, and political and cultural events all affect the reliability of historical records. An accurate history depends on its keeper as well as on its historian, coral paleoclimatology depends on the skill of the scientist, instrumental limitations, and assumptions about linear coral skeletal growth.

Third, unlike documentary records, paleoclimate research gives us the opportunity to apply a hypothetico-deductive approach to investigate interannual climate change. Long-term records of ENSO-related climate variability in tropical and extratropical regions allow us to test hypotheses about ENSO mechanisms. A scientific method also fosters reproducibility of results; paleoclimatologists can return to a tropical region, even to the same coral reef, and reanalyze the paleoclimate history using the same or new methods.

In summary, corals provide a unique opportunity to expand upon observational and historical records for identifying trends and patterns of change. Charles et al. (1997) demonstrated this point in their quantitative comparison of Indian and Pacific Ocean interannual climate variability over the past 150 yr using corals from the Seychelles, Tarawa Atoll, and the Arabian Sea. They discovered that interannual climate variability in the Indian and equatorial Pacific have been linked for more than a century, that interdecadal shifts in Asian monsoon were important in pre-anthropogenic periods, and that the QBO has governed Indian Ocean SST variability for at least a century. Likewise, in a comprehensive analysis of historical ENSO records, Whetton and Rutherford (1994) suggested that coral paleoclimate records might surpass available historical records for establishing long-term climate history over the past few centuries. I now turn to unique aspects of coral biology that makes them so useful.

Corals: Monitors of Tropical Paleoclimate

The paleoclimatology of tropical interannual and seasonal climate variability relies heavily on the record derived from massive coral colonies (figure 7-2). Corals have become the sine qua non of short-term climate variability related to ENSO variability, equatorial upwelling, changes in the subtropical gyres, and North Atlantic trade winds (Druffel et al. 1990; Dunbar and Cole 1993). As many as 92 different coral colonies had been studied as of the 1993 review of Dunbar and Cole (1993). Through chemical and growth-banding signals entrapped in their calcium carbonate ($CaCO_3$) skeletons, these corals provide interannual, seasonal, and even monthly trends in SSTs, rainfall, oceanic salinity,

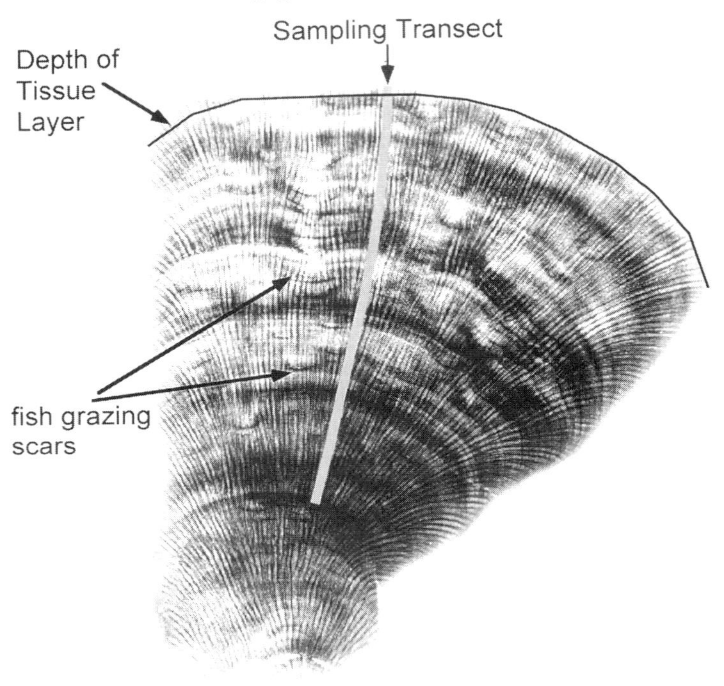

FIGURE 7-2 Photograph of *Porites lobata* coral colony in cross section showing transect for geochemical sampling of seasonal and interannual oceanic variability. Also shown are fish-grazing scars and the depth in the coral skeleton of the organic tissue layer. Courtesy of Braddock Linsley.

vertical mixing, and anthropogenic environmental changes. As we saw in chapter 6, corals also yield significant information on centennial-scale climate events such as the Little Ice Age.

Three aspects of coral biology render them ideal as a monitor of climate change. First, their colonial nature and life history as sedentary reef builders allow colonies to live long in a single place, recording its history. Second, coral grows through the process referred to as skeletogenesis, which provides both a means to date climate records through annual growth-band counting and a record of physical and climatic events that alter growth rates. Third, coral skeletal chemistry is altered by changes in the oceanic environment caused by climate changes; these changes become incorporated into the permanent skeletal record. We consider each of these attributes in some detail.

Corals, Colonialism, and Life History

Corals belong to the phylum Coelenterata—the skeletonized cousins of jellyfish. Thousands of individual coral organisms make up a single colony. Each individual in the colony is a metazoan (called a polyp) with an extremely simple body plan. Coral polyps consist of two folds of cells, differentiated into simple tissue, but lacking a central nervous system, and having relatively few types of specialized cells. Despite their simplicity of body plan, hermatypic reef-building coral have played a key role in earth history because they build immense $CaCO_3$ reefs. Coral colonies build reefs through a mutually beneficial symbiotic relationship with dinoflagellate algae called zooxanthellae. Zooxanthellae live in the coral tissue and require light to photosynthesize. Vigorous coral skeleton growth is possible only with active zooxanthellate photosynthesis and growth. Thus, hermatypic corals must live in shallow water, usually in depths above 150 m; many taxa live at depths less than 15 m. Hermatypic coral growth is also limited by water temperature and salinity—most species are unable to live in regions where temperatures drop below 16–17°C, and most species inhabit waters of normal marine salinity.

Colonial organisms, through an asexual mode of reproduction, have a means of perpetuating their colony's existence. As a consequence of colonialism, coral growth continues indefinitely, often over several centuries. The sedentary life style of the colony renders it a stationary monitor of periodic swings in oceanic conditions at a single location as long as the coral colony survives. A colony is the consummate paleoclimatological tropical-weather monitoring station.

As a result of these attributes, several tropical areas have coral-derived climate records covering the past 50–400 yr. These regions include Australia, the Indian Ocean, Indonesia, Bermuda, the eastern and western Pacific Ocean, and the Caribbean. Still, climatologists require an increasing density of coral records to examine broad tropical climate trends. Expanding the density of coral paleoclimate records will continue to be a major priority in future years.

Coral Skeletogenesis, Growth, and Chronology

Coral skeletogenesis—their ability to secrete aragonitic skeletons—makes them a massive reservoir of the earth's $CaCO_3$. Skeletogenesis also confers upon corals their paleoclimatological value. The growth of corals in general, and the formation of density bands in particular, are complex, actively researched topics. Because the study of ENSO events is often concerned with seasonal and annual changes, we must examine the processes governing coral growth in some detail.

The coral skeleton in the form of annual density growth bands provides its own chronology with which to analyze and interpret the paleoclimatic signal (figure 7-2). Like annual tree ring dendrochronology, sclerochronology provides a means of dating paleoclimate events through the use of growth bands. Annual and even seasonal paleoclimate events can be resolved using growth-band chronology.

Early studies showed that coral growth varied seasonally, producing annual growth bands, suggesting that density-band couplets might allow high-quality age models (Ma 1937; Knudson et al. 1972). In a survey of almost 1500 specimens of 47 coral genera and subgenera, Weber et al. (1975) showed that the average thickness of coral skeletal growth bands is positively correlated with mean annual water temperature. Other factors being equal, the maximum density of the coral skeleton often occurs when seawater temperature is above average.

Although the general aspects of coral annual growth bands were established by the 1970s, knowledge of the mechanisms, rates, and patterns of growth; taxonomic variability; and effects of environmental factors on corals was very limited. A more thorough understanding of coral growth has recently come about through systematic field investigations of coral growth. Most notable are studies by J. M. Lough and D. J. Barnes. In a series of papers on corals from the Great Barrier Reef off Australia, they made significant strides to place corals on a par with tree-ring dendroclimatology in terms of their value for interannual paleoclimate reconstruction. Focusing mainly on the genus *Porites,* Lough and Barnes examined the influence of rainfall, cloudiness, ENSO events, and other environmental factors on coral growth (Lough and Barnes 1989). Experiments were aimed at understanding an array of issues of coral growth: the seasonal origin of high- and low-density banding and the growth rate of *Porites* (Lough and Barnes 1990), growth band density variability in an onshore-offshore gradient 16–120 km from shore (Barnes and Lough 1992), variation in the skeleton-building tissue at different depths (Barnes and Lough 1992), the empirical and modeling evidence for different types of growth (Barnes and Lough 1993), and modeling density-band formation (Taylor et al. 1993b). These papers along with those by Druffel et al. (1990), Gagan et al. (1994), and others (see Lough and Barnes 1989) form the basis for the following discussion coral growth banding.

The manner in which corals secrete their shell is critical for deciding upon a sampling strategy and developing an age model for a coral colony. Skeletogenesis in massive corals has three main components. *Extension* is a the outward development of skeletal material growing at the outer edges of the colony. Extensional growth is usually fairly equally distributed over the entire colony surface. The uniform growth of the coral aragonite skeleton is an extracellular activity, whereas the growth of the coral animal's tissue is intracellular. High-density bands form during periods of growth associated with reduced skeletal extension, that is, at times of reduced intracellular tissue building and reduced extracellular calcification. Although there are exceptions, low- and high-density bands usually reflect high and decreasing light levels, respectively, because light influences the productivity of the zooxanthellae, which in turn mediate calcification and tissue growth.

The second process is *thickening* of the existing skeleton. Barnes and Lough (1992, 1993; Lough and Barnes 1992) found that in massive corals such as *Porites* and *Pavona* from the Great Barrier Reef, skeletal thickening occurs throughout the depth of the tissue layer that represents the previous 4–13 months worth of skeletal growth. Subannual growth can vary by as much as 30% of the annual mean, with the thickness of the coral tissue layer ranging from 2 to 10 mm. This phenomenon obviously could compromise the integrity of the subannual paleoclimate signal, because thickening of existing skeleton through continued calcification in living tissue could smooth any environmental variability derived from the geochemistry of the coral skeleton.

Gagan et al. (1994) specifically addressed this concern in a study of isotopic variations in coral skeletons from Pandora Reef in the Great Barrier Reef. They studied a 6-yr record (1978–1984) of oxygen and carbon isotopes from *Porites lutea.* What made the study unique is that by analyzing nearly weekly sampling, they could identify with uncanny precision a brief (2-week) mid-winter SST cooling event that occurs near Pandora Reef each winter. The mid-winter isotopic event provided a time marker to date the coral skeleton at key points within the growth bands independent of the density banding itself. Gagan et al. (1994) concluded that the signal distortion of the short-term paleoclimate events due to tissue layer calcification and variation in the intraannual coral extension was negligible. Thus, in practice, the value of coral growth as a climate monitor of seasonal-scale events is not diminished due to growth-rate changes over the course of a year. The good correlation of some coral records to environmental data also suggests that skeletal-thickness effects are small.

The third growth process is periodic and abrupt *tissue uplift* of the lower tissue margin, occurring about every 30 days associated with the formation of skeletal structures called *dissepiments.* Dissepiments are thin, horizontal bulkhead structures connecting the vertically growing coral skeleton. Dissepiments are preserved as part of the coral even after tissue growth has stopped, and the spacing between them records environmental events influencing growth. Variation in the depth of the tissue layer of *Porites* could also influence paleoclimate interpretation of geochemical records. For example, Barnes and Lough (1992) found that thickness varied with the size of the colony and the location and environmental conditions (probably nutrients). Relatively little tissue thickness variation, however, was found among different species of coral. By establishing the relationship of the tissue layer to the dissepiment, Barnes and Lough (192) established that dissepiments could be used for identification of extreme paleoclimate events.

The growth bands formed during these three processes represent seasonally—and therefore environmentally mediated—changes in the rate of growth and degree of skeletal thickening. In general, in the Great Barrier Reef *Porites,* high-density bands form in the summer and the low-density bands in the winter, although there are occasional exceptions. Seasonal light and temperature variability are generally considered the main factors controlling the thickness of density bands, but other factors play a role. Lough and Barnes (1989) found that in *Porites solida* from the Great Barrier Reef minimum skeletal density can be related to rainfall, cloud cover, and atmospheric pressure, whereas high density is related to spring climatic conditions. Wellington and Glynn (1983) and Wellington and Dunbar (1995) studied corals from the Pacific Ocean off Central America and showed that light also affects coral growth banding. The low density band forms in the dry season at Cano Island in the Gulf of Chiriqui when the light level is high. In Galapagos corals, the high density band forms in the months January through April.

Corals grow at widely different rates not only among different taxonomic groups, but also within relatively small regions and even single colonies. For example, Lough and Barnes (1990) measured growth rates of 5–15 mm/yr in 40 colonies of four species of *Porites* from three separate stations located 16, 60, and 120 km from shore. They observed a trend of increasing average density as one moved from onshore to offshore locations, which suggested a decrease in annual growth rate as one moves farther offshore. However, there were no statistically significant common patterns of density and growth in their among-reef and within-reef comparisons. The paleoclimate message from this regional study was to use a single coral colony as a monitor of regional climatic variability because even a neighboring colony may show a different density-banding pattern.

Although coral banding is clearly an annual growth phenomenon, a factor called the "soft year" can complicate the use of annual growth bands for paleoclimate research. The soft year was described by Buddmeier et al. (1975); it refers to the fact that equivalent points within an annual density pattern may not represent the same time of the year. Lough and Barnes (1990) established the existence of the soft year by collecting 40 colonies of *Porites* at the same time in November when the high density band had just formed. They were able to establish intra-annual skeletal variation related to the soft year. To minimize the effects of this factor in subannual paleoclimatology, careful and consistent orientation of the coral for geochemical sampling and growth-band measurement is required.

Not only does water temperature affect the growth and formation of density bands, but temperature can also cause parts of the colony to die while other polyps survive. Fitt et al. (1993), for example, showed that for some Caribbean corals, growth hiatuses of up to one year could be induced by temperature-related stress, after which the coral resumed normal growth. This punctuated growth is important for identifying the impact of individual extreme ENSO events that potentially can disrupt the growth of a colony. Paleoclimatologists need to recognize growth cessation because, should growth stop for a year, an error factor of ± 1 yr must be added to the paleoclimate chronology for a colony. This small adjustment could affect the correlation of interannual climate trends with records from other regions.

In summary, coral growth bands signify differences in the density of the skeleton related to seasonal growth variability due to multiple factors. The process of skeletogenesis is important for using coral growth bands as chronological tools and also as a measure of ecological stress related to temperature and other factors. Understanding of the relationships between water conditions, light and temperature, and the density of growth bands is improving steadily. The three main sources of variability in coral growth—regional environmental variability of colonial growth, taxonomic factors, and intracolony variation—are now well established for several genera.

Developing a Coral Chronology

Paleoclimatologists use annual bands as a simple means to date coral paleoclimate records by counting backward from the coral surface, or from the time the colony died, as one would count tree rings from the center toward the periphery. Several guidelines and caveats must be followed and acknowledged. The most important strictures for obtaining an accurate chronology are colony orientation, careful site selection, systematic x-ray and gamma-radiation densitometry, and cross-checking of density banding with isotopic markers calibrated to known events. Often coral chronologies based on density banding are used to develop an age model that is further refined through the use of seasonal carbon and oxygen isotopic variations (e.g., McConnaughey 1989; Cole et al. 1993a). Thus, under the proper circumstances of growth and sampling strategy, corals can provide not only annual resolution but also seasonal and even monthly paleoclimate records. Fossil Quaternary coral colonies dated by the uranium series method (chapter 4) can also yield a floating paleoclimate record of interannual oceanic variation during past interglacial or glacial periods.

Coral Skeletal Chemistry

In addition to the excellent chronology obtained from growth banding, the second attribute of the coral skeleton for recording climate-related oceanic and atmospheric environmental change is the geochemical signal of chemical isotopes and trace elements entrapped in its skeleton.

Oxygen isotopes are extremely valuable for recording both SSTs (figures 7-3, 7-4) and, indirectly, the impact of precipitation variability on sea-surface salinity (SSS). The calibration of oxygen-isotope ratios to oceanic parameters is described in many studies (Fairbanks and Dodge 1979; Druffel 1985; McConnaughey 1989; Linsley et al. 1994; Carriquiry et al. 1994; Wellington et al. 1996; Dunbar et al. 1996; Leder et al. 1996).

Three main factors influence the oxygen isotopic ratios in coral skeletons—temperature of the seawater, which can vary seasonally, especially in the eastern tropical Pacific Ocean; the isotopic composition of the seawater, $\delta^{18}O$, which also varies regionally and seasonally owing to the effects of evaporation and precipitation; and biologically mediated factors (Dunbar et al. 1996), which we called *vital effects* in previous chapters. In cases where SST variability is significant and seawater $\delta^{18}O$ variability is minimal, most studies have found a consistent relationship between isotope ratios and temperature in coral skeletons. For example, Druffel (1985) and McConnaughey (1989) reported isotope-temperature relationships of 0.22‰ per 1°C and 0.21‰ per 1°C, respectively, whereas Gagan et al. (1994) reported an isotope-temperature relationship of 0.18‰ per 1°C. Applying this relationship

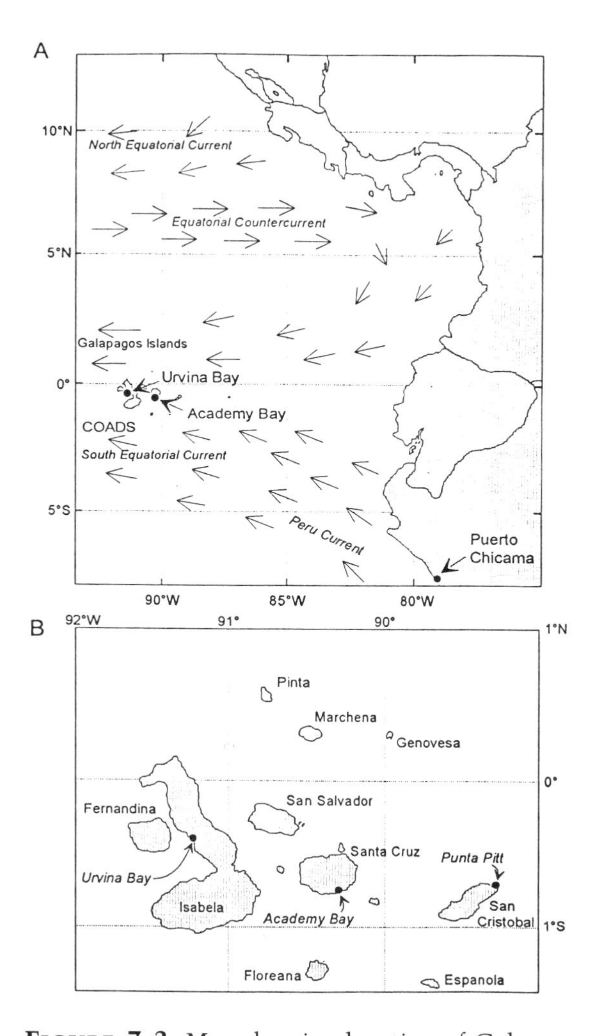

FIGURE 7-3 Map showing location of Galapagos Islands and ocean currents related to paleoclimatic reconstruction of long-term El Niño-Southern Oscillation history from Galapagos corals. A: Eastern Pacific. B: Enlargement of Galapagos area. COADS (Comprehensive Ocean Atmospheric Data Set) is site of instrumental oceanographic data collected from a 2° by 2° square centered on 1°S 92°W. Courtesy of R. Dunbar and American Geophysical Union. From Dunbar et al. (1994).

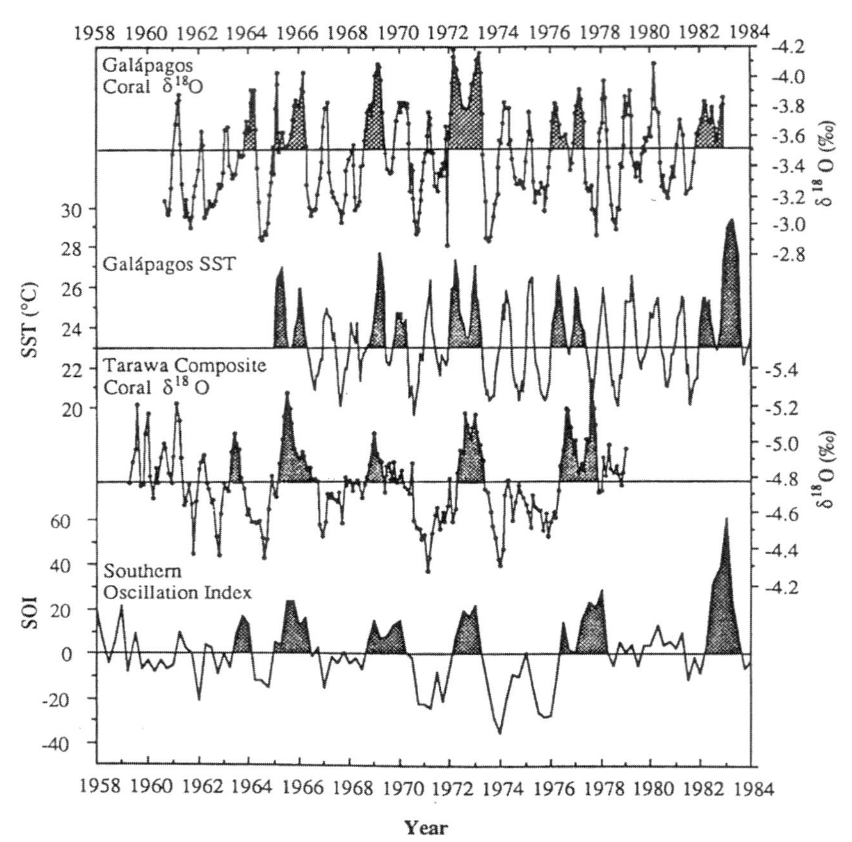

FIGURE 7-4 Coral oxygen-isotope records from Tarawa Atoll and Punta Pitt, San Cristobal Island, Galapagos, compared with measurements of ENSO-related SSTs and Southern Oscillation Index (SOI), show relationship between eastern and western Pacific Ocean ENSO patterns and serve as a means to calibrate isotopic measures of coral skeletons for long-term ENSO reconstruction. Courtesy of J. Cole and American Geophysical Union. From Cole and Fairbanks (1990).

to a paleoclimate study, oxygen isotopes can be used to reconstruct ocean temperature changes with a high precision (about ± 0.5°C). In studies on the genus *Porites* from Cano Island, Costa Rica, Carriquiry et al. (1994) found that a strong isotopic depletion (i.e., a negative isotopic anomaly) characterized skeletal isotope values from El Niño events. The strongest isotopic excursion was from the extreme oceanic warming that occurred during the 1982–1983 El Niño event.

In contrast to Cano Island, annual SST variability in the Gulf of Chiriqui off Panama is only 1–2°C. However, a strong annual cycle of oxygen-isotope variability recorded in Chiriqui corals reflects the influence of seasonal ITCZ migrations and rainfall variability on SSS as well as oxygen-isotope ratios. Dunbar and Wellington (1981) first described the isotope-SSS relationship as $\Delta^{18}O$ SSS = 0.12‰ for corals from the Gulf of Panama. Linsley et al. (1994) later reported a 10‰ $\delta^{18}O$ change in seasonal precipitation in Panama, with the most depleted values found during the most intense rains of the wet season owing to the way oxygen isotopes fractionate during rapid precipitation events. They broke down the contribution of salinity (80%) and temperature (20%) variability to the total 0.9‰ $\delta^{18}O$ annual signal in Gulf of Panama corals. As a general rule, when ocean salinity decreases due to rainfall, both seawater and coral $\delta^{18}O$ decrease.

Several recent field studies have improved the calibration of the $\delta^{18}O$ record of corals with SST. They were designed specifically to better understand the potential impact of growth (Lough and Barnes 1990, 1992), metabolic effects, and sampling density on the quality and reliability of the coral isotope record. Three species of *Pavona* and *Porites* were examined from the Galapagos (Wellington et al. 1996) and the genus *Montastrea* from Florida (Leder et al. 1996). The sampling interval ranged from 10 to 24 per year for the Galapagos and from 50 to 55 samples for the Florida corals. Wellington et al. (1996) used several sophisticated analytical methods, including comparison of instrumental and satellite SST measurements and biannual dye, to mark the coral colony growth.

Important conclusions came out of these studies. First, the $\delta^{18}O$ of coralline aragonite is a reliable indicator of SST over a 100-km area in the Galapagos; it faithfully records past seasonal and monthly changes in SST from the region. Seasonal SST variability was successfully tracked in all regions examined, and some local SST variability was also identified. Second, 20 samples taken over the 13–18 mm/yr annual skeletal growth period was sufficient to identify seasonal changes in SST. Third, interannual SST changed 1.48°C during a 1993–1994 calibration experiment; this produced a $\delta^{18}O$ change of 0.31‰ that accounted for 95% of the variance. The $\delta^{18}O$ of seawater, although annually variable, was not a major factor in obscuring the coral $\delta^{18}O$-SST relationship in the Galapagos. Finally, interspecific variability in the $\delta^{18}O$ among the three different coral species did not severely affect the SST signal; all species successfully tracked secular trends in SST. Regional calibration studies such as those by Wellington, Leder, and colleagues are critical for establishing improved precision of SST variability.

The final source of potential isotopic variability in coral skeletons stems from vital effects. The skeletons of many corals, like those of foraminifera, are secreted with $\delta^{18}O$ values several parts per mil less than the $\delta^{18}O$ of the surrounding seawater. This biologically mediated disequilibrium (McConnaughey 1989) is believed to be constant both in individuals from any single genus and also along the axis of growth for a single coral colony. Thus, coral vital effects are usually not a major problem within a single study (McConnaughey 1989; Linsley et al. 1994), but biological processes pose additional challenges as the field of coral scleroclimatology expands and interregional and intergeneric isotope records are compared.

The $^{13}C/^{12}C$ ratios in tropical coral aragonite yield a more ambiguous signal than that of oxygen isotopes for studying past ENSO variability (Swart et al. 1996b). The $^{13}C/^{12}C$ ratio is generally related to both endogenous factors such as coral ecology and physiology, acting in concert with exogenous factors such as light intensity (Carriquiry et al. 1994). The most important endogenous factor is photosynthetic activity of the zooxanthellate algae that live symbiotically in the corals (Fairbanks and Dodge 1979; Swart 1983, Wellington and Dunbar 1995). Because higher solar insolation, itself a function of sunlight and clouds, generally leads to a carbon isotopic enrichment in the coral tissue and skeleton, the carbon isotopic signal is in a sense a proxy for the rate of photosynthesis. Photosynthesis rates vary both seasonally and with water depth, and relationships reflecting this variation have been found in coral skeletons.

Shen et al. (1992a) showed promising results in establishing an inverse correlation between the $\delta^{18}O$ and $\delta^{13}C$ in corals from Punta Pitt, eastern Galapagos Islands. The carbon isotopic variation was considered to be related to cloud cover that varies during rainy and dry seasons. Wellington and Dunbar (1995) were also able to show high coherency and positive correlation between $\delta^{18}O$ and $\delta^{13}C$ in coral time-series records from several tropical sites, supporting the idea that cloud cover and light intensity influence the $^{13}C/^{12}C$ ratios.

At other sites, however, factors unrelated to light intensity can overwhelm the seasonal-insolation shifts and distort the carbon-isotope record. For example, water clarity variability due to upwelling and plankton blooms may cause carbon isotope variability at Urvina Bay in the Galapagos (Wellington and Dunbar 1995). Carriquiry et al. (1994) concluded that seasonal variability in Costa Rican coral

$^{13}C/^{12}C$ ratios is related to the photosynthesis-respiration ratio and to endogenous factors mediated by solar activity. Of the most important environmental factors, they found that ecological stress, irradiance intensity, turbidity (mainly due to runoff), and resuspension of sediment can all influence the carbon-isotope signal. The complex biological factors involving coral symbiosis, growth, and metabolic activity clearly were at least as important as exogenous solar insolation and other environmental effects.

In summary, carbon isotopes are not yet routinely used in coral paleoclimatology except in situations where they aid in developing a chronology. Wellington and Dunbar (1995:19) reviewed the complex hydrographic, climatic, and biological factors that control coral skeletal $\delta^{13}C$ and concluded "ENSO events generally have no predictable effects on $\delta^{13}C$ with the exception of the 1982/83 event at Cano Island." Additional research on the uptake of carbon isotopes by coral is needed.

The third category of geochemical tools in coral skeletons are elemental concentrations (see Shen and Sanford 1990). During coral growth, coprecipitation of minute amounts of cadmium (Cd), barium (Ba), and manganese (Mn) are incorporated into the coral skeleton, probably through substitution for the calcium (Ca) ion in the crystal lattice of the coral aragonite. Trace-element substitution often occurs in direct relation to climate-related variables. The coprecipitation of strontium from seawater is a case in point. The amount of strontium incorporated into coral skeletons from three separate genera collected from widely spaced tropical areas in the Pacific and Caribbean appears to be strongly controlled by water temperature (Smith et al. 1979). Shen and Boyle (1988) and Lea et al. (1989) also demonstrated the utility of barium for evaluating nutrient variability and water temperature in corals. Lea et al. (1989) found that in the coral *Pavona clavus* from the Galapagos Islands, $BaCO_3$ replaced $CaCO_3$ in relation to water temperature. Low barium content was found in warm water, high barium in cooler water. They concluded the Ba:Ca ratios changed in response to the degree of upwelling and entrainment of nutrient-rich water into the surface waters, a signal then taken up by the corals.

The coprecipitation of cadmium in corals from the Galapagos seems to be a function of seasonal upwelling related to oceanic currents. Manganese uptake on the other hand is controlled mainly by atmospheric dust and the effects of sunlight on surface ocean. During a typical non-ENSO year, the Panama Current off South America brings oligotrophic nutrient-depleted and cadmium-depleted waters southward. During ENSO events, the boundary between the cool upwelling nutrient-rich water and the warm Panama Current waters is shifted farther southward, and surface waters become more cadmium depleted.

The Equatorial Undercurrent also affects seawater trace-element concentrations in complex ways that ultimately can lead to differential elemental uptake by corals from different regions. The Equatorial Undercurrent influences the western Galapagos more than it does the eastern Islands. Delaney et al. (1993) examined the seasonal signature preserved in coral manganese and cadmium records from eastern and western Galapagos corals. A colony of *P. clavus* (UR-LL-86) from Isabela Island in the western Galapagos showed that for the interval 1946–1953, lower Mn:Ca ratios characterized intensified seasonal upwelling during the non-ENSO years 1946–1950. The Cd:Ca ratios from the same Isabela Island coral showed less distinct seasonal cycles than did the Mn:Ca record for the western Galapagos. In the eastern Galapagos off Hood and San Cristobal Islands, Linn et al. (1990) and Shen and Sanford (1990) found the opposite effects for cadmium and manganese. Eastern Galapagos Cd:Ca ratios in *P. clavus* for the interval 1965–1979 were higher after seasonal upwelling events and showed a stronger signal than did Mn:Ca. Thus, the proximity of the western Galapagos to the Equatorial Undercurrent appeared to cause a stronger manganese signal in this region; conversely, the nutrient-depleted Panama Current surface waters exert a greater influence on the cadmium record of the eastern Galapagos.

The incorporation of small amounts of radioactive uranium into coral skeletons, which often remain a closed chemical system after death, is another geochemical attribute of the coral skeleton. Although not a climate proxy, uranium uptake makes fossil corals amenable to accurate dating. When thermal ionization mass spectrometry (TIMS) is used, $^{230}Th/^{234}U$ ages have a small age uncertainty $< 1\%$ for samples 1000 yr to 100 ka). Therefore, fossil corals can provide seasonal and interannual paleoclimate records from glacial periods or past interglacial intervals.

Strontium is taken up into coral skeletons from seawater and its abundance has been used for paleoceanographic research. Beck et al. (1992) calibrated strontium-calcium (Sr:Ca) ratios in Pacific Ocean corals to SST and oxygen-isotope data and hypothesized that strontium uptake in corals is mediated mainly by water temperature. We have seen in earlier chapters that the Sr:Ca technique has been instrumental in debates about glacial-age climate in tropical regions (see Guilderson et al. 1994). However, the Sr:Ca method hinges on assumptions that biological factors are minimally important in controlling strontium uptake and that strontium content in seawater is relatively invariable. De Villiers et al. (1995) demonstrated that biological effects related to photosynthesis and metabolic processes and to seawater strontium variability can in some cases lead to uncertainties in SST estimates of 2–3°C.

Alibert and McCullough (1997) calibrated Sr:Ca ratios in *Porites* from the Great Barrier Reef for application to the study of ENSO-generated SST variation. They determined a Sr:Ca–temperature relationship as follows:

$$\text{Sr:Ca} \times 103 = 10.48\ (\pm\ 0.01) - 0.0615\ (\pm\ 0.0004) \times T,$$

where T is temperature in °C. Importantly, they found that, although environmental changes (e.g., rainfall) can affect coral extension and skeletal density, these processes did not affect Sr:Ca partitioning in *Porites.* They applied their Sr:Ca SST method to reconstruct ENSO-temperature variability since 1964 and found significant cooling associated with El Niño episodes in 1965, 1972, and 1982–1983 and a net 1.3°C SST increase since 1965.

Uranium also offers promise as a potential tool for reconstructing past SST variations using coral skeletons. This topic was recently reviewed by Shen and Dunbar (1995).

From the preceding pages, it is easy to extol the virtues of corals as powerful tools for studying interannual tropical climate history. Corals, of course, have several drawbacks limiting their use in paleoclimatology. For example, they are not available in extratropical regions. Also, errors in counting growth bands can produce problems of interpretation of longer-term paleoclimate records, although evidence that coral growth records global cooling episodes, such as that resulting from the 1991 Mt. Pinatubo volcanic eruption (Crowley et al. 1997), increases confidence that long-lived coral chronology can be advanced. Corals' main benefit to paleoclimatology—the sensitivity of their growth to environmental stress from salinity, temperature, and nutrient depletion—can also be a major weakness in that growth cessation can occur, as it did during the extreme 1982–1983 ENSO event (Druffel et al. 1990). Moreover, there is limited understanding about certain aspects of coral physiology and metabolism, such as the way ultraviolet light and low nutrient levels evoke a physiological response that can influence growth. Coral physiology, growth, and skeletal isotopic disequilibrium are biological processes that epitomize the complications that arise in paleoclimatology because of poorly understood organism-environment connections (see the discussion in chapter 3). Furthermore, some geochemical indicators record local and regional events and are not necessarily related to global or oceanwide climate anomalies (Delaney et al. 1993). Regional and global climate events should be replicated across a wide region and/or be evident in multiple climate indicators at any single site.

One final practical limitation, common to ice cores and dendroclimatology as well, is that the sheer number of sample analyses needed in each study impedes extensive geographic and temporal coverage of coral records. This is especially so if one wishes to establish intra-annual—that is, seasonal or monthly—variation as well as interannual and interdecadal trends. Climatologists demand as fine a network of sites as possible to compare instrumental and paleoclimate trends. So the dilemma arises: how many samples are required to adequately identify trends in ENSO or QBO without sacrificing the statistical validity of the study.

Quinn et al. (1996) addressed the question of sampling strategy in a study of oxygen isotopes in 40 yr of coral growth sampled at monthly intervals using corals from Amedee, New Caledonia, and the coral from Tarawa Atoll studied by Cole and Fairbanks (1990). Amedee is located in the Eastern Coral Sea (22°S, 16°7'E) in an open ocean site far from coastal influences. They studied a colony of *P. lutea* with an average growth rate of 1.23 cm/yr; for the 40-yr interval of study, they sampled every 1.03 mm per density band couplet. Taking advantage of the sequential (i.e., not random) growth of the colony, they took a single linear sampling pathway along the colony's axis so as not to compromise the temporal sequence.

Quinn's group found a conspicuous pattern of 40 annual oxygen-isotope cycles corresponding precisely to the exact number of density-band couplets. A correlation of $r = 0.88$ existed between Amedee SST and oceanic oxygen-isotope ratios for a 20-yr period for which local temperature measurements were available. The isotopic trends in the Amedee coral indicated annual SST variations averaged about 5.3°C.

Spectral analyses of the isotopic data also revealed important variations in the frequency domain. Spectral peaks occurring at 6.7, 2.9, 2.3, and 1.0 yr provided strong paleoclimate evidence for periodic influence at ENSO (2.9–6.7 yr), QBO (2.3 yr), and annual (1.0 yr) frequency bands.

To determine the minimum number of samples needed to identify ENSO and QBO events with confidence, Quinn et al. (1996) combined the monthly isotopic measurements into bimonthly and quarterly groupings. They discovered that even with this lumping procedure, the same ENSO, QBO, and annual signals generally appeared. They judged that the bimonthly sampling gave essentially the same results as the monthly sampling and thus was adequate for identifying all three frequency signals; the quarterly samples yielded no loss of information about interannual and decadal time scales, but quarterly sampling was not refined enough to identify the full amplitude of annual cycles.

The significance of the study by Quinn et al. is that the character and the percent variance explained by ENSO, QBO, and annual periods changed little with reduced sampling means. One practical ramification stemming from this conclusion is that future researchers can be less stringent about how many samples are analyzed, especially if the goals of a particular study involve reconstruction of ENSO or lower-frequency climate events.

Tropical ENSO Paleoclimate Records

We now turn our attention to some of the more important ENSO-related paleoclimate records that emerge from corals. As mentioned earlier, 92 coral paleoclimate records were available as of the 1993 summary of Cole and Dunbar; many more have been obtained since. however, fewer than 10 true multicentury coral records have been published, and relative to the paleoclimate reconstructions described in earlier chapters, coral paleoclimatology is in its infancy. Dunbar et al. (1996) reviewed records from 10 key coral sites in the eastern Pacific from Baja, California, to the Galapagos and the Gulf of Panama. Coral oxygen-isotope records from eastern Pacific sites range from 25 yr at Bartolome Island (Galapagos) to 374 yr for Urvina Bay (Galapagos, see later). Here I concentrate on four regions of the Pacific Ocean that together give a cohesive picture of the past few centuries of tropical ENSO-related climate history: ITCZ-related precipitation variability in the eastern Pacific off Central America, SST variability in the Galapagos region, trade-wind and precipitation oscillations in the central Pacific near Tarawa Atoll, and the evolution of ENSO SST and thermocline variability in the western Pacific near New Caledonia and Australia's Great Barrier Reef. I also briefly discuss coral records from Barbados and the Indian Ocean, as well as the interannual records of ENSO in extratropical areas obtained from dendroclimatology.

The Eastern Pacific Climate Record: ITCZ and ENSO

The Pacific Coast of Central America along the western shores of Costa Rica and Panama is the backdrop for studies of climate variability related to migration of ITCZ that occurs during ENSO events.

Research has been focused on areas in the Gulf of Chiriqui (Chiriqui Lagoon), off the Pacific coast of Panama, and off Cano Island, Costa Rica.

Gulf of Chiriqui

The Gulf of Chiriqui (8°N, 82°W) harbors some of the most abundant and diverse coral colonies in the eastern Pacific Ocean and has been the site of several key paleoclimate studies. Large colonies of the genera *Pavona* and *Porites* can reach 4 to 5 m in height. The ITCZ is a primary source of seasonal climate variability in the Gulf of Chiriqui region, characterized by small annual oceanic temperature variations of about 1–2°C and a rather pronounced wet season. Seasonal precipitation changes in the Gulf of Chiriqui region influence the $\delta^{18}O$ of seawater and corals and are used to reconstruct oceanic salinity changes due to ITCZ-related precipitation anomalies.

Climate-related isotopic variation at the Gulf of Chiriqui is quite significant. Rainwater depleted in $\delta^{18}O$ predominates when the ITCZ migrates northward to the Gulf of Chiriqui (8–10°N latitude) in August and September from its position near the equator in March and April. Salinity variability can account for as much as a 1.1‰ shift in the oxygen isotope records in Chiriqui corals (Linsley et al. 1994). Freshening of seawater by runoff from the nearby land area also exerts a significant influence on ocean salinity and the $\delta^{18}O$ of surface waters.

Detailed investigation of corals from Secas Island in the Gulf of Chiriqui have helped to decipher ITCZ history for the period spanning the past three centuries (figure 7-5). Building on earlier studies by Druffel et al. (1990), Linsley et al. (1994) performed an in-depth study of subseasonal, seasonal, and interannual variability in ITCZ in the eastern Pacific for the period 1707–1984. They analyzed two corals: *Porites lobata,* a reef coral that grows at rate of several millimeters to centimeters per year, and *Pavona.* The Chiriqui record provides an exceptionally detailed record because growth-banding is evident in the coral skeleton. Low-density bands form in *Porites* and *Pavona* during the dry season, and the high-density bands represent growth during the ITCZ-influenced wet season of May through December. Such a rapid growth rate and accurate growth-band chronology afforded an excellent opportunity to obtain a centennial-scale paleoclimate record.

Linsley et al. sampled the coral core at an average frequency of 9.9 samples per year, with a range of 7–17 samples for any 12-month period; they estimated that perhaps only a single year might be missing from the entire three-century record and the long-term climate record had an accuracy of about ± 2 years. To set the annual calendar

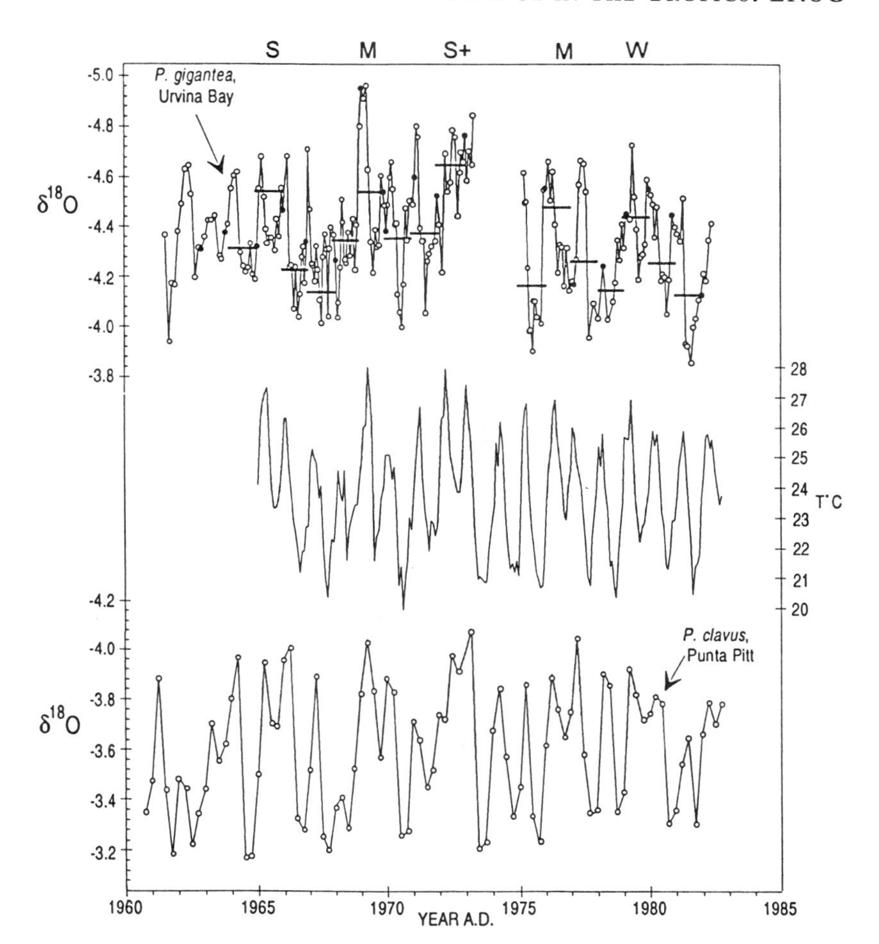

FIGURE 7-5 Long-term tropical climate variability in ENSO-related precipitation since 1700 A.D. obtained from coral oxygen-isotope variation from Secas Island, Gulf of Chiriqui, Panama. Rightmost panel shows original detrended and 2-yr filtered data. RC = reconstructed climatic components. Courtesy of B. Linsley and the American Geophysical Union. From Linsley et al. (1994).

within each annual growth band's light-dark couplet, Linsley assumed that the lowest $\delta^{18}O$ value measured in a couplet represented the peak wet season in the month of November.

El Niño events are manifested in the Chiriqui coral as relatively minor shifts of approximately 0.2–0.4‰ due to SST changes, accounting for about 20% of the total oxygen-isotope signal observed in the coral record. These isotopic shifts are much smaller in amplitude than those due to the larger seasonal ITCZ-related precipitation variability, which account for 80% of the modern isotopic variability measured in seawater isotopes from Chiriqui.

One major conclusion reached by Linsley et al. (1994) was that the ITCZ has remained a fundamental part of the tropical climate system for at least the past three centuries. The isotope record from 2730 samples covering the years 1707 until 1984 showed that the ITCZ behavior exhibits remarkably stable and regular patterns over centennial time scales. For at least 277 years, the ITCZ probably moved northward to the Gulf of Chiriqui region near 8–10°N latitude during each Northern Hemisphere summer, when maximum rainfall extends northward from near the equator. These results confirmed that seasonality has characterized ITCZ migrations for centuries and has also constrained the latitudinal extent of annual ITCZ migration over long time scales.

In a second part of the study, Linsley and colleagues evaluated the periodicity in the Chiriqui isotope record using statistical methods to remove the annual and long-term signals. This approach resulted in several significant periodicities over the 277-year $\delta^{18}O$ record. Listed in decreasing order of the percent variance explained, these are 9, 3–7, 17, and 33 yr.

The 9-year band (ranging from 7.5–11.8 yr), is suspiciously close to the well-known sunspot 11-year cycle found in other climate proxy records (see chapter 6). Linsley and colleagues found the 11-year period was strongest during the 1800s. However, they were reluctant to assign a definite causal mechanism, such as solar variability, to this pattern because they could not establish a rigorous correlation with trends in sunspot and solar irradiance. They also cited a lack of a plausible physical mechanism linking small irradiance changes in the upper atmosphere to processes operating at the earth's surface. Nonetheless, the decadal-scale period characterizing ITCZ variability over several centuries represents an important component of tropical climate evolution.

The 3- to 7-yr periodicity accounted for 30% of the variance and seems to manifest an ENSO signal. Compared with the Galapagos record discussed later, however, the Chiriqui ENSO events are not strong components of the local climate record. Sea-surface temperature variability during ENSO events at Chiriqui is smaller than it is at other eastern Pacific tropical sites. At Chiriqui the opposing effects of SST and precipitation on the coral $\delta^{18}O$ signal combine to dampen the ENSO signal. The Chiriqui record, for example, does allow the identification of certain large ENSO events and shows a decrease in ENSO amplitude during the 1920–1930 period and again during the 1820s. In sum, firm correlations between the Chiriqui ENSO record could not be established with coral records from Tarawa Atoll or the Galapagos, although the Chiriqui record has now been correlated with coral records from New Caledonia and Madang, New Guinea (Quinn et al. 1998).

Long-term centennial-scale climate information related to the Little Ice Age (LIA) also emerged from the Chiriqui record. For instance, the average seasonal rainfall during and after the Little Ice Age did not change appreciably; hence, this widely discussed climate event may not have altered the ITCZ regularity or intensity. Although seasonal changes during the Little Ice Age were probably of the same magnitude as those today, large-amplitude interdecadal oscillations in $\delta^{18}O$ occurred over the period including the Little Ice Age, suggesting that interdecadal climate variability at that time was far greater than it was over the past century.

Gulf of Panama

Like the Gulf of Chiriqui, the Gulf of Panama also shows strong SSS changes during the wet-season months (May through November) when the ITCZ moves to its more northerly location. Salinity can be reduced to 20–25‰. However, rainfall anomalies during El Niño events such as those at Chiriqui do not reach the Gulf of Panama region, which remains relatively dry throughout the year.

Druffel et al. (1990) demonstrated convincingly that ENSO events are recorded in the coral isotopic record from Uraba Island in the Gulf of Panama and from Uva Island in the Gulf of Chiriqui. The seasonal ranges of $\delta^{18}O$ in Uraba and Uva Islands corals were 1.9‰ and 1.1‰, respectively, reflecting cooling due to seasonal upwelling in the Gulf of Panama. Furthermore, times of peak isotopic enrichment or depletion are offset by 4–6 months in the two areas because of these seasonal variations. Druffel and colleagues found that the seasonal range of $\delta^{18}O$ is suppressed during El Niño events (especially the events of 1965, 1969, and 1982) owing to reduced enrichment during the upwelling season and less depletion during the wet warm season. Cautioning that coral mortality during the extreme events might lead to difficulty in recognizing older events, and that seasonal effects might complicate interannual records, Druffel and colleagues nonetheless demonstrated strong evidence of the sensitivity of coral skeletal geochemistry in these regions to precipitation and temperature oscillations associated with ENSO.

Shen et al. (1991, 1992a) used trace-element geochemistry to document ENSO-related upwelling changes over the past few decades in the Galapagos Islands and at Contadora Island in the Gulf of Panama. In the Galapagos (see later), they were able to use Cd:Ca ratios as a "paleofertility" indicator of nutrient changes in Galapagos corals. Off Contadora Island in the Gulf of Panama, however, trace-element records derived from corals are more complex than they are in the Galapagos because of the influence of metals derived from riverine runoff, nearshore remobilization from sediments, and particulate desorption. Despite these complications, *Pavona gigantea* showed seasonal variation in Mg:Ca ratios, but it was more difficult to identify clear ENSO signals in the Contadora Island corals than it was in the Galapagos specimens.

Galapagos (Urvina Bay, Punta Pitt)

As they have for Charles Darwin and scores of other naturalists, the Galapagos Islands off the coast of South America have provided the raw material for important discoveries about tropical climate history. Galapagos coral paleoclimate records have added significantly to our understanding of the dynamics of tropical climate change over the past few centuries. Located directly on the equator, the Galapagos experience wide seasonal swings of SSTs. During the wet, warmer season (January through May), southeast trade winds weaken and the ITCZ migrates south, bringing warm, less-saline water. In contrast, during the dry, cool season (June through December) the Peru Coastal Current carries cooler (20–24°C), more-saline water from the southeast to mix with the South Equatorial Current.

Oceanic conditions in the Galapagos also change dramatically during ENSO events. El Niño events can cause temperature anomalies of 1.5–4.0°C, depending on the strength of the particular event. The SST maximum during an El Niño is 1.0–2.0°C warmer than usual; during the cool season after an El Niño, SST remains above normal levels.

Two main regions of the Galapagos have yielded paleoclimate records: Urvina Bay on the western side of the western Isabela Island, and Punta Pitt on the north coast of the eastern island of San Cristobal. Between these two islands lies the central Galapagos island of Santa Cruz, where long-term oceanographic and meteorological data from Academy Bay serve as a source of calibration data for coral geochemistry.

Several studies have firmly established that the oxygen-isotope record from Urvina Bay is primarily a record of SST variability (Shen et al. 1991, 1992a; McConnaughy 1989; Druffel et al. 1990; Dunbar et al. 1994; Wellington and Dunbar 1995; Wellington et al. 1996) (figure 7-6). McConnaughy (1989) showed that oxygen isotopes in Galapagos corals could be used to identify El Niño events. Low $\delta^{18}O$ values in the Galapagos characterized warm SSTs recorded during ENSO events in 1963, 1965, 1967, 1972–1973, and 1976. Druffel et al. (1990) studied coral skeletons dating back to 1929 from four sites in the Galapagos,

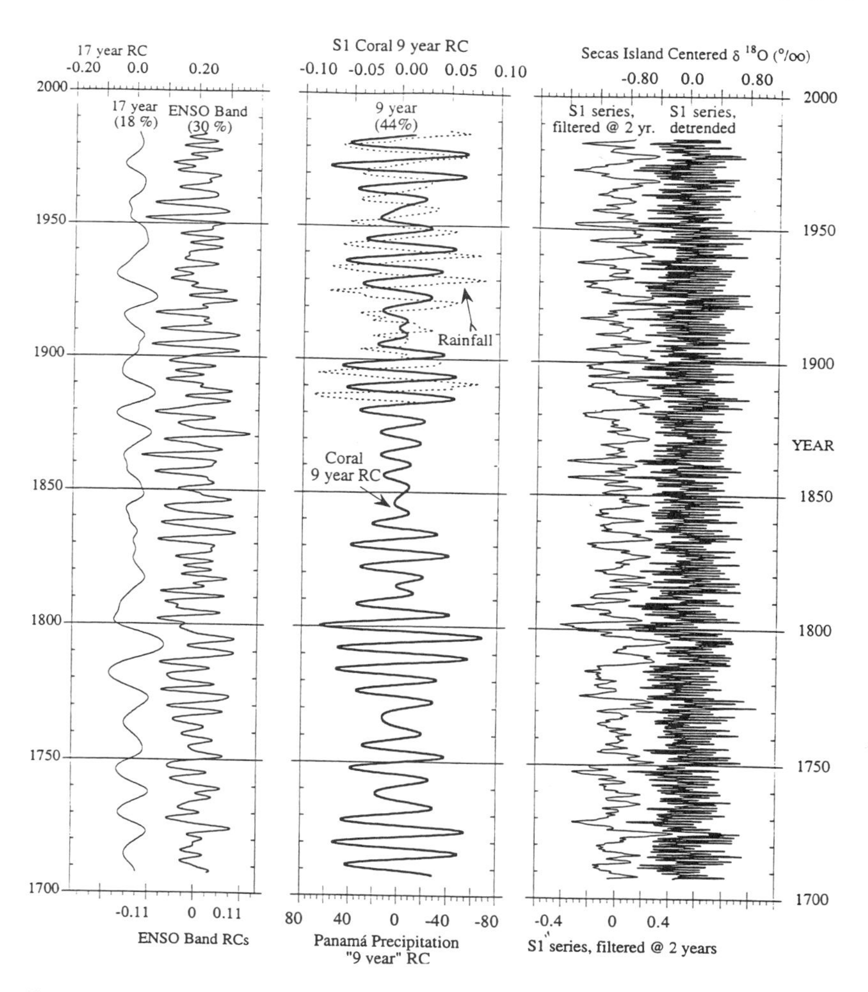

FIGURE 7-6 Calibration of coral oxygen-isotope trends from two Galapagos corals from Urvina Bay and Punta Pitt to measured temperature record. Courtesy of R. Dunbar and American Geophysical Union. From Dunbar et al. (1994).

emphasizing the 25-yr period between 1962–1983. They found that the $\delta^{18}O$ of corals from the Galapagos reflected higher average water temperature and lower average salinity during ENSO events. The seasonal range of $\delta^{18}O$ was actually reduced during most ENSO events. Wellington and Dunbar (1995) studied a 247-mm-high *Pavona gigantea* (specimen UV-87) that died from bleaching during the 1982–1983 El Niño. El Niño events are expressed in UV-87 as negative $\delta^{18}O$ anomalies of 0.2–0.3‰ relative to the non-El Niño baseline. Stronger El Niño events exhibited greater amplitudes for the oxygen-isotope anomaly than did weaker ones.

As mentioned earlier, Shen et al. (1991, 1992a) successfully used Cd:Ca ratios to document ENSO-related upwelling changes in the Galapagos Islands as well as at Contadora Island in the Gulf of Panama. At Punta Pitt on San Cristobal, Shen and colleagues showed a strong concurrence between Cd:Ca ratios and SST anomalies due to seasonal upwelling of cold water to the surface during the period 1965–1980. They also showed that a curve of Punta Pitt coral Mn:Ca ratios flattens out during the 1965, 1969, 1972, and 1976 ENSO events. In a second Galapagos coral from Urvina Bay, Shen showed that Cd:Ca ratios vary in phase with temperature anomalies but that Mn:Ca ratio changes are out of phase because cadmium and manganese have the opposite gradients in eastern Pacific surface waters.

Armed with a strong correlation between historical ENSO events and with coral oxygen-isotope ratios well established, researchers could consider the paleoclimate record from the Galapagos as a long-term record of tropical SST variability. In a seminal paper, Dunbar and colleagues (1994) studied a 367-yr climate record preserved in two species of *Pavona* from Urvina Bay on the western edge of Isabela Island (figures 7-7 and 7-8). One specimen was a 3-m-high colony of *Pavona clavus* (UR-86); the other was the small colony of *P. gigantea* (UR-87) mentioned earlier. The chronology for UR-86 was established by counting back growth bands from the year the colony died in 1953 to the year 1586. To cross-check their density-band age estimates, Dunbar et al. (1994) also ran several mass spectrometric determinations of ^{230}Th and identified a large Mn:Ca excursion, discovered earlier by Shen et al. (1991), caused by a major volcanic event in 1825.

Dunbar and colleagues used both oxygen isotopes and growth rate as paleoclimate proxies. By calibrating the $\delta^{18}O$ of UR-87 to instrumental records of SST for the period 1965–1983, they showed that SST accounted for 80% of the variance in the $\delta^{18}O$ record at interannual and decadal scales. For comparison, recall that at Tarawa Atoll in the western Pacific and at Chiriqui Lagoon in Panama, most of the oxy-

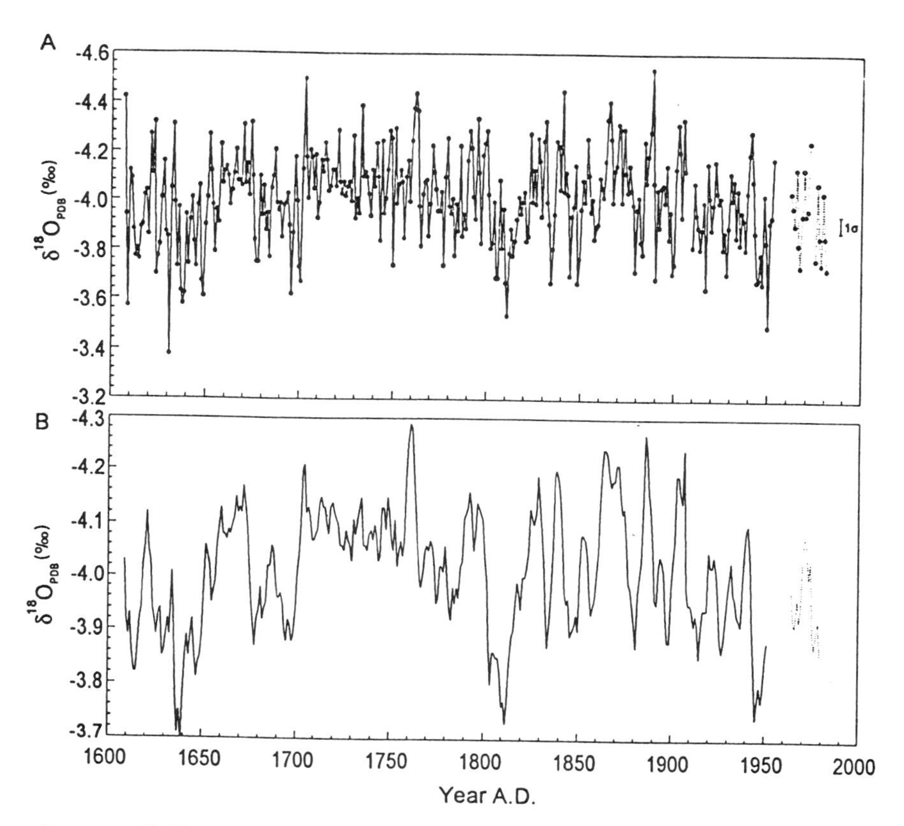

FIGURE 7-7 Centennial-scale variability in coral oxygen isotopes from Urvina Bay, Galapagos Islands. A. Original data (solid line coral = specimen UR-86; dotted line = UR-87). B. Five-point moving average of the $\delta^{18}O$ data. Courtesy of R. Dunbar and the American Geophysical Union. From Dunbar et al. (1994).

gen-isotope signal was controlled by precipitation-related factors. Dunbar's group also found a wide range of environmentally mediated growth rates in Galapagos corals, varying from a minimum of 5 mm/yr to a maximum of 22 mm/yr, averaging 12.9 mm/yr.

How well did the Urvina Bay UR-86 coral match the historical El Niño record of Quinn (1992)? Briefly, four of the largest El Niño events in the historical record, from the years 1642, 1728, 1743, and 1891 A.D., were faithfully recorded by strong isotopic excursions. Of the 30 largest isotopic excursions deviating from the baseline by 0.23–0.48‰, 20 match El Niño years identified in Quinn's ENSO chronology, and

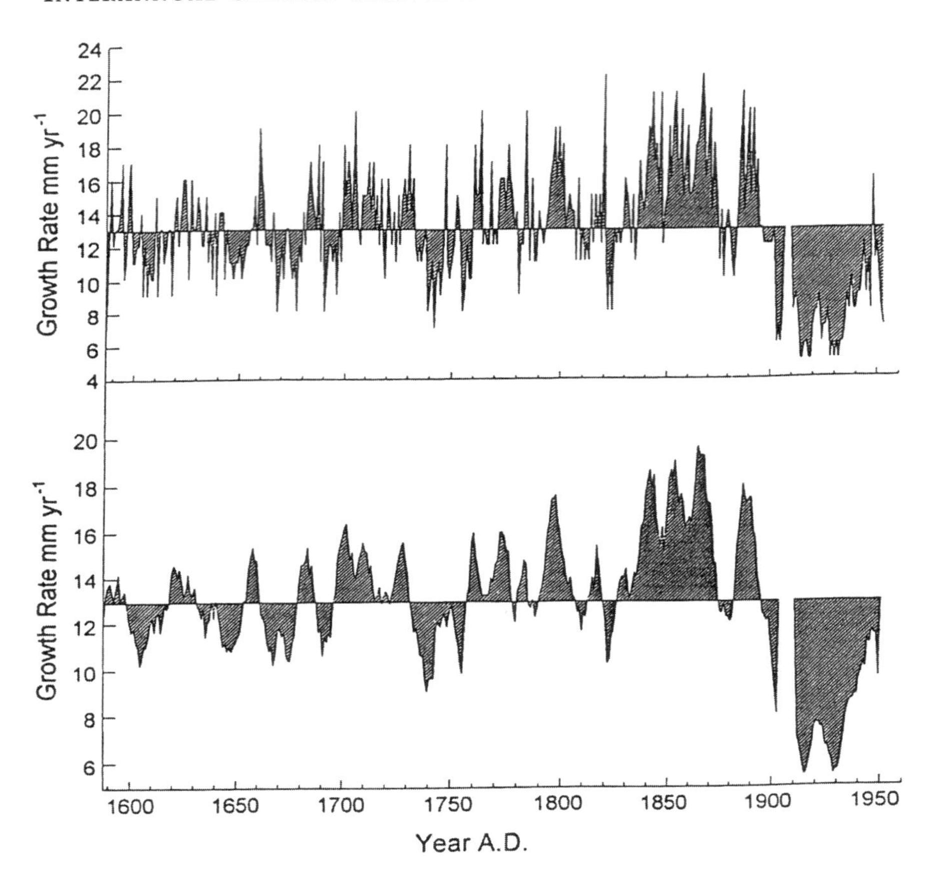

FIGURE 7-8 Centennial-scale variability in coral growth rate from Urvina Bay, Galapagos Islands, coral UR-86. Lower panel is 5-point moving average of original data in upper panel. Superimposed on long-term trends in growth rate such as high rates during the mid to late 1800s are 23-yr and 11-yr high frequency cycles of coral growth. Courtesy of R. Dunbar and the American Geophysical Union. From Dunbar et al. (1994).

an additional 8 excursions occur in a year immediately adjacent to one of Quinn's historical El Niños. Of the 100 largest oxygen isotope excursions, 88 match exactly or within one year of Quinn's ENSO years. Moreover, 9 of the mismatches occurred in the 1600s, a period when Quinn's historical data were probably too spotty to be reliable. Thus, among their many important conclusions, Dunbar and colleagues provided compelling evidence that $\delta^{18}O$ oscillations from Galapagos corals record most of the ENSO events identified independently on the basis of documentary data by Quinn (1992). The Galapagos corals provide a unique quantitative archive of oceanic changes during these events (figure 7-4).

A second major contribution from the Urvina Bay coral was its record of nearly four centuries of decadal and centennial tropical climate variability. As we saw in the previous chapter, the period since about 1600 A.D. is critical because it spans the much-disputed Little Ice Age and the subsequent "global warming" of the past century (Diaz and Pulwarty 1994). The patterns of *Pavona* isotopes show that interannual SST variability during most of the 1600s and 1800s was slightly greater (about 2.5°C) than that during the twentieth century. The lowest isotopic variability occurred during the 1700s, a time when low $\delta^{18}O$ values signify generally warmer SSTs. Thus, a period of tropical climatic stability split the two major intervals comprising the Little Ice Age, 1600–1660 and 1800–1825 A.D. In fact, these relatively cool periods in the Galapagos correspond to some North American tree-ring records for seventeenth- and early nineteenth-century climate anomalies (Jacoby and D'Arrigo 1989). The bipartite nature of the Little Ice Age in the Galapagos, with periods of cold during the 1600s and again during the 1800s, indicates that climatic anomalies such as those recorded in North America and Europe can also be recognized in the eastern Pacific Ocean.

What do the Urvina Bay corals say about the much-publicized warming of the past century? Dunbar et al. found that, in contrast to many instrumental, tree-ring and other paleoclimate records indicating climatic warming since the Little Ice Age ended in late 1800s (e.g., Wigley 1995; Jones 1994; Nicholls et al. 1996), eastern Pacific SSTs did not rise between 1880 and 1940.

The seemingly anomalous behavior of ENSO in terms of its strength and frequency since the 1970s has led researchers to ask how this behavior compares with long-term ENSO behavior. The UR-86 record suggests that variability in ENSO frequency is a common feature of the Galapagos climate. During the 1600s, ENSOs occurred every 4.6–7 years, in the early to mid 1700s, every 3.0–4.6 yr, and during the early to mid 1800s, about every 3.5 yr.

Other significant climate shifts that are preserved reflect a change in ENSO band frequencies during the mid 1850s, when a strong 33- to 35-yr periodicity disappeared, shifting to a 17-yr period. The Galapagos growth-band and oxygen-isotope record also shows evidence of 11- and 22-yr cycles, raising the solar-climate issue discussed in chapter 6. Despite the temptation to speculate that these cycles were related to

11-yr sunspot and 22-yr solar magnetic cycles, Dunbar et al. (1994) suggested mechanisms related to short-term solar forcing could plausibly account for these changes. Nonetheless these solar cycles may modulate annual and ENSO climate events in unknown ways.

Finally, the Urvina Bay *Pavona* record also showed that the devastating 1982–1983 ENSO event was the strongest, at least in terms of Urvina Bay isotopic record, since 1586.

The Pacific Record of Tarawa Atoll

Interannual paleoclimate records from Pacific atoll corals contrast sharply with the eastern Pacific record showing strong SST variability in the Galapagos. Tarawa Atoll sits astride the equator at 1°N and 172°E in a region of the Pacific that experiences extremely high SSTs, among the warmest oceanic temperatures anywhere. Although SSTs do not vary much at Tarawa during ENSO events as they do in the eastern Pacific, Tarawa experiences extreme swings in interannual precipitation patterns (Ropelewski and Halpert 1987). Precipitation varies when the Indonesian low-pressure system migrates eastward during El Niño events, bringing with it heavy rainfall. Rainfall during ENSO years exceeds 300 mm/month and, during the greatest extremes, reaches as much as 500–800 mm /month. These anomalies are the equivalent of 300–400% of rainfall in a typical, non-ENSO year (Cole and Fairbanks 1990, 1993a).

This ENSO rainfall anomaly has a severe effect on oceanic environments, leaving its distinct signature on the isotopic record of coral skeletons in Tarawa Atoll. Oxygen isotope ratios of rainwater vary greatly during the two climatic extremes. The $\delta^{18}O$ values resulting from relatively low-intensity rains during non-ENSO years are isotopically heavy (nonshaded areas in figure 7-4). In contrast, rainwater $\delta^{18}O$ values are reduced to about –8 to –10‰ during strong ENSO events, when rainwater is isotopically lighter. Enough rain falls during ENSO events to alter the isotopic composition of the seawater in which the corals grow and to lower the surface salinity. Near-surface water stratification results in the concentration of isotopically light, low-salinity water in the surface layers where the corals secrete their skeleton. Comparing Tarawa oxygen isotopes to SOI values obtained through observational measurements, Cole et al. (1993a) found that most ENSO events between 1960 and 1980 were preserved as isotopic minima. The strong 1982–1983 El Niño event was not recorded in the isotopic record from Tarawa because the foci of precipitation was located east of Tarawa (Cole et al. 1993a), reflecting the need for an array of coral sites to properly assess interannual climate variability.

Tarawa's coral record yields a 96-yr history of rainfall variation. Cole et al. (1993a) compared the coral oxygen isotope record with instrumental records of the Southern Oscillation and found the $\delta^{18}O$ record shares up to 85% of the variance of instrumental measures. Also, between 1930 and 1950, when the ENSO cycle was somewhat dampened, the annual variation in coral $\delta^{18}O$ reached its highest levels. This may be the result of either greater influence of seasonal changes in rainfall or the effects of cool surface waters penetrating the Tarawa region from the east (Cole et al. 1993a).

In general, Cole et al. (1993a) concluded that the last century's ENSO-related interannual tropical climate changes from Tarawa Atoll correlated with those in *Pavona* from the Galapagos, thousands of kilometers to the east. The Tarawa-Galapagos correlations convincingly show that the precipitation and temperature anomalies of El Niño events extend back at least through the past century. Cole and coworkers found, however, that the coastal South American historical record of El Niño episodes of Quinn et al. (1987) is occasionally decoupled from the Tarawa record and Pacific-wide ENSO extremes.

Trade Winds and ENSO in Tarawa

Shen et al. (1992b) developed a technique whereby the uptake of manganese into the coral skeleton is used as an indicator of trade-wind activity, and they applied it to the same Tarawa corals analyzed for oxygen isotopic variability by Cole et al. (1993a). Shen et al. (1992b) found that manganese from a 30-cm living colony of *Hydnophora microconos* oscillated in phase with ENSO events during the period 1960–1977, with particularly high ratios (50–80% above background) during the six months of an ENSO event. The 1965, 1972, and 1976 El Niño events were especially marked by their manganese peaks. Moreover, the manganese record paralleled closely the periodic changes in oxygen-isotope ratios from the corals, which became isotopically negative during El Niño events. Shen et al. reasoned that, unlike in the eastern Pacific, where seasonal vertical mixing leads to variability in manganese content of the seawater, the Tarawa seawater manganese reflects mainly remobilization of manganese temporarily trapped in the atoll's sediments. They interpreted these trends to signify that the transient appearance of strong west winds along the equator in the equatorial Pacific caused remobilization of manganese from sediments and at the same time abundant rainfall in a typically dry area.

Thus, oxygen isotopes and manganese concentrations in corals provide complementary evidence for precipitation and wind-strength variability during El Niño events.

Great Barrier Reef, Australia, and New Caledonia

Several studies have been carried out on corals from the Great Barrier Reef region of eastern Australia (e.g., Druffel and Griffin 1993; Gagan et al. 1994; Cole 1996). Among these researchers, Druffel and Griffin (1993) came to several important conclusions about long-term ENSO behavior in their study of $\Delta^{14}C$, $\delta^{18}O$, and $\delta^{13}C$ records from a 323-yr-old coral colony from Abraham Reef on the southwestern tip of the Great Barrier Reef. First, they found that interannual variation in surface ocean $\Delta^{14}C$ occurs in which low $\Delta^{14}C$ values coincide with strong ENSO events. Either the shoaling of the thermocline or advection of $\Delta^{14}C$-depleted water is the cause of this variability. In their long-term record of $\Delta^{14}C$—between 1635 and 1875—most, but not all, ENSO events recorded in the Abraham Reef coral $\Delta^{14}C$ record matched those documented historically by Quinn et al. (1987). Druffel and Griffin found evidence for low $\Delta^{14}C$ values during El Niño episodes between 1635 and 1875, but ENSO activity did not explain the observed variability between 1875 and 1920, a period these authors referred to as the "anomalous period." Furthermore, a 6-yr periodicity in $\Delta^{14}C$ and $\delta^{18}O$ that was apparent for the time 1653–1795 was absent in the latter part of their record between 1797 and 1957.

Druffel and Griffin also found unusually large-amplitude variations in $\Delta^{14}C$ between 1680 and 1730, as much as three times the amplitude of ENSO events of the 1950s. By comparing the coral $\Delta^{14}C$ record to tree-ring $\Delta^{14}C$, a measure of atmospheric radiocarbon, Druffel and Griffin discerned that the coral radiocarbon variability did not result from changes in atmospheric production. Instead, they argued that changes in vertical ocean mixing and/or surface advection were probably responsible for large-scale $\Delta^{14}C$ variations in the corals during this period.

The most significant general conclusion stemming from Druffel and Griffin's study is that ENSO behavior in the western Pacific is not stable over multicentennial time scales. There are shifts in the amplitude and the frequency of ENSO events probably related to long-term changes in oceanic circulation, mixing, and/or advection.

At Amedee Lighthouse, New Caledonia, T. M. Quinn et al. (1998) studied a 335-yr stable isotope record of *Porites*. This record showed important decadal and centennial changes in ENSO behavior in the western Pacific Ocean. During El Niño events, SSTs and rainfall decrease in this region of the Pacific. The seasonal coral oxygen-isotope record is controlled mainly by oceanic temperature, although salinity may also be an important factor in interannual $\delta^{18}O$ variability. Quinn et al. (1998) also found several spectral peaks in the coral $\delta^{18}O$, including a strong 3.6-yr periodicity, a frequency commonly encountered in other oceanic, atmospheric, and paleoclimatic records from the Pacific. The Amedee *Porites* oxygen-isotope curve also recorded large ENSO-related excursions during years that often matched those documented by W. H. Quinn (1992) for South America. However, the coral and documentary events did not always match in magnitude, and Crowley et al. (1997) have argued that some western Pacific cooling events may be forced by major volcanic eruptions rather than ENSO. More generally, Quinn et al. were cautious about making conclusions about which ENSO-related frequencies were the most significant in this region until further studies could be carried out.

In addressing questions about centennial and decadal climate variability, Quinn and colleagues discovered strong evidence for twentieth-century warming: Western Pacific SSTs reconstructed from the Amedee coral for the twentieth century were 0.3°C warmer that SSTs from between 1657 and 1900 A.D. In addition, they found that significant shifts occurred in western Pacific climate over decadal time scales. For example, the period 1940–1960 was relatively quiescent in terms of ENSO activity, and significant shifts in coral oxygen isotope variability occurred in 1940–1941, 1976–1977, and 1988–1989, coincident with shifts recorded in instrumental records. Their time-series analyses also revealed a concentration of variance centered at 15.4 yr.

Caribbean-Atlantic Area

Florida

Smith et al. (1989), Hudson et al (1989), and Swart et al. (1996a) studied a 160-yr old *Solenastrea* coral from Florida Bay. Although the emphasis in these studies was on the impact of human activities and hurricanes on the coral record, before about 1940 the fluorescent (Smith et al. 1989) and stable-isotope (Swart et al. 1996a) records of *Solenastrea* showed a 5-yr periodicity that is very likely related to ENSO-driven rainfall anomalies.

Barbados and the Amazon River

Altough Ba:Ca ratios have not yet been applied to reconstructing long-term trends in ENSO-related climate variability, Shen and Sanford (1990) used this ratio from corals from Barbados in the Caribbean in an exploratory study to test its use as an indicator of discharge from the Amazon River, 2000 km away, that occurred during seasonal and interannual events. For the period 1966–1978, a good correspondence between Ba:Ca maxima and salinity minima seemed to reflect Amazon discharge, which varies seasonally, reaching its maximum during the months of May through June. However, a 2–4 month lag existed between peak Amazon discharge and the Barbados salinity minimum and Ba:Ca maximum. Although this lag may be related to complex oceanic circulation patterns, Shen and colleagues recommended a cautious interpretation of historical discharge trends related to ENSO events until longer time series become available.

Extratropical Impacts

Paleoclimate records from tree rings, ice cores, and sediments can yield valuable information on long-term variability in ENSO cycles in extratropical regions, although until recently, most research on late Holocene ice-core and sedimentary records have focused on decadal and centennial time scales. This section describes a few representative studies specifically relevant to ENSO variability and its impact.

In western North America, Meko (1992) found that tree-ring records for the period 1700–1964 produced strong seasonal rainfall anomalies at frequencies of 2.8–10.2 yr. ENSO-band oscillations were evident in moisture-related tree growth patterns in trees from Arizona and New Mexico. The ENSO signal, however, was somewhat weaker in southern California trees, and an ENSO influence almost disappeared in more northerly regions of Wyoming, Montana, and Oregon. Meko's study also revealed interdecadal (> 20-yr) variability that may be related to low-frequency climate factors.

Swetnam and Betancourt (1992) measured fire-scar frequency on trees from Arizona and New Mexico for the period 1700–1905. They concluded there was an inverse correlation between the prevalence of fires (measured by fire-scar frequency) and tree growth, which itself is a function of winter-spring precipitation. The correspondence, especially evident for the interval 1740 through the 1760s and the 1840s through 1860s, reflects high tree-ring growth during low wildfire activity and low growth during times of high fire activity. The preliminary results also suggested that the area burned was greatest during years when archival records (measured by a high SOI) indicated non–El Niño conditions. Large burn areas are usually due to drier conditions. Conversely, the lowest fire activity occurred during El Niño years (periods of low SOI values), when wetter conditions persisted.

Thompson et al. (1992) summarized ice-core records of ENSO-related atmospheric anomalies from the Dunde ice cap, Tibetan Plateau, China, and the Quelccaya ice cap from the Peruvian Andes. Historical El Niño events can be identified in the Quelccaya record, where they are associated with reduced net ice accumulation, higher concentrations of insoluble and soluble dust, and less negative $\delta^{18}O$ values for many El Niños. Significantly, low-frequency, decadal-scale variability also characterizes the low-latitude ice core records of the past 1500 yr, and Thompson and colleagues believe that these non-ENSO oscillations may in fact be similar in character to El Niños in terms of their effect on South American atmospheric conditions. Low-latitude ice cores will continue to be an important source of data on ENSO-related atmospheric change.

Several studies suggest that long-term decadal variability in the strength and frequency of ENSO events and modulation of ENSO activity has occurred over the past few centuries owing to other forcing factors as yet not identified with certainty. Lough (1992:215) summed up such a conclusion for tree-ring records since 1600 A.D. obtained from the arid southwestern United States and northern Mexico: "[there is] evidence of changes in the character of the SOI over the past few centuries, or at least, a modulation of the North American teleconnection pattern." A similar conclusion was reached by Anderson (1992a) on the basis of historical and Nile flood records since 622 A.D. These records were derived from multiple proxies, including marine and lacustrine sedimentary, isotopic, and micropaleontological records. Anderson et al. (1992:419) concluded "more than 400 yr of historical record, tree-ring record, and varved marine sediments contain evidence for long-term changes in the frequency and amplitude of ENSO's primary oscillation or in effects related to ENSO." The link between ENSO and lower-frequency forcing such as that due to solar activity (Anderson 1992b, Anderson et al. 1992) will continue to be an active field for future paleoclimate research.

Although brief, this summary of extratropical ENSO-related continental paleoclimate records indicates great potential for determining long-term interannual and interdecadal regional variability in climate changes originating at least in part from tropical ENSO-related anomalies.

GODDARD INSTITUTE OF SPACE STUDIES GENERAL CIRCULATION MODEL OF ISOTOPIC RESPONSE TO INTERANNUAL CLIMATE

IN THIS AND previous chapters, evidence has been presented for large-scale climatic perturbation of earth's hydrologic system over seasonal-to-centennial time scales. Thus climatically induced variability in the oxygen-isotope composition of rainwater, oceanic surface waters, atmospheres, and ice have been shown to reflect changes in oceanic circulation, SST, ice volume, precipitation rate, SSS, and sources of precipitation, among others. Oxygen-isotope oscillations are in turn incorporated into paleoclimate records in corals, ice cores, foraminiferal isotopes, and other archives.

During ENSO events, seasonal-to-decadal climate change leaves its signature in oxygen-isotope variability of tropical corals, in which the $^{18}O/^{16}O$ ratio varies seasonally, interannually, and decadally. Unraveling the local, regional, and global factors contributing to the tropical coral isotope signal can be complex, and sometimes the sources of the various signals remain uncertain. Moreover, ENSO-scale events clearly have impacts on extratropical regions, as evidenced from tree-ring and ice-core records, but these effects are still poorly understood. It would be useful in the quest to understand the patterns and causes of interannual climate variability to examine the global scale of variability in oxygen isotopes and see if the expected signal is recorded in the paleoclimate record.

Cole et al. (1993b) did just this when they asked the overarching question: What is the global-scale oxygen-isotopic response to ENSO-like interannual climate variability? To find an answer, they conducted an investigation that combined general circulation modeling (GCM) with paleoclimate data from the tropics and extratropics. This study entailed running climate simulations using an isotopic tracer version of the Goddard Institute of Space Studies (GISS) GCM to simulate interannual variability in oxygen isotopes. To concentrate on the ENSO-forced isotopic variation, they prescribed the model with January and July SSTs from the extreme El Niño event of 1982–1983 and the La Niña event of 1955–1956 and examined anomalies in isotopic variation as output. They focused their paleoclimate-model comparison on the interannual isotopic variation from four climate zones critical for understanding ENSO and its teleconnections:

1. Pacific warm pool—Tarawa Atoll (Cole and Fairbanks 1990; Cole et al. 1993a)
2. Tropical South America—Quelccaya ice cap (Thompson et al. 1984, 1986)
3. Tibet—Dunde ice cap (Thompson et al. 1988, 1989)
4. Western Europe (Rozanski et al. 1993)

Their results yielded several provocative insights into the causes of isotopic variability in particular and interannual paleoclimatology in general. First, they discovered for the western Pacific region that the GCM reproduced the inverse relationship between the amount of recorded variation and the isotopic composition of rainfall. This indicates that western Pacific coral isotope records should generally reflect precipitation-related factors as interpreted by prior investigators (e.g., Cole and Fairbanks 1990). Similarly, in South America, the GCM output suggests precipitation anomalies appear to control the isotopic signal preserved in the region of the Quelccaya ice cap paleoclimate record. These South American paleoclimate trends are related to air masses forming over the Amazon Basin. Conversely, the paleoclimate-GCM simulation results indicated that temperature anomalies during ENSO events appear to control the isotope record in Europe.

In Tibet, the Asian isotopic record is more uncertain. Isotopic variability is correlated neither with local climatic anomalies related to temperature nor with any apparent changes in the source of atmospheric water vapor. Cole's group felt that the Tibetan Dunde ice cap paleoclimate record preserves a continental record of climate change in Asia but not necessary a global signal.

Cole et al. (1993b) also contrasted the dominant influence of precipitation anomalies on isotope record of ENSO-forced climate changes with the temperature-dominated record during the last glacial period that was also simulated by the GISS GCM. In the broadest terms, the model simulations support the results from many empirical paleoclimate studies showing that temperature and ice-volume factors overwhelm the precipitation signal during the large-amplitude climate changes that accompany glacial-interglacial cycles (chapters 4 and 5). The interannual temperature-related isotope variability, on the other hand, can be seriously complicated by changes in water-vapor sources and transport mechanisms.

In summary, the research on interannual ENSO-related climate variability that has taken place during the past decade has developed rapidly. Although they are not nearly as mature as research on orbital and suborbital climate change, the fields of coral physiology, growth, and ecology, as well as the calibration of geochemical proxies to the available instrumental records, have progressed swiftly. The strategies and proxy tools are now available to paleoclimatologists to extend the observational record of oceanic and atmospheric parameters centuries and perhaps millennia back in time. Coral tropical paleoclimatology is providing climate modelers with additional means to conduct ENSO "hindcasting" and forecasting with increasing predictability. Among the major challenges confronting ENSO investigators, one challenge requiring many more long-term records, is decoupling of the high-frequency ENSO signal from low-frequency, decadal-scale forcing.

VIII

Sea-Level Change

. . . in speculating on catastrophes by water, we certainly anticipate great floods in the future, and we may, therefore presume that they have happened again and again in past times.

Sir Charles Lyell, 1830.

BY LAND OR BY SEA?

For centuries Scandinavian naturalists have noted a remarkable phenomenon—their country's land area was increasing and their coasts were rising. Skerries were becoming islands, waterways were closed, and submerged forests were rising. Nils-Axel Morner of Sweden (1979a) describes how in 1743, Anders Celsius, inventor of the thermometer and famed for his temperature scale, calculated a rate of coastal rise (or relative sea-level fall) of 1.2 cm/yr along Swedish coasts. The rising land gave considerable support to the eighteenth-century theory of Neptunism, which held that sea level was continuously falling in a one-way direction. Indeed, this trend of falling sea level continues today in Scandinavia at a rate of 1–2 cm/yr.

On Smith Island, located in the heart of Chesapeake Bay, the opposite has been happening. When English colonists settled in the seventeenth century, Smith Island was ideal for grazing cattle, and bay waters off its shores were bountiful with finfish and shellfish. Over the past three centuries, however, its coasts, like those of other Chesapeake Bay islands, have suffered from severe erosion and shoreline retreat. Smith Island has been reduced in area from 11,000 acres in 1849 and to 7000 acres of mostly marshland by the 1980s. Tide gauges, first established in the Chesapeake and its tributaries in Baltimore (1903), Annapolis (1929), and Solomon (1938), Maryland, as well as Washington, D.C. (1931), record vertical coastal submergence at rates between 3.17 and 3.56 mm/yr (Hicks and Hickman 1988; Emery and Aubrey 1991). In a century, this amounts to more than a 30-cm sea-level rise relative to the land surface. Coasts along Chesapeake Bay are retreating in a linear direction as well. Dolan et al. (1983) concluded that the average rate of retreat of Chesapeake Bay coastlines was 0.7 m/yr, a rate much higher than that of other coasts of the United States.

The Chenier Plain and Mississippi Delta of the Louisiana coast, a complex nexus of swamps, fresh and saltwater marshes, alluvial plains, and barrier islands, is another region undergoing coastal submergence. About 3950 km^2 of Louisiana coastal wetlands—a region the size of the state of Rhode Island—have been lost between 1930 and 1990. Louisiana has 40% of the total wetlands in the United States and has suffered 80% of the nation's wetland loss over the past century. As a result, concern about future coastal land loss in Louisiana and elsewhere has become a national priority (Boesch et al. 1994).

What factors are causing Sweden's coasts to rise, Chesapeake Bay islands to shrink, and Louisiana coastal wetlands to disappear? Can the much-publicized global sea-level rise account for coastal submergence in Chesapeake Bay and in Louisiana, and a sea-level fall in Scandinavia simultaneously? In previous chapters the topic of sea-level change was described with reference to several subtopics of paleoclimatology: emerged coral terraces and the orbital theory of climate change, low sea level during the last glacial maximum (LGM), and glacial meltwater that caused sea level to rise during deglaciation. In this chapter, I focus on the broader topic of sea-level change and the multiple factors that complicate efforts to understand sea-level history and its intimate relationship with climate change.

Four important themes permeate sea-level and climate change and coastal evolution. First, a clear distinction must be made between the concepts of eustatic and relative sea-level change. The concept of *eustatic,* or global, sea-level change is credited to Austria's Eduard Suess (1885–1909), who theorized that worldwide ocean-level changes caused observed changes in sea level. Eustatic changes in ocean level are sometimes informally referred to as absolute or global sea-level changes. Today researchers attribute eustatic sea-level variability to three major processes: steric expansion (or contraction) of upper ocean layers due to temperature and density changes; glacio-eustasy, the growth and decay of ice sheets and glaciers and concomitant volumetric changes in the world's ocean water; and tectono-eustasy, changes in ocean basin volume (table 8-1).

Geologists also recognize, however, that observed sea-level change along a coastal region is a manifestation of *both* eustatic sea-level change *and* other processes that cause the land to rise or fall. Observed changes in sea level for any particular region are therefore referred to as *relative* sea-level changes, and the sea-level history for that region is called a *relative sea-level curve.* The most important

TABLE 8-1. Some Factors Producing Observed Sea-Level Change and Their Approximate Time Scales

Processes	*Time Scale (yr)*	*Comments*
"Absolute" or eustatic sea-level change		
Steric expansion and contraction	1–100	
Glacio-eustasy		
Alpine glacial growth and melting	10–1000	Faster response to forcing than ice sheets; glacial surges extremely rapid
Ice-sheet growth	1000–1,000,000	Stepwise growth toward full glacial interval
Ice-sheet decay	100–10,000	Rapid ice-sheet decay under some conditions
Tectono-eustasy	100,000–10,000,000	Ocean-volume changes
Relative sea-level change		
Ocean dynamic processes	1–10	Kelvin and Rossby waves
Geoidal changes	1–100	Gravitational variation
Subsidence and uplift	100–10,000,000	Regionally variable; depends on sedimentation tectonics and local geology
Tectonic uplift and subsidence	<1–1,000,000	Regionally variable; coseismic vertical movements most rapid
Isostatic processes		
Sedimentation and erosion	1000– > 10,000	Accumulation and removal of sediment loads
Hydro-isostasy	1000– > 10,000	ocean basin sinking lags behind ice-sheet melting
Glacio-isostasy	1000– > 10,000	Slow restoration to gravitational equilibrium after deglaciation due to high mantle viscosity

processes that affect relative sea-level records include geoidal factors (spatial and temporal variability in gravitational effect on the ocean surface), coastal subsidence or uplift due to sediment compaction or ground-water removal, local and regional tectonic uplift or subsidence, and isostatic processes related to sedimentation and erosion, hydro-isostasy, and glacio-isostasy (table 8-1). Although a eustatic sea-level rise may occur as a global and nearly synchronous event, completely separating the eustatic component of a relative sea-level curve from those parts caused by local and regional factors is difficult if not impossible. Historical coastal changes in Scandinavia, Chesapeake Bay, and Louisiana serve as textbook examples of how coastal processes act in concert with changing ocean level to affect observed trends in relative sea level. Consequently most researchers abide by the maxim, there is no single "eustatic" sea-level curve. To successfully attribute causes to any observed sea-level change, it is necessary first to decouple the effects of land-based processes like subsidence and uplift from those related to changes in ice or ocean-basin volume.

A second aspect of sea-level and climatic change is the concept that large-scale ice-volume change, the major cause of eustatic sea-level oscillations over tens of thousands of years, is itself an active agent of climate change. As discussed in earlier chapters, the areal and volumetric extent of continental and sea-ice volume influence planetary albedo, atmospheric circulation, oceanic salinity and temperature, and biogeochemical cycling in the oceans. In the example of the postglacial 125-m sea-level rise between 20 and 5 ka, this inundation of the world's continental shelves affected ocean-margin sedimentation, upwelling, biological productivity, and nutrient budgets, ultimately influencing earth's global elemental cycling and climate.

Third, all modern coastlines—whether barrier islands, coral reefs, mangroves, estuaries, deltas, or glaciated fjords—have one feature in common. Their geomorphology, sedimentology, and ecology, to varying degrees, have been influenced by postglacial sea-level rise and the redistribution of load from continental ice sheets to ocean basins. In tropical regions, notches forming overhanging cliffs testify to Holocene erosion of emerged reefs that formed during previous high stands of sea level (figure 8-1). In estuaries like Chesapeake Bay, drowned river valleys such as the ancient Susquehanna River became flooded during postglacial sea-level rise (Colman and Mixon 1987). Postglacial sea-level change left its mark on the terrestrial fauna and flora of many islands, especially in shallow marginal seas like those around Indonesia, where islands harbor terrestrial species that migrated across land bridges connecting them when sea level was lower.

FIGURE 8-1 Exposed coral-reef terrace along the coast in the city of Santo Domingo, Dominican Republic, corresponding to marine isotope stage 5e. Region is relatively stable tectonic area such that the fossil reef's elevation about 3–4 m above modern sea level represents high global eustatic sea level during the past interglacial period, about 135–120 ka.

One striking example of mammalian migration during low sea level and subsequent isolation was the crossing of the Bering Land Bridge by *Homo sapiens* at least by 12 ka, just before the rapid sea-level rise that inundated the Bering Strait region.

Finally, we must distinguish between sea-level change and coastal land loss due to processes unrelated to either relative or absolute sea-level rise or fall. Wave and storm erosion, coastal ocean currents, storm surges, changes in sediment budget (i.e., variations in sand supply), local geology (slope failure and landslides), local climate (temperature and precipitation) and vegetation are processes that affect coastal zones (Pilkey et al. 1989; Fletcher 1992). Natural processes are often exacerbated by human activities such as construction of jetties, groins, or channels; the extraction of oil, water, or gas; and irrigation and land-use changes (Jelgersma 1996). Experts believe, for example, that near Grand Isle, Louisiana, a relative sea-level rise of about 1 cm/yr accounts for only a fraction of the total coastal land loss. The remainder is attributed to other factors related to land-use practices (Boesch 1994). Others disagree as to the effects of anthropogenic factors. Whereas Sahagian et al. (1994) estimate that global sea level rose 0.54 mm/yr owing to groundwater mining, wetland drainage, deforestation, and impoundment of freshwater by dams, Gornitz et al. (1994) contend that the total anthropogenic contribution to sea-level rise over the past 60 years due to these processes and irrigation was a net fall in sea level of 1.63 mm/yr. We will not discuss coastal processes or anthropogenic activity causing coastal change further, except to note that they may contribute to and/or accelerate the impact of naturally occurring sea-level change. Many authors discuss the combined effects of coastal processes and sea level to explain regional sea-level trends (e.g., Milliman and Haq 1996).

After a brief section on early concepts of sea-level change, this chapter presents sections on geological evidence for sea-level change; the major processes that cause eustatic and relative sea-level change; Cenozoic, late Quaternary and Holocene sea-level history; and historical sea-level change of the past century.

EARLY CONCEPTS OF SEA-LEVEL CHANGE

PRE–TWENTIETH CENTURY theories to explain sea-level changes have a long and diverse history closely linked with theories about the earth. Some of these are described in engaging books entitled *Eustasy: The Historical Ups and Downs of a Major Geological Concept,* edited by Robert H. Dott, Jr. (1992); *Sea Levels, Land Levels and Tide Gauges,* by K. O. Emery and David Aubrey (1991); and *Phanerozoic Sea-Level Changes,* written by Anthony Hallam (1992). Generations of leading scientists have tackled the still perplexing problem of the nature and causes of sea-level change, and it is judicious to mention a few notable examples. Early ideas that sea level was not stable came from evidence derived from marine fossils preserved in sediments exposed far above modern sea level and from observations of emerging and submerging modern coastlines. Many early interpretations of marine fossils and sea-level change were also deeply rooted in theology. For example, evidence for changing sea level was considered to be a sign of the biblical Noachian flood, a theory advocated by early seventeenth Natural Theologists such as Thomas Burnet (see Gould 1987; Dott 1992).

In the early 1800s, the Diluvialist, Catastrophist, and Neptunist schools of science also used varying lines of evidence for catastrophic submergence of the land by the ocean to support their theories. *Diluvial* is a term coined by William Buckland (1823) to explain superficial gravel deposits that he attributed to a great catastrophic flood. Buckland's diluvial period of relatively high sea level had been preceded by a long "antediluvial" geological era of lower sea level and a "postdiluvial" (or alluvial) period succeeding it.

Georges Cuvier promoted the theory of Catastrophism to account for the alternating series of fossiliferous marine and nonmarine sedimentary strata and the extinction of fossil species in the Paris Basin, France. Cuvier saw relative sea-level changes that were induced by catastrophic floods. Each flood, Cuvier thought, was caused by an inversion of the sea floor, an idea that led to the "sea-floor inversion" theory of eustasy (Dott 1992).

Neptunism is a school of thought traced to the seventeenth-century work of France's Benoit de Maillet (Carozzi 1992). Carozzi (1992:19) summarizes de Maillet's Neptunist ideas as follows: "The present-day diminution of the sea, as inferred by de Maillet, on the earth's surface is the sea-level fall episode of an eternally repeated cosmic and complete eustatic cycle produced by a complex interchange of water and ashes between celestial bodies within and between the various vortices." In eighteenth-century Scotland, James Hutton, and later Abraham Gottlob Werner at Freiberg in Saxony, among others, fostered various versions of Neptunism (see Berggren 1998; Strahler 1987; Dott 1992), which flourished in the early nineteenth century. Dott (1992:3) characterized Neptunism as signifying "one-way eustatic fall of sea level." Thus according to Dott, Neptunism explained the prevalence of marine fossiliferous strata exposed on land and the rising coastlines of Scandinavia as evidence for vertical uplift by the land rather than changes in the level of the oceans.

During the nineteenth century, Charles Lyell also held as a basic tenet of his Uniformitarian view of earth history that gradual uplift and subsidence of the land was a preferred mechanism to explain sea-level changes. Lyell believed that the mechanism he found so prevalent in many parts of the modern earth—volcanic activity—was the major cause of observed changes in ocean level evident from fossil marine organisms now found far from modern oceans. Citing the work of early naturalists of the seventeenth and eighteenth centuries, as well as his own observations, Lyell postulated that volcanic activity had displaced large land areas either upward or downward relative to the level of the sea. Lyell also strongly favored a theory of glacial drift to explain the existence of diluvium and glacial erratic boulders as the result of icebergs "drifting" into a region rather than the product of continental ice-sheet movement.

As we saw in chapter 4, the glacial theory of Louis Agassiz, which he developed in the mid-nineteenth century to explain glacial geomorphic features and sediments, was instrumental in the development of the orbital theory of climate change. Agassiz's glacial theory also revolutionized the study of sea-level change by providing a new mechanism—continental ice-sheet growth and decay—to explain the existence of glacial features and to account for large-scale, rapid sea-level changes. With the general acceptance of Agassiz's theory during the latter part of the nineteenth century, glaciology and glacial geology became intricate parts of research on glacio-eustatic sea-level history. As I discuss below, glacio-eustasy is the dominant factor in explanations of large-scale (50–100 m), rapid (< 100 ka) sea-level changes occurring during late Cenozoic glacial-interglacial transition.

In sum, the earliest notions about the possibility of sea-level change emanated from the same types of evidence—rising land surfaces, fossiliferous sediments, and glacial deposits—that are still used now, as we describe in the next section.

GEOLOGICAL, GEOCHEMICAL, GEOPHYSICAL, AND BIOLOGICAL EVIDENCE FOR SEA-LEVEL CHANGE

WE CAN DIVIDE the types of evidence for sea-level change into four categories: geological, geochemical, geophysical, and biological. Geological evidence comes from two broad subfields: geomorphology and stratigraphy. Geomorphological evidence for ancient shorelines that now lie either above or below present sea level is plentiful along coastal regions. Wave-cut escarpments, relict barrier islands, and fossil coral reefs are three such features. Stratigraphic evidence for sea-level oscillations occurring over various time scales include data derived from sedimentary sequences showing multiple marine transgressive and regressive episodes, punctuated by either subaeriel erosion or nonmarine deposition.

Geochemical evidence comes mainly from the stable oxygen isotopic record of deep-sea benthic and planktic foraminifers described in chapter 2. Oxygen-isotope variability, which is a function of global ice volume and other factors such as oceanic temperature and salinity and the vital effects of foraminiferal species, has been applied primarily in research on orbital-scale sea-level variability (see chapter 4).

Geophysical evidence stems mainly from two widely divergent subfields: seismic stratigraphy (Vail et al. 1977; Haq et al. 1987, 1988) and geophysical modeling of earth's isostatic response to glacial ice-sheet growth and decay (Peltier 1993, 1994). Seismic stratigraphy uses the distinct seismic signatures of different types of sedimentary deposits and their bounding unconformities, supplemented by geochronological methods of biostratigraphy and magnetostratigraphy, to detect transgressive and regressive sedimentary sequences of continental margins (also called onlap-offlap sequences). Transgressive-regressive sedimentary sequences are believed by many to result largely from eustatic sea-level changes and regional tectonic and sedimentary processes. Seismic stratigraphy has been the major impetus behind the development of a chronostratigraphic eustatic framework, often referred to as a eustatic sea-level curve, for much of the Mesozoic and Cenozoic (Haq et al. 1987, 1988, see below).

Geophysical research on earth's crustal and mantle rheology and deformation also plays an important role in sea-level research. A global model of "paleotopography" heavily dependent on radiocarbon-dated late Quaternary sea-level curves known as the ICE-4 model was developed by Richard Peltier to test models of earth's crustal properties and examine global sea level and isostatic motions over the past 20 ka. Peltier's models of glacio-eustatic sea-level change and accompanying crustal isostatic response to deglacial unloading of ice sheets contributes heavily to interpreting late Quaternary relative sea-level curves.

Biological evidence for sea-level change comes from a wide variety of organisms that today live at or near sea level, whose fossil record helps to establish the location of ancient coastal zones. The overarching principle in the reconstruction of relative sea-level history from biological evidence is as follows: the more a species' ecological tolerances restrict its range to within the intertidal or shallow subtidal zones, the more suited it is as an indicator of past sea level. In the book entitled *Sea-Level Research: A Manual for the Collection and Evaluation of Data,* edited by Orson van de Plassche (1986), experts on various biological groups devote entire chapters to marine mollusks, corals, coralline algae, gastropods, buried forests, botanical macrofossils, foraminifers, ostracodes, diatoms, and even human-constructed shell middens to describe the rich array of biological evidence available to scientists conducting sea-level research. Among the best sea-level indicators are tidal marsh grasses (*Spartina alterniflora*), corals (*Acropora palmata*), mollusks (*Crassostrea virginica*), and benthic marsh foraminifera (*Trochammina*), among others. As with other biological indicators of climate-related phenomena, our knowledge of sea-level history rests on biological principles just as much as it does on the physics of glaciers or the earth's crust.

Finally, tide gauges are means to examine historical trends in sea level extending back about a century (figure 8-2). Tide-gauge measure-

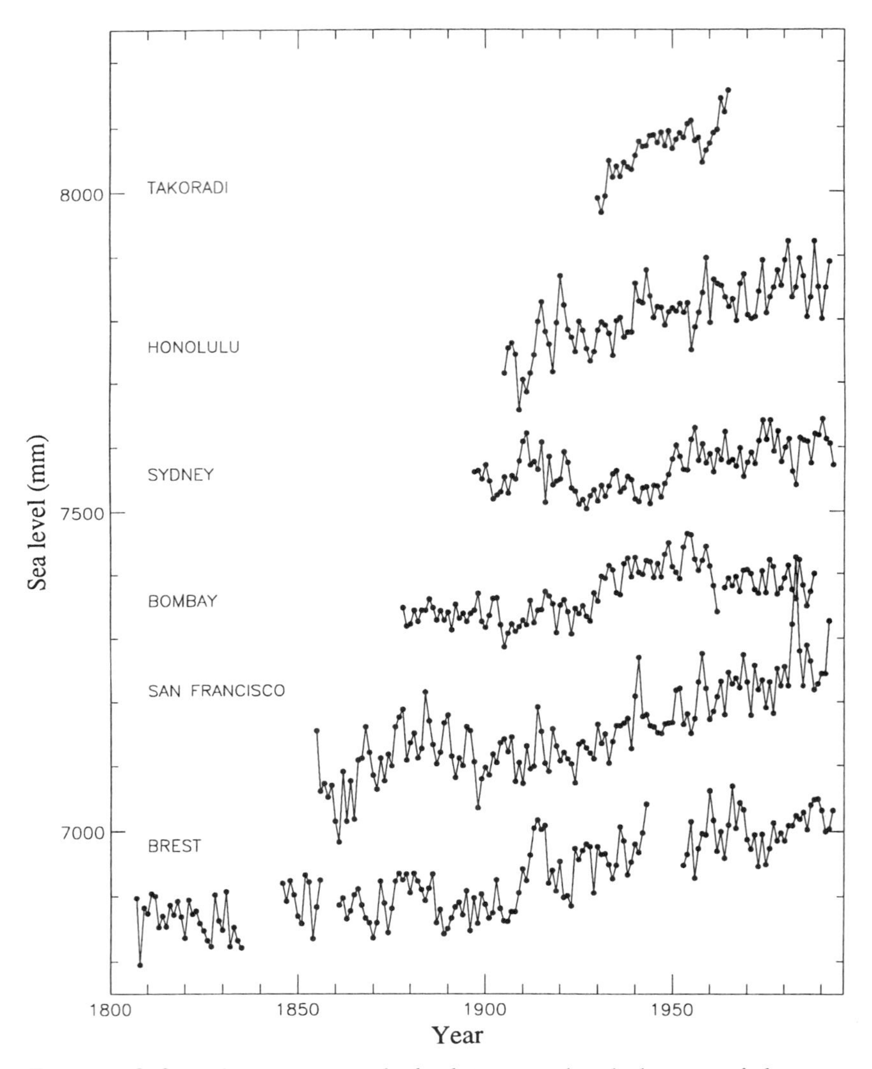

FIGURE 8-2 Tide-gauge record of relative sea-level changes of the past century along coastal regions having different isostatic and tectonic processes. Courtesy of World Meteorological Association.

ments of sea level are, however, complicated not only by tectonic and isostatic factors that cause vertical motions of the land surface, but also by piling up of water due to prevailing winds and ocean currents, changes in instrumental procedures, and other factors. Emery and

Aubrey (1991:162) conducted the most thorough assessment of the value of tide-gauge records as a measure of global sea level and came to the following conclusion:

> We can find no justification for using a few long-term tide-gauge stations as representative for regional or global averages. The longer term data appear anomalous compared to the better regional and global estimates derived from more but shorter term data. Although the reasons for this anomaly are only hypothesized here, it appears that the long-term station data prior to about 1930 cannot be used to evaluate changes in rates of eustatic sea-level rise.

In the following section on processes affecting sea level, we will see that each particular type of evidence plays an important role in the study of sea-level history.

PROCESSES AFFECTING SEA-LEVEL CHANGE

SCIENTISTS RECOGNIZE several major processes that cause eustatic and relative sea-level changes over different time scales (table 8-1). The most important are steric changes (ocean expansion due to atmospheric and near surface oceanic warming), glacio-eustasy (sea-level rise and fall due to melting and growth of glaciers and ice sheets on land), tectono-eustasy (continental plate motions leading to ocean-ridge volume and global sea-level changes manifested in marine transgressions and regressions), ocean dynamic factors (i.e., Rossby and Kelvin waves), geoidal changes, local coastal subsidence and uplift (sediment accumulation and removal) and tectonics, and isostasy (including glacio-isostasy and hydro-isostasy, vertical crustal motions due to gravitational effects of load redistribution of glacial ice and meltwater in the oceans). The following sections summarize each of these as they pertain to coastal processes and sea-level variability (e.g., Nummendal et al. 1987; Morner 1979b; Cronin 1982; Tooley and Shennan 1987; Pirazzoli 1991; Peltier 1993; Fletcher 1992; Fletcher and Wehmiller 1992; Milliman and Haq 1996).

Eustatic Sea-Level Change

The concept of eustasy was formally introduced by Suess (1885–1909), who theorized that changes in sea level were due to global changes in the level of the world's oceans. Suess attributed subsidence of the crust to the cooling and shrinking of the earth, causing sea level to fall, and the continuous deposition of sediments into oceans, causing sea level to rise. In the ensuing century, parallel theories were developed by Chamberlin (1898, 1916), Stille (1924), Umbgrove (1939), Graubau (1940), and other geologists in Europe and North America to account for cyclic stratigraphic sequences on the world's continents (see Dott 1992; Hallam 1992).

Eventually, Suess's idea that sedimentation led to global sea-level rise was discarded because marine geological and stratigraphic investigations (Hallam 1963) showed it was too minor to cause sea-level to rise and fall the > 100 m observed in stratigraphic and geologic records. Instead, ocean volume changes associated with plate tectonic theory became, and remain today, the preferred mechanism to explain first-order and perhaps second-order sea-level cycles over tens of millions of years. Eustatic sea-level changes are viewed today as resulting mainly from three types of processes: steric (thermal) effects, glacio-eustasy, and tectono-eustasy.

Steric Effects—Thermal Expansion of Oceans

Steric height, a term used in oceanography, is defined as the vertical distance (i.e., depth) between two surfaces of constant pressure. This measure is used by oceanographers to account for density differences in the oceans. The main steric effect causing eustatic sea-level change is believed to be temperature change in upper ocean layers. During the past century, steric sea-level rise is believed to have been caused by increasing atmospheric temperatures because higher temperatures warm the world's sea-surface waters and decrease their density, causing the ocean to expand (Wigley and Raper 1987). In one study, Nerem (1995) concluded from satellite data that the global mean rate of sea-level rise was 4–6 mm/yr. This rate is higher than rates calculated from post-1900 tide-gauge records (Emery and Aubrey 1991). Nerem computed that an estimated 6- to 9-cm rise in sea level would accompany a 1°C increase in sea surface warming. This estimate suggests that about one half of the total historical 10- to 25-cm sea-level rise of the past century was due to thermal expansion of the ocean surface. However, there is considerable uncertainty in this estimate. One source of uncertainty lies in measuring the depth below the ocean surface that is warmed. In one study of 42-yr records of ocean temperatures off coastal California, Roemmich (1992) found a temperature increase of 0.8°C in the uppermost 100 m and smaller but significant warming as deep as 300 m. Roemmich concluded that steric expansion could account for anywhere between 30% and 100% of the observed sea-level rise from adjacent California tide-gauge records (corrected for isostatic adjustment). He emphasized, however, that because spatial and temporal variability in steric oceanic effects over decadal time scales are poorly documented, the Pacific trends cannot be extrapolated to other oceanic regions.

Despite these uncertainties, current estimates of the contribution of steric effects to sea-level rise of the past century, as determined by based on observational and modeling studies, is generally estimated at about 2–6 cm, with a best estimate of 3.5–3.8 cm (Meier 1993; Warrick and Oerlemans 1990; de Wolde et al. 1995; Cubasch et al. 1995; Warrick et al. 1996). These measures are equivalent to a rate of sea-level rise between 0.2 and 0.6 mm/yr, which roughly matches that expected from oceanic warming of 0.07°C/yr. Estimates from modeling studies depend primarily on estimates of atmospheric temperature rise, while those based on tide-gauge records depend on the length of tide-gauge records selected. The remaining portion of the historical sea-level rise most likely results from melting of alpine glaciers (Meier 1984, 1993). Unknown contributions include meltwater Antarctica and Greenland (see later) and possibly anthropogenic factors such as groundwater mining, deforestation, wetland loss, creation of reservoirs, and irrigation (Gornitz et al. 1994).

Glacio-eustasy

Glacio-eustasy refers to the theory that the melting and growth of large continental ice sheets are the cause of observed sea-level changes over 10 ka to 1 Ma. To fully appreciate the concept of glacio-eustasy and how difficult it is to determine sea-level history due to the growth and melting of glaciers, we must first review the earth's cryosphere and the processes governing glacial and ice-sheet behavior.

Earth's Modern Cryosphere

Although the existence of an ice-covered continent was known in the early nineteenth century, it was not until early twentieth-century expeditions to Antarctica by Ernest Shackleton, Robert Scott, and Roald Amundsen did naturalists begin to unlock the mysteries of the immense quantity of ice stored on this unexplored continent. Ironically, by the time of early Antarctic explorations, European and North American scientists already knew as much or more about the late Pleistocene ice sheets that covered Europe and North America than about the scope and history of the modern Antarctic Ice Sheet. Indeed,

Maclaren (1842) and later Penck (1882) had already postulated a drop in global sea level of about 100 m due to formation of the late Pleistocene ice sheets, a figure remarkably close to modern estimates of 125 m.

The earth's modern cryosphere—that part of the global water budget stored in ice—is a dynamic and critical part of the entire climate system (figure 8-3). As discussed in chapter 2, Antarctica holds 60% of the world's fresh water and 91% of its ice, an equivalent to 13.9×10^6 km^2 in area and 30.1×10^6 km^3 in volume. Melting of the grounded ice portion of the Antarctic Ice Sheet would cause sea level to rise about 73 m; melting of the Greenland Ice Sheet would raise sea level an estimated 7.4 m; and the world's mountain glaciers and small ice caps would cause a sea-level rise of 0.3 m.

Processes Governing Budgets of Ice Sheets and Glaciers

At any one time, an ice sheet or a glacier has either a positive or negative mass balance depending on whether it is growing or melting. Multiple factors govern the mass balance of an ice sheet or a glacier and thus how it will affect eustatic sea level. These factors include basal melting, subglacial lithology and glacial till distribution,

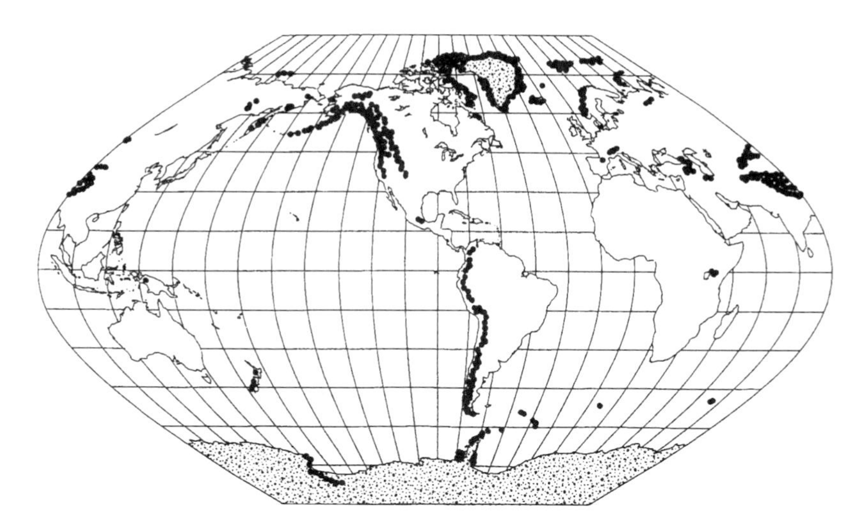

FIGURE 8-3 Modern distribution of ice sheets (stipled) and alpine (black dots) ice. Courtesy of Intergovernmental Panel on Climate Change (IPCC), World Meteorological Organization. From Warrick et al. (1996).

oceanic temperatures surrounding the margins of an ice sheet, the position of sea level with respect to grounded marine-based ice sheets, summer ablation, winter snow accumulation, ice-stream behavior, and long-term equilibrium factors related to the past deglaciation.

Glaciers and ice sheets are thus complex, dynamic masses of ice, each with its own budget. Glaciologists can measure mass balance either volumetrically in kilograms per year or in terms of the sea-level equivalent in meters or centimeters per year. In simplest terms, a glacier with a positive mass balance means its winter accumulation of snow, converted eventually to ice, is greater than its summertime ablation (or melting) and/or its loss of mass due to calving icebergs in the case of marine-based glaciers. Glaciers with positive mass balance are increasing their ice mass and, other factors being equal, would cause a eustatic sea-level fall. Glaciers with negative mass balances exhibit the opposite relationship—summertime melting or calving exceeds winter accumulation—these glaciers would contribute to a net sea-level rise. In addition to seasonal ablation and accumulation, the mass flux at the grounding line—the boundary between grounded and floating ice—is a key measurement in terms of the potential impact of ice sheets on sea level. Meier (1993) reviewed the factors that influence the mass balance of glaciers and came to the general conclusion that a 1°C temperature increase would cause a sea-level rise of about 0.3–0.5 mm/yr; a 5% increase in precipitation would cause sea level to fall about 0.1 mm/yr.

Glaciologists continue to debate whether the mass balance of Antarctic and Greenland are presently positive or negative, a topic of critical importance for predicting future sea-level change (Warrick et al. 1996). To estimate Antarctic Ice Sheet mass balance, Bentley and Giovinetto (1991) used observational glaciological data to estimate accumulation to be 1660×10^{12} kg/yr and mass flux at the grounding line to be 1260×10^{12} kg/yr. These results gave a positive mass balance of about 400×10^{12} kg/yr. A positive mass balance for Antarctica might result from atmospheric warming of about 0.5 °C, which would cause increased precipitation and snow accumulation. Conversely, Huybrechts (1990, 1994) computed a negative mass balance of -351×10^{12} kg/yr for Antarctica using a numerical model of Antarctic ice. This study suggested an annual net loss of mass through calving. Jacobs (1992) studied iceberg calving and basal melting of Antarctica and suggested that, at least for the 1980s, Antarctica had a net negative mass balance and may have contributed to recent sea-level rise. Paterson (1993) agreed with this conclusion, stating that loss of grounded

Antarctic ice might be occurring at a rate of 235 gross tons/yr. Paterson suggested, however, that mass-balance estimates are too uncertain to use them to estimate sea-level change.

Ice streams are important features of the West Antarctic Ice Sheet as they flow into the adjacent Ross Ice Shelf and Ross Sea at rates more rapid than the rates of most glaciers. Bindshadler (1997) estimated the migration rate of Ice Stream B to be 488 m/yr, a rate characteristic of surging glacial ice. If this rate is maintained, Bindshadler estimates that the lifetime of the West Antarctic Ice Sheet, which holds the equivalent of about 5–6 m of sea level, would be somewhere between 1200 and 6000 years.

The mass balance of Greenland over the past century is also unclear (see Meier 1993; Paterson 1993). Zwally et al. (1989) used satellite data from 1978–1986 to show the equivalent ice growth of 0.45 ± 0.25 mm/yr. Budd and Smith (1985) and Fortuin and Oerlemans (1990) also suggested Greenland has a positive mass balance and thus contributes to a lowering of the net historical sea-level trend. A review of ongoing research on Greenland and Antarctic Ice Sheet behavior is beyond the scope of this chapter, but major improvements in understanding the mass balance of different segments can be anticipated in the near future.

Small glaciers and ice caps also contribute to sea-level change. Meier (1984, 1993) and Trupin et al. (1992) used monitoring data from dozens of glaciers from 31 regions of the world to provide a range of estimates of 0.18–0.46 mm/yr for the sea-level rise during the past century due to melting of small glaciers and ice caps. However, most of their data came from Northern Hemisphere glaciers, and additional work is needed. Warrick et al. (1996) estimated that sea level rose 0.60 mm/yr owing to small glacier melting for the period 1985–1993. As we saw in chapter 6 in reference to climate change since the Little Ice Age, historical records indeed show that small glaciers have been receding around the world during the past century, and estimates suggest that they have contributed between 5 and 15 cm to the total sea-level rise since the late nineteenth century.

Paterson (1993) summarized available evidence for the contribution that modern ice sheets and glaciers make to the total 1.8 mm/yr sea-level rise of the past century as follows: smaller glaciers, about 0.46 ± 0.26 mm/yr; Greenland Ice Sheet, about 0.29 ± 0.15 mm/yr; Antarctica, about 0.65 ± 0.61 mm/yr. The remaining 0.4 ± 0.2 mm/yr was due to thermal expansion of the oceans.

To summarize, although significant progress has been made in glaciology, determining the glacio-eustatic contribution to the past century's sea-level rise has a large degree of uncertainty (1.2 mm/yr ± 0.6 mm/yr), stemming in part from spotty historical data on most of the world's small glaciers and a limited but improving understanding of the budget of the Greenland and Antarctic Ice Sheets. Similarly, modeling of the glaciological response to climate change is still inadequate to hindcast past or forecast future behavior with certainty. Bindschadler thus states (1997:409): "The idealized `steady-state' glacier can only be found in textbooks. Climate is constantly changing on many time scales, forcing ice masses to respond on a separate set of time-scales." In response to the title of Meier's (1993) paper, "Ice, Climate, and Sea Level: Do We Know What is Happening?," unfortunately the answer is not yet.

Tectono-eustasy: Plate Tectonics and Ocean Volume Changes

The plate tectonics revolution in earth sciences in the 1960s changed the way geologists viewed the earth and contributed to a resurgence of the concept of tectono-eustasy as a major force to explain long-term marine transgression and regressions over time scales of 1 to 10 Ma. In addition, sea-floor spreading was a plausible mechanism to explain how marine sediments could be deposited inland many hundreds of kilometers and then how deposition would cease for the following period, and finally how a stratigraphic unconformity would develop during intervening periods of marine regression and global low sea level. One theory of tectono-eustasy holds that changes in ocean-ridge volume influence global sea level because when oceanic ridges spread quickly, young ridge material is more buoyant and low-lying coastal regions become flooded (Pitman 1978; Donovan and Jones 1979; Kominz 1984). Continental collisions might in theory cause sea level to fall. Global sea-level changes resulting from ocean volume changes occur slowly—at a rate of only ~1 cm/1000 yr—relative to those caused by glacio-eustasy and other processes. Later, I briefly describe global Phanerozoic sea-level curves based on seismic stratigraphy and biostratigraphy (Vail et al. 1977; Haq et al. 1987; Vail 1992) and the ongoing debate about the relative contribution of tectono-eustasy and glacio-eustasy to sea-level history.

Relative Sea-Level Changes

Sea level varies locally and regionally owing to complex processes superimposed upon eustatic sea-level changes. We divide them into the categories of oceanic dynamics, geoidal variations, and isostatic factors.

Ocean Dynamics and Short-Term Sea-Level Variability

Sea level varies not only spatially but also over very short time scales due to ocean-atmosphere affects. For example, geostrophic flow of ocean currents can cause differences in sea level of up to 1000 mm. Additionally, wind fields over the oceans and changes in surface currents create planetary-scale waves called Kelvin and Rossby waves. Sea-level variability up to 500 mm due to Kelvin and Rossby waves occurs over seasonal and interannual time scales related to oceanic-atmospheric dynamics such as the El Niño–Southern Oscillation (ENSO) phenomenon (chapter 6). Strong westerly winds cause Kelvin waves to propagate eastward in the equatorial Pacific. Rossby waves, the oceans' equivalent to large-scale meanders of the atmospheric jet stream, are large-scale dynamic responses to wind and buoyancy (i.e., heating and cooling). They are of longer wavelength than Kelvin waves (hundreds to thousands of kilometers) and their amplitude is relatively small (< 10 cm) (Chelton and Schlax 1996). Rossby waves propagate slowly (< 10 cm/s) westward, and upon reaching the western edge of the Pacific Ocean, they create a new Kelvin wave, which then moves eastward again.

Satellite altimetry has led to major advances in the study of Rossby and Kelvin waves and the variation in elevation of the ocean surface they cause. The TOPEX-Poseidon satellite, launched in August 1992, circles the earth between 66°N and 66°S, using radar altimetry to study short-term sea-level variation by measuring the height of the ocean surface relative to a reference ellipsoid (Fu et al. 1996). The satellite data show an extremely large sea-level rise in the eastern Pacific related to the warm pool flow during years of relatively strong El Niño signal. TOPEX-Poseidon results tend to confirm the historical trend in sea level measured by tide gauges (Cheney et al. 1994). Nerem (1995) studied 2.5 yr of TOPEX-Poseidon data to examine the interannual and seasonal sea-level record. After making complex corrections and calibrations for tides, tropospheric and ionospheric conditions, atmospheric pressure, and other factors, he obtained an estimate of a net rate of sea-level rise of 5.8 ± 2.5 mm/yr. Although this estimate is sure to undergo revision, and the 2.5-yr interval of measurement is obviously too short to say whether this is a long-term trend, satellite monitoring of sea level will continue to be a critical element for tracking future sea-level trends.

Gravitational Factors—the Geoid

It should be apparent by now that the term *sea level* is like the term *solar constant*—an oxymoron; at any one time, the surface of the ocean in different regions of the earth lies at different heights. A reference ellipsoid called the earth's geoid is used to measure short-term spatial variability in sea level. The geoid is defined as earth's equipotential surface, and it owes its existence to differing gravitational attraction in different regions of the globe. The significance of an irregular geoid for sea-level studies is that sea level is not at all *level*. Therefore, one may, in theory, measure a period of high sea level along one coast at a particular time but a different sea level simultaneously on another coast.

The fact that sea level variation due to geoidal effects would complicate the correlation and interpretation of past sea-level curves from different areas was recognized by Fairbridge in 1961 in a seminal paper on eustasy. Fairbridge was among the first to suggest that geoidal changes might be responsible for disagreement in postglacial sea-level curves derived from the earliest (1960s) radiocarbon-dated sea-level indicators. Geoidal effects have also been discussed in detail by Morner (1979b), who argued that there can be no "global" sea-level curve; only regional eustatic curves can be constructed. Over time scales greater than decades, geoidal effects are difficult to identify in the geological record of sea level because of chronological limitations of correlation. Thus, since concern arose about correlation of sea-level curves and geoidal complications in the 1960s and 1970s, researchers seldom try to address these processes in studies of long-term sea-level history.

Isostasy

Isostasy is the theory that the earth's crust and mantle behave like a visco-elastic material in response to the load redistribution that occurs when thick wedges of sediments accumulate or are removed or when ice sheets melt or grow. This section discusses several different types of isostasy as they relate to sea-level history.

Local Coastal Uplift and Subsidence

It is useful to examine isostatic processes along a coast in relation to other processes affecting local sea-level records. Three major geological processes lead to coastal submergence that can produce a local or regional sea-level change. The first encompasses *tectonic* processes that characterize convergent-plate margins along coastal regions where one plate is superimposed upon another and fault movement can displace large crustal blocks downward, causing subsidence of a coastline. Coseismic zones can experience large earthquakes in which underthrusting can displace land upward. For example, on Sado Island in the Sea of Japan, a great 1802 earthquake led to tectonic uplift, elevating the coast several meters. Tectonic uplift is also critical for the interpretation of emergent flights of coral terraces like those along New Guinea and Barbados. Chappell et al. (1996) showed 1- to 4-m-scale events along the Huon Peninsula probably caused episodic uplift, leading to new coral reef accretion.

In contrast to tectonic activity, the second process involves the accumulation of sedimentary wedges along tectonically passive continental margins. Suess (1885–1909; see Dott 1992) believed that sedimentation delivered from continents to the world's oceans was so great as to be the cause for rising sea level because he thought the ocean basins were filling up. Although sedimentation is too small to affect sea level globally, sedimentation can cause local vertical subsidence in coastal regions. The Mississippi Delta of coastal Louisiana is a prime example where sediment from a large portion of North America is eventually deposited on the continental shelf and slope of the northern Gulf of Mexico. The weight of the sediment accumulated in continental-margin sedimentary basins causes the earth's crust to subside isostatically, forming thick sedimentary prisms that are often kilometers thick.

A third way coasts can subside and cause a relative sea-level rise is through the removal of groundwater from subsurface geological formations near coasts. Natural processes such as karst formation—the dissolution of carbonate rock by the action of ground water—can lead to increased permeability and porosity and cause subsidence. Human-induced subsidence caused by groundwater removal has also caused subsidence but the effects are local compared with the overall coastal submergence due to other factors described earlier.

Epeirogeny

Epeirogeny is another term applied to isostatic subsidence due to loading and unloading of sediments. Epeirogeny is pertinent to a major puzzle for scientists seeking to find a suitable mechanism to explain sea-level changes occurring over 1- to 5-Ma time scales. These are referred to by Haq et al. (1987) as third-order sea-level cycles. Third-order sea-level changes appear to occur too slowly to be related to glacio-eustatic processes, which occur over 10- to 100-ka time scales (ice decay and growth), but too rapidly to be caused by mid-ocean ridge-volume changes (Hallam 1992).

Mechanisms proposed to explain 1- to 5-Ma sea-level cycles include intrabasin stress from sediment loading, chaotic mantle convection, and polar wandering. Cathles and Hallam (1991) dismissed these factors and instead recently proposed a new mechanism to account for third- and higher-level cycles of sea level deduced from the stratigraphic record. They suggest that plate elevation changes of up to 200 m result from changes in stress and strain in the earth during collisions between the earth's lithospheric plates. Furthermore, they argue that the creation of new rifts between lithospheric plates can produce quick rupture, then an elastic response, with density changes large enough to cause 50 m of subsidence. When a new rift forms, seafloor depression occurs, then isostatic equilibrium allows mantle flow back into the formerly depressed region. A eustatic sea-level change of 18-45 m can occur, depending on the size of the plates involved. More remarkably, they proposed that these changes can occur in less that 30 ka—on a par with the rate of sea-level change during glacial events. They cite as evidence the correspondence of couplets of black shales and periods of extinction of marine organisms, invoking a complex linkage between the area of ocean bottom and overpopulation, food exhaustion, and extinction of benthic organisms. Whether stratigraphic and geophysical evidence supports this hypothesis remains to be seen, but it serves as a reminder of the complexities of sea-level change.

Glacio-isostasy and Hydro-isostasy

We turn now to the topic of Quaternary sea-level and ice-sheet history and the theories of glacio-isostasy and hydro-isostasy. Glacio-isostasy is the theory that the earth's crust and mantle respond to the load redistribution that occurs when ice sheets melt or grow. In its simplest form, the last great ice sheets that covered large parts of the continents about 21 ka caused glacio-isostatic depression in glaciated regions proportional to the thickness of the ice. During deglaciation, the load from ice is removed and the land "springs" back over several thousand years. It is useful to examine the roots of modern thinking on glacio-isostasy because of its significance for studying sea-level history.

The aforementioned rising land and falling sea level in Scandinavia posed a perplexing geological problem that attracted the interest of some of the natural sciences' greatest luminaries (see Morner 1979a).

After Celsius' proposed ideas in the eighteenth century to explain the rising land in terms of evaporation and plant uptake of moisture, Sweden's Carl von Linne, father of botany, identified 77 beaches that he deduced may have been the result of the biblical deluge. The Englishman John Playfair was one of the first to introduce the idea that crustal movements affect sea level, and when Sir Charles Lyell later visited Sweden, he became convinced that the evidence for elevated relict shorelines was firm. Lyell proposed that the rising land of Scandinavia supported his hypothesis that vertical land motions cause sea level to change. Charles Darwin agreed with Lyell on the topic vertical movements of the land; his first major work on the theory of tropical coral reefs described a general subsidence of the Pacific Ocean he believed explained the origin of Pacific reefs.

The hypothesis that ice from the past ice ages depressed the earth's crust is credited to Jamieson (1865), who by incorporating the new glacial theory of Agassiz (1840) recognized the profound effect ice would have on a nonrigid crust. Soon afterward, early maps of Scandinavian isobases, ice margins, and land-sea distribution, meticulously compiled by De Geer (1888–1890, 1892), demonstrated convincingly not only the great magnitude of the glacio-isostatic uplift, but also the dome-like shape of the uplift inherited from the shape of the Fennoscandinavian Ice Sheet. De Geer showed that the thickest part of the ice sheet must have been centered in central Scandinavia, whereas thinner ice had covered coastal regions where uplift was of smaller magnitude.

De Geer also applied his theory to interpreting glacial lake and marine shoreline levels in North America (1892). Indeed, North American Pleistocene geology has its own laboratories for isostasy, the most famous being relict lake levels around Great Salt Lake in Utah, where ancient Lake Bonneville occupied a much larger area during the last glacial. Dutton (1871) first coined the term isostasy in his studies of glacial Lake Bonneville (see also Gilbert 1890), although there had been a long history of observations of isostasy in Europe before this time.

Fennoscandinavian uplift as well as vertical crustal changes in other glaciated regions is now explained by the theory of glacio-isostasy. This theory holds that the earth behaves as an elastic material in response to short-term stress—such as earthquakes—but more like a fluid in response to long-term stress changes such as centrifugal force. The earth's asthenosphere behaves like a liquid and the lithosphere like a buoyant mass floating on the asthenosphere. Like a rubber ball when squeezed, it regains its original shape once pressure is released. The earth's gravitational response in its mantle to redistribution of massive loads of ice and water at the earth's surface leads to differential adjustment depending on the location of the regions relative to the ice sheet. Over time scales of 10–20 ka, like those that characterize glacial-interglacial cycles, both viscous and elastic responses are in evidence. In a formerly glaciated region such as Fennoscandinavia, uplift is a reflection of this glacial-isostasy. The greater the thickness of the ice sheet, the greater the depression and subsequent uplift. Relative sea-level changes due to glacio-isostasy are therefore reflected in varying rates of sea-level rise or fall. Coastal submergence in nonglaciated regions occurs at rates of about 2 mm/yr, coastal emergence in formerly glaciated regions at rates of 1.2 cm/yr. Glacio-isostasy is incorporated into modern geophysical models of earth's late Quaternary paleotopography, discussed later.

Ice is not the only mass to weigh down the earth's crust and mantle. As ice sheets melt, ocean volume rises glacio-eustatically and the shifting of the load from continental ice to the ocean water in theory causes a vertical adjustment of the ocean basins. The ocean basin response to this loading is described as the process of hydro-isostasy (figure 8-4). Daly (1910; see Bloom 1967; Hopley 1982) postulated hydro-isostatic effects in his glacial theory of coral reefs, and other early European workers (e.g., Hogbom 1921; Nansen 1922) described its effect in Scandinavia. Bloom (1967) pointed out that many early workers thought the idea of hydro-isostasy was speculative for lack of definite evidence. Bloom tested the theory of hydro-isostasy in 1967 with data from radiocarbon-dated shorelines from Massachusetts, Connecticut, New Jersey, and Florida. Reasoning that there would be differential subsidence depending on the depth of water and breadth of the continental shelf (greater subsidence where there was a greater load of deglacial water), Bloom compared sea-level curves from these areas for the last 12 ka and found evidence supporting hydro-isostatic loading. Currently, hydro-isostatic effects are secondary in importance to glacio-isostatic effects, which are now incorporated into geophysical models used to predict the vertical response to redistribution.

Paleotopography

Isostatic response of the earth's crust and mantle to glaciation has provided an ideal natural experiment for geophysicists to probe the nature of the deep earth's interior. Several generations of geophysical models have used radiocarbon-dated shorelines to characterize the

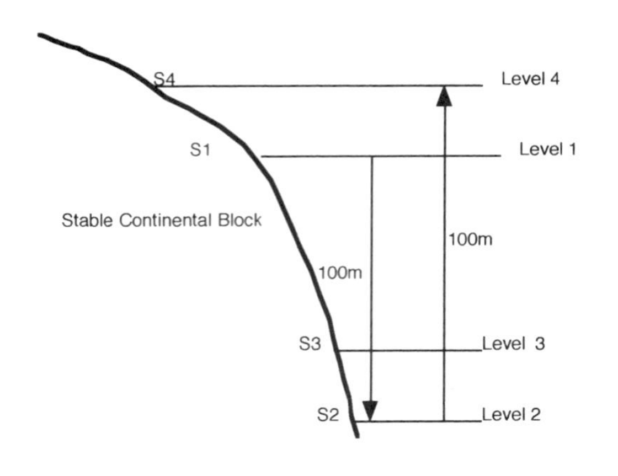

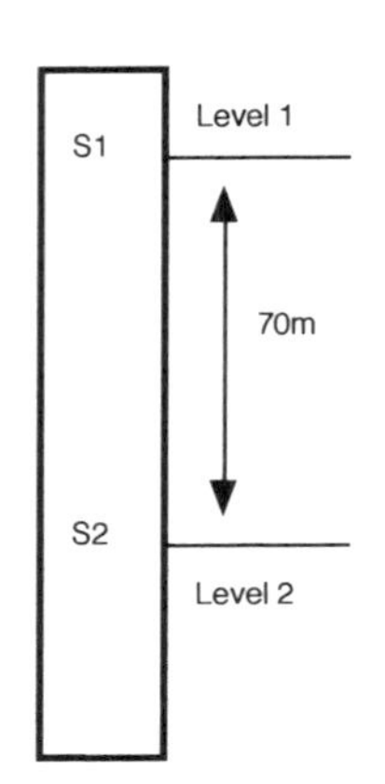

Contrast between Observed Relative Sea Level Changes along a Continental Margin and an Oceanic Island due to potential effects of Hydroisostasy

FIGURE 8-4 Generalized diagram illustrating the theory of hydro-isostasy. On left, a nonglaciated, nontectonic coast, Level 1 is sea level during an interglacial period; ice-sheet growth on continents lowers sea level to Level 2; Slow isostatic rise of sea floor along continent raises level to Level 3; rapid postglacial sea-level rise occurs as ice sheets decay, and sea level comes up to Level 4. Delayed isostatic sinking lowers the level back to Level 1. These combined glacio-eustatic changes and isostatic adjustments produce shorelines that are 100 m apart, 30 m more than the actual change in eustatic sea level. On right, mid-oceanic island behaves like a dipstick such that the island (in contrast to the continent) rises isostatically with ocean sea floor when sea level is lowered. Ocean islands thus in theory record actual change in sea level. See Bloom (1967) and text for discussion.

earth's interior and to identify the vertical response to load redistribution along coastal regions of the world. Walcott (1972) recognized three regions, each in theory having a distinct response to deglaciation: glacial uplifted areas, submerging forebulge in peripheral areas, and emergence along coasts due to tilting. Peltier (1974), Cathles (1975), and Peltier and Andrews (1976) expanded on Walcott's work in using a more realistic spherical visco-elastic earth. Clark et al. (1978) and Peltier et al. (1978) later constructed the first global isostatic model of sea-level changes that included both the oceans and continents. They identified six regions, each responding differently to deglacial load redistribution. They generally found supporting evidence for their model from patterns of sea-level rise or fall from the world's coastal regions. For example, in the east coast of North America, the coastal submergence was predicted as part of the collapsing forebulge, which is now believed to explain the relatively rapid rate of sea-level rise in the Chesapeake Bay region.

The latest set of geophysical models relevant to sea-level change have been developed by Richard Peltier and colleagues (Peltier 1988, 1994, 1995; Tushingham and Peltier 1991). The models have evolved to incorporate relative sea-level curves from the world's coastlines. As a means of fine-tuning these models, contamination from glacio- and hydro-isostatic responses has been removed from historical tide-gauge records (Peltier and Tushingham 1989). The culmination of geophysical modeling of the earth's ice-age surface is called the ICE-4 model (Peltier 1994), which reconstructs global "paleotopography" for the LGM. As we see below, this geophysical model plays a critical role in evaluating field evidence for glacial and postglacial global ice-volume and sea-level history.

CENOZOIC SEA LEVEL: TECTONO-EUSTASY VERSUS GLACIO-EUSTASY

IN 1977 PETER VAIL and colleagues (Vail et al. 1977) of Exxon published a monograph on global eustatic sea-level changes used to interpret and correlate stratigraphic sequences of sedimentary basins around the world. Basing their sea-level curve on extensive seismic and biostratigraphic information accumulated during years of petroleum exploration, Vail and colleagues postulated that regional marine transgressions and regressions from passive continental margins could be used to erect a global sea-level curve (see also Vail and Hardenbol 1979). The sea-level curve was later referred to as a curve of onlap and offlap sequences in recognition of the fact that these sequences represented relative sea-level curves. The Vail curve stimulated a new generation of research into the long-dormant field of long-term eustatic sea-level history.

In brief, the Vail model of eustasy, resting on foundations built by L. Sloss, among others, held that periodic stratigraphic unconformities signified periods of regression and erosion and a hiatus in the deposition of marine sediments. Vail then applied this concept that seismic stratigraphy was a means to recognize sequences to develop a distinct set of criteria to identify strata within a sequence. Each complete sea-level cycle formed a distinct depositional sedimentary sequence—a comformable succession of genetically related strata bounded by unconformities. By convention, the sequence began with a sea-level low stand. Next, relative sea level rose again, forming a sedimentary sequence. During the third phase of a cycle, sea level dropped until it reached a low stand, which is reflected in the boundary unconformity. Vail classified sea-level cycles by the length of their period. For example, first-order cycles range over periods of hundreds of millions of years; third- and higher-order cycles occur over periods with frequencies of less than 5 Ma years. A revised eustatic sea-level curve covering the Mesozoic and Cenozoic was generated by Haq et al. (1987, 1988) and is now known as the *Haq sea-level curve.* Haq and colleagues also subscribed to the belief that global sea-level events could be recognized using sedimentary packages and dated and correlated using magnetostratigraphic and biostratigraphic data available for most sedimentary basins along continental margins.

In the same year that Vail et al. published their landmark paper, Anthony Hallam (1977) also published an important paper describing stratigraphic and paleontological evidence for long-term Phanerozoic patterns of marine inundation as a record of large-scale sea-level changes.

Since these influential papers, the Vail and Haq sea-level curves, in particular the chronology of post-Triassic coastal submergence and emergence and the mechanisms controlling long-term sea-level changes, have been actively researched and widely debated. Two major criticisms permeate the controversy. The first issue is geologists' ability to date the events accurately enough to correlate among widely separated sedimentary basins and thus demonstrate the synchroneity of sea-level events. This problem applies especially to the correlation of third-order cycles, which in some sedimentary basins may represent regional changes in sediment supply or other processes unrelated to eustatic sea level (Christie-Blick et al. 1988). Miall (1997) was strident in criticizing the Vail-Haq sea-level curves because of limitations inherent in global correlation of sedimentary sequences and sea-level records. Miall demonstrated that random processes could produce the same sea-level patterns as those obtained from seismic records.

The second issue in the Vail and Haq sea-level curve controversy involves finding a mechanism that could explain the relatively rapid third- and higher-order events. Matthews (1988), for example, raised the issue as to whether the third- and fourth-order sea-level cycles were the result of glacio-eustasy. Morner (1981), in contrast, emphasized his long-standing opinion that geoidal and other effects render efforts to find a "global" sea-level curve futile. He stated that a global sea-level curve is "an illusion" because of geoidal effects—and he recommended that only regional eustatic curves be developed (p. 344). Haq and colleagues have defended both the age models used to correlate sea-level changes globally and the ability to correlate third-order cycles, vis-à-vis the number of sequences and the characteristic stacking.

Cenozoic deep-sea oxygen isotope records, ocean drilling on continental margins, and attempts to determine Antarctic ice-sheet history have recently shed new light on reconstructions of long-term sea-level history. The most important outcome of this research is that the Cenozoic Era was characterized by glacio-eustatic sea-level changes resulting from both North and South Polar regions. Antarctic ice-sheet oscillations have occurred since at least the earliest Oligocene (Barron et al. 1991), and Northern Hemisphere ice volume has fluctuated since at least the late Miocene, 7–10 Ma (Jansen et al. 1990; Jansen and Sjøholm 1991; Larsen et al. 1994). What follows is a brief summary of these findings.

Deep-Sea Isotope Records

Deep-sea oxygen isotope records of foraminifera were discussed in chapters 3 and 4 in terms of their value as a monitor of continental ice volume. Early views that the steadily heavier oxygen values of the Cenozoic represented colder oceanic temperatures have been modified (Matthews and Poore 1980; Prentice and Matthews 1988) to suggest there is an ice-volume component to some foraminiferal isotopic records. Building on Quaternary deep-sea isotope research (Shackleton 1967; Shackleton and Opdyke 1973), Prentice and Matthews (1991) argued instead that the Cenozoic oscillations in deep-sea tropical planktic foraminiferal oxygen isotope values reflected a polar ice-volume history as far back as 40 Ma. Prentice and Matthews (1988) also developed the Snow Gun hypothesis to identify a mechanism whereby Antarctic ice growth might occur during periods of relative warmth, like the Miocene and Pliocene, driven by warming of deep water by low-latitude marginal seas. The Snow Gun theory held that, in contrast to ice-sheet growth related to cold deep-water formation of Quaternary glacial periods, warmer pre-Quaternary deep-sea waters may have contributed to Antarctic ice-sheet development.

Continental Margin Record

Sea-level history derived from stratigraphic and oxygen isotopic records from emerged and submerged sedimentary records from continental margins also indicate high-amplitude Cenozoic sea-level fluctuations. Miller et al. (1991) and Wright and Miller (1992) studied Oligocene and Miocene sea level along the eastern United States and found 12 $\delta^{18}O$ spikes of > 0.5‰ in Oligocene and Miocene foraminifera. They interpreted the isotopic excursions as a sign that sea level fell between 20 and 80 m during low stands. Pekar and Miller (1996) found 10 major unconformities in records from Ocean Drilling Program (ODP) legs off New Jersey that matched periods of low sea level in the Haq curve. The amplitude of Oligocene sea-level changes were, however, only 24–34 m, about one half the amount postulated by Haq and colleagues. McGinnis et al. (1993) offered a mechanism to explain the smaller magnitude of Tertiary sea-level events obtained from continental margin ODP legs and the Haq curve. They suggest that the early-late Oligocene (38–34 Ma) sea-level fall, estimated by Haq to be as much as 150 m, is overestimated because part of the observed relative sea-level fall is the result of flexural isostatic rebound of the lithosphere along the continental margin due to deep-sea erosion and retreat of the continental slope.

Miller et al. (1996) integrated evidence from sequence stratigraphy from ODP cores and onshore coastal plain deposits in New Jersey with oxygen isotope curves to draw an improved sea-level curve for the Oligocene–middle Miocene that generally supported the Haq model for global Cenozoic third-order sea-level oscillations. They found an excellent correspondence among seismic reflectors, increases in $\delta^{18}O$ of foraminifera (indicating reduced ice volume), and coastal plain unconformities, all indicators of lowered sea level. Miller and colleagues argued that between 36 and 10 Ma, multiple sea-level cycles occurred over million-year time scales. About 10 cycles punctuated the period 12–24 Ma, one occurring about every 1–2 Ma. Data from ODP Leg 166 of the Bahama Banks lent similar support for eustatic Cenozoic sea-level change.

The Glacial Record

Are the alleged glacio-eustatic sea-level oscillations hypothesized from isotopic and continental margin stratigraphic records supported by direct evidence for ice-volume instability from polar regions

(Mangerud et al. 1996)? Quite likely. It is now generally accepted that continental-scale ice existed in Antarctica since the earliest Oligocene, about 40 Ma (Barron et al. 1991; Ehrmann et al. 1992) and in Greenland since at least 7 Ma (Larsen et al. 1994). Extensive stratigraphic, ice-rafting, and micropaleontological evidence has been obtained from ODP legs such as Leg 119 from the Kerguelen Plateau region and other sites near Antarctica (Barron et al. 1991). However, even as Antarctic and Greenland ice-sheet history begins to emerge from glaciological, isotopic, stratigraphic, lithological, and ice-rafted debris (IRD) evidence in deep-sea sediments, sea-level history due to polar ice-volume instability remains elusive. To illustrate this complexity, I examine the debate surrounding late Neogene Antarctic Ice Sheet stability.

The debate revolves around how dynamic the Antarctic ice sheet has been for the past 6 Ma or so. The stabilist view (Sugden et al. 1993; Kennett and Hodell 1993) holds that Antarctic ice volume has remained fairly stable since at least the Miocene. This conclusion is based mainly on geomorphological, glaciological, and chronological evidence from several regions of Antarctica (Denton et al. 1993), glacial modeling (Huybrechts 1993), deep-sea ice rafting data, and stable-isotope data. Kennett and Hodell (1993) reviewed evidence for ice rafting around Antarctica, a line of evidence signifying the existence of a large ice sheet. They concluded (p. 213) that "the Subantarctic IRD records an expansion of the Antarctic cryosphere during the latest Miocene through earliest Pliocene, relative stability during the early Pliocene, and further expansion during the late Pliocene."

The dynamicist view is based on stratigraphic and paleontological evidence from several regions of Antarctica, suggesting one third to one half deglaciation during parts of the Pliocene over the past 5 Ma (Webb et al. 1984; Webb and Harwood 1991; Wilson 1995). Independent evidence for relatively high sea level during the Pliocene (Haq et al. 1987; Dowsett and Cronin 1990; Wardlaw and Quinn 1991) suggests there may have been partial deglaciation of Antarctica.

The long-term history of the Greenland Ice Sheet has emerged within the past few years from deep-sea sedimentology and isotopic records. Greenland ice began building up in the late Miocene, perhaps as much as 8–10 Ma. Jansen et al. (1990) found ice-rafted sediments, diamictons, and dropstones as old as 6 Ma in ODP sediments from the Norwegian Greenland Sea. Jansen and Sjøholm (1991) estimated glacio-eustatic sea level drops near 5.0 and 3.9 Ma to have been as much as 60–100 m based on combined IRD and oxygen-isotope data from the Norwegian Sea. After about 2.75 Ma, there was intensification of Northern Hemisphere glaciation with European glaciation beginning around 2.57 Ma (Jansen and Sjøholm 1991).

Additional evidence for dynamic Northern Hemisphere ice comes from Greenland, where Larsen et al. (1994) discovered glacial till, diamictons, and dropstones in deep-sea cores from off the southern coast of Greenland. They concluded that cooling around Greenland started after 10 Ma and full glacial conditions developed by 7 Ma. These events may have led to a glacio-eustatic sea-level drop of 12 m. These and other studies indicate a significantly more complex and dynamic late Miocene and Pliocene glacial history of the Northern Hemisphere than researchers recognized just a decade ago and provide a mechanism for pre-Quaternary low-amplitude glacio-eustatic sea-level changes (Mangerud et al. 1996). Evidence for sea ice in the Arctic Ocean extends back at least until 12–14 Ma (Thiede et al. 1996), but sea-ice would not have affected eustatic sea level.

In closing this brief discussion on long-term sea-level history, it should be emphasized that some geologists remain skeptical that long-term Phanerozoic eustatic cycles are glacio-eustatic in nature (Dott 1992). Similarly, there is perhaps even more skepticism surrounding the idea that pre-Cenozoic sea-level cycles reflect orbital-scale sea-level changes (Sloss 1991; see chapter 4). Moreover, disputes about the long-term contribution of Antarctic and Greenland ice to eustatic sea level will likely continue into the foreseeable future. Regardless of the outcome of these debates, there is growing evidence that since their initial formation > 35 Ma and > 7 Ma, respectively, Antarctic and Northern Hemisphere polar ice sheets have undergone dynamic oscillations that were probably large enough in many instances to cause sea-level oscillations of at least 10–30 m over time scales of 100 ka to 1 Ma. The possibility that Cretaceous sea-level fluctuations are glacio-eustatic in origin remains a viable hypothesis as well (Stoll and Schrag 1996). Still, it remains a major challenge to decouple glacio-eustatic sea-level changes related to climate change from those produced by other mechanisms and to reconstruct long-term sea-level changes.

QUATERNARY SEA-LEVEL HISTORY

IN CHAPTER 4, I described the contribution that Quaternary sea-level history has made to testing the basic tenets of the orbital theory of climate change. The age and elevation of tectonically uplifted coral reef terraces in Barbados and New Guinea record swings in sea level over tens to hundreds of thousands of years that according to many researchers match the timing of ice-volume fluctuations inferred from deep-sea oxygen isotope curves and times of low solar insolation. During the past 500 ka, each 100-ka glacial-interglacial cycle underwent a ... bout 100 m below the present level, and a high sea ... s of the current level. The dominant 100-ka ... the period of orbital eccentricity was super... h stands occurring about every 20 ka, associ... sional period. During most of the past 500 ka, ... below its current level because the volume of ... its was greater than it is now. During the late ... eistocene (2.5–0.5 Ma), eustatic sea level oscil... ital frequency but at a lower amplitude, perhaps ... l. 1989; Ruddiman et al. 1989; Cronin et al. 1994; ... er, the paleoclimatological record from emerged ... al stratigraphy, and deep-sea isotopes have led to a ... not universal, acceptance of the influence of or... te Cenozoic global climate, the cyclic waxing and ... s, and glacio-eustatic oscillations.

... o Cenozoic orbital-scale cyclic ice-volume and sea... reconstructing sea level during the LGM and the pro... vel rise and ice-sheet decay during deglaciation are two ... portant topics in paleoclimatology. The next two sec... ea-level controversies surrounding sea level and global ... ring the LGM and the subsequent deglaciation. The ev... ative sea-level changes over the past 20 ka is scattered ... e literature, in part because regional field studies have ... ried out on so many of the world's coastlines. Recognizing the ... portance of correlating relative sea-level curves, the International Geological Correlation Program (IGCP) has sponsored four successive international projects over the past 20 years to examine late Quaternary relative sea-level history and to sort out eustatic, geoidal, and isostatic effects. These projects have been coordinated by leading sea-level researchers—Arthur Bloom, Paolo Pirazzoli, Orson van de Plassche, and David B. Scott—and have resulted in many field excursions and publications on regional sea-level history (e.g., Bloom 1992).

In addition to extensive field studies, late Quaternary sea-level research depends heavily on accurate chronology of sea-level events. Excellent summaries of Late Quaternary relative sea-level obtained from continental margins using conventional ^{14}C can be found in Bloom (1977) and Pirazzoli (1991). In addition, recent improvements in the chronology of sea-level history have come from accelerator mass spectrometric (AMS) ^{14}C and uranium-series thermal ionization mass spectrometry (TIMS) dating (Bard et al. 1990a,b, 1992, 1993).

Thick Ice and Thin Ice During the Last Glacial Maximum

The thickness of late Quaternary ice sheets has been debated for at least a century and a half. During the formative days of Agassiz's glacial theory, Maclaren (1842) had already estimated that sea level would be lower by 350 feet because of ice locked up on the continents, even excluding ice in the south polar region. Later, Penck (1882) also estimated a glacio-eustatic sea-level drop of 100 m. These were remarkably accurate estimates foreshadowing most currently accepted estimates, which range from 100 to 130 m.

Currently, the problem of how far sea level fell during the LGM about 21 ka remains intimately related to conflicting estimates of continental ice volume and past sea-level positions at the continental shelf edge. Two opposing viewpoints can be expressed in terms of the thickness of the last great ice sheets (in order of decreasing volume): the Antarctic, Laurentide of eastern Canada and the midwestern and northeastern United States (figure 8-5), Fennoscandinavian, Barents-Kara Sea in the eastern Arctic Ocean, Greenland, Cordilleran in the Pacific northwest extending into southern Alaska, Innuitian in the Canadian Arctic island archipelago, northern British Isles, and Icelandic. It should be noted that the extent of the Barents-Kara Sea ice sheet has been a topic of some debate (e.g., Denton and Hughes 1981; Grosswald 1993, see below) and that there were other smaller ice caps, especially at high elevations, I have not listed.

Historically, one school of thought has adhered to a "thick ice scenario" in which, at their greatest extent, LGM ice sheets held the equivalent of as much as 130–160 m of sea level. The other theory holds to a "thin ice scenario," which calls for thinner ice sheets and a much smaller fall in eustatic sea level of about 100 m (table 8-2). The evidence upon which each ice-volume–sea-level scenario rests, however, comes from a spectrum of complex topics that include coral reef accretion and growth, ice mechanics, earth rheology, stable-isotope fractionation, foraminiferal ecology, salt marsh accretion, and others.

Sea level and ice volume during the LGM and the succeeding deglaciation has been reconstructed through four main avenues of investigation. First is the combination of glacial geology, carried out by extensive field mapping of glaciated and periglacial regions in middle and high latitudes (e.g., Andrews et al. 1993b, 1995b, references

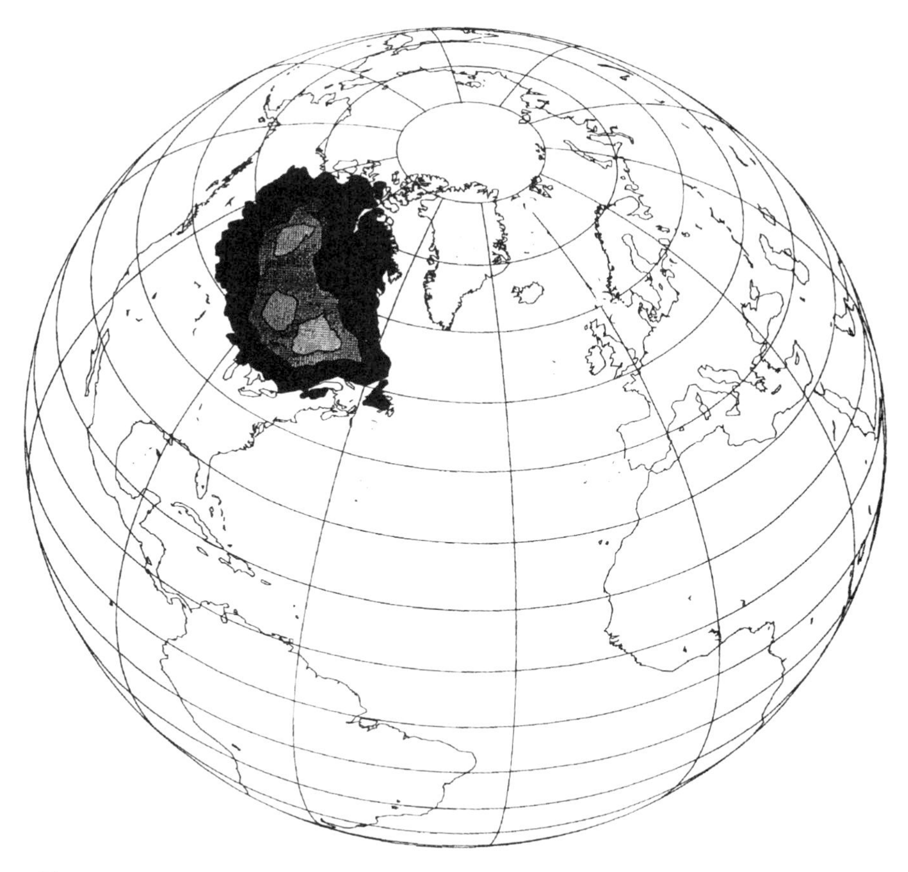

FIGURE 8-5 Late Quaternary North American Laurentide Ice from about 14 ka, about the time of meltwater pulse 1a. Courtesy of Peter Clark and the American Geophysical Union. From Clark et al. (1996a).

TABLE 8-2. Selected Estimates of Global Sea Level During the Last Glacial Maximum

Estimate	*Study*	*Sea-level change (m)*
CLIMAP maximum	Denton and Hughes 1981	163*
CLIMAP minimum	Denton and Hughes 1981	127*
Barbados	Fairbanks 1989	125†
ICE-4	Peltier 1994	105

*CLIMAP Project members (1981) used these studies for their LGM estimates.
†Includes for tectonic uplift about 0.34 mm/yr, or 7 m since the LGM.

therein; Denton and Hughes 1981; Dyke and Prest 1987; Sibrava et al. 1986), and glaciological modeling (Denton and Hughes 1981; Clark et al. 1996b) that addresses glacial mechanics and the way ice sheets respond to various boundary conditions and forcing. The global LGM ice reconstruction of Denton and Hughes (1981) was used in the CLIMAP project, and it provided two extreme ice reconstructions. The maximum ice sheet reconstruction called for 97.8×10^6 km^2 total volume of ice, of which 65.3×10^6 km^2 would lower sea level; the rest was floating marine ice sheets. The minimum LGM reconstruction is estimated at 84.2×10^6 km^2 total ice volume, of which 51.3×10^6 km^2 contributed to a sea-level drop. These maximum and minimum ice configurations were the equivalent to 163 and 127 m of sea-level drop, respectively, which after taking into account isostatic effects would produce eustatic sea-level drops of 117 and 91 m. The Denton and Hughes Laurentide Ice Sheet reconstruction held that there was a single dome located over Hudson Bay where surface elevations exceeded 3800 m. This thick ice scenario was used by CLIMAP (1981) and other researchers modeling glacial-age climate.

The CLIMAP ice sheet reconstructions and sea-level estimates for the LGM, however, were not always accepted (see Andrews et al. 1986) and have since been modified. One important factor requiring reevaluation of ice sheet thickness is the nature of the subglacial bed throughout glaciated regions. Clark et al. (1996b) recently revised long-standing concepts about the thickness of the Laurentide Ice Sheet in North America that characterize the CLIMAP reconstructions by applying a different viewpoint on subglacial sediments. Specifically, they noted that geological field evidence indicates that large parts of the land surface below the Laurentide Ice Sheet consisted of relatively wet, easily deformable glacial till, referred to as "soft bed" till by Clark et al. (1996b). In contrast to areas where the subglacial substrate is more rigid and ice flow is influenced mainly by ice deformation, the wet, soft till substrate influences ice flow both by ice deformation and by sediment (i.e., till) deformation. By applying this twofold classification of hard bed and soft bed substrate to crystalline and sedimentary bedrock regions, respectively, to a numerical reconstruction of the Laurentide Ice Sheet, Clark et al. (1996b) found evidence for a much thinner Laurentide Ice Sheet than previous studies had indicated. Their paleo-topography showed four ice domes instead of one, one on either side of James Bay, one in the Keewatin district west of Hudson Bay, and one over the Foxe Basin. They depict the overall features of the ice sheet as follows (p. 681):

> In general, the ice-sheet surface topography is best described as a broad plateau with elevations 2000–2500 m asl [above sea level] centered over the area of crystalline bedrock and surrounded on its southern, western, and northwestern sides (corresponding to areas of low-viscosity till) by a low brim of ice ranging in elevation from 500–1500 m asl.

This thin ice sheet scenario calls for a Laurentide Ice Sheet volume that is 75% thinner than the estimated maximum volume of Denton and Hughes (1981). Obviously it would result in a much smaller sea-level drop during the LGM. Whereas these ice sheet reconstructions involve numerous assumptions and potential sources of uncertainty, other methods of ice volume reconstruction discussed next tend to support the thin ice model.

The second approach to reconstructing glacial sea level is to determine past sea-level positions directly from ancient shorelines using stratigraphic, geomorphological, and paleobiological study of submerged shorelines and coastal environments preserved at various elevations below modern sea level (e.g., Bloom et al. 1974; Pirazolli et al. 1989; Tooley and Shennan 1987). During the 1960s and 1970s, radiocarbon dating of shells, peats, and other material from outer continental shelves produced the earliest sea-level curves (Fairbridge 1961; see Pirazolli 1991), but relatively few dates were found that were old enough to define sea level during the LGM (Field et al. 1979). For example, in the Newman et al. (1980) compilation of 4000 ^{14}C dates, only about 40 are older than 14 ^{14}C ka, and of these, only a dozen or so are from water depths > 80 m below sea level. The remainder of the dates come from subsequent periods of deglaciation, when sea level had already risen from its glacial low stand. Such is also the case for Pirazzoli's (1991) compilation of 900 sea-level curves from the world's coastal regions.

In addition to problems of tectonic and isostatic subsidence and uplift in many regions, several other problems prevented early workers from establishing an accurate sea-level position for the LGM from continental margin shoreline sediments. One of the problems was that some types of dated organic material, such as fossil mollusk shells, did not come from the original water depth at which they lived (that is, postmortem sedimentary processes had transported them into deeper water). Another dilemma was whether the dated species (e.g., the oyster *Crassostrea virginica*) had lived precisely at sea level or many meters below mean sea level and thus provided only a minimum depth for a paleo–sea-level position.

This situation changed radically when submerged coral reefs from the Caribbean Island of Barbados, radiocarbon dated by Fairbanks (1989), and Muruao Atoll in French Polynesia, dated by the uranium-series TIMS method by Bard et al. (1992, 1993), provided a much better source of direct paleo-shoreline data from the LGM. Corals provided a stable sea-level datum from which to evaluate LGM sea level. Fairbanks (1989) obtained a transect of cores from 50–130 m depth off Barbados that penetrated fossil reefs of the coral *Acropora palmata,* the same species of coral used successfully to document high sea-level events from emergent reefs associated with orbital climate change (see chapter 4). *A. palmata* is a well known species of coral ideally suited to paleo–sea-level studies because its ecology limits its depth range to < 5 m in most regions (Lighty et al. 1982). The deepest submerged fossil *A. palmata,* lying at 113 m below sea level, was dated by radiocarbon at 17.1 ka and later by uranium series dating (Bard et al. 1992). This remains the single best-dated LGM sea-level datum available. Assuming tectonic uplift was about 7 m since the LGM, and given the < 5-m depth range of *A. palmata,* Fairbanks (1989) estimated glacial eustatic sea level was about –121 ± 5 m. A sea-level fall of 121 m lies toward the lower end of estimates derived from glaciological evidence used in the CLIMAP studies.

A third method used to estimate LGM sea level is geophysical modeling that investigates the earth's short-term vertical isostatic response to redistribution of load due to continental ice growth and decay (Clark et al. 1978; Tushingham and Peltier 1991; Peltier 1994). Peltier's geophysical model to describe "ice age paleotopography" is called the ICE-4 model. ICE-4 calculates paleotopography for the earth during the LGM using a visco-elastic model of the earth's response to ice-sheet melting. The model essentially uses hundreds of relative sea-level curves from around the world to constrain earth's crustal and mantle isostatic adjustment to deglaciation and create an inverse curve. It is constructed from the earlier ICE-3 model (Tushingham and Peltier 1991), which was tuned to sea-level curves mostly from glaciated areas and mostly covering the past 8 ka. The ICE-4 model described by Peltier (1994, 1995) is significantly improved in that it is tuned to the uncorrected Barbados LGM sea-level depth of 118 m (Fairbanks 1989).

The ICE-4 model estimates eustatic sea level at 21 ka to be about 105 m below present. The 105-m eustatic sea-level estimate is 13 m below the uncorrected depth of the 21 ka radiocarbon dated Barbados *A. palmata* coral because of geoidal and isostatic effects (Peltier 1995). Although the ICE-4 model has been criticized for underestimating eustatic sea level because it does not take into account presumed uplift of Barbados during the past 21 ka (Edwards 1995), Peltier responded that isostatic disequilibrium could easily account for the difference. Moreover, even if the uplift of 7 m were incorporated into the ICE-4 model, the estimated ice-sheet thickness would change by only 200 m, an inconsequential amount in terms of the use of ice-age paleotopography in studies of atmospheric general circulation modeling of glacial climates. The critical conclusion derived from the ICE-4 model is that LGM ice sheets were significantly thinner—about 35% thinner than many prior estimates and as much as 1000 m thinner for portions of the Laurentide Ice Sheet—than the CLIMAP LGM ice-sheet reconstructions.

A fourth, less direct method of estimating glacial sea level comes from the deep-sea foraminiferal oxygen-isotope record of ice-volume history (figure 8-6). As discussed in chapter 2 and elsewhere in this

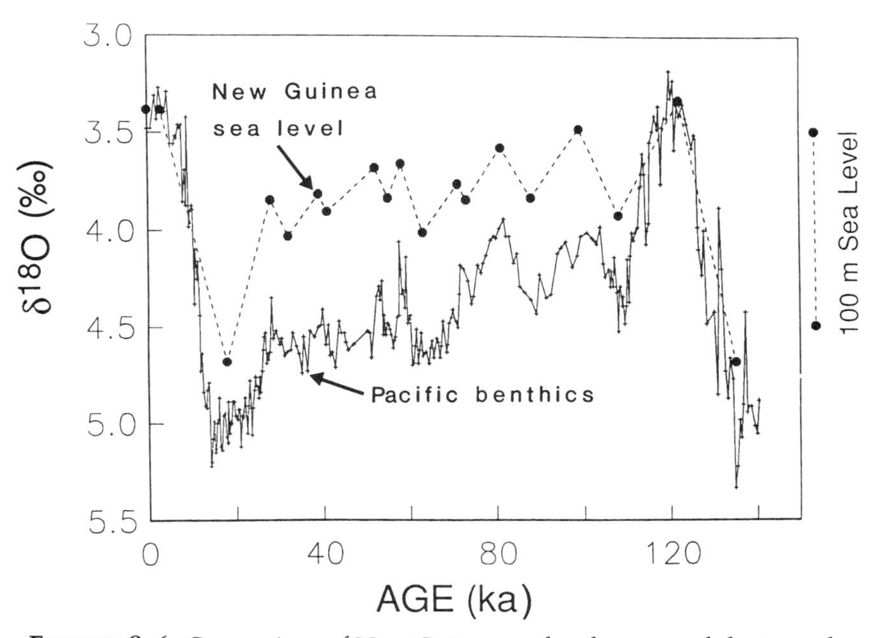

FIGURE 8-6 Comparison of New Guinea sea-level curve and the ice-volume curve obtained from deep-sea Pacific Ocean benthic foraminiferal oxygen isotopes for the past 130 ka. The discrepancy between the two records most likely results from the contribution of deep-sea temperature changes to the oxygen isotopic signal. Courtesy of A. Mix, with permission from the Geological Society of America. From Mix (1992).

book, oceanic oxygen isotopic composition reflects in part continental ice-sheet volume because of differential fractionation of the light and heavy oxygen isotopes, so that the $^{18}O:^{16}O$ ratios of deep-sea foraminifers record changes in ice volume (Shackleton and Opdyke 1973; Chappell and Shackleton 1986; Schrag et al. 1996). The oxygen isotopic method of estimating ice volume has the advantage of being unaffected by isostatic and tectonic effects. Using the calibration of 0.11‰ per 10 m of sea level (Fairbanks and Matthews 1978), and all other factors being equal, a typical 1.5–1.7‰ oxygen isotopic shift between glacial age and modern foraminifers would imply a fall in sea-level change during the LGM of as much as 150–170 m. However, the isotope signal in foraminifers is also affected by other factors, the most important being ocean temperature changes. Given the evidence discussed earlier that tropical sea-surface temperatures (SSTs) dropped during the LGM and that deep-sea bottom water temperature fell at least 1.5–2.0°C globally (Labeyrie et al. 1987; Chappell and Shackleton 1986) and as much as 2.5–3.0°C in the Atlantic Ocean (Dwyer et al. 1995), estimating *eustatic* sea-level change from foraminiferal $\delta^{18}O$ requires care. In general, reevaluation of the ice volume signal from oxygen isotopes that takes glacial-age reductions in surface and deep bottom water into account would place the isotopic LGM ice volume estimates more in line with those derived from new glacial geological interpretations, the Barbados coral record, and geophysical modeling.

In sum, while the above discussion is a great oversimplification of the complex topic of LGM ice reconstruction, it serves to characterize the two opposing points of view and the types of paleoclimate evidence garnered to support each. Recent evidence from glaciological data, coral paleo–sea level, geophysical modeling, and revised oxygen-isotope results suggests that glacial-age ice sheets may have been substantially thinner than was thought 15–20 years ago. This concept has serious implications for the rate of sea-level rise during subsequent deglaciation.

Late Quaternary Deglaciation and Sea-Level Rise

The period between 21 and 6 ka, when late Quaternary ice sheets melted, can be divided into four stages based on the Barbados sea-level curve of Fairbanks (1989) and other sea-level records referenced later (figure 8-7). The first, early deglaciation about 21.0–14.5 ka, was a time when the Barents Ice Sheet may have disintegrated. The second, about 14.5–12.5 ka, is called Meltwater Pulse (MWP) 1a and includes a rapid pulse of melting about 14.2–13.8 ka. Most researchers believe a

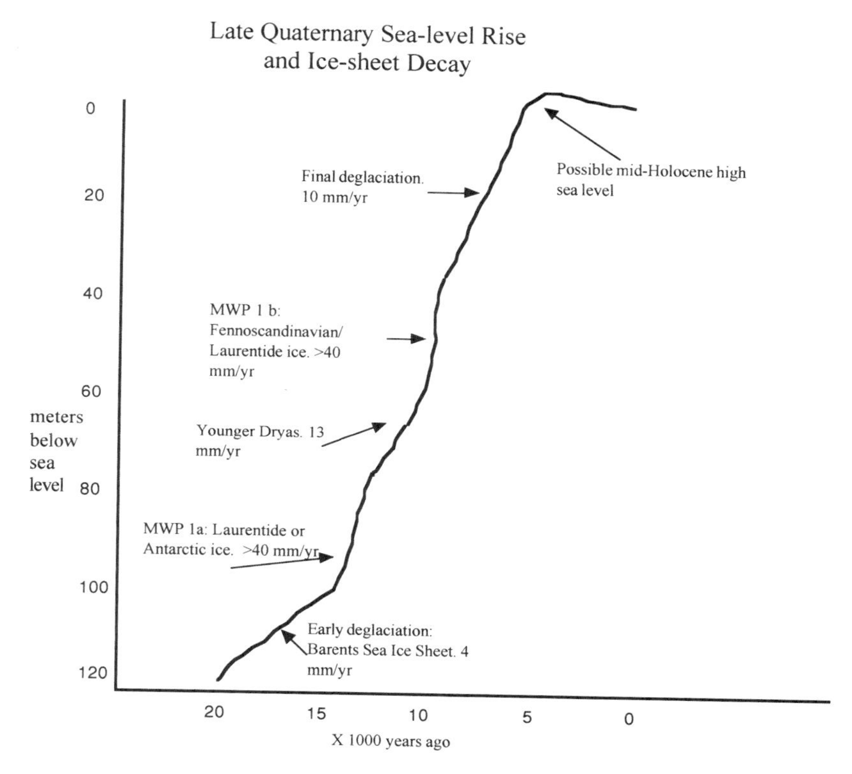

FIGURE 8-7 Late Quaternary sea-level curve compiled from several sources, including Barbados, Caribbean, and Pacific Ocean coral records of sea-level changes and continental ice sheet history. Sea level is characterized by slow, steady rise from the glacial maximum until about 14.5 ka, rapid rise of about 40 mm/yr from 14.5 to 13.5 ka (Meltwater Pulse [MWP] 1a) and at about 10 ka (MWP 1b), slow rise during the Younger Dryas between MWP 1a and 1b, slow rise to the present level or slightly above it by about 6 ka, and possibly a slight fall since 6 ka.

large portion of the Laurentide Ice Sheet collapsed during MWP 1a, although recently Clark et al. (1996a) suggested that maybe parts of Antarctica may have played a role. The third period corresponds to the Younger Dryas, 12.5–11.5 ka, when the rate of sea-level rise slowed appreciably. The fourth stage was Meltwater Pulse 1b, centered about 11.6–11.1 ka, when the Fennoscandianvian Ice Sheet and probably a portion of the Antarctic Ice Sheet melted.

The most reliable data linking the decay of a particular ice sheet to a particular sea-level event comes from detailed geological studies at judiciously chosen locations along the dynamic margins of late-glacial ice sheets. Some of the more critical regions include the Hudson Strait area of Canada studied by John Andrews and colleagues (Andrews and Tedesco 1992; Kaufman et al. 1993; Andrews et al. 1986, 1993a,b, 1995a,b), the Barents Ice Shelf examined by Elverhoi et al. (1993), and the southern Laurentide Ice Sheet margin in Canada (e.g., Hillaire-Marcel and Occhietti 1980; Dyke and Prest 1987; Teller and Kehew 1994; Stoner et al. 1996). I draw on these and other studies to construct the following sea level–ice sheet history.

Early Deglaciation—Barents Ice Sheet

The Barbados record suggests that sea level rose relatively slowly, a total of about 20 m between about 21 and 14.5 ka. What was the first ice sheet to collapse after the last glacial period? There is widespread evidence that the Barents Ice sheet, which lay partially grounded on the continental shelf of the eastern Arctic Ocean, may have decayed first, contributing about 10 m to the initial postglacial sea-level rise. Geophysical, glacial geological, and stratigraphic data suggest that a large ice sheet lay grounded on the shallow Barents Sea and that it may have began to melt during the earliest stage of deglaciation. Elverhoi et al. (1993) obtained radiocarbon dates on shells taken from glacial tills on the Barents Sea floor and confirmed that the inception of the last ice sheet to occupy the area was about 22–23 ^{14}C ka. This supported earlier work that step-wise ice sheet decay began about 15 ^{14}C ka.

In a remarkable application of his model of the earth's geophysical response, Peltier (1988) also found indirect evidence for the existence of a Barents Ice Sheet. Peltier reasoned that the load redistribution of mass during deglaciation would alter two components of earth's planetary rotation: the nontidal component of acceleration of the earth's axial rate of rotation and the drift of the rotation pole with respect to surface geography. Basically, the earth's rotation speeds up as the ice sheets melt and load is transferred to the oceans. Peltier used modern satellite geodetic data to infer that both rotational components did not fit with the accepted current rate of sea-level rise of about 0.5 mm/yr (e.g., Meier 1984). To account for the discrepancy, an additional mass of ice, unknown to exist on the continents, was necessary. Peltier suggested that during the last glacial period a large ice mass in the general vicinity of the Barents Sea would account for the modeling results.

Jones and Keigwin (1988) provided support for the existence of a

Barents Sea Ice Sheet in a study of oxygen-isotope records from the Fram Straight, the only deep-water oceanic gateway to the Arctic Ocean, situated between Spitzbergen and northern Greenland. They found an isotopic excursion dated by AMS radiocarbon dates that preceded the characteristic deglacial two-step isotopic spike of Termination 1 by 1–2 ka. They hypothesized that this represented a significant meltwater event, when fresh water discharge from the Barents Ice Sheet caused lighter isotopic ratios in the planktonic foraminifer, *Neoglobigerina pachyderma.* Because this isotopic spike was not accompanied by a large rise in sea level, however, they suggested that it might be due to ice sheet decay from an ice mass grounded wholly or partially below sea level.

Additional evidence for a large ice sheet in the eastern Arctic Ocean came from the Yermak Plateau, where Vogt et al. (1994) discovered iceberg plowmarks along the ocean bottom in 450–850 m water depth. They interpreted these as recording a 400- to 600-m-thick Arctic Ocean ice sheet calving from a Barents and Kara Sea source area. The estimated weight of these "megabergs," as Vogt et al. refer to them, was 10^7–10^8 tons. Iceberg keel marks formed on the ocean floor during the past ice age were also found off Norway, Greenland, and Baffin Island, Canada. Although the iceberg keels are not well dated, Polyak et al. (1995) documented independent stratigraphic and micropaleontological evidence in the Barents Sea itself for early deglacial ice sheet disintegration that was followed by a larger two-step deglaciation. Taken together, these studies suggest a dynamic breakup of a marine-based ice sheet.

A final line of evidence for early Barents Ice Sheet decay comes from the ingenious use of lithologic provenance by Bischoff (1994), who determined the sequence of late glacial oceanic sedimentation on the Vøring Plateau off the west coast of Norway. During the glacial maximum, most glacially derived sediment in the eastern Norwegian Sea originated in the Fennoscandinavian ice sheet to the east. As deglaciation began, there was a dramatic change in ice-rafted drop-stone lithology in the marine sediments dated at 15.0–14.5 ^{14}C ka. This age coincides with the isotopic shift of Jones and Keigwin (1988) and also with the stratigraphic and faunal changes of Polyak et al. (1995). Vøring Plateau sediment shifted from crystalline rocks to coarse sand and gravels nearly identical to those sediments that cover the Barents Sea shelf, implying a change to the Barents Sea as the drop-stone source area. Bischoff estimated that sedimentation rates increased 10-fold during initial deglaciation (in calendar years about 18–17 ka) and conjectured that this lithological transition was remarkably swift, occurring in just 400 yr. Following this catastrophic ice decay, there was a more gradual 4.5-ka-long disintegration of the remaining ice sector in the Spitzbergen-Svalbard area.

In summary, considerable evidence indicates that an ice shelf was present in the Barents Sea area and that much of it probably melted catastrophically relatively early in the postglacial period, perhaps followed by more gradual disintegration of the remainder.

Meltwater Pulse 1a

Fairbanks (1989) proposed that sea level rose 24 m in less than 1000 years during MWP 1a based on the radiocarbon-dated Barbados coral *A. palmata*—the first rapid postglacial rise in sea level signifying massive continental ice-sheet decay. Meltwater Pulse 1a discharged about 14,000 km^3/yr into the North Atlantic, an amount equivalent to all North Atlantic river discharge between 0–90°N latitude. Blanchon and Shaw (1995) further analyzed the Barbados sea-level record and argued that one could distinguish three rates of sea-level rise in the Barbados stratigraphic record based solely on coral ecology and taphonomy. During relatively slow sea-level rise of < 14 mm/yr, the occurrence of monospecific *A. palmata* assemblages track sea-level rise because this species' ecology allows it to accrete its skeleton at rates below this threshold. When sea level rises at rates > 14 but < 45 mm/yr, the Barbados stratigraphy shows that *A. palmata* corals are displaced into a mixed coral assemblage that is more characteristic of a water depth range of 5–10 m. At exceptionally rapid rates of sea-level rise, > 45 mm/yr, sea level rises too fast even for a mixed coral reef framework to develop.

Blanchon and Shaw applied this model of coral paleobiology and taphonomy to identify *A. palmata* reefs that were drowned by rising sea level. They found three horizons at 80, 50, and 15 m below sea level in Fairbanks's Barbados core where sea level rose at a rate exceeding 45 mm/yr. These were dated near 14.2–13.8 ka (MWP 1a), 11.5 ka (MWP 1b), and 7.6 ka. Each date signifies a catastrophic, that is, within a few centuries, sea-level rise equivalent to 13.5, 7.5, and 6.5 m, respectively. Where did the meltwater come from?

The origin of MWP 1a has long been held to represent meltwater from the Laurentide Ice Sheet. The evidence for this conclusion comes from glacial geology on the continent (Teller and Kehew 1994), oxygen-isotope evidence for meltwater events in the deep-sea record (Keigwin et al. 1991), and ice-rafted material in the North Atlantic (Bond et al. 1992). Drainage of freshwater outflow via the Mississippi

River was believed responsible for isotopic excursions recorded in Gulf of Mexico foraminifers (Leventer et al. 1983) and the Bermuda Rise (Keigwin et al. 1991).

Clark et al. (1996a), however, recently argued against the Laurentide Ice Sheet as a source for MWP 1a for at least two reasons. First, the amount of ice available in the sector of the ice sheet draining via the Mississippi River outlet was insufficient to cause the observed isotopic excursion. Second, the rate of ice ablation was implausible given current knowledge about ice sheet sensitivity to temperature change. Clark's group argues that evidence from glacial IRD and stable isotopes from the two other outlets of the Laurentide Ice Sheet, the Hudson Strait (Andrews et al. 1993a, 1994) and the St. Lawrence River (Hillaire-Marcel et al. 1993; de Vernal et al. 1996; Stoner et al. 1996), indicate no major IRD pulses were synchronous with MWP 1a and the corresponding period of rapid sea-level rise. Moreover, Heinrich events in the North Atlantic Ocean were not synchronous with MWP 1a, as might be expected if they were caused by Laurentide Ice Sheet decay (Bond et al. 1992, 1993). These data led Clark and colleagues to radically alter the concept of MWP 1a from that of a Laurentide Ice Sheet event to an Antarctic Ice Sheet event. They theorized that Antarctic ice had, in fact, contributed to most of the meltwater discharge about 14–13 ka. In doing so, they add a new dimension to the reconstruction of late Quaternary sea-level history with implications for thermohaline circulation and mechanisms of late Quaternary climate change as well.

Younger Dryas Sea Level

The Younger Dryas stadial is represented in the sea-level record at Barbados in the form of a much reduced rate of sea-level rise and a meltwater equivalent about one fifth the volume of MWP 1a (Fairbanks 1989). A sea-level curve based on radiocarbon-dated corals from a core from the Huon Peninsula, New Guinea (Chappell and Polach 1991), does not contradict the inferred slower Younger Dryas rate of sea-level rise. Fairbanks argued that the Younger Dryas was essentially a period of reduced meltwater discharge sandwiched between two periods of massive meltwater discharge and reduced oceanic surface salinities. Fairbanks suggested that this relationship is the opposite of that between meltwater discharge, surface salinity, and climate cooling in the North Atlantic region proposed by Broecker et al. (1988b) as an explanation for the Younger Dryas cold snap (see chapter 5). This distinction is critical for explanations of the Younger Dryas event, but all that is important here is to note that the rate of sea-level rise was substantially reduced from that just prior the Younger Dryas. Blanchon and Shaw (1995) estimated Younger Dryas *A. palmata* accretion and sea-level rise to be 13 mm/yr, much slower than the catastrophic 45 mm/yr rate during MWP 1a, but more rapid than the earliest deglaciation rate of 4 mm/yr 20–15 ka.

There is abundant glacial evidence for ice-sheet readvances and reduced rates of glacial retreat around the time of the Younger Dryas cold episode. For example, Polyak et al. (1995) found a period of non-deposition of glacio-marine sediments reflecting a pause in the decay of the remnants of the Barents Sea ice sheet near the Younger Dryas time. A detrital carbonate event signifying ice-stream discharge of Heinrich event–like icebergs is also evident in Hudson Strait (Andrews and Tedesco 1992; Andrews et al. 1993a,b, 1995a).

Meltwater Pulse 1b

Sea level rose a total of about 28 m during MWP 1b, centered at about 11.5–11.0 ka (Fairbanks 1989) at a rate exceeding 45 mm/yr (Blanchon and Shaw 1995). Chappell and Polach (1991) also recorded rapid sea-level rise during coral reef accretion in New Guinea. There are several possible ice sheet sources for MWP 1b. One is the northern region of Labrador where Kaufman et al. (1993) documented several major advances of Laurentide ice between 14 and 9 ka. The most rapid and well-dated advance is the abrupt Gold Coast advance of the Labrador dome of Laurentide ice. Kaufman and colleagues cite evidence that the ice advanced as much as 300 km along a 200-km-long calving ice sheet margin into the Hudson Strait area. The entire advance-retreat cycle was completed within only about 300 years. They estimated that this advance might account for as much as one fourth of the total sea-level rise during MWP 1b. The Gold Coast advance apparently did not significantly affect North Atlantic ocean salinities and circulation, either because it did not produce a sufficient quantity of icebergs or their ocean trajectories were too far to the south. The remainder of the MWP 1b sea-level rise most likely came from Fennoscandinavia and/or eastern North America through the St. Lawrence River.

Three points about the evidence for this sequence of late Quaternary postglacial sea-level history need to be emphasized. First, widely divergent opinions remain about which ice sheet contributed to each stage of the postglacial sea-level rise; we have only briefly touched upon some of the more recent theories. The uncertainty in deglacial ice sheet–sea level chronology stems in part from a relatively poor chronology of Antarctic late glacial history compared with that from the Northern Hemisphere. The chronology is also uncertain because meltwater events from the deep sea isotope record are subject to various interpretations.

Second, the chronology of the postglacial sea-level record does not yet approach the near-calendar year chronology available from some European terrestrial and Greenland ice core records (chapters 5, 9). There are only a limited number of well-dated postglacial sea-level records from submerged shorelines for the interval spanning the glacial maximum through early deglaciation before about 10 ka. Even in the case of the Barbados curve, rates of sea-level rise must be interpolated between relatively few radiocarbon dates for the periods of most rapid deglaciation. The lack of calendar-year chronology makes it very difficult to precisely correlate sea-level history with the high-resolution paleoclimate records now available from ice cores, lakes, and tree-ring records. Moreover, the lack of a high-resolution sea-level chronology means there are no firm estimates of the fastest rates at which sea level can rise, impairing the interpretation of sea-level change in the context of very rapid (centennial-scale or faster) climate change.

Third, improvements in separating out the temperature, salinity, and vital effects from the ice-volume signal in stable-isotope ratios would greatly improve the development of a "continuous" sea-level curve from high-resolution deep-sea sequences.

Despite the many uncertainties, research on postglacial sea level–ice volume history is progressing rapidly in terms of linking glacial and stratigraphic records from ice margins to paleoceanographic and paleo-atmospheric records. There is compelling paleoclimatological evidence that postglacial sea level rose at an exceptionally rapid, catastrophic rate of at least 45 mm/yr, about 50 times faster than the rate of sea-level rise that has occurred over the past century that is attributed to alpine glacial melting.

Holocene Sea-Level History

We conclude our discussion of late Quaternary sea level with brief comments on sea-level history of the last 8 ka. Pirazzoli (1991) compiled several hundred late Quaternary sea-level curves from around the world, most covering the last 8 ka, which constitutes a large part of the global relative sea-level database. While each relative sea-level curve is subjected to local and regional isostatic and tectonic effects, together they comprise the foundation upon which the most sophisticated geophysical model of global paleotopography rests (Peltier 1994).

There are two key features of Holocene sea-level history. First, eustatic sea level probably reached its present level or slightly higher by about 6 ka (figure 8-7). Second, the probability is strong that mid-Holocene eustatic sea level was briefly a meter or two higher than the present sea level, although separating isostatic and eustatic effects remains an impediment to conclusively demonstrating how much global ice volume was reduced.

A recent study of Holocene sea level documented from coral-reef growth on Abrolhos Island, of the Easter Group off Western Australia, has improved our understanding of this early- to mid-Holocene interval. This region is ideal for sea-level study because it is a nontilted, stable continental margin that fits into a far-field category (thus affected relatively little by glacio-isostatic effects) relative to the location of major Northern Hemisphere ice sheets. Eisenhauer et al. (1993) studied two cores, one from Morley Island with a coral growth rate of 7.1 m/ka, the other a coral from Soumi Island, which grew at 5.8 m/ka. They constructed an extremely accurate Holocene sea-level curve and found that relative sea level reached a peak high stand 6.5–4.7 ka, followed by relative fall in sea level. The Abrolhos sea-level record gave support for the postulated mid-Holocene high sea level predicted by geophysical models (e.g., Clark et al. 1978; Peltier 1988), perhaps due to a hydro-isostatic response vis-à-vis mantle flow outward from the region after about 5 ka.

Eisenhauer's group also contrasted the Abrolhos record with those from the Huon Peninsula (Chappell and Polach 1991) and Barbados (Fairbanks 1989). They found that the Abrolhos and Huon curves were generally similar to each other, and both differed significantly from the Barbados record. The most critical conclusion was that the Abrolhos and Huon curves reached the present sea level by about 6 ka, but at this same time, sea level at Barbados was still 10–12 m *below* the present level and still rising.

What factors caused this critical difference? Eisenhauer et al. (1993) conjecture that geoidal changes might account for a sea-level rise in Barbados after 6 ka because Barbados is located in the Northern Hemisphere, relatively close to the glacial-age ice sheets. Mitrovica and Peltier (1991) constructed a geophysical model that explains how such geoidal changes might influence a relative sea level toward the end of the deglaciation. They predicted that water would be siphoned off from far-field areas toward near-field areas after a phase of ice melting.

The siphoning effect was proposed to be the result of the redistribution of meltwater. In other words, extra water was pulled from Southern Hemisphere areas such as Abrolhos toward Northern Hemisphere regions such as Barbados, which caused the apparent relative sea-level fall in the former region and the sea-level rise in the latter. Additional confirmation of these complex relative sea-level trends should be sought in other regions and in independent ice-volume records from isotopic data.

Recent studies of Holocene sea-level history along eastern North America have also yielded insights into relative-sea level during the middle Holocene. Gayes et al. (1992) and Scott et al. (1995b) studied sea-level history at Murrells Inlet, South Carolina, where relative sea level oscillated in comparison with modern sea level as follows: From a level –3 m about 4600–5200 yr BP, sea level rose to +1 m about 4280 yr BP, then fell to –3 m at 3600 yr BP. Since 3600 yr BP, sea level rose steadily at a rate of 10 cm/100 yr. Scott et al. (1995a) found supporting evidence from Chezzetcook Inlet, Nova Scotia for a high rate of relative sea-level rise between 5500 and 3500 yr BP. The rise rate of 67 cm/100 yr supported the results of Murrells Inlet studies showing that there was not a simple monotonic sea-level rise since 6 ka. Scott et al. (1995a,b), while acknowledging the complexity of factoring out the isostatic contribution to east-coast sea level, nonetheless suggested that perhaps 2 m of the mid-Holocene sea-level high stand may have been a eustatic event.

Despite its importance as a baseline against which the past century's 10- to 25-cm sea-level rise must be measured, the sea-level record for the past 3–4 ka is poorly known. This stems in part from the same problems that plague most late Holocene decadal and centennial-scale climate research. There is a low signal-to-noise ratio in reconstructed sea-level curves. Sea-level oscillations were minor during the past few millennia relative to massive (> 100 m) ice-volume oscillations of late Quaternary glacial-interglacial cycles. Still, even relative sea-level changes 0.5–1.0 m during the late Holocene could have ramifications for ice-sheet and glacier dynamics in relation to centennial-scale climatic variability during a warm interglacial period. A few recent studies have suggested that late Holocene sea level might be unstable.

Varekamp et al. (1992) conducted an intriguing study documenting sea-level oscillations during the past 1500 yr near Clinton, Connecticut. Using salt-marsh foraminifera to estimate past sea-level positions, and radiocarbon and anthropogenic metal input to establish a firm chronology, they found two periods of accelerated rate of sea-level rise. One dated at 1200–1450 A.D. coincides with the Medieval Warm Period–Little Ice Age transition; the other, which occurred during the late nineteenth century, corresponds to beginning of the past century's "global warming." In data from the Little Ice Age, about 1400–1850 A.D., they found virtually no relative sea-level rise at all. Because coastal Connecticut is situated in an isostatically subsiding region, the observed "stability" during the Little Ice Age suggests that a rise in eustatic sea level may have actually occurred.

Varekamp and Thomas (1998) reviewed the evidence for sea-level variability over the past few millennia based on salt marsh records and concluded that rates of relative sea-level rise for the past 2000 years were 1–2 mm/yr for the eastern United States, that higher rates have occurred in recent centuries, and that relatively high rates of sea-level rise precede twentieth-century increases in mean annual global temperatures. Although evidence on the timing and scale of late Holocene eustatic sea-level events remains inconclusive, microstratigraphical studies like those described by Varekamp and Thomas are bound to be repeated in other coastal regions.

Tanner (1992) used a somewhat unorthodox proxy, kurtosis of beach sands from sand ridges on St. Vincent Island, Florida, to determine that sea level has risen and fallen rapidly seven times during the last 3 ka at rates of 10 mm/yr. The total range of change was 1–3 m. Tanner's proxy method and age model, however, are suspect and more reliable relative sea-level records come from coastal marshes and coral-reef records.

In summary, when relative sea-level records are reconstructed from paleoclimatological methods, all coastlines exemplify to one degree or another the complex processes confronting inhabitants of coastlines of Scandinavia, Chesapeake Bay, and Louisiana. Multiple processes can cause observed sea-level changes along any and all coastlines and uncertainty remains when attributing cause to reconstructed sea-level trends. Only through additional relative sea-level records (including sorely needed records from the LGM and early deglaciation), better glaciological budgets, and improved geophysical and glacial models will the many factors that control sea-level change be fully decoupled.

HISTORICAL SEA-LEVEL CHANGE

WE CONCLUDE THIS chapter with a brief summary of sea-level history obtained from tide-gauge records of the past few centuries and possible causes of historical sea-level trends. As we have mentioned, sea level has risen approximately 10–25 cm over the past century

(Nicholls et al. 1996). Many questions arise from this conclusion. How fast is it rising now? Is the historical rate of rise faster or slower than pre–twentieth century rates? What has caused the historical sea-level rise?

The earliest tide-gauge devices were put in place in Amsterdam in 1682, Venice in 1732, and Stockholm in 1774 (Pirazzoli 1991) to record changing sea level in these early urban centers of Europe. As it became recognized that sea-level change varied from one coastal region to the next, it also became clear that a wider network of sites was needed to evaluate local versus broader sea-level trends. Consequently, more and more tide gauges were established to better quantify sea level. Now the Permanent Service for Mean Sea Level at Bidston Observatory in England maintains a set of 420 tide-gauge records at least 20 years old from which tide-gauge estimates of sea-level change are derived (Raper et al. 1996).

We saw in figure 8-2 the complex tide-gauge records of sea-level history obtained from six regions. The most obvious aspect to this figure is that the trend in each relative sea-level curve is different from the others owing to the many factors that affect relative sea level discussed earlier.

Like many trends derived from monitoring records of climate-related phenomena, sea-level trends obtained from tide gauges suffer from the tradeoff between the length of the record and the geographic coverage available. The longest recorded tide gauges yield longer trends for a particular coast but tell nothing about sea-level change in other regions (Gornitz 1995a,b). Tide gauges in place for the past few decades give a broad spatial coverage (though still predominantly in the Northern Hemisphere) but they cannot be used to establish how sea level varies over longer time scales. Global sea-level trends cannot be established without broad geographic coverage of the world's coastlines because of the local processes influencing sea level. Consequently, the literature contains estimates of the total sea-level rise over the past century that vary depending on the length and location of the tide-gauge records used.

In recent years, many papers have reconstructed sea level using tide-gauge records in order to better explain the patterns and causes of historical sea-level trends (e.g., Gornitz and Lebedeff 1987; Emery and Aubrey 1991; Douglas 1991; Fletcher 1992; Gornitz 1995b; Warrick et al. 1996; see Raper et al. 1996 for review). By removing the effects of vertical land movements using geological (Gornitz et al. 1995b) and geophysical (Peltier and Tushingham 1989) methods, scientists have reached general agreement on two issues. First, the tide-gauge record is sufficient to evaluate the amount of global sea level only for perhaps the past 60–100 years. Before about 1930, the record is too limited. Second, there is a general consensus that global sea level has been rising along the world's coasts during the past century. However, estimates of the total sea-level rise and thus the net rate of rise vary widely. For example, Emery and Aubrey (1991) and Nicholls et al. (1996) concluded that total global sea level has risen somewhere between 10–25 cm over the past century. Raper et al. (1996) suggested that, once noise and interannual variations due to oceanographic and atmospheric effects are removed, the range of sea level–rise estimates run from a low of 5 cm to a high of 30 cm for the past century (or 0.5 to 3.0 mm/yr). A rate of 1–2 mm/yr is an oft-cited average rate of rise for the past century (see Douglas 1991, 1992; Gornitz 1995b).

An average rise of sea level of 1 to 2 mm/yr over the past century leads to the question, do historical tide-gauge records offer any evidence for an acceleration in the *rate of sea-level rise* during the past 100 years. Douglas (1992) studied short and long-term tide-gauge records to address this question. He found that 23 tide-gauge records from 10 different coastal regions for the interval 1905–1985 showed sea level rose at a rate of 0.11 (±0.012) mm/yr. A less extensive tide-gauge data set available for the period 1850–1991 covering a smaller geographical area than the 1905–1985 data set showed that no statistically significant acceleration in the rate of sea-level rise had occurred after 1850. In another study of the rate of recent sea-level rise, Maul and Martin (1993) showed there was likely no acceleration in sea-level rise in the Florida Keys since 1846. Any acceleration in sea-level rise must have preceded the Florida historical records available to them, a point that underscores the need to reconstruct late Holocene sea-level trends from geological evidence.

A similar question might be: is the past century's rate of sea-level rise rapid compared with rates of sea-level change for the past 2000 years? Varekamp et al. (1992), studying late-Holocene sea level from Connecticut, found evidence that sea level had varied by as much as a few tens of centimeters over the past 2000 years but that there had not been a steady rise or fall during that interval. Regarding the correlation of relative rates of sea-level rise and atmospheric temperatures, Varekamp and Thomas (1998:75) concluded:

> The period since 1800 A.D. (including the modern global warming for which climatologists have speculated on an anthropogenic cause) is associated with the highest rates of RSLR [relative sea-level

> rise], but no acceleration in RSLR is evident...with the rapidly rising temperatures of the last 100 years, in agreement with tide gauge records.

Thus, the evidence is good that global sea level has risen over the past century and that the rate of relative sea-level rise has oscillated over the past 2000 years, but the evidence about the correlation of relative sea-level rise with global temperature is less conclusive. We are then drawn to ask what factors might be causing ocean levels to rise over the past century? The two most likely processes are thermal expansion of the oceans and glacio-eustatic processes related to decreasing glacier and/or ice-sheet volume. Various types of climate, ocean, and coupled general circulation models predict a sea-level rise about equal to that observed in the tide-gauge record and resulting from thermal expansion and glacial melting. For example, de Wolde et al. (1995) estimate thermal expansion of the world's oceans to be about 2–5 cm with a best estimate of 3.5 cm. Hydrographic surveys monitoring oceanic temperatures confirm to a limited extent the expected increase in near-surface oceanic temperatures, which is very likely associated with surface-temperature rise (Raper et al. 1996). Moreover, to a first approximation, observations of mountain glaciers indicate glacial recession over the past 100 yr, generally on the order of magnitude expected from the tide-gauge data (Meier 1984, 1993). Thus independent theoretical and empirical evidence tend to corroborate at least the trend and general magnitude of historical sea-level rise. About one half the past century's sea-level rise is attributed to thermal expansion, and one half to melting of alpine glaciers, but these estimates have large errors associated with them, mainly owing to geoidal and isostatic processes. Moreover, the great uncertainty surrounding the contribution to historical sea level of the modern ice sheets in Greenland and Antarctica, which have not reached an equilibrium state since deglaciation ended the last ice age, must be reiterated.

If we accept that thermal expansion and glacial recession over the past century are the primary immediate causes of historical sea-level rise, then we are led to ask which factors caused the oceans to warm (and expand) and alpine glaciers to retreat? How important are natural and human factors to the sea-level record of the past century? The available evidence is still too limited to conclusively relate sea-level rise to atmospheric and oceanic warming that might be due to human activities. As we saw in chapter 6, the retreat of alpine glaciers after the cool climate of the Little Ice Age of the fifteenth through the nineteenth centuries reflects one of several low-amplitude climatic cycles that characterize the Holocene interglacial period. Holocene oceanic sea-surface temperature change has also been documented (Keigwin 1996). With regard to sea level, it is still uncertain whether centennial-scale sea-level oscillations of 10–25 cm occur during earlier parts of the Holocene interglacial or during previous interglacial periods. Thus scientists are left to explain exactly when and why historical sea level began to rise and whether the past century's rise is part of cyclic low-amplitude sea-level history or an unprecedented event in the Holocene interglacial period.

SUMMARY

IN THIS BRIEF survey of the patterns and processes of sea-level change, we have seen that many factors influence the local relative sea-level history of any particular region. The most important processes include oceanic thermal expansion, geoidal variability, isostatic and tectonic uplift and subsidence of the land, ocean-atmospheric dynamics, and glacio-eustatic effects. All of these processes tend to complicate efforts to reconstruct past and project future sea-level trends. In any discussion of past, present, or future sea-level trends these factors must be taken into account as potentially significant causes of the observed sea-level variability.

Despite these complications, researchers can often separate out the primary cause of observed sea-level changes, because each process operates over distinct spatial and temporal scales. The geological record is an invaluable source in ongoing efforts to understand sea level and associated climatic change at all time scales over which these processes operate.

IX

Paleo-atmospheres: The Ice-Core Record of Climate Change

The carbon dioxide is critical because of its peculiar thermal capacity by virtue of which it retains the heat of the sun to a relatively extraordinary degree, a capacity which is shared by water vapor. . . . Whenever, therefore, there is a notable percentage of carbon dioxide in the atmosphere, it performs a most important function in conserving the heat of the sun and raising the temperature of the lower atmosphere and the Earth's surface.

T. C. CHAMBERLIN, 1898

ATMOSPHERIC CHANGE: HUMAN AND NATURAL FACTORS

IN 1973 Charles D. Keeling published a landmark paper showing a startling rise in atmospheric carbon dioxide (CO_2) concentrations at Mauna Loa, Hawaii, from 312 to 330 parts per million volume (ppmv) between 1958 and 1972. At that time, scientists had only limited evidence about pre-industrial atmospheric CO_2 content, but it was clear the progressive rise in CO_2 was probably due mainly to fossil fuel emissions, cement production, and deforestation (Keeling 1973; Keeling et al. 1989). Although CO_2 is referred to as a trace gas and constitutes only 0.035% of earth's modern atmosphere, concern grew about the environmental consequences. Carbon dioxide holds the potential to warm the atmosphere near the surface because CO_2 is a radiatively active gas, which means its molecules can absorb various long wavelengths of terrestrial (i.e., thermal) radiation reemitted from earth. Carbon dioxide also has a long atmospheric residence time (50–200 yr), such that the rate at which it is removed from the atmosphere by natural processes is much lower than the rate at which humans are producing it. The radiative forcing of a molecule of gas like CO_2 and the residence time of the gas are used to calculate what is called the *global warming potential* (*GWP*) of the gas for a given period of time. Carbon dioxide's GWP is standardized to the value 1; the GWP of methane (CH_4) is 21; nitrous oxide (N_2O) is 290; and chlorofluorocarbons (CFCs) range between 3000 and 8000. All of these greenhouse gases will also have an impact on future climate. Earth's atmospheric chemistry and the history of its greenhouse gas content lie at the heart of the global warming debate. In fact, global warming from greenhouse gas emissions is arguably the most pressing international environmental issue today. This issue's prominence stems from discoveries like Keeling's and from paleoclimatological discoveries from ice cores described in this chapter.

Before Keeling's discovery, there had been well-founded speculation that greenhouse gases from human sources might alter global climate. The influence of CO_2 on climate can be traced back at least to the Swedish chemist Svante Arrhenius, who in 1892 calculated that a doubling of atmospheric concentrations would warm the earth by 5–6°C. Arrhenius also theorized that during past geological eras atmospheric CO_2 varied mainly because of volcanic activity. American geologist T. C. Chamberlin (1897, 1898, 1899) also recognized atmospheric CO_2 as an important factor in climate. He proposed that higher weathering rates and reduced marine calcium carbonate production would combine to reduce atmospheric CO_2 levels. Chamberlin also believed that volcanoes played a large role in elevating CO_2 concentrations in the atmosphere, reasoning that periods of global warmth such as the Cretaceous had characteristic features that should lead to high CO_2 concentrations and warm climate. For example, he posited that low continental elevations, high marine carbonate production, and low weathering rates of continental rocks were all factors controlling the global carbon budget. During the early part of the twentieth century, other scientists attempted to estimate past CO_2 concentrations using indirect lines of evidence, but a direct means to measure ancient CO_2 concentrations was not available (Revelle and Suess 1957; Revelle 1985).

In 1980, unambiguous evidence emerged from Antarctic ice cores that pre-industrial, nineteenth century concentrations of atmospheric

CO_2 were significantly below modern levels. Air trapped in slowly accumulating polar ice preserved CO_2 concentrations of 280–290 ppmv from pre-industrial times (informally defined here as before the late nineteenth century). These levels were a full 70 ppmv below current levels; a century of human activity was responsible for more than a 25% increase in CO_2 concentration over natural levels. This percentage equates to a total volume of 25–50 gigatons (~2.13 gigatons = 1 ppmv) of carbon emitted into the atmosphere during the past century at an annual rate of increase of 6 gigatons/yr, or 1.8 ppmv/yr (table 9-1) (Houghton et al. 1996).

Equally as startling as the postindustrial twentieth-century rise in greenhouse gas concentrations was the discovery from glacial-age ice that CO_2 concentrations during the last glacial period were only about 200 ppmv. During the last deglaciation, atmospheric concentrations rose 80 ppmv, reaching their interglacial levels of 280–290 ppmv over about 10,000 yr. Likewise, it was discovered that two other radiatively active gases, CH_4 and N_2O, also oscillated naturally over glacial-interglacial time scales (table 9-1). These discoveries about natural and human-induced fluctuations in potentially climate-altering atmospheric gases sent shock waves throughout the paleoclimate community that still reverberate.

TABLE 9-1. Atmospheric Changes in Radiatively Active Species of Trace Gases

	Concentration			*Twentieth Century*	
Species	*Glacial*	*Pre-industrial**	*Current*	*Annual concentration change*	*Annual % change*
Carbon dioxide (CO_2)	200 ppmv	280–290 ppmv	365 ppmv	1.5 ppmv	0.4
Methane (CH_4)	300–400 ppbv	700 ppbv	1730 ppbv	10 ppbv	0.6
Carbon monoxide (CO)	—	90 ppbv	0.6 ppbv		0.7
Nitrous oxide (N_2O)	—	275	312 ppbv	0.8 ppbv	0.3
Chlorofluorocarbons (CFCs)	—	0.1–0.5 ppbv	0.01–0.02 ppbv		

*"Pre-industrial" informally refers to the time before the late nineteenth century. Houghton et al. (1996) and other studies often plot trends since 1850 A.D.

Sources: Graedel and Crutzen (1993); Houghton et al. (1996); Battle et al. (1996)

THE ICE-CORE RECORD OF PALEO-ATMOSPHERES AND CLIMATE CHANGE

THE BRANCH OF paleoclimatology concerned with the study of greenhouse gas evolution, other atmospheric gases, chemical species, particulate material (dust) from many sources, and wind, as well as their role in climate change, is the field of paleo-atmospheric science. In prior chapters, we introduced selected topics about earth's changing atmosphere in the context of climate change over various time scales. Among them, evidence for increased aridity and elevated atmospheric dust content during Quaternary glacial periods (chapter 4), the role of atmospheric water vapor in the tropics as a forcing mechanism of rapid climate change during deglaciation (chapter 5), and global isotopic variability of water vapor during short-term climate events (chapters 6, 7). We also encountered geological proxies such eolian (wind blown) sediment in deep-sea sediment cores, which researchers use to reconstruct atmospheric parameters.

In addition, various paleoclimatological methods are used to reconstruct paleo-CO_2 concentrations through indirect means. These include carbon-isotope measurements from tree rings (Peng et al. 1983), deep-sea microfossils (Shackleton and Pisias 1985), fossil peats (White et al. 1994), and carbonate material in paleosols (figure 9-1) (Cerling 1991, 1992; Retallack 1990). Changes in leaf stomatal density represents another innovative technique used to infer changes in atmos-

FIGURE 9-1 Tertiary paleosols exposed in the Badlands region of South Dakota are examples of sediments from which paleo-atmospheric CO_2 concentrations are estimated using carbon-isotope methods for periods in the geological record older than the oldest polar ice.

pheric CO_2 (Van der Burgh et al. 1993; McElwain and Chaloner 1996). All play important and expanding roles in paleoclimatology.

Ice cores, however, have become a Rosetta stone for paleoclimatology because they preserve an archive of "fossil" atmospheres. The discovery of natural oscillations in greenhouse gases from fossil air trapped in polar ice ranks as one of the most important advances in the field of climate and earth science. Indeed, ice cores provide paleoclimatologists with a quantitative, accurate baseline of natural variability in atmospheric trace gases over thousands to hundreds of thousands of years with which twentieth century emission trends from human activities can be compared.

Ice cores provide a wealth of paleoclimate data extending far beyond the history of trace gases. Direct measurements of CO_2, CH_4, and N_2O concentrations are only a part of the ice-core record. As many as 50 chemical species and physical properties have been measured in ice cores (Grootes 1995; Bales and Wolff 1995), including stable isotopic composition of both the ice matrix itself and trapped molecular oxygen within the ice, cosmogenic isotopes (^{10}Be [beryllium]), insoluble particulate matter (dust), ice and air geochemistry (soluble and insoluble anions and cations), electrical conductivity, among other climate proxies. From these proxies, inferences can made about past winds and atmospheric-circulation changes, sea-ice dynamics, rapid atmospheric-temperature change, bipolar climate change, solar activity, global biogeochemical cycles, terrestrial vegetation and marine phytoplankton activity, volcanic activity, biomass burning, wetland evolution, and many other factors. Ice-core paleoclimatology intersects almost every aspect of climate history described in the preceding chapters of this book.

This chapter is devoted to the paleoclimatic record of atmospheric change over the past few hundred thousand years derived mainly from ice cores. The first part of this chapter focuses on the history and principles of ice-core paleoclimatology. It includes short sections on the brief but spectacular history of ice-core research and the major ice coring projects, on the climate proxies found in the ice, and on dating and correlation tools applied to ice cores. In these sections, important syndepositional and postdepositional processes that occur during the snow-firn-ice transformation are described.

The second part of this chapter contains three sections describing the contribution of the ice-core record to orbital climate change over Quaternary glacial-interglacial cycles, millennial-scale climate changes during the last glacial period and deglaciation (Dansgaard-Oeschger events and the Younger Dryas), and rapid climate change over decades to centuries. These are the three themes covered in chapters 4, 5, and 6, respectively; records from polar ice and low-latitude ice caps in China and Peru were already discussed in chapter 6 in reference to short-term climate and atmospheric change of the past few millennia. Here I will expand on evidence for abrupt climate changes that occurred during the last deglaciation.

Throughout the chapter, the reciprocity between the physical and biological processes that influence atmospheric composition on regional and global scales serves as a backdrop against which climate change at various time scales should be viewed. Physical and chemical processes control precipitation and the accumulation rate of snow and ice. They influence the transport and deposition of chemical impurities trapped in the ice, the isotopic ratios of oxygen in the ice, and many other properties of glacial ice. These processes act in concert with biological processes such as respiration, photosynthesis, dimethylsulfide production by marine organisms, and wetland CH_4 production to mediate many of the global signals preserved in ice. Although one cannot cover in a single chapter even a small part of the exponentially growing field of global biogeochemical cycling, one can still easily gain an appreciation for the integrated nature of paleoclimatology from a discussion of the principles and application of ice-core research.

"ONE THOUSAND CENTURIES": THE CAMP CENTURY CLIMATE RECORD

IN ONE OF THE first important ice-core studies, Willi Dansgaard and colleagues (1969) broke the proverbial ice in a paper entitled: "One thousand centuries of climate record from Camp Century on the Greenland Ice Sheet." This title seems somewhat unusual because earth scientists usually refer to time during the Quaternary period in terms of thousands (or hundreds of thousands) of years, using conventions such as ka (kiloannum), ka BP, or exponents (10^3–10^4 yr). They do not refer to "centuries" that passed 50,000–100,000 yr ago. Dansgaard's usage of "centuries" conveyed a message that Camp Century's consequential climate record should be spoken of in that vernacular.

Dansgaard's study of the Camp Century isotopic record yielded an exceptional history of climate change over multiple time scales and is a quintessential ice core investigation. Using new discoveries about the oxygen isotope–temperature relationship, they studied the oxygen isotopes from 1600 samples from a 12-cm diameter, 1390-m long, ice core taken by the United States Army Cold Regions Research Laboratory in 1966. The age of the ice at 1390 m core depth was estimated at about 100,000 years; the core thus gave a continuous climate history for "1000 centuries," spanning most of the last glacial-interglacial cycle. Dansgaard's group made the preliminary interpretation that an observed 13-ka cycle was somehow related to earth's precession. In light of the formative stages of knowledge about orbital climate change derived from coral reef–sea level and deep-sea foraminiferal studies at the time, Dansgaard's age model and climate inferences from the Camp Century record were remarkable achievements.

Camp Century was also an important precursor to contemporary ice-core research on rapid climate change occurring over millennial time scales. Dansgaard and colleagues were able to identify the classical European climatostratigraphic stages—the Bølling, the Allerød, and the Younger Dryas events—in the Camp Century isotope curve. Atmospheric-temperature change over Greenland added a whole new component to the understanding of deglaciation, which up to this point had been mostly studied from glacial geology and palynology.

Even in the Holocene section of Camp Century ice, a period considered by many to have a relatively stable climate, Dansgaard's group identified multiple isotopic oscillations and tantalizing evidence for notable short-term climate events. They suggested that during the past 1000–1400 yr, oxygen isotopic ratios varied, probably because of solar variability, with a period of about 120 years. As discussed in chapter 5, climate instability in the Holocene and previous interglacial periods has since been identified in several paleoclimate records. Moreover, solar influences on climate are receiving increasing attention as a potential cause for short-term variability (chapter 6). As we will see later, the three scales of climate change—orbital, millennial, and centennial-decadal—shown by Dansgaard's Camp Century studies continue as the subject of ice-core research almost 30 years later.

A Brief Summary of Ice-Core Programs for Paleoclimatology

Investigations of polar ice sheets and low-latitude, high-elevation ice caps pose complex logistical problems and require large international cooperative research programs. Ice cores sites from programs devoted largely to understanding climate history obtained over the past 40 years are listed in table 9-2. These come mainly from the Greenland and Antarctic Ice Sheets and smaller glaciers whose general locations are shown in figure 9-2. The Greenland Ice Sheet extends more than 2000 m from its surface to underlying bedrock; the oldest ice at its

TABLE 9-2. Summary of Major Ice Cores Used in Paleoclimatology

Site	*Core depth (m)*	*Year*	*Elevation (m)*	*Accumulation ($gm/cm^2/yr$)*	*Latitude*	*Longitude*
Greenland						
Camp Century	1390	1966	1885	32	77°10'N	61°08'W
Dye 3	2037	1981	2480	50	65°11'N	43°49'W
GRIP	3028	1992	3238	Variable	72°34'N	37°37'W
GISP2	3053	1993	3208	Variable	72°36'N	38°30'W
Antarctica						
Byrd	2163	1968	1530	16	80°01'S	119°31'W
Dome C	905	1978	3420	3.4	74°39'S	124°10'E
Vostok	3700	1980/90s 3490	2.3		78°28'S	106°48'E
Law Dome DE08	234	1987	1300	116	66°43'S	113°12'E
Taylor Dome		1990s	2400		77°48'S	96°24'E
Arctic Canada						
Various sites: Devon Island, Agassiz Ice Cap, Barnes, Mt. Logan						
Low Latitudes						
Quelccaya, Peru	160	1984	5670	1.5 m/yr	13°56'S	70°50'W
Dunde, Qinghai-Tibet Plateau	139	1987	5325	0.4 m/yr	38°06'N	96°24'E
Guliya, Tibet Plateau	306	1990–1992	6710	0.14–0.26 m/yr	35°17'N	81°29'E

Sources: From Robin 1983; Raynaud et al. 1993; Sowers et al. 1992; Thompson et al. 1990; Bradley 1989; Grootes 1995; Thompson 1996.

base is more than 130 ka old and may be as much as 200 ka. One of the earliest international programs was the Greenland Ice Sheet Project (GISP1), initiated by the United States, Denmark, and Switzerland in the early 1950s as an integrated field and laboratory investigation to study the three-dimensional geophysical and geochemical character of the Greenland Ice Sheet (Langway et al. 1985). The Dye-3 Greenland ice core obtained in 1971 was a 10.2-cm-diameter cylinder of ice initially reaching a depth of 372 m; by the summer of 1979, new drilling efforts to reach bedrock commenced, and in 1981, using advanced deep drilling methods, the investigators reached bedrock 2037 m below the ice-sheet surface. Integrated studies involving physical stratigraphy, mechanical properties, chemical microparticles, gases,

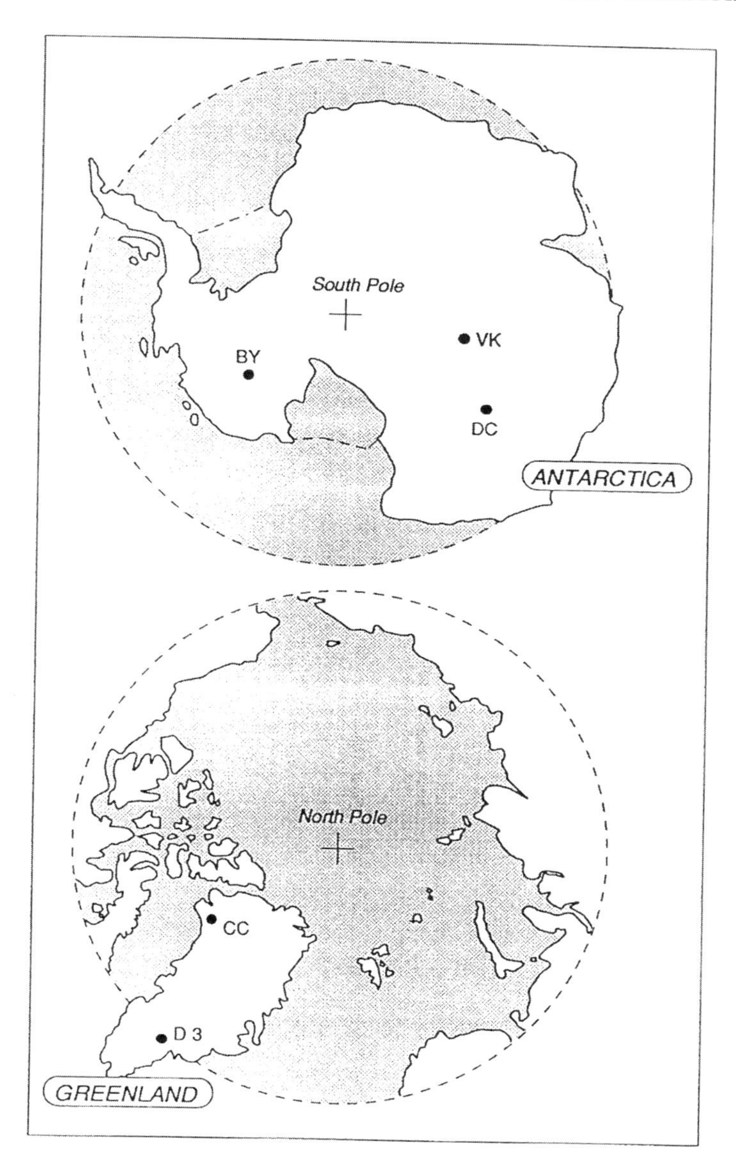

FIGURE 9-2 Map showing the locations of major ice-core sites in Greenland and Antarctica. VK = Vostok; DG = Dome C; BY = Byrd; D 3 = Dye-3; CC = Camp Century. The Summit cores (GISP and GRIP) are from central Greenland. Courtesy of R. Delmas and the American Geophysical Union. From Delmas (1992).

stable isotopes, and radioactive isotopes led to numerous fundamental discoveries about the Greenland Ice Sheet.

During the past decade, two new Greenland ice-core programs were initiated in central Greenland. They were only about 25 km apart near a site called Summit. One, the Greenland Ice-Core Program (GRIP), was led by a European group; the other, the Greenland Ice-Sheet Project (GISP2), was sponsored by the United States. A recent volume of the *Journal of Geophysical Research* was devoted entirely to results from the GISP2 and GRIP projects ("Greenland Summit Ice Cores," *Journal of Geophysical Research*, volume 102, no. C12, pp. 26315–26886). The GRIP and GISP2 members successfully recovered long-term records of atmospheric and climatic change back to at least 110 ka. Many discoveries stemming from the GRIP and GISP2 programs are highlighted later.

The Antarctic Ice Sheet has also provided numerous ice cores for paleoclimate research. Antarctica is colder and its ice sheet is larger and thicker than Greenland's. Antarctic ice reaches 3700 m thick at the famous Russian Vostok ice-core station. Early coring at Vostok began in 1974; by 1980–1982, drilling had reached a depth of 2083 m, and soon afterward, coring was extended to a depth of 2546 m (Jouzel et al. 1987, 1993). As of 1996, about 3350 m of ice had been penetrated (Petit et al. 1997). Vostok ice accumulates more slowly than at GISP and GRIP sites, so its temporal resolution is not as good. Yet because Vostok is so cold (–55°C), it preserves a 400-ka record of atmospheric trace gases largely unaltered by melting and postdepositional processes. A history of research activity on the Vostok ice core is provided by Robin (1983), Oeschger and Langway (1989), Grootes (1995), and Vostok Project Members (1995).

Several shorter polar Greenland and Antarctic ice cores have figured prominently in research on climate and atmospheric changes over past 20 ka (Jouzel et al. 1995; Mayewski et al. 1996), the past century (Battle et al. 1996), and the past millennium (Fisher et al. 1996). International collaborative research on polar ice is also being conducted by joint European-Japanese teams (Clausen et al. 1996). Salient results of some of these studies are outlined in later sections; details are available in the reviews listed in table 9-2.

In low-latitude regions, small ice caps found at high elevations are a primary source of atmospheric paleoclimate information unavailable from other sources. Teams led by Lonnie Thompson and Ellen Mosley-Thompson of Byrd Polar Research Center, of Ohio State University, have studied the Quelccaya ice cap (Thompson et al. 1984,

1985) and the Huascarán ice cap (Thompson et al. 1995b), both in Peru; the Dunde ice cap, Qinghai, China (Thompson et al. 1989); and the Guliya ice cap, Tibet (Thompson et al. 1995a, 1997). Because alpine glaciers are not as old or as thick as polar ice sheets, the paleoclimate record obtained from them is limited mainly to climate changes occurring since the last glacial maximum (LGM). The alpine ice-core record of the past 2000 years of decadal and centennial climate variability is especially noteworthy.

CLIMATE PROXIES FROM ICE CORES

CRYSTALLINE POLAR ICE has ample space between its molecules of water (H_2O) to trap chemical impurities and gases in the form of fossil air and provide a more direct measure of past atmospheres. The preservation of fossil air, atmospheric chemicals and particulates, and crystalline precipitation, albeit via complex processes that can obscure the original signal (see later), is a feature of the ice core record that makes it a unique archive of paleo-atmospheric conditions. In this section I outline the main ice-core proxy methods. Figure 9-3 schematically illustrates the steps that occur in transition of snow to firn to polar ice that result in such an exceptional preservation of past atmospheric conditions.

The primary paleoclimate indicators measured in ice cores include stable isotopic ratios of the ice lattice, stable isotopes of the trapped gases within the ice itself, relatively unreactive trace gases (e.g., CO_2) that give a global climate record, glaciochemical signatures derived from soluble (HNO_3, HCl, H_2O_2, NH_3) and insoluble particulate matter (e.g., cations such as NH_4^+, Ca^{2+}, Mg^{2+}, and anions such as NO_3^-, Cl^- SO_4^{2-}), electrical conductivity (a measure of ice acidity, itself a function of anion-cation concentrations), and cosmogenic isotopes (table 9-3).

Ice-core climate proxies can be classified in several ways (for summaries see Delmas 1992; Raynaud et al. 1993; Bales and Wolff 1995; Grootes 1995). One way is to group ice-core properties by their chemical characteristics. Bales and Wolff, for example, distinguish between reversibly and irreversibly deposited chemical species. Reversible species are those, such as acidic species, that continue to interact chemically with air during the snow-firn-ice transition, the process by which snow becomes ice. Irreversible species, such as some particulate aerosols, generally do not interact with air once deposited in the snow. Emphasis in this scheme is placed on the dynamics of atmospheric chemical characteristics, such as global oxidation capacity.

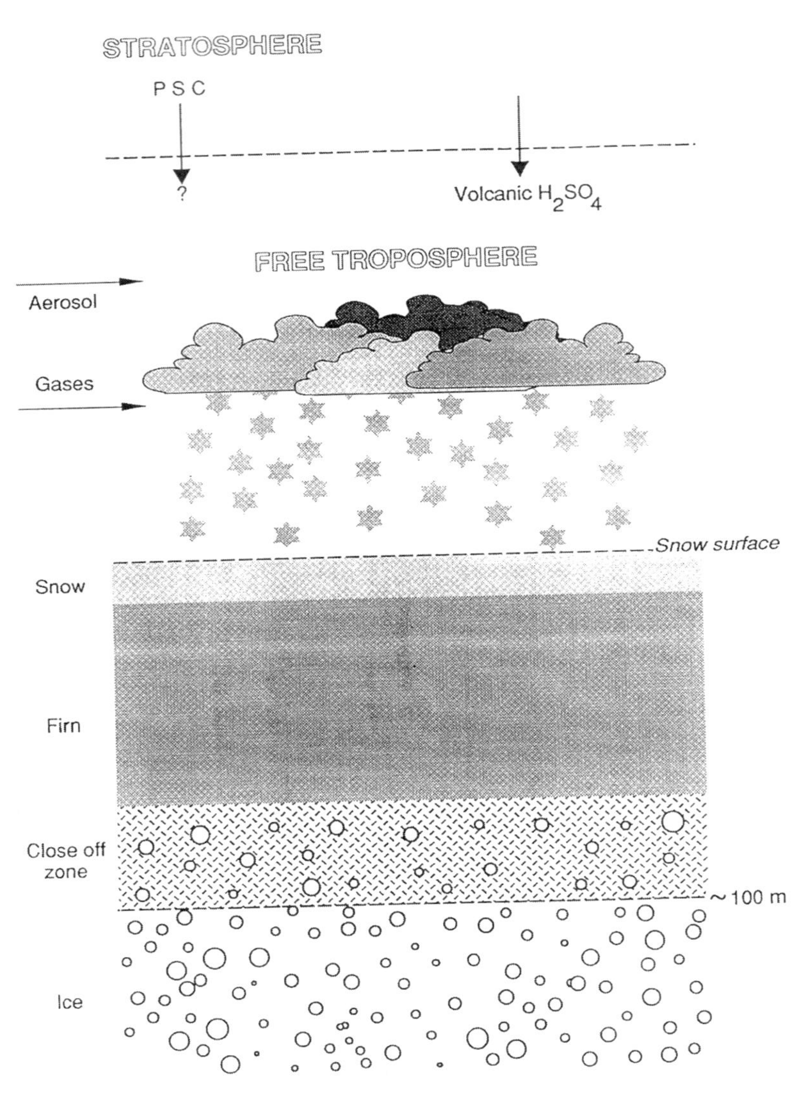

FIGURE 9-3 Schematic diagram showing the steps by which atmospheric aerosols and gases become incorporated into glacial ice and ultimately serve as proxies for paleo-atmospheric reconstruction. Diagram also depicts the approximately 100-m thick upper layer of ice sheets in which the snow-firn-ice transition takes place, closing off the trapped air from the atmosphere. This occlusion zone is the reason the trapped air is younger than the ice itself. Courtesy of R. Delmas and the American Geophysical Union. From Delmas (1992).

TABLE 9-3. Major Climate Proxies Measured in Ice Cores

Measure	*Climate signal*
Atmospheric gas content in occluded air	
CO_2 concentration	Biological systems, ocean pump
CH_4 concentration	Wetlands, oceans, biomass, animals, continental shelf hydrates, permafrost
N_2O concentration	Biogeochemical nitrogen cycles from marine and terrestrial activity
Stable isotopes in ice	
$\delta^{18}O_{ice}$	Atmospheric temperature
deuterium, δD (^2H:^1H)	Atmospheric temperature
d, deuterium excess ($\delta D - \delta^{18}O_{ice}$)	Ocean surface conditions (humidity, sea-surface temperature, wind velocity at source)
Stable isotopes in gas	
$\delta^{18}O_{air}$	Ice volume, oxygen cycle
$^{13}C/^{12}C$ in CO_2	Relative size of carbon reserves, different sources and sinks
$^{13}C/^{12}C$ in CH_4	Source of methane, different sources
$^{15}N/^{14}N$ in N_2	Gravitational and thermal fractionation of air in firm before close-off
Unstable cosmogenic isotopes	
$^{14}C/C$	Cosmogenic production changes, carbon reservoir changes, dating (< 30–40 ka)
^{10}Be and ^{36}Cl	Cosmic ray production, snow accumulation, dating
Insoluble particulate matter (dust)	
Ca^{2+}	Dust, aridity, wind atmospheric transport
SO_4^{2-}, NO_3^-	Sea salt, volcanic eruptions, ash
Cl^-	Sea salt, wind transport
NH_4^+	Ammonium, summer biomass burning
Electrical conductivity measurements (ECM)	Dust, air circulation, volcanic events

See Sowers et al. 1991, 1992, 1993; Raynaud et al. (1993); Delmas (1992, 1995); Grootes (1995).

One can simply divide ice core properties into chemical, physical, mechanical, and geophysical types of measurements.

Another way to classify ice core proxies is to focus on the process of the climate system that a particular proxy measures. For instance, calcium concentrations measure continental sources of dust, whereas chloride concentrations measure sea-salt (marine) sources and changing sea-ice conditions. Methane concentrations measure wetland activity due to hydrological changes on land. Cosmogenic isotopes such as ^{36}Cl and ^{14}C measure solar processes related to climate. Chemical species that are water soluble (e.g., deuterium [D], $\delta^{18}O$) generally give a record of local and regional response to climate because they reflect atmospheric processes occurring at or near the site of precipitation. Conversely, those chemical species (such as CO_2) that have long atmospheric lifetimes (years to centuries) and are well mixed throughout the atmosphere give a more global climate signal. Table 9-3 groups proxies by the type of measurements taken and also gives the most important aspect of climate each proxy is used to reconstruct. Brief summaries of each proxy method follow.

Stable Isotopes and Atmospheric Temperature

Oxygen ($^{18}O/^{16}O$) and hydrogen isotopic (deuterium, $^{2}H:^{1}H$, D:H) ratios of glacial ice are the main sources of information on atmospheric temperature change. Note that this refers to the $\delta^{18}O$ of *ice* (designated $\delta^{18}O_{ice}$) as opposed to the $\delta^{18}O$ of molecular oxygen, O_2, trapped within air bubbles (designated $\delta^{18}O_{air}$), which is used mainly as a means to correlate ice-core records to other climate records (see below). Deuterium and $\delta^{18}O_{ice}$ each have a linear relationship with temperature. In early work, Dansgaard et al. (1973) proposed a relationship of 0.63‰ per 1°C for a temperature-$\delta^{18}O_{ice}$ coefficient. More recent studies have revised the temperature-$\delta^{18}O_{ice}$ calibration (Dansgaard et al. 1993; Grootes et al. 1993; Cuffey et al. 1995). Kapsner et al. (1995) expressed the following relationship:

$$T(K) = [(\delta^{18}O_{ice} + 18.2)/53] + 273,$$

which is equivalent to 0.53‰ per 1°C. Cuffey et al. (1995) and Alley et al. (1997) used the calibrations 0.33‰ per 1°C to estimate Holocene atmospheric temperatures at Summit, Greenland (see also Johnsen et al. 1995).

Isotopically derived temperature estimates generally reflect the local atmospheric temperatures above the atmospheric inversion layer where precipitation is formed. However, factors other than temperature can affect the oxygen isotope ratio; these include sea-surface conditions, cloud temperatures, the season when the precipitation fell, and changing source area and storm tracks of the moisture (Jouzel et al. 1982; Steig et al. 1994; Cuffey et al. 1995; Charles et al. 1995).

Oxygen-isotope ratios can vary seasonally by 20‰, in contrast to an 8–10‰ isotopic shift between the LGM and the Holocene. Secular changes in moisture source can also alter the isotope-temperature calibration over time (Kapsner et al. 1995).

To be certain that variations in $\delta^{18}O_{ice}$ signify regional or hemispheric climate events related to temperature, isotopic changes should be reproduced in different regions of the same ice sheet. Grootes et al. (1990) studied oxygen isotopes in four cores from a region about 500–700 miles off the Ross Sea, for the period covering the past 1400 years. They demonstrated that oxygen isotope fluctuations from four separate cores could be explained not only as a temperature signal, but also as a result of variation in topography of core sites, summer vs winter accumulation rate, and firn formation. Grootes's study provided a clear example that some ice-core proxy records are dominated by local factors and are not representative of a regional or global climate signal.

Despite complications resulting from regional, seasonal, and source-area factors, stable isotopes can provide a reliable paleotemperature history. Analyses of $\delta^{18}O_{ice}$ records from multiple cores from Greenland and Antarctica indicate that similar patterns of isotopic change characterize each region during late Quaternary climate changes (e.g., Dansgaard et al. 1993; Grootes et al. 1993). Grootes (1995:551) summarized the oxygen isotope–temperature relationship as follows:

> . . . the isotopic composition of precipitation reflects primarily the temperature difference between the ocean surface at which the water vapor formed and the place of precipitation (air temperature at condensation level some distance above the surface). Thus the isotopes do not provide a simple, direct local temperature record, although the precipitation temperature often dominates.

Not surprisingly, at the scale of glacial-interglacial oscillations, ice core isotopic records show evidence for significantly colder temperatures during the LGM. The scale of isotopic shift between the LGM and the Holocene, however, varies among different ice cores. For example, at Camp Century, northern Greenland, the net change was 11‰; in Antarctica about 5–7‰; in the Dunde ice cap, China, about 2‰; and in the Devon ice cap, Canada, about 8‰. These differences reflect the degree to which climate cooling during the LGM varied owing to regional and local factors.

Deuterium-isotope ratios are sometimes preferred as estimates of paleotemperature because they are influenced less than oxygen isotopes by kinetic isotopic effects (Jouzel et al. 1989). Steig et al. (1994) concluded that high-amplitude deuterium isotope shifts in Greenland ice deposited during the past century might be attributed to interannual and seasonal changes in the extent of sea-ice conditions instead of solely temperature changes. Researchers take the additional step of calculating a value known as "deuterium excess," $d = \delta D - \delta^{18}O_{ice}$ (Johnsen et al. 1989). The value *d*, or the difference between the two isotopes, is sensitive to sea-surface temperatures in the source area of the moisture, relative humidity, and wind speed. Deuterium and oxygen isotopes together can track both local temperature over the ice sheet and ocean-surface conditions near the source area, important properties that vary during periods of rapid climate change such as the Younger Dryas (Johnsen et al. 1989).

Trace Gases

Carbon Dioxide

Carbon dioxide (CO_2) is a water-soluble gas currently present in the atmosphere in a concentration of 0.035%, a very small proportion relative to other atmospheric gases such as oxygen and nitrogen. For contrast, the planet Venus's atmosphere has a CO_2 concentration of 98%. Carbon dioxide has a long residence time in earth's atmosphere, between 50 and 200 yr, depending upon terrestrial and oceanic sources and sinks that exchange with the atmosphere. This long residence time makes measurements of "fossil" CO_2 concentrations from ice cores excellent indicators of global atmospheric changes. Moreover, the radiative properties of CO_2 molecules make changes in CO_2 suspect as a forcing factor in the amplification of natural long-term climate change (Lorius et al. 1990).

Changes in atmospheric concentrations of CO_2 also reveal clues about the earth's total carbon budget. Changes in the sinks and sources of atmospheric carbon involve major reorganizations of the terrestrial and marine biosphere. Indeed, the confirmation from ice-core records that atmospheric concentrations of CO_2 gases during glacial periods were 25% lower than concentrations during interglacial periods spawned an enormous effort by the scientific community to explain the mystery of reduced glacial atmospheric CO_2 levels. The critical role of the oceans in sequestering carbon through ecosystem (productivity), chemical (nutrient), and/or ocean-circulation changes were soon offered as competing hypotheses (Broecker 1982; Kier and Berger 1984; Knox and McElroy 1984; Wenk and Siegenthaler 1985; Boyle 1988b; Broecker and Peng 1989). Research on the oceans' "biological pump" as a dynamic reservoir of carbon has continued since.

Methane

Methane (CH_4) is the second major atmospheric trace gas directly linked to biological activity. The primary naturally occurring sources of atmospheric CH_4 are tropical and mid- to high-latitude wetlands, where the gas is produced by anaerobic bacteria (Senum and Gaffney 1985). Prather et al. (1995) estimated that wetlands currently produce an average of ~115×10^{12} g/yr within a range of between 55×10^{12} g/yr and 150×10^{12} g/yr. Estimates of tropical wetland CH_4 flux based on studies of the Amazon Basin are 60×10^{12} g/yr (Bartlett and Harriss 1993). Other minor natural sources of CH_4 include ancient fossil (coal, lignite, natural gas) and abiotic (e.g., volcanic) sources, termites, animals, marine gas hydrates (clathrates), and permafrost. Anthropogenic contributions stem mainly from intestinal organisms (enteric fermentation) and agriculture (rice fields).

The primary sink for CH_4 is oxidation by the hydroxyl radical (OH) in the troposphere through a series of four chemical steps, ultimately producing formaldehyde. The OH radical itself forms from solar radiation and has the effect of cleansing the atmosphere of many chemical species. The concentration of CH_4 in 1990 in the high northern latitude atmosphere was about 1725 ppbv, continuing an increase that began in pre-industrial times (Dlugokencky et al. 1994). Recent trends in atmospheric CH_4 and its sinks and sources are a major research area in atmospheric chemistry.

Changes in CH_4 concentrations in ice-core records has been considered a proxy indicator of tropical and high-latitude wetland activity, depending on the time interval (Chappellaz et al. 1990; Brook et al. 1996). Temperature and precipitation appear to control the formation of CH_4 such that, other factors being equal, the warmer and wetter the regional climate, the greater the production of CH_4. Researchers estimate that glacial-age OH concentrations may have been 10–30% lower than current levels. Methane and related wetland activity are closely tied to climate changes associated with 21-ka precessional changes and to short-term climate changes during glacial, deglacial, and Holocene periods.

Nitrous Oxide

Nitrous oxide (N_2O) is a long-lived atmospheric gas produced by the earth's oceans and soils at a rate of about 9 (6–12) × 10^{12} g/yr. In the atmosphere, N_2O is removed via photodissociation from sunlight. Nitrous oxide also varies in gases obtained from ice cores. Ice cores show N_2O increasing about 8% since the pre-industrial period due to human activity and about 30% during the last glacial-interglacial climate transition (Leuenberger et al. 1992). Machida et al. (1995) documented the past century's rise in N_2O, and Battle et al. (1996) conducted more detailed studies of N_2O changes over the past century. By modeling complex physical and chemical processes affecting air in firn and sampling 120 m of firn at the surface of the Antarctic Ice Sheet at the South Pole, Battle's group determined that atmospheric N_2O increased slowly (0.06%/yr) until about 1958, when it rose more rapidly (0.22%/yr). The rapid rise of N_2O was unexplained; it might have resulted from changing conditions in the oceans and/or natural soils, as well as from increased use of fertilizers. The paleoatmospheric record of N_2O and its use in paleoclimate studies is not as advanced as that of CO_2 and CH_4, although lower N_2O levels during glacial periods may reflect lower soil activity and or reduced N_2O from seawater.

Glaciochemical Species

A major group of chemical species derived from aerosols found in ice cores include anions and cations in the soluble fraction of snow. Aerosol particles have a range of particle radiuses: windblown dust (including volcanic dust) ranges from 1 to 20 μm, sea salt from 8 to 12 μm, and pollen grains from 10 to 100 μm. Aerosol species affect the acidity of the snow and ice, and generally each chemical species has a different source. Some important anion species and their most likely chemical sources are nitrate (NO_3^-) from nitric acid (HNO_3), sulfate (SO_4^{2-}) from hydrogen sulfate (H_2SO_4), and chloride (Cl^-) from sea salt.

Sulfate is a good example to illustrate the many factors that can affect the deposition of glaciochemical species in snow. Three ultimate sources of sulfate are sea salt, dimethyl sulfide produced by marine organisms through complex pathways, and volcanic material. These oceanic, biological, and geological sources all vary independently of one another, leading to multiple sources of natural variability in atmospheric sulfate concentrations. Complicating matters more, seasonal factors affect sulfate deposition. In Antarctica, maxima in sulfate levels are reached during spring and summer, minima in winter. These differences are due in part to atmospheric circulation and chemical processes. All these factors ultimately influence sulfate concentrations found in Antarctic snow and those measured in paleoclimate studies of ice cores (Delmas et al. 1982; Delmas 1995).

Despite this complexity, processes affecting many glaciochemical species are understood well enough to apply them to understanding climate change. Ammonium (NH_4^+) is another soluble trace gas that is a byproduct of biomass burning; it can serve as a useful indicator of fire history on continental areas near source areas of air masses passing over Greenland (Mayewski et al. 1993; Taylor et al. 1996). Later we will see several other examples of glaciochemical proxies.

Explosive volcanic eruptions also produce nitrates and sulfates found in glacial ice. Fine sulfurous ash can remain in the stratosphere for 6–18 months; coarser particles settle within a few months (Hammer 1977). By subtracting the sea-salt background component from the sulfur in ice cores, one can obtain a measure of excess sulfur and identify historical volcanic events, which can be used for cross-checking age models in cores (Mosley-Thompson et al. 1993). Many papers have demonstrated the existence of volcanic ashes in Greenland (Hammer 1984) and Antarctic (Delmas et al. 1985) ice cores. Langway et al. (1988) demonstrated interhemispheric correlation of major volcanic events, and Clausen and Hammer (1988) showed that even a single event, such as the massive Tambora eruption in 1815, can have variable impacts on the sulfate record throughout various parts of the Greenland Ice Sheet.

Mass Accumulation Rate

Climate change can lead to changes in precipitation that affect snow accumulation such that the rate of deposition of the ice itself can serve as a sensitive paleoclimate indicator. The mass accumulation rate (MAR) is not exactly the same as the precipitation falling on the ice because of three complicating factors (Mosley-Thompson et al. 1993): deflation, redeposition, and sublimation, all of which can result in the removal of mass (precipitation) from the ice surface. To use accumulation and ice-layer counting as a chronological and climatological tool, Mosley-Thompson and colleagues devised the following equation to estimate the original layer thickness $L(t)$ at time (t) and express the relationship between ice layers and accumulation rate:

$$L(t) = L_0 e^{(-bT/H)},$$

where L_0 is current ice layer thickness, b is current accumulation rate (both in ice equivalent), H is the thickness of the ice sheet at the core site, and T is the age of the particular ice layer. Variable ice accumulation rates controlling $L(t)$ values were calculated for Greenland and Antarctic cores and other dating means and revealed little correspondence in decadal-scale changes in climate. Mosley-Thompson et al. (1993) attributed discrepancies to differences in regional precipitation patterns.

A positive correlation between net accumulation and $\delta^{18}O$ has been obtained in several Greenland ice core records and provides strong evidence that accumulation is a valuable indicator of climate change, especially during the last glacial period, deglaciation, and the Holocene (see Meese et al. 1994).

Processes Affecting Ice Core–Gas Concentrations

Measuring ice core properties requires complex, meticulous procedures and a deep understanding of processes that can compromise the original chemical signal. While this statement is true for all measured properties, fossil gas concentrations trapped in air bubbles pose particular challenges because of their low concentrations and the significance of obtaining accurate historical concentrations of greenhouse gases. Although the methodological aspects of ice-core research are beyond the limits of this chapter, a few critical points require discussion. These topics are discussed in detail in Sowers et al. (1997).

Ice-Air Uncertainty

It is useful to begin this section with a discussion of the processes that occur during and after entrapment of the air that can affect the interpretation of the leads and lags between atmospheric CO_2 and global climate change. The age of trapped air is younger than the ice in which it is enclosed because the firn that constitutes developing ice does not close off the air at the surface of the ice (figure 9-3). Rather, gas bubbles are sealed off about 50–100 m below the ice surface. The consequence of this process, called *air occlusion,* is that the age of the ice is greater than the age of the air.

The magnitude of the ice-air difference depends on the accumulation rate and temperature. The more rapid the accumulation rate and the higher the temperature, the more rapid is densification and the smaller the difference in ice age and gas age. This produces a smaller age uncertainty for any particular level of ice analyzed for the concentration of certain gases at particular times (table 9-4) (Schwander and Stauffer 1984). An example of a high-accumulation-rate ice core is the DE08 core from Antarctica. High accumulation rates there lead to only a 35-yr age difference. In contrast, at Vostok, the age difference can be as much as 4 ka, or 5–10% of the age for sections of ice deposited before the last glacial period (Barnola et al. 1991).

A related concern is that because air bubbles close at different times, the measured gas concentrations are actually an integration of air from a certain interval of time. This limitation is most important during periods of rapid climate change when differing occlusion rates might blur the climate signal.

To further complicate matters, ice accumulation, which is a function of temperature and precipitation, can change over time. Climatically induced changes in accumulation rates mean that the ice-air age uncertainties themselves can vary as one moves down through the core from intervals of glacial to interglacial ice. Sowers and Bender (1995), for example, estimated an age error of about 600 yr for GISP2 ice formed at the beginning of the last deglaciation (about 17 ka) but only 300 yr for ice formed near the end of deglaciation (about 8 ka).

Gravity is an important process potentially influencing the paleoclimate signal because of the way it affects molecules of different weights in the incipient trapped air during the transition from firn to ice (Craig et al. 1988; Schwander 1989). CO_2 is heavier than O_2 and N_2, so it sinks a greater distance in the firn. The CO_2 concentration at the time when air bubbles become completely sealed off from the

TABLE 9-4. Ice-Air Differences in Greenland and Antarctic Ice Cores

Location	*Core*	*Annual accumulation (meters water equivalent)*	*Difference between ice age and mean age of air*
Antarctica	Dome C	0.036	1700
Antarctica	South Pole	0.084	950
Antarctica	Byrd	0.16	240
Greenland	Crete	0.265	200
Greenland	Camp Century	0.34	130
Greenland	Dye 3	0.5	90

Sources: From Schwander and Stauffer (1984), other sources cited in text.

atmosphere could be greater in the base of the firn column (figure 9-3). This process could lead to slightly biased gas concentrations when the atmospheric concentration of CO_2 changes before gas bubbles are completely enclosed at the firn-ice transition. Nevertheless, Schwander and Stauffer (1984) estimated that gravitational effects on gas concentration accounted for only about 1% of the initial atmospheric concentration. The gravitational effect accounted for only about 2 ppmv in Vostok ice (Barnola et al. 1991). In sum, gravitational processes affecting trace gases that occur during the transition from snow to firn to ice are reasonably well known and any corrections needed are now routine.

Procedures and Processes Potentially Affecting Trace Gases

Extraction and analytical methods for gases in ice have been particularly important for establishing an accurate measure of pre-industrial levels of CO_2 (Neftel et al. 1982, 1985; Sundquist 1985). Early studies melted the ice before chemical analyses. The dry extraction method of crushing the ice in a vacuum gave consistently better results in studies in the Vostok ice core (Barnola et al. 1987; Raynaud et al. 1993). For CH_4, Blunier et al. (1993) found that wet and dry extraction methods in analyses of CH_4 removed from central Greenland ice deposited over the past 1000 years generally yielded consistent results.

A high degree of analytical accuracy is necessary to measure and evaluate the climatic significance of long-term changes in CO_2. Gas chromatography and laser infrared spectrometry have overall errors of about 3% (the equivalent of 10 ppmv of the total 300 ppmv). Paleo-CO_2 measurements from the Vostok ice core have been studied by comparing data from different laboratories (Barnola et al. 1991), which has resulted in an analytical accuracy of 5 ppmv for the past 145 ka of Vostok ice. This value compares to the total range of CO_2 variability of about 80 ppmv between glacial (190–200 ppmv) and pre-industrial interglacial (270–280 ppmv) periods.

Several syn- and postdepositional processes can affect the chemical signatures preserved in the ice. These include physicochemical adsorption of gases occurring during early stages of ice formation, chemical processes within the ice over long time periods, changing concentrations due to gravity and molecular diffusion within incipient and trapped air, and alteration of air composition by hydrate formation at great depth.

Physical and/or chemical adsorption of gases onto the surface of the snow or ice might occur because CO_2 has a higher solubility, so enrichment can occur when it is adsorbed onto firn crystals. By measuring CO_2 and CH_4 from both ice found in recent air and from ice from an Antarctic core, Etheridge et al. (1992) showed that only minor differences exist between the two. In cold dry areas like Antarctica, adsorption accounts for less that 10 ppmv variation.

Molecular diffusion can also affect the trace-gas signal from ice cores. If the CO_2 becoming trapped in firn does not remain in equilibrium with the atmosphere and mixing is not complete within the firn, then a true atmospheric record cannot be obtained. Schwander (1989) suggested that in high-accumulation regions such as Siple, Antarctica, the close-off of the atmosphere within the newly formed ice occurs at a firn-ice density of 0.8 g/cm. However, Barnola et al. (1991) cautioned that some slow but significant mixing can continue even at higher ice densities at which accumulation is slow. This uncertainty surrounding the gas-ice age difference is about 35 years in high-accumulation-rate areas (e.g., Siple, Antarctic Peninsula) and as much as 2500 years in deep parts of slowly accumulating ice (e.g., Vostok).

Physicochemical changes might in theory also occur after the air is trapped, as well as during the period of entrapment. For example, CO_2 molecules might interact with the surrounding ice or with other occlusions during storage and transfer. Tests of CO_2 concentrations taken at different times after initial core recovery show little or no difference in CO_2; thus, this is only a minor factor.

Carbon dioxide concentrations are also affected by alkaline and acidic impurities in the ice. Grootes and Stuiver (1987) discussed how molecular diffusion can influence the gas concentration through post–ice formation processes. They reasoned that, because the nature of the impurities in Greenland ice are quite different from those in Antarctic ice, one would expect that the CO_2 record might be differentially altered. Delmas (1993) showed that HNO_3, NO_3^-, and SO_4^{2-} introduce a natural artifact into the CO_2 concentration variability measured in Greenland ice cores during millennial Dansgaard-Oeschger events. Such is not the case in Antarctica, where dust concentrations are 10 times lower than those over Greenland.

Two final processes can cause measured CO_2 concentrations to differ from those originally trapped in the air even where the effects from atmospheric impurities are minimal. These processes are air hydrate formation, which can occur under extreme pressure due to the thickness of the ice, and ice fracturing, which occurs in brittle zones (250–1400 m below the Antarctic surface) found during drilling. Raynaud et al. (1993) pointed out that the patterns of CO_2 and CH_4 obtained from the ice-core records themselves are inconsistent with the hypothesis that hydrate formation and fracturing influence the record. Specifically, they point out that the CO_2 and CH_4 records for both the last deglaciation, 15–10 ka, and the penultimate deglaciation, 145–130 ka, are very similar, even though the former is derived from ice at depths having no hydrate formation and the latter corresponds to the brittle zone where air hydrates do form. They also indicate that both the Byrd and Vostok Antarctic CO_2 and CH_4 records of atmospheric changes during the last deglaciation are generally similar, even though the glacial-deglacial transition is located in the air-hydrate zone at Byrd and but not at Vostok. Raynaud et al. (1993:928) concluded:

> . . . the good agreement for the glacial interglacial changes of CO_2 and CH_4 recorded in different types of ice (with and without air hydrates, or fractures or cracks, as well as different snow accumulation rates, ice structures, and so on) on the same core (Vostok) or among different cores supports the notion that, overall, the long-term trace-gas record from ice cores accurately reflects atmospheric changes.

In sum, most processes that might in principle compromise the orignal CO_2 concentration in Vostok and Byrd, Antarctica, ice are well enough understood so that the measured fossil CO_2 concentration does not differ appreciably from the actual paleo-atmospheric concentration.

THE DATING AND CORRELATION OF ICE CORES

GLACIAL ICE DIFFERS considerably from other geological (mostly sedimentological) and biological records in that it lacks some properties used in conventional geological age dating and chronology. For example, the magnetic polarity of the ice cannot be measured as it can be in sediments; consequently the paleomagnetic time scale that supports Cenozoic paleoclimate research is lacking. Too much ice is required for radiocarbon dates, and uranium-series dating, amino acid racemization, thermoluminescense, and potassium argon methods are not applicable to ice cores. Conversely ice cores have the exceptional advantage that snow accumulates rapidly; areas of high-accumulation rate often preserve annual layers, producing unprecedented climate records from the last glacial period, the deglaciation, and the Holocene. This section briefly describes the fundamental methods used to date glacial ice and correlate the ages with each other and with other paleoclimate records.

Dating an ice core is generally carried out through the use of one or multiple methods usually grouped into five general categories: seasonal trend dating and annual layer counting, marker horizons (volcanic events), cosmogenic isotopes, ice-flow modeling (combined with annual layer counting), and the stable oxygen isotope record of O_2 (Hammer 1989).

Seasonal Trend Dating

Seasonal and annual paleoclimate records can be obtained from ice if one can recognize seasonal variations using unique properties of glacial ice. For example, one can search for chemical, visual, electrical, or physical signatures that reflect seasonal changes in temperature, atmospheric composition, winds, or other factors at the site (figure 9-4). Seasonal-trend dating of ice cores is similar to using coral-growth banding and skeletal chemistry to study interannual tropical paleoceanography (chapter 7). Like coral banding, seasonal ice-core dating requires enormous numbers of sample analyses, as many as 8–15 per year of ice. Electrical conductivity measurement (ECM), which is a proxy of ice core acidity, is one such method. ECM can yield as many as 15 measurements per year (Taylor et al. 1996).

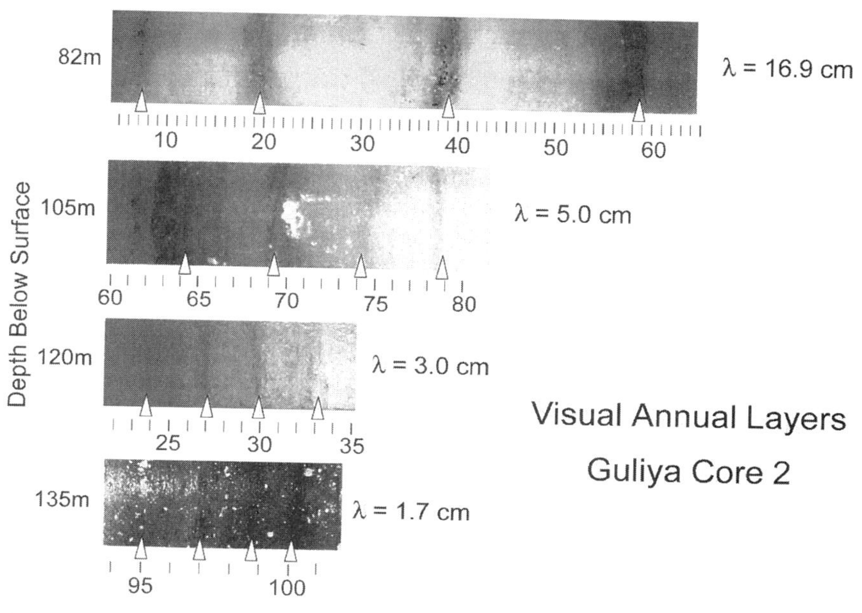

FIGURE 9-4 Annual layers of ice in South American ice core. Courtesy of E. Mosley-Thompson.

The identification of seasonal signatures must be based on carefully determined relationships between the measured property and its variability during modern seasonal changes (Hammer 1989). The atmospheric proxy for continental aridity—dust—illustrates this requirement. In prior chapters, eolian (wind-blown) deposition of fine sediment in the ocean was shown to be a meaningful record of changes in terrestrial ecosystems and aridity over continents. Higher dust concentrations generally mean greater aridity during glacial periods (chapter 4) and, for some regions (East Antarctica and the Peruvian Andes), greater aridity during short-term coolings such as the Little Ice Age (chapter 6).

Dust particles in ice cores are often separated into distinct size classes—those > 0.63 μm but < 2 μm and those > 2.0 μm diameter (Thompson et al. 1989; Mosley-Thompson et al. 1993). In the atmosphere over polar regions, dust content is relatively low and dust concentrations in the polar ice are orders of magnitude lower than those over nonpolar ice caps (e.g., Peruvian Quelccaya and Chinese Dunde ice caps, Mosley-Thompson et al. 1991).

Thompson et al. (1989) showed that dust concentrations increase during the dry season and decrease during the wet season over the Dunde ice cap in the Tibetan Plateau of China. These oscillations produce visible stratigraphic markers in deeper parts of the ice. Combined with oxygen isotope ratios, Thompson's group erected a near-annual layer chronology back to more than 4 ka. Similar integrated annual chronologies have been constructed using dust, isotopes, and volcanic-marker horizons (see below) in the Quelccaya ice cap in Peru (Thompson et al. 1985, 1986), in central Greenland (Meese et al. 1994), and at Siple Station in Antarctica (Mosley-Thompson et al. 1993).

Excellent seasonal signals are also obtained from glaciochemical signatures such as nitrate, sulfate, and $\delta^{18}O$ isotopes from Greenland and the Antarctic Peninsula. These records show well-defined seasonality trends in atmospheric temperature and chemistry (see Mosley-Thompson et al. 1990, 1991, 1993). Sulfate concentrations, for example, are at a maximum in Antarctic ice during spring and summer and at a minimum during winter. This fluctuation signifies atmospheric chemical and circulation changes (Mosley-Thompson et al. 1993) and sea-salt and non–sea salt biogenic emissions of sulfurous compounds from marine organisms (Legrand 1995). At Siple Mosley-Thompson's group used oxygen isotopes, sulfate, and nitrate to estimate an annual accumulation rate of 0.5 m/yr for an 18-yr interval (1792–1800) during the Little Ice Age at an ice-core depth interval between 121–130 m.

Seasonal trend dating using isotopes to count individual years backward from the ice surface has its limitations. Among the most problematic is slow ice-accumulation rate. Grootes et al. (1990) among others noted that seasonal fluctuations could not be detected in ice cores recovered from low–snow accumulation areas such as Vostok, Antarctica.

Local variation in a particular ice-core isotopic record can also obscure the annual signal. Processes known as sastrugi (also called wind-scouring), snow drifting, and simply irregular noise in the geochemical record of the ice can produce irregularities. These complications led Grootes et al. (1990) to advocate that regional climate interpretations should be based on a suite of cores rather than just a single core. ECM measurements in glacial ice can also be compromised by winter scour and other processes so, for example, the ECM record should be compared with other proxies of climate (e.g., Fisher and Koerner 1994). In summary, season-trend dating has been extremely useful in paleoclimatology, especially in documenting the past few centuries of climate history that can be compared with other high-resolution records. Annual layer counting can have a relatively small error margin—only 1% in the upper parts of cores—but one that increases with depth.

Marker Horizons

A second dating method, often used in conjunction with seasonal trends and layer counting, is the use of marker stratigraphic horizons such as volcanic ash. Hammer et al. (1980) pioneered the study of ice-core acidity in relation to volcanism, and recent studies by Clausen and Hammer (1988) and Langway et al. (1988), as well as the review by Clausen et al. (1995), expand on the record of volcanic events in Greenland and Antarctic ice.

Many volcanic ashes, especially those produced by historical eruptions, have well known ages based on archival records. These ashes can be identified as sulfate spikes rising above the background level of H_2SO_4 in the upper sections of ice cores, providing an independent means, based on layer counting, to confirm ages. One noteworthy example is the volcanic ash from the famous 1815 Tambora eruption, an event that led to the "year without summer" in Europe because of the cooling effects of volcanic particulate matter in the atmosphere. Mosley-Thompson et al. (1990) found a sulfate spike at 113 m depth in the Siple Station core corresponding to the years 1817–1819, when the atmosphere carried Tambora volcanic particulates around the world. Tambora is also recorded in Greenland ice cores (Clausen and Hammer

1988). Another well-known volcanic event is the February 19–March 6, 1600 A.D., eruption of Huayanputina, Peru, which stands out as a spike in large airborne particles in the Quelccaya ice core (Thompson et al. 1986). Many eruptions of historical Icelandic volcanoes, and prehistorical volcanoes, during the past millennium are recorded in Greenland (Hammer 1984; Clausen et al. 1997; Zielinski et al. 1997).

Prehistorical ashes are also useful in ice core research, especially those that are well dated from their occurrence in varved lake sediments through layer counting and radiocarbon. In Greenland ice cores, for example, the Vedde ash, well known from marine and continental sediments and dated at about 10.5 ^{14}C ka, is useful for study of the Younger Dryas event.

Cosmogenic Isotopes

A third dating technique is the use of the cosmogenic radioactive isotopes carbon (^{14}C), beryllium (^{10}Be), and chlorine (^{36}Cl) produced in the atmosphere by high-energy cosmic rays. The measurment of these isotopes in ice cores provides a method to trace cosmogenic isotopic production, changes in the geomagnetic field and atmospheric chemical processes (Stuiver and Braziunas 1989; Beer et al. 1988, 1990; Raisbeck et al. 1992). Cosmogenic carbon and beryllium isotopes behave differently in the atmosphere. ^{14}C has a longer atmospheric lifetime of 50–200 yr and circulates between the oceans and the biosphere as it is incorporated into organic material via the global carbon cycle. ^{10}Be is removed from the atmosphere within weeks of its formation and thus records changes in production.

The rationale for using ^{10}Be as a dating tool is that, if ^{10}Be is produced in the atmosphere at a constant rate, then changes in the accumulation rate of ice should be reflected in changing ^{10}Be concentrations. After analyzing three Greenland ice cores, Beer et al. (1988) concluded that atmospheric mixing and transport had minor effects on ^{10}Be concentrations, that atmospheric production dominated the signal, and that ^{10}Be appeared to vary over 11-yr cycles, modulated by sunspot cycles. The variation was due to changes in either primary cosmic ray flux or the intensity of the earth's geomagnetic field. They also argued that the correspondence between ^{10}Be concentrations and paleotemperature at 80- to 90-yr periods reflected Gleissberg solar cycles (chapter 6).

Raisbeck et al. (1987, 1990, 1992) and Yiou et al. (1985) showed that Antarctic ^{10}Be concentrations of glacial age ice were twice those of interglacial ice over longer time scales. They argued this difference indicated lower snow-accumulation rates during glacial periods. Beryllium trends then are especially useful to supplement chronologies derived from glaciological data and isotopic stratigraphy. Raisbeck et al. (1987) also found that certain beryllium spikes in the Vostok ice record probably record periods when cosmogenic production of ^{10}Be was not constant. One ^{10}Be event near 60 ka did not correspond to climatic events recorded by other indicators, such as shifts in oxygen isotopes and dust records. It might signify a 1- to 2-ka period of rapid change in ^{10}Be production in the atmosphere due to solar modulation or a supernova event.

Ice-Flow Modeling

A fourth method used to develop an age model for deeper intervals of the ice core involves ice-flow modeling. Glaciologists view ice cores as bodies of ice that flow according to physical laws of stress and strain, not necessarily as stratigraphic records of climate. There is a large literature on empirical and theoretical studies of ice sheet flow, or ice rheology (Paterson 1978; Peltier 1995; see *Annals of Glaciology* 1996, volume 23). In the case of ice-flow modeling for ice-core age models, the study of ice flow must take into account the dynamic nature of glacial ice. In particular, researchers must estimate the way in which ice from upstream of an ice-coring site can influence the ice-core record at the site itself by the way it flows. Ice-flow data are combined with snow-accumulation rates derived from paleotemperature estimates of layer counts to produce an integrated age-depth model at some sites.

Ice flow modeling is most important for understanding deeper ice layers, which are under greater pressure to flow and thus distort the record. The longest ice-core climate record from Vostok (Jouzel et al. 1993; Petit et al. 1997) had several age models derived mainly from ice-flow modeling. Jouzel et al. (1993) called the Vostok age model the extended glaciological model (EGM).

The effects on entrapped gases of physical ice mixing at great depths can be substantial. Souchez et al. (1995) studied the lowermost 7 m of basal silty ice in the GRIP core and showed exceptionally high CO_2 (130,000 ppmv) and CH_4 (5000 ppmv) concentrations, covarying with oxygen isotope and total gas content. They concluded that neither anaerobic bacterial production of these gases nor molecular diffusion could produce such a trend. Instead, mechanical mixing at great depth ultimately related to paleo-soil gases depleted in oxygen and enriched in CO_2 and CH_4 was a better explanation.

In Greenland, recent studies of the GRIP and GISP2 cores serve as a prime example of the importance of understanding ice flow for correlating paleoclimatological events. In early studies of GRIP, Johnsen et al. (1992) reported isotopic evidence for a surprisingly large amplitude of Eemian interglacial climate variability. The concept of a highly variable interglacial climate challenged conventional views that Quaternary interglacial periods were climatically stable (chapter 5). Other studies were soon conducted to see if high-amplitude climate variability existed in other Eemian records from the deep sea (Keigwin and Jones 1994; McManus et al. 1994), including analyses of the nearby GISP2 ice core (Taylor et al. 1993a). Marine and GISP2 records indicated a much lower level of Eemian climate variability than the GRIP core had indicated. Many researchers have reached the conclusion that the GRIP and GISP2 isotopic records older than about 110 ka may be compromised owing to ice flow or other factors (Grootes et al. 1993; Bender et al. 1994). Johnsen et al. (1997) recently concluded on the basis of 70,000 oxygen isotopes analyses from the GISP core that (p. 26,387) "abrupt and strong climatic shifts are also found within the Eemian/Sangamon Interglaciation," although problems still remain with the core's stratigraphic continuity.

Oxygen Isotope Stratigraphy of Trapped Oxygen

A fifth method of dating and correlating ice core records to the standard Quaternary time scale is the use of the isotopic signal of trapped molecular oxygen ($\delta^{18}O_{atm}$). Bender et al. (1985, 1994) and Sowers et al. (1993) showed that it was possible to construct an oxygen-isotope curve similar to that derived from deep-sea foraminifera (chapter 4) from molecular O_2 trapped in the ice. Knowing that the residence time of oxygen is 2–3 ka, Bender and colleagues reasoned that the $\delta^{18}O_{atm}$ of oxygen molecules trapped in ice can serve as a proxy for ice volume just as oxygen isotopic ratios in marine foraminiferal calcite change because of ice-volume changes. Isotopic ratio variability due to ice-volume changes would in theory be translated from the seawater to the atmospheric-oxygen reserve via the action of photosynthetic (oxygen-producing) marine organisms, and then into the polar ice bubbles, a series of processes simplified as follows:

$$\delta^{18}O_{seawater} \rightarrow \text{photosynthesis} \rightarrow \delta^{18}O_{atm} \rightarrow \text{polar ice} \rightarrow \delta^{18}O_{ice} \text{ (ice bubble).}$$

The isotopic fractionation of oxygen is, however, complex, and several factors must be taken into account before oxygen-isotope stratigraphy can be applied to date climate events (Bender et al. 1985; Sowers et al. 1991; Sowers and Bender 1995). These include hydrological (changes in evapotranspiration and evaporation and precipitation) and ecological (photosynthesis and respiration) factors that can influence the $\delta^{18}O$ signal. The Morita-Dole effect is the general term given to the difference between the $\delta^{18}O$ of seawater and that of the atmosphere. To isolate the ecological factors that influence the $\delta^{18}O$ signal, Bender, Sowers and colleagues calculated the ratio for the average isotopic composition of photosynthetic oxygen. Photosynthetic oxygen of marine phytoplankton is about 0‰, the same as the $\delta^{18}O$ of the seawater in which they live. Selective uptake of light oxygen during respiration by marine phytoplankton leads to a range of ocean $\delta^{18}O$ values from –7‰ to –25‰ (average of –20‰). Terrestrial $\delta^{18}O$ values are a function of precipitation, humidity, and plant physiology, and terrestrial plants produce a range of $\delta^{18}O$ values between 4‰ and 8‰. The modern $\delta^{18}O_{atm}$ is about 23.5‰ and represents an approximate 1:1 contribution of gross photosynthesis from terrestrial and marine sources.

On a glacial-interglacial scale, changes in global biological productivity can alter the $\delta^{18}O_{atm}$ of past atmospheres such that marine and continental isotopic effects might influence the isotope signal preserved in ice-core molecular oxygen. Changes in the ratio of marine to continental primary productivity, for example, can alter the ice core $\delta^{18}O_{ice}$ because $\delta^{18}O$ of leaf water is heavier than that of sea water. A 30% change in marine and terrestrial productivity such as that occurring during a glacial-interglacial transition is the equivalent of a 0.5‰ isotopic change in atmospheric O_2 versus seawater O_2. Sowers et al. (1993) argued that, with the exception of one period near 110 ka, hydrological and biological factors appear to have remained near present values over the last glacial-interglacial cycle, so the dominant signal in the ice-core oxygen-isotope record from trapped air represents an ice-volume signal.

Ice-core isotope stratigraphy was applied to revise the age of the Vostok core for the last full glacial-interglacial cycle of the past 135 ka (Sowers et al. 1993) and to correlate Greenland (GISP2) and Antarctic (Vostok) climate records for the past 100 ka (Bender et al. 1994). Sowers et al. obtained an excellent fit between the $\delta^{18}O_{atm}$ and $\delta^{18}O_{foram}$ from the SPECMAP time scale, supporting the hypothesized relationship between $\delta^{18}O_{atm}$ and global ice volume. They caution that the correlation is tentative because they observed subtle differences between $\delta^{18}O_{atm}$ and $\delta^{18}O_{foram}$ especially during cold intervals. These differences are likely due to secular changes in processes controlling the oxygen isotopic composition, including changes in the global average of respiratory isotopes from the continental biosphere or changes in the ratio of marine to continental productivity. Nonetheless, the GISP2 and Vostok oxygen-isotope stratigraphy puts the ice core record of climate and CO_2 into a common temporal framework. More generally, the ice core $\delta^{18}O_{atm}$ stratigraphy has been a major breakthrough in paleoclimatology because it has enabled the correlation of climate records from the two poles with each other and with deep-sea marine climate records, allowing the study of phasing between the ocean and atmosphere.

A final age-related factor is that ice in the polar regions is relatively young. The oldest ice at Vostok is only ~500 ka, at Greenland ~200 ka. This factor, of course, precludes direct measurement of paleo-atmospheric greenhouse gas concentrations from periods of global warmth such as the mid Cretaceous (Barron and Washington 1985; Barron et al. 1995), the Eocene (Sloan et al. 1995), and the Pliocene (Dowsett et al. 1994). Several ice cores nonetheless cover periods of major climatic transition during glacial terminations of the past four glacial periods, times of interglacial climatic warmth, and periods of extremely rapid climate change during the Younger Dryas.

In summary, limitations of dating and climate proxy indicators of ice cores are similar to those inherent in all paleoclimatology, albeit stemming from processes unique to the snow-firn-ice transition, glaciological flow, and chemical processes occurring in glacial ice. The temporal resolution of climate history that can be achieved from ice cores varies widely depending on many complex factors. In some ice cores seasonal and annual signals can produce calendar-year resolution with relatively small error margins (< 50–200 years). Individual events such as volcanic eruptions can also be pinpointed with exceptional accuracy in many cores, especially in the Greenland Ice Sheet, supplementing layer counting and seasonal trends.

In other regions, the age of ice can be determined with estimated errors no better than several thousand years. Large age uncertainty characterizes the deepest layers of some east Antarctic ice cores where ice flow at depth due to immense pressures can distort the original stratigraphy and ice accumulation is slow. Moreover, the processes described above that control the inclusion of gases and dust into the trapped air also influence the age and correlation of the paleoclimate signal. Still, the rapid development of correlation methods of the past few decades have produced exceptionally reliable paleo-atmospheric–paleoceanographic correlations, and discoveries will surely accelerate in the near future.

GLACIAL–INTERGLACIAL CLIMATE, CARBON DIOXIDE, AND METHANE

THE HISTORICAL Mauna Loa atmospheric CO_2 record generated by Keeling (1973, see Keeling et al. 1989) raised weighty questions about future climate and inspired interest in natural CO_2 variability. What were pre-industrial concentrations of atmospheric CO_2? What is the natural variability of atmospheric CO_2 during the Holocene? During the last glacial period? What role does CO_2 play in global glacial-interglacial climate change?

Polar ice cores answer these basic questions about earth's greenhouse gases back through the past 400 ka. The following sections review CO_2 and CH_4 variability over the past 1000 yr, during the LGM, and over orbital time scales. Before discussing the ice core trace gas record, however, I examine the long-term atmospheric evolution with particular reference to atmospheric CO_2. Indeed, knowledge of the early evolution of earth's atmosphere has advanced considerably over recent years, and many lines of geological evidence indicate that the atmosphere has changed significantly over earth's 5 billion year history (Holland 1984). Kasting (1993) provides a useful summary of Precambrian atmospheric evolution. Before about 2 billion years ago (Ba), the atmosphere lacked free oxygen. Atmospheric oxygen levels probably increased considerably about 2 Ba and again near 800 Ma, coincident with major evolutionary changes in earth's biosphere. Carbon dioxide levels are also believed to have been substantially different during the Precambrian. One line of reasoning holds that because the sun's luminosity was much lower (roughly 30% at 4.6 Ba), an enhanced greenhouse effect is needed to explain a Precambrian climate in which water existed in liquid form. This problem has been called the "faint early sun paradox" (see Hoyt and Schatten 1997). Although alternative explanations to greenhouse gas composition have been offered to explain the faint sun paradox (e.g., different albedo), it is most readily explained if earth's early atmosphere was compositionally different from the current atmosphere (Kasting 1993).

Although Phanerozoic (540 Ma to present) evolution of earth's atmosphere was not as dramatic as that during the first 4 Ba of the planet's history, it was nonetheless substantial, intricately linked to climate change and the global carbon budget. Berner (1990, 1997) reviews our understanding of Phanerozoic CO_2 levels estimated from published geochemical and paleontological proxies and from Berner's own carbon-cycle model called GEOCARB (Berner 1994).

Mora et al. (1996) found that middle to late Paleozoic CO_2 dropped by a factor of 10 between 450 and 280 Ma, as indicated by solid carbonate and organic material carbon isotopes. Between the Silurian and the Pennsylvanian, atmospheric CO_2 fell from 12–16 times to about 2–3 times modern concentrations. The steepest decline was tied to the evolution of terrestrial ecosystems (see also Berner 1997). Leaf stomatal density changes are also believed to signify physiological manifestation of declining Paleozoic CO_2 levels (McElwain and Chaloner 1996).

Several studies have concentrated on post-Paleozoic atmospheric CO_2 levels. These include those by Barron et al. (1993) for the Cretaceous; Sloan et al. (1995) for the Eocene; Ruddiman and Kutzbach (1989), Raymo et al. (1988), Cerling (1991, 1992), and Cerling et al. (1993) for the Cenozoic; and Shackleton and Pisias (1985) and White et al. (1994) for the Quaternary. During the Cenozoic, as earth's climate shifted from one of global warmth into one of colder temperatures, many complex factors influenced the atmosphere. One potentially important factor was the uplift of the Tibetan and North American plateaus, which may have led to elevated continental weathering rates and drawdown of atmospheric CO_2. Cenozoic high mountains and plateaus contrasted strongly with low continental relief during the Cretaceous. The evolution of certain plant types during the late Miocene (7 – 5 Ma) that were capable of a distinct photosynthetic pathway for carbon (called C4 plants) may have resulted from lower atmospheric CO_2 levels (Cerling et al. 1993). Leaf stomata also support the hypothesis of declining CO_2 levels during the Cenozoic (Van der Burgh et al. 1993).

Atmospheric Carbon Dioxide and Methane: the Past 1000 Years

Short-term variability in atmospheric trace gases is best determined from ice-core records. Before the advent of ice-core paleoclimate research, estimates of pre-industrial atmospheric concentrations of CO_2 were in the vicinity of 290 ppmv (e.g., Bray 1959). If we extrapolate the Mauna Loa trend of increased CO_2 due to fossil fuel combustion back, we obtain a pre-industrial estimate of 297 ppmv. Nonetheless, early estimates included a large degree of uncertainty, and no firm data were available about natural short- and long-term variability of CO_2 and CH_4 (Sundquist 1985).

Neftel et al. (1982, 1985) provided early evidence that pre-industrial CO_2 levels in the Siple ice core, in which the ice-air age difference is less than 100 yr, were 280 ± 5 ppmv. As methods to extract gases improved and additional ice cores were analyzed, other studies confirm that pre-industrial eighteenth century levels of CO_2 were near 270–280 ppmv (see Oeschger et al. 1985). Raynaud and Barnola (1985) suggested that during the last 1000 years, before the early 1800s, atmospheric CO_2 may have varied only by about 10 ppmv (see also Etheridge et al. 1996). Additional studies of high-accumulation-rate ice-core records from, for example, D47 and D57 of Antarctica (Barnola et al. 1995), the South Pole (Siegenthaler et al. 1988), and Siple (Friedli et al. 1986) have since confirmed that short-term variability is about ±10 ppmv around a mean value of about 280 ppmv. One possible exception is the small excursion occurring between about 1200 and 1400 A.D., when CO_2 rose about 10 ppmv above background late-Holocene levels (Barnola et al. 1995). The causes of low-amplitude late Holocene oscillations are not known, in part because of the ± 3–5 ppmv analytical error (Raynaud et al. 1993), but the oscillations most likely result from small imbalances in global carbon cycling. In general, ice core data have provided compelling evidence that atmospheric CO_2 variability during the latest part of the Holocene exhibited a much lower amplitude than the human-induced rise of the past century or that over glacial-interglacial time scales.

Atmospheric concentrations of CH_4 have also increased during historical times far higher than the relatively stable background levels of about 700–900 ppbv in evidence from high-accumulation-ice core records. For example, Stauffer et al. (1984) studied the CH_4 from the Siple core and found concentrations of 800–900 ppbv back to the year 1800. At another high-accumulation site (DE08 on the Law Ice Dome in Antarctica) where the air-ice age difference was only about 35 yr, Etheridge et al. (1992) showed CH_4 levels from 1841 through 1978 increased from 823 to 1481 ppbv. Nakazawa et al. (1993) studied Greenland and Antarctica CH_4 records back to 1300 and 1600 A.D., respectively, confirmed a pre-industrial global mean of about 720–740 ppbv, and discovered slight differences in the Northern and Southern Hemisphere CH_4 concentrations related to latitudinal distribution of terrestrial CH_4 production.

Methane concentrations have varied over the past 1000 yr of pre-industrial history more than CO_2 concentrations. For example, Blunier et al. (1993) found oscillations of 70 ppbv of CH_4 (about 10% of the pre-industrial mean) occurring before 1500 A.D. that may be associated with the Medieval Warm Period. These variations were caused by either changes in atmospheric chemistry, climatically driven changes in wetland emissions, agricultural emissions, or a combination of these processes. Holocene CH_4 oscillations suggest a possible link between high-frequency, low-amplitude climate changes and wetlands (see below).

Atmospheric Carbon Dioxide and Methane During Glacial-Interglacial Cycles

Having established the natural interglacial CO_2 and CH_4 concentrations for the late Holocene, we might ask what was the long-term glacial-interglacial variability and what was the role of greenhouse gases in high-amplitude Quaternary orbital climate changes. To address this problem many questions must first be answered. What were glacial-age concentrations of CO_2 and CH_4? Do atmospheric CO_2 changes lead to climatic warming that accompanies deglaciation, and do they initiate or contribute to atmospheric temperature rise? Or does CO_2 lag behind atmospheric temperature change? In the former case, CO_2 would be considered a plausible forcing mechanism to explain large-scale climate changes. In the latter, CO_2 changes might reflect complex biogeochemical feedback responses, which may amplify or, in the case of falling CO_2 levels, dampen climate change. A third possibility is that CO_2, atmospheric temperature, and other climate proxies changed in phase during glacial-interglacial climate change. The ice core record of climate provides empirical evidence to address these critical topics.

Atmospheric Gases During the Last Glacial Period

Unequivocal evidence has accumulated from ice-core records that CO_2 and CH_4 concentrations were significantly lower during the LGM than during the pre-industrial Holocene. Delmas et al. (1980) and Neftel et al. (1982) securely established that the CO_2 concentrations in the atmosphere of the LGM 20 ka were substantially lower than pre-industrial concentrations. These pioneering studies carried out in Grenoble, France (Delmas et al. 1980), and Bern, Switzerland (Neftel et al. 1982), documented CO_2 trends for the past 20 ka and 40 ka, respectively. Both studies discovered glacial CO_2 concentrations near 180–200 ppmv. Later studies of the Dye-3 Greenland core (see Oeschger et al. 1985) and Byrd, Antarctica, core (Neftel et al. 1988) provided additional evidence that glacial CO_2 levels were reduced 25% below Holocene levels.

Carbon dioxide changes for the glacial period between 50 and 20 ka were also documented at Byrd, Antarctica (Neftel et al. 1988). The

Byrd record showed that relatively minor oscillations ±10 ppmv around a mean of 190 ppmv characterized these changes. Other climatic indicators from the Byrd ice core such as CH_4 and N_2O (Stauffer et al. 1988), oxygen isotopes (Johnsen et al. 1992), and carbon isotopes (Leuenberger et al. 1992) indicate that CO_2 and climate are closely linked. Early evidence for a CO_2 maximum about 30–40 ka from Dye-3, Greenland (Stauffer et al. 1984), were not supported by the Byrd ice core record. The Dye-3 gas chemistry record for this interval is likely compromised, perhaps because of high dust concentrations (Delmas 1993; Sowers and Bender 1995).

Like the CO_2 record, the CH_4 record from Byrd also shows extremely low glacial-age concentrations (400 ppbv) during the last glacial period compared with late Holocene levels (Stauffer et al. 1988). Global atmospheric CH_4 concentrations appear to vary considerably during the glacial period 70–20 ka. Chappellaz et al. (1990, 1993) for GRIP and Brook et al. (1996) for GISP2 showed that CH_4 concentrations were elevated (up to 500–600 ppbv) several times between about 30 and 80 ka, a period when CO_2 concentrations measured at Vostok remained level. These natural high-frequency (100–150 ppbv) fluctuations in CH_4 are related to Dansgaard-Oeschger temperature fluctuations (see chapter 5 and below).

Carbon Dioxide and Climate Change During the Last Glacial-Deglacial Transition

The record of the last deglaciation in the GISP2 and Vostok ice cores provides evidence for important interhemispheric lead and lag relationships between CO_2 concentrations and the global climate system (figure 9-5). A common chronology for the two polar regions was developed from the oxygen-isotope stratigraphy discussed earlier. Sowers et al. (1991) and Sowers and Bender (1995) showed that the deglacial decrease in $\delta^{18}O$ in trapped oxygen over the interval 15–8 ka is large enough, about 1.5‰, to allow correlation of climate records between the two hemispheres. The ice-air age difference at GISP2 is about 240 and 630 yr for the Holocene and last glacial period, respectively. For the Byrd ice core, the correlation uncertainty is ±600 yr between 15 ka and 8 ka and ±1500 yr between 30 ka and 15 ka (Sowers and Bender 1995).

With this chronology, Sowers and colleagues examined atmospheric CO_2 and CH_4 changes during the 7-ka period of deglaciation and discovered a salient feature about the interhemispheric relationship

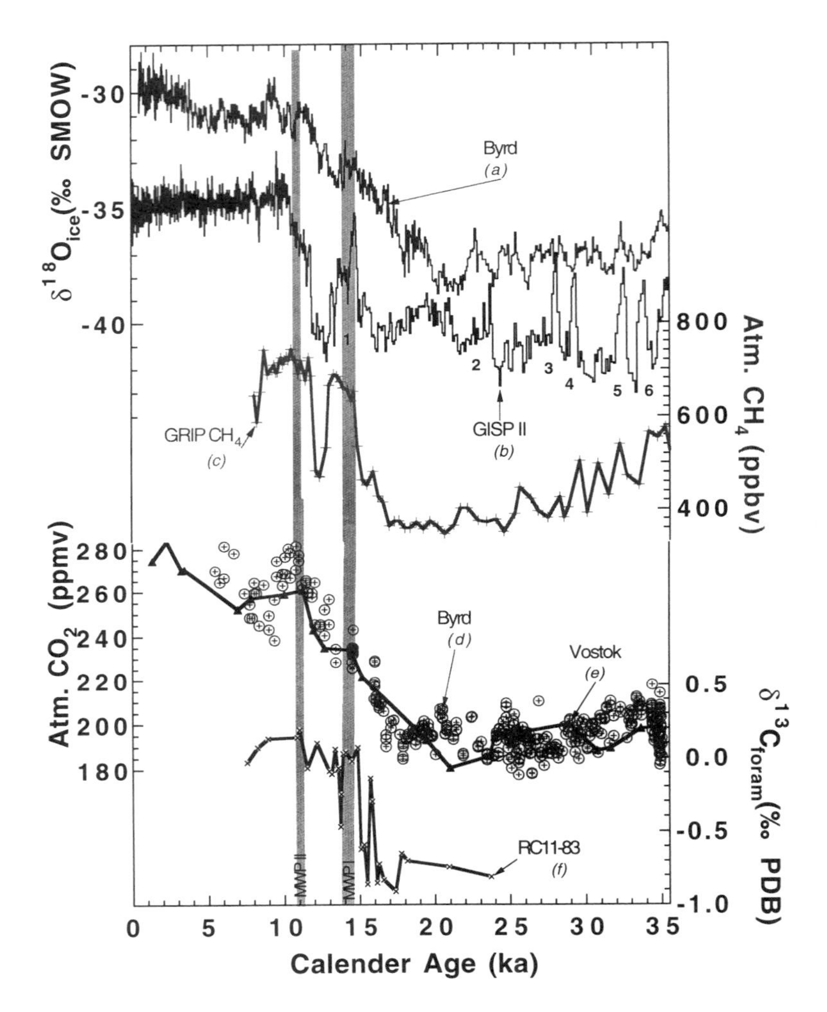

FIGURE 9-5 Comparison of deglacial climate record from GRIP, GISP2, and Byrd ice cores and deep-sea core RC11-83 carbon isotopes. Curves show that Antarctic deglaciation recorded by oxygen isotopes and CO_2 concentrations precedes by several thousand years climate changes in tropical and Northern Hemisphere ice marked by rising Greenland ice-core CH_4 concentrations and postglacial sea-level rise as recorded in Meltwater Pulses 1a and 1b. MWP I = MWP 1a; MWP II = MWP 1b; SMOW = standard mean ocean water. Courtesy of the American Association for the Advancement of Science and Todd Sowers.

between greenhouse gases and climate. Global atmospheric concentrations of CO_2 and CH_4 began to rise 2–3 ka *before* the onset of atmospheric warming recorded in the Greenland ice-core record. Sowers and Bender (1995) came to this conclusion by demonstrating in the GISP2 record that the first time $\delta^{18}O$ rises above the glacial level is at 14.7 ka, during the Bølling interstadial. Moreover, all Greenland ice-core climate records preserve this event at about the same time. Sowers and Bender also garnered a large amount of climate-proxy evidence from North Atlantic deep ocean circulation, sea-surface temperatures, and European terrestrial climate records to augment the GISP2 atmospheric record and demonstrate convincingly that Northern Hemisphere climate began to warm about 14,700 yr, as we discussed in chapter 5. In contrast, the Byrd deglacial temperature rise of 7–10°C, marked by the first time Byrd $\delta^{18}O_{ice}$ rose above glacial levels, was 18 ka. This deglacial inception was a full 3.7 ka *before* the Greenland shift of comparable magnitude (figure 9-5).

Raynaud et al. (1993) reviewed the Vostok record of CO_2, CH_4, and atmospheric temperature over longer time scales, also using the oxygen-isotope stratigraphy as a correlation tool. They came to several important conclusions that were largely dependent on the phase relationships between the Vostok ice-core record and the deep-sea ice-volume record first established by Sowers et al. (1991). First, during the past two deglaciations, CH_4, CO_2, and Southern Hemisphere atmospheric temperatures changed approximately simultaneously, although trace gases may have lagged behind temperature by up to 1000 yr (owing to ice-air age uncertainty). Second, there is unambiguous evidence that atmospheric CO_2 rose as much as 35 ppmv above glacial levels 4–7 ka before the major decay of large Northern Hemisphere ice sheets began. This means that sea-level changes did not induce changes in the concentrations of atmospheric greenhouse gases. Third, changes in deep-sea Southern Ocean temperatures and the carbon-isotope records, which both precede deep-sea foraminiferal oxygen isotope shifts (Sowers et al. (1991), support phasing of Vostok greenhouse gases and ice volume. Finally, as first suggested by Lorius et al. (1990), Raynaud et al. (1993:932) concluded:

> The most striking feature of the CO_2 and CH_4 records . . . is the close correlation between greenhouse gases and climate over the last climatic cycle. This correlation, as well as the phase relationships among greenhouse gases and climatic parameters, suggest that those greenhouse gases have participated, along with orbital forcing, in the glacial-deglacial changes.

Glacial-Interglacial Climate and Carbon Dioxide and Methane Over Tens to Hundreds of Thousand of Years

The Vostok ice core record has provided unique and unprecedented evidence for atmospheric evolution corresponding to large-scale global climate cycles that have dominated earth's climate over the past 400 ka. Long-term records of CO_2 and CH_4 encompassing the past two glacial-interglacial cycles have been established in a series of landmark studies of the Vostok ice core (Barnola et al. 1987, 1991; Chappellaz et al. 1990; Raynaud et al. 1988; Jouzel et al. 1993; Petit et al. 1997). The Vostok trace gas and paleotemperature records are shown in figure 9-6. Leuenberger et al. (1992) also established trends in N_2O from the Byrd ice core. Because of the importance of the Vostok record, a few comments about the nature of the Vostok ice core site are in store. Vostok is situated in East Antarctica, about 320 km downstream of another site known as Dome B. Antarctic ice flows from Dome B toward Vostok. The flow model used to date Vostok ice depends heavily on understanding the style and rate of ice flow between the two sites, a flow that can create a complex ice stratigraphy. For instance, the Holocene at Dome B starts at 500 m ice depth; at Vostok the equivalent horizon is at 300 m. Thus the combination of ice flow and slow snow accumulation rates at Vostok leads to a large age uncertainty at Vostok, especially deeper in the core. The deepest interval (2546 m) studied by Jouzel et al. (1993) contained an age uncertainty of ±20 ka for ice dated at 220 ka near the penultimate interglacial period (marine isotope stage 7).

Despite these limitations, three facets of long-term climate change unveiled by the Vostok record stand out. First, Barnola et al. (1987, 1991) established that atmospheric CO_2 concentrations were about 190–200 ppmv during glacial periods corresponding to marine isotope stages 6 and 4–2. Reduced atmospheric CO_2 levels are a fundamental feature of late Quaternary glacial periods and, to a first approximation, correlate with periods of global cooling indicated by many other climate records. Temporal variability of CO_2 concentrations during glacial periods is relatively small (±10 ppmv).

Second, the Vostok record shows that CH_4 concentrations also varied over 100-ka glacial-interglacial cycles. Concentrations during the penultimate glacial period, marine isotope stage 6, were slightly lower (about 400 ppbv) than those of the last glacial period (Chappellaz et al. 1990). Furthermore, Chappellaz et al. (1990) showed that during both deglaciations and glacial inceptions, CH_4 rose and fell in phase with atmospheric temperature changes recorded in isotope records.

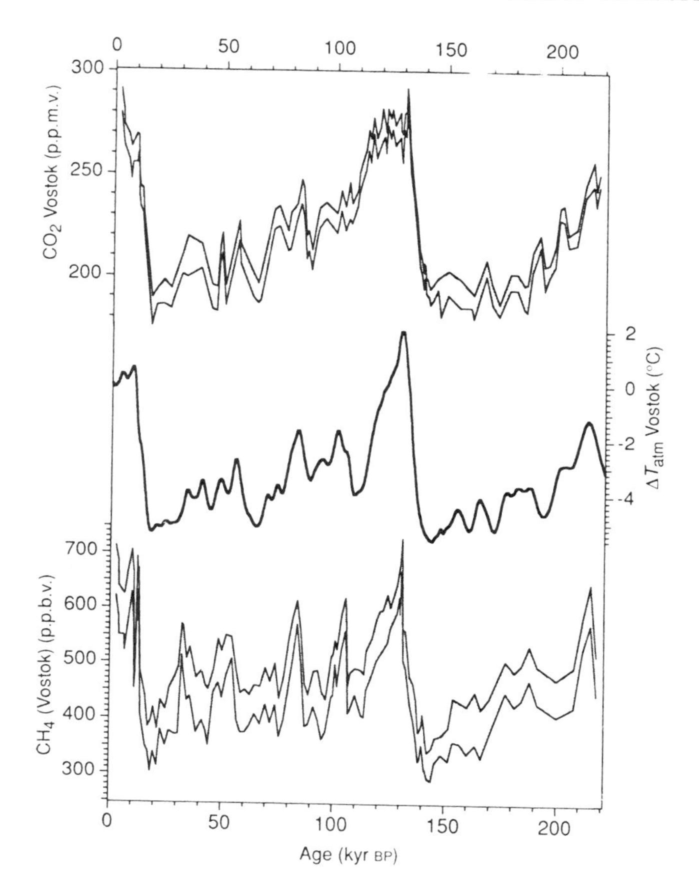

FIGURE 9-6 The record of CO_2, CH_4, and atmospheric temperature at Vostok for the past 240 ka covering the past two glacial-interglacial cycles based on the Vostok Antarctica ice-core record. Courtesy of J. Jouzel with permission from Macmillan Magazines. From Jouzel et al. (1993).

Third, Vostok shows that levels of CO_2 during previous interglacials were equal to or very slightly higher (300 ppmv) than those of the Holocene. Vostok CO_2 trends supported evidence from the Byrd core that concentrations tend to rise to these levels in phase (< 1000 yr) with deglacial temperature rise during glacial terminations. The rate of natural CO_2 increase during deglaciation was about 12 ppmv/1000 yr, compared with a rate of rise due to human activities of 80 ppmv/100 yr for the past century.

Barnola et al. (1987) showed that, in contrast to periods of global warming, during global cooling between about 130 and 110 ka, temperature and CO_2 share no correspondence (figure 9-6). As earth enters into a glacial period, CO_2 concentrations fall at least several thousand years *after* atmospheric temperatures fall. In a further investigation of the phase relationship between Vostok CO_2 and paleotemperature, Barnola et al. (1991) used a new firn densification model for the penultimate deglaciation (Termination 2, the isotope stage 6/5 transition) and the transition into the last glacial about 120–110 ka (isotope stage 5e–5d transition). They affirmed that during climate warming, CO_2 concentrations rise simultaneously or lag only slightly (by < 1000 yr) behind atmospheric temperature, but during climatic cooling, atmospheric temperature falls about 4.5 ka before CO_2 levels fall. Temperature also leads CO_2 entering the major glacial phase at about 80 ka. A new age model for Vostok (Sowers et al. 1993; Sowers and Bender 1995) also highlights the temperature-CO_2 asymmetry. These results suggest that CO_2 plays a different role during periods of climate warming versus cooling and that complex biogeochemical feedbacks are required to explain these patterns (Jouzel et al. 1989, 1993).

Orbital Theory and Ice Core Climate Records

Is the orbital theory of climate change supported by the ice core record, and if so, what roles if any do CO_2 and CH_4 play in concert with changes in solar insolation? As discussed in chapter 4, high-latitude forcing of climate is an earmark of the orbital model of Imbrie et al. (1992). Obliquity cycles occurring at a 41-ka period are dominant at high latitudes, whereas a precessional influence should manifest itself in low-latitude climate signals. We also saw that nonlinearity in the climate system may enhance the 100-ka cycle over the past 600 ka. Whereas the ice-core record is too short to examine eccentricity with any confidence, it does bear upon the obliquity and precessional cycles.

In their early study of the oxygen-isotope record from Camp Century, Greenland, Dansgaard et al. (1969) first uncovered the importance of ice cores for testing for orbital influence on climate. They noted that oxygen-isotope maxima occurred at about 6, 19, 32, and 45, as well as perhaps 59 and 74 ka, suggesting a 13-ka cycle that might be due to the dual influence of eccentricity and precessional cycles. The age model on the Camp Century core was, however, inadequate to carry these correlations any further.

The Vostok and GRIP and GISP2 ice cores have yielded detailed data of the frequency spectra of atmospheric temperature, CO_2, CH_4, and dust that support the orbital theory. Lorius et al. (1985) noted that a striking alternation of cold- and warm-interval temperature swings (up to 10°C) at Vostok for the past 150 ka matched minimum and maximum insolation values for 80°S latitude. Although they did not have enough samples to conduct spectral analyses, they visually identified 40-ka cycles that suggested obliquity forcing. They calculated that insolation changes of 7% during a full obliquity cycle would cause a 4°C temperature change, much lower than the estimated 10°C.

Jouzel et al. (1987), Barnola et al. (1987), and Genthon et al. (1987) further examined the long-term Vostok ice core record for evidence of orbital influence over the past 160 ka. Jouzel et al. (1987) measured the temperature record from deuterium and found support for a 40-ka frequency and a weaker 20-ka frequency. Within the dating uncertainty, these patterns suggested obliquity and precessional influence and supported orbital theory. Barnola et al. (1987) concluded that the Vostok record of CO_2 oscillated at a 21-ka frequency, with a weak 41-ka signal. Cross-spectral analyses comparing the deuterium and CO_2 records showed that during climate cooling, CO_2 lagged behind the deuterium temperature curve.

Using the deuterium-based temperature and CO_2 records from these two studies, Genthon et al. (1987) conducted a series of spectral analyses to examine how atmospheric temperatures in the south polar region responded to forcing from CO_2 and high-latitude Northern Hemisphere insolation. This study was important because it attempted to quantify the contribution of both CO_2 and insolation to the south polar climate. Genthon et al. (1987) found that the relatively weak forcing due to insolation changes were amplified by CO_2, an idea suggested by Pisias and Shackleton (1984) based on deep-sea core isotopic records. Although Genthon's group could not conclusively show interhemispheric phase relationships, they still found that the CO_2 amplification hypothesis best explains the apparent long-term synchroneity in Northern and Southern Hemisphere climate patterns, and they considered it a preferred explanation of the 100-ka cycles of the past 600 ka.

Petit et al. (1990) studied the evidence for orbital influence in the Vostok dust record covering the past 185 ka. Sampling the upper 2202 m of ice at 8-m spacing, Petit et al. found evidence that dust concentrations at Vostok varied at 21.6-ka and 18.5-ka frequencies—indicating that precessional insolation cycles very likely exerted influence on continental aridity and/or wind intensity. The Vostok dust record indicated that glacial periods in the Southern Hemisphere corresponding to marine isotope stages 2, 4, and 6 had high dust levels and those corresponding to marine stages 1, 5, and 7 had low dust contents. Ocean sedimentary dust records from the Indian Ocean (e.g., Clemens and Prell 1991) and the China loess record (Ding et al. 1994; Rutter et al. 1996) are two examples of corroborating evidence for greater glacial aridity related to precession.

Jouzel et al. (1993) later found a strong positive correlation between reconstructed atmospheric temperature and CO_2 and CH_4 for the past 220 ka at Vostok. They measured five parameters—dust, deuterium, CO_2, CH_4, and the oxygen-isotope ratio of O_2—down to a depth of 2546 m of ice. Spectral analyses using both the original time scale of Lorius et al. (1985) and a newer, slightly revised age model revealed a strong 41-ka peak. The 41-ka power was dampened during the uniformly cold interval between 140 and 200 ka (marine isotope stage 6). Chappellaz et al. (1990) also showed that the Vostok CH_4 record for the past 160 ka had a 21-ka precessional cycle related to changes in wetlands.

In a recent study of Vostok deuterium and ECM data covering four complete 100-ka glacial-interglacial cycles, Petit et al. (1997) produced a new Vostok paleoclimate record back to more than 400 ka (through marine isotope stage 11). This record bears a striking resemblance to the deep-sea oxygen-isotope curves and lends credence to the idea that the 100-ka eccentricity cycle that dominates mid- to late Quaternary climate records also exists in the longest ice-core record to date.

The Greenland ice-core record extends back only about 110 ka but still yields support for orbital influence on climate. Bender et al. (1994) noted that, in both Greenland and Vostok records, the warm interstadials corresponding to marine isotope substages 5a and 5c (Dansgaard-Oeschger interstadials 21 and 23) corresponded to maxima in summer insolation at 60°N. Several other extended interstadials corresponded to moderate insolation. Brook et al. (1996) also uncovered a precessional frequency for their 110-ka record of CH_4 from the GISP2 core. Low-frequency CH_4 cycles at GISP2 also generally match those at Vostok for the period 20–110 ka and are positively correlated with high-latitude June insolation values.

Brook et al. (1996) made the additional discovery that CH_4 varies over millennial time scales, apparently in phase with Dansgaard-Oeschger interstadial events. Orbital influence might also play an indirect role in these high-frequency oscillations. In general, Brook et al. (1996) support the hypothesis that low-latitude tropical wetlands controlled the long-term glacial atmospheric CH_4 concentrations, as suggested by Chappellaz et al. (1993). However, they also note that high-latitude CH_4 variability may help to explain the GISP2 interstadial events because CH_4 production might increase during interstadials in ice-free areas near ice-sheet margins. Brook et al. concluded that that short-term CH_4 patterns are probably not *directly* related to orbital-scale climate forcing but that insolation changes modulated high-frequency millennial-scale climate events.

In summary, the Vostok ice core records provide considerable empirical evidence that climate over the past 400 ka changed at frequencies that correspond to oscillations in the earth's orbital parameters of obliquity, precession, and eccentricity. Greenland ice cores indicate precessional influence on tropical atmospheric moisture and wetland CH_4 production over the past 110 ka and perhaps on high-latitude wetland activity as well.

MILLENNIAL-SCALE ATMOSPHERIC VARIABILITY AND CLIMATIC CHANGE

SOME OF THE earliest research on ice cores exposed the significance of interhemispheric symmetry and synchroneity of climate change. Epstein et al. (1970), for example, considered that the Byrd ice core record of the Wisconsinan glacial climates (75–10 ka) indicated that changes in Antarctic climate were synchronous with the much better known Northern Hemisphere continental climate record. The chronology available at that time was insufficient to be conclusive.

Rapid millennial-scale climatic changes have recently been studied in Greenland and Antarctica with special reference to the Younger Dryas and other climatic reversals during the last deglaciation (21–6 ka) and during Dansgaard-Oeschger interstadial events of the last glacial period (110–20 ka).

Younger Dryas in the Northern Hemisphere

As we discussed in chapter 5, the Younger Dryas event is the most intensely studied rapid climate reversal. In their classic study of the Camp Century ice core, Dansgaard et al. (1971) identified the classical European chronology including the Oldest Dryas, Bølling, Older Dryas, Allerød, Younger Dryas, and Preboreal climatic sequences and presented isotopic evidence that at the end of the Younger Dryas cold snap (12.5–11.5 ka), climate warming resumed over about a century.

The exceptional deglacial chronology available from GRIP and

GISP2 annual layers has added a vast new dimension to our understanding of this event. Alley et al. (1993) used several means to identify annual layers of ice in GISP2: visible summer melt layers, ECM oscillations due to dust and acids (see Taylor et al. 1993a), laser-light scattering of dust, and stable isotopic chemistry. Over selected intervals 1600–1800 m deep in the ice, these methods yielded an age uncertainty of < 1% near the Younger Dryas. The age of the Younger Dryas-Preboreal transition was determined to be 11,660 (±250 yr) at GISP2. At GRIP, it was 11,550 (±70 yr) (Taylor et al. 1993a; Alley et al. 1993; Sowers and Bender 1995).

The climate over Greenland indeed changed rapidly at the termination of the Younger Dryas. At GISP2, the end of the Younger Dryas is marked by changes in oxygen isotope ratios that signify a rapid climate warming of at least 4°C over an approximately 20- to 50-yr period (Severinghaus et al. 1998). Atmospheric dust concentrations fell even more rapidly, reaching near interglacial levels within just 20 yr at the Younger Dryas termination. A third indicator of this abrupt warming is the estimated doubling in ice-accumulation rate over only a few years (Alley et al. 1993).

Mayewski et al. (1993) affirmed that the Younger Dryas in Greenland began and ended rapidly, within only 10–20 yr, based on glaciochemical evidence from GISP2. Concentrations of calcium (Ca^{2+}) change in response to changes in the intensity of atmospheric circulation and/or in aridity in the calcium-source area. Calcium concentrations are extremely high during the cold, arid Younger Dryas stadial and much lower during warm interstadial climates. Mayewski and colleagues made several startling discoveries. First, decadal-scale fluxes in Ca^{2+} occurred *within* the 1000-yr-long Younger Dryas. Second, a massive flux of sulfate occurred within the Bølling-Allerød warm period near 13,713–13,531 yr, signifying volcanic sources or open, ice-free sea-surface conditions perhaps related to polynyas. Third, an increase in ammonium and nitrate near 12,859–12,786 yr may have been related to destruction of continental biomass after the Bølling-Allerød. Based on multivariate analyses of glaciochemical data, they concluded that the Younger Dryas stadial could not be explained solely as a shift in wind strength but rather it marked both a major shift in the size of the polar cell over the North Atlantic region (including at least three separate expansions of the polar cell during the Younger Dryas stadial) and changes in the source strength of chemical species.

Similarly, Chappellaz et al. (1993) found strong evidence that there was a sharp drop of more than 200 ppbv in CH_4 concentration during the Younger Dryas cold snap, whereas CO_2 levels may have only paused briefly in their deglacial rise to Holocene levels.

Another study of rapid late glacial climate change came from analyses of ice-accumulation rates in GISP2 (Meese et al. 1994). Accumulation rates at GISP2 increase rapidly but in a step-like manner. The rise is punctuated by at least four plateaus at 11,195, 10,650, 9950, and 9250 yr, reflecting brief, temporary cooling events. Meese and colleagues also identified two periods of exceedingly high snow accumulation: the early Holocene altithermal period (10–7 ka) and the interval 620–1150 A.D., which overlaps with the Medieval Warm Period.

Whereas many studies have focused exclusively on the Younger Dryas event, Stuiver et al. (1995) developed an unparalleled continuous bidecadal oxygen-isotope record for the last 16.5 ka at GISP2. Their study confirmed the classical climatostratigraphic sequence of Oldest Dryas, Bølling, Older Dryas, Allerød, Younger Dryas, and Holocene at GISP2. They also concluded that the deglacial climate transition was not merely a shift in temperature; rather it consisted of more complex atmospheric changes, supporting the ideas of Dansgaard et al. (1989) and Alley et al. (1993). Stuiver et al. (1995) showed that a net 5‰ shift in oxygen isotopes occurred during the Younger Dryas–Preboreal warming in Greenland ice and that the rate of isotopic change was about 0.25‰/yr. This rate was much faster than that during the Allerød–Younger Dryas cooling. They postulated that two distinct opposing climate modes characterized the North Atlantic region. One was a two-branched system characteristic of cold climatic intervals in which the high-level jet stream has northerly and southerly branches. The other is more typical of current conditions, in which there is a single jet stream. Stuiver's group offered compelling isotopic evidence from the deglacial ice-core record that the two-branched system ended nearly instantaneously at 14,670 yr and again at 11,650 yr.

The Magnitude of Rapid Climate Change

Having discussed multiproxy evidence for the rate of Allerød–Younger Dryas cooling and Younger Dryas–Preboreal warming, we should consider in more detail the magnitude of atmospheric cooling and warming. Dansgaard et al. (1989) found the Younger Dryas–Preboreal warming was 7°C at the Dye-3 Greenland core. They also suggested on the basis of the deuterium excess record that the retreat of sea ice during this transition may have led to cooler oceanic temperatures in the water-vapor source area in the adjacent Nordic Seas. Large-scale temperature oscillations during the Younger Dryas event are also in evidence from the GRIP (Johnsen et al. 1992) and GISP2 (Alley et al. 1993) ice cores. For example, Johnsen et al. (1992) suggested that a 10°C drop in temperature occurred during the Younger Dryas at GRIP. Mayewski et al. (1996) postulated that at the onset of the Younger Dryas at GISP2, the chloride content, which has a marine source and is a good indicator of glacial conditions, suddenly rose to 75% of its glacial value in only about 20 years.

Severinghaus et al. (1998) used a new tool, isotopes of nitrogen (^{15}N:^{14}N ratios) trapped in polar ice bubbles with oxygen isotopes to revise the scale of glacial and Younger Dryas atmospheric cooling over Greenland. The ratio of heavy-to-light isotopes (either nitrogen or argon) trapped in ice will vary when surface temperatures change abruptly (see Sowers et al. 1992). ^{15}N:^{14}N and ^{40}Ar:^{36}Ar ratios also vary in polar ice when a large thermal gradient in the firn causes the heavier isotopes to migrate down the firn while the lighter ones generally move up the firn column. A spike in the nitrogen-isotope ratio can therefore supplement the oxygen-isotope record when one is determining the rapidity and scale of climate change. Applying this reasoning to the Younger Dryas event in Greenland, Severinghaus et al. (1998) found a distinct nitrogen isotope spike at the Younger Dryas event. By measuring the thickness of ice separating nitrogen and oxygen isotopic spikes, they estimated the rate of temperature change at the Younger Dryas termination was abrupt (less than 50–100 yr). Moreover, the nitrogen- and argon-isotope method suggested that the Younger Dryas was 14°C (± 3°C) colder than the present, which agrees with the value based on the borehole temperature reconstruction of Cuffey et al. (1995). Thus the total warming at the end of the Younger Dryas and the early part of the Preboreal was about 14°C.

Borehole temperature measurements from Greenland also led Cuffey et al. (1995) to suggest that the last glacial cooling was as much as 20°C (Cuffey et al. 1995). A 20°C temperature drop is also double previous estimates of about 10°C. The borehole method does not allow resolution of brief events such as the Younger Dryas, but if correct, it supports the more general idea that atmospheric temperatures above Greenland are subject to extremely large-amplitude swings.

In summary, the climate of the North Atlantic region near Greenland was extremely unstable during deglaciation, reflecting changes in atmospheric polar cell size, sea-ice conditions in Nordic Seas, temperature, moisture sources, and other properties.

Millennial Climate Change in Antarctica

Finding the Younger Dryas equivalent in the Southern Hemisphere has been a thorn in the side of paleoclimatologists for years (Rind et al. 1986; chapter 5). Recently, however, a cooling event punctuating the last deglaciation has been identified in Antarctic ice cores. It has been named the Antarctic cold reversal (ACR) (Jouzel et al. 1992, 1995).

Jouzel et al. (1995) examined several ice-core records of the deglacial Antarctic record. They compared the climate record from the Dome B ice core with other cores from Dome C, Vostok, and Byrd. The Dome B core has a high accumulation rate (2.8 mm/yr yielding a 3-yr temporal sampling resolution), making it suitable for examination of rapid climatic events. Jouzel argued that asynchroneity between Northern and Southern Hemisphere high-latitude climate records characterized the last deglaciation, Termination 1. Using the synchronous peaks in atmospheric dust content that demarcate the LGM and the postglacial decline in dust concentrations to Holocene levels (14.6 ka) as trans-Antarctic time markers, they postulated that the ACR, which was defined at Dome by deuterium changes, predated the Younger Dryas by about 1200 years.

Although the ACR represents a well-defined climatic reversal during Termination 1, three important differences distinguish it from the Northern Hemisphere's Younger Dryas. First, the Younger Dryas–age cooling was three times as great in Greenland (≥ 10°C) as in the Antarctic (only 3°C near the surface) as measured by deuterium. Second, whereas Younger Dryas dust in Greenland returned to near-glacial concentrations, in Antarctica dust did not increase during the ACR. Third, and most telling, the peak cooling during the ACR occurred between 13.5 and 12.5 ka, about 1000 yr before the well-dated cooling in Greenland. This led to the conjecture of interhemispheric climatic asymmetry. Jouzel et al. (1995) also argue that a pre–Younger Dryas age for the ACR is consistent with evidence that the initial large-scale deglacial warming began at 17 ka in Antarctica but not did not begin until 14.7 ka in Greenland (Sowers and Bender 1995). Jouzel et al. (1995) thus believe that warming in the Southern Hemisphere leads that in the Northern Hemisphere during the last deglaciation.

The 13.5-ka ACR cooling event, however, remains an enigma if one accepts the chronology of deglacial events from some other Southern Hemisphere paleoclimate records. For example, a 13.5-ka Southern Hemisphere cooling does not match the glacial record from New Zealand, where an 11.5-ka-old glacial advance is documented (Denton and Hendy 1994), nor does it match the Chilean glacial and palynological record (Lowell et al. 1995). In a study of the deglacial transition of dust (calcium) and sea-salt (chloride) records from the Taylor Dome, Antarctica, Mayewski et al. plotted the ACR as correlative with the Younger Dryas episode, in contrast to the ideas of Jouzel et al. (1995). They found high chloride concentrations at Taylor Dome during several climate reversals that interrupted deglaciation, and they tentatively correlated these with the Northern Hemisphere's Oldest Dryas, the intra-Allerød oscillation, as well as the proposed Younger Dryas–ACR correlation (figure 9-7). However, synchroneity between the Taylor events and Northern Hemisphere events has yet to be demonstrated.

Rapid Climate Change in Low Latitudes

Using multiple paleoclimate proxy tools (chloride, nitrates, sulfates, pH, and oxygen isotopes) to measure atmospheric dust and temperature, Thompson and colleagues (1989) examined the 40-ka record of ice in the Dunde ice cap. Dunde is a 60-km^2 area located high in the Qinghai-Tibetan Plateau in China, just south of the Gobi Desert. The upper 70 m of Dunde ice allow an annual resolution; the lower 70 m extend the record back to the last glacial stage, about 40 ka.

Thompson et al. found strong evidence that major shifts in dust and temperature occurred during the last glacial transition into the Holocene. Specifically they showed that the glacial period had higher precipitation rates and stronger winds. Precipitation decreased as glacial lakes dried up as the Holocene began. This low-latitude region also experienced modest summer cooling during the last glacial period.

Younger Dryas and Antarctic Cold Reversal Events: Interhemispheric Asymmetry?

It is useful to summarize the Northern and Southern Hemispheric differences during the Younger Dryas–ACR sequence:

1. Younger Dryas atmospheric cooling was as much as 10°C over Greenland, perhaps much more, but much less in Antarctica—3°C near the surface and 2°C at the inversion layer in some parts of Antarctica.
2. Dust concentrations fell quickly during deglaciation in Greenland, but much more slowly in Antarctica.
3. During the ACR, calcium (dust) levels remained low near the Bølling-Allerød and Holocene values in Antarctica, while in

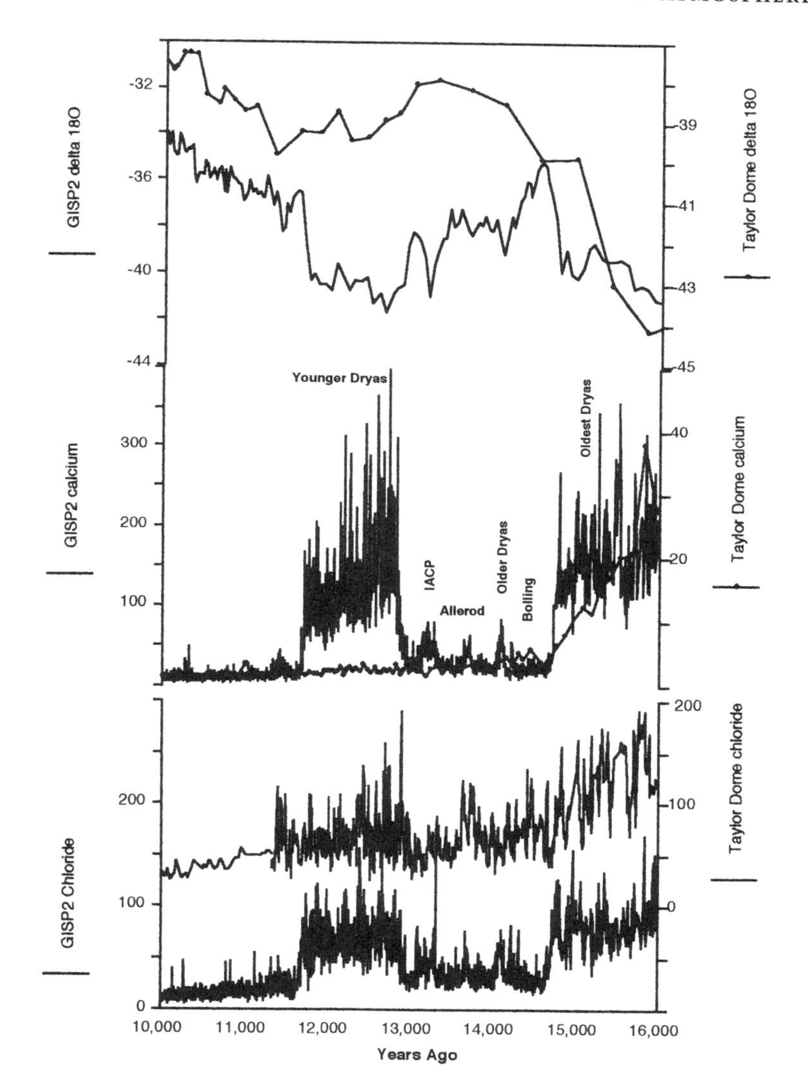

FIGURE 9-7 Comparison of rapid climate changes during deglaciation 16–10 ka, as measured from Greenland (GISP2) and Antarctica (Taylor Dome) ice cores based on oxygen isotopes of the ice, calcium, and chloride glaciogenic proxy indicators of atmospheric conditions. ACR is the Antarctic Cold Reversal that may actually precede the Northern Hemisphere Younger Dryas cooling by 1000 years. Courtesy of P. Mayewski with permission from the American Association for the Advancement of Science. From Mayewski et al. (1996).

Greenland Younger Dryas calcium rose substantially to near glacial levels.

4. Chloride dust, indicating marine sources, rose to 75% of glacial levels during the Younger Dryas in Greenland but to 54% of glacial levels during the ACR at Taylor Dome, Antarctica.
5. In the Southern Hemisphere, climatic warming both during the initial deglaciation and during the Younger Dryas–Preboreal transition proceeded more gradually than in the Northern Hemisphere during the same events.
6. The phasing of hemispheric deglaciation and climate reversals is not clear; maximum cooling during the ACR may have preceded Younger Dryas cooling in the Northern Hemisphere by as much as 1000 years, and the earliest deglacial warming may have begun in the Southern Hemisphere about 17 ka—a full 2500–3000 yr before warming began at 14.5 ka in Greenland, northern Europe, and the North Atlantic.

The interhemispheric synchroneity, or lack thereof, of the Younger Dryas stadial is a critical issue to resolve if paleoclimatologists wish to understand the cause of this event. Since the discovery of the ACR climate reversal in the deglacial ice-core record of Antarctica, investigators have proposed three possible relationships with the much more firmly dated North Atlantic, European, and Greenland Younger Dryas. The ACR could correspond to the Younger Dryas (asynchronous cooling), to the Bølling-Allerød interstadial (interhemispheric asymmetry, with the Southern Hemisphere cooling leading that in the Northern Hemisphere), or to another short-lived Northern Hemisphere climate oscillation (such as the Killarney oscillation [Levesque et al. 1993, 1994]) or a deep-sea isotopic event (Keigwin et al. 1991). If interhemispheric symmetry holds, then atmospheric processes likely influence global climatic reversals. If asymmetry prevails, deep-oceanic transport is likely involved, causing the 1000-yr lag time. The resolution of this issue will surely come from integrated ice-core, deep-sea, and continental paleoclimatological reconstructions.

Abrupt Holocene Climatic Oscillations

The Younger Dryas and ACR event(s) is (are) not the only rapid climatic reversal that punctuated the past 15 ka. Near the end of the deglacial episode, multiple indicators document another abrupt climate cooling over the North Atlantic region dated precisely at 8.2 ka (Alley et al. 1997). This event can be identified in GISP2 calcium and chloride records, ice accumulation rates, fire activity, atmospheric temperatures, and CH_4 concentrations (Meese et al. 1994; Taylor et al. 1997; Alley et al. 1993; Brook et al. 1996). Blunier et al. (1995) also discovered CH_4 and oxygen-isotope excursions in the GRIP core associated with climate cooling at about 8.2 ka. This sharp early Holocene event coincided with continental records for extensive low-latitude aridity in Africa and Tibet and thus seemed to be synchronous, at least across the Northern Hemisphere.

O'Brien et al. (1995) showed that the 8.2-ka event was one of several cooling events during the late glacial–Holocene dated at > 11.3, 5.0–6.1, and 2.4–3.1 ka that are present in glaciochemical signatures at GISP2. They reflect a quasiperiodic 2500- to 2600-yr pattern of cooling over Greenland within the Holocene interglacial period (chapters 5, 6). These Holocene events signify more than just regional events over the North Atlantic–Greenland area (Blunier et al. 1995; O'Brien et al. 1995): They correlate with worldwide glacial advances mapped in early studies by Denton and Karlen (1973, 1976). Based on Holocene atmospheric conditions implied by the glaciochemical data, O'Brien's group assert these periodic cold intervals are reminiscent of Maunder-Spoerer–type solar oscillations of the past few centuries, except they recur at a much lower frequency (chapter 6). Their causes, however, will not be known with certainty until records are obtained from other regions.

Dansgaard-Oeschger Interstadials in Ice Cores

Dansgaard-Oeschger interstadial climate events of the glacial interval between 110 and 20 ka were introduced in chapter 5 as another type of rapid climate event. Chappellaz et al. (1993) and Brook et al. (1996) discovered millennial scale Dansgaard-Oeschger oscillations in atmospheric CH_4 concentrations during the last glacial period in the GRIP and GISP2 ice cores in Greenland, respectively. The Greenland cores have the distinct advantage over the Vostok core in that their ice-air age difference is relatively small; consequently, millennial-scale gas variability can be observed. Chappellaz et al. (1993) noted the ice-air age difference at GRIP of only 210 yr near the surface and 750 yr at the LGM, compared with about 2500 yr at Vostok.

Brook et al. (1996) reached four main conclusions in their study. First, they confirmed the data from the GRIP core (Chappellaz et al. 1993) that CH_4 changes on millennial time scales. Second, they argued that these abrupt shifts correspond to Dansgaard-Oeschger climate events, which, as we saw in chapter 5, are very likely global

ʾlimate changes. Third, they determined that CH_4 varied in phase ith Northern Hemispheric summer insolation, a clear link with orbital influence mentioned above. Fourth, they postulated that the tropical terrestrial biosphere played a major role in causing rapid changes in atmospheric CH_4 during the late Quaternary.

Are Dansgaard-Oeschger events interstadial warmings global in scale or are they confined to the North Atlantic region? Using the deuterium record of Vostok and the calcium and $\delta^{18}O_{ice}$ from GISP2 as climate proxies, and the $\delta^{18}O_{air}$ for correlation, Bender et al. (1994) compared the strength and timing of Vostok and GISP2 interstadial events between 105 and 20 ka. They found that the Antarctic record showed evidence of Dansgaard-Oeschger events, but there were significant differences between the two records. For example, in Greenland Dansgaard-Oeschger interstadials occur rapidly; in Antarctica, interstadial warming and return to glacial conditions are both relatively gradual events. Greenland events are also more numerous than those in the equivalent interval of the Vostok record. The GISP2 record had 22 interstadial events (including interstadials 21 and 23 corresponding to marine isotope stages 5a and 5c, about 85 ka and 105 ka, respectively), but only those Greenland interstadial events lasting more than about 2000 years were found in Antarctica. This fact suggested to Bender et al. (1994) that interstadial events of longer duration begin in the Northern Hemisphere and spread to the Southern Hemisphere, whereas shorter events do not affect the southern regions.

What caused high-amplitude Dansgaard-Oeschger events? Bender's group estimates that greenhouse gas changes in CH_4 and CO_2 were insufficient in magnitude to cause more than about 20% (1°C) of the total 3°C warming at Vostok during each Dansgaard-Oeschger event. Moreover, orbital solar insolation changes might have been factors only in the higher-amplitude events (e.g., near 85- and 105-ka insolation maxima). The covariance between the paleoceanographic record from the Santa Barbara Basin off California (Behl and Kennett 1996) and Greenland Dansgaard-Oeschger events suggests atmospheric processes are involved. To account for interhemispheric symmetry and amplitude of Dansgaard-Oeschger events, oceanic circulation changes and partial deglaciation probably in the Northern Hemisphere and atmospheric processes are considered possible causes.

The sum of ice-core evidence indicates that millennial-scale events are a feature of both polar regions during both interglacial and glacial intervals. Ice-sheet and ocean-circulation changes perhaps originating in the North Atlantic regions may be important for transferring interstadial climate changes to low latitudes and high southern latitudes.

The precise phasing of these events remains clouded because the chronology for the Vostok core is not as precise as that for Greenland cores.

THE "FLICKERING SWITCH" OF CLIMATE

HOW FAST CAN a major climate change take place? Ice cores are excellent archives to document rapid, large-amplitude atmospheric changes occurring over a century or less. Several ice cores, notably GISP2 and GRIP, have both annual (or subannual) chronology and a suite of proxy indicators sensitive to rapid atmospheric change. Evidence from dust, isotopes, glaciochemical data, and ECMs from several ice cores support extraordinarily abrupt climate shifts when the earth is experiencing glacial, interglacial, and transitional climatic states (figure 9-7).

Taylor et al. (1993a) narrowed their search for rapid climate change to an exceptionally refined temporal scale. Using electrical conductivity (ECM) combined with oxygen isotopes and glaciochemical signals at GISP2, they unearthed a hierarchy of climate changes at subcentury time scales between 42 and 10 ka, including rapid atmospheric shifts *within* the Younger Dryas cold snap. ECM gauge the ability of the ice to conduct an electrical current, itself a function of the amount of acidic and basic material in the ice. The more acidic the ice, the stronger is the electrical current. The more volcanic activity (producing sulfates) and/or nitrates (produced by changes in atmospheric circulation), the more acidic is the ice. Ice acidity can, however, be neutralized by airborne alkaline chemicals such as ammonium and dust blown off continents. These atmospheric constituents, preserved in polar ice in the ECM record, provide a highly sensitive barometer of the atmosphere over Greenland.

ECM records, supported by the oxygen-isotope and glaciochemical data, preserve millennial-scale stadial and interstadial events between 42 and 10 ka, including the Bølling–Allerød–Younger Dryas–Preboreal deglacial sequence. High ECM values characterize the interstadials and low values the Younger Dryas stadial. A second temporal level of climate instability was also discovered within the Bølling-Allerød period of warming, in which century-scale ECM oscillations occurred.

A third category of climate change discovered by Taylor et al. (1993a), which they termed the "flickering switch," were short-lived (10- to 20-yr duration), abruptly terminating (in < 5 yr) reversals in ECM during times when regional climate was in transition between cold and warm states. Even more remarkable, they noted that the ECM fluctuations differ in the level of "spikiness" during three distinct types of warm climatic regimes: Dansgaard-Oeschger interstadials (e.g., about 38–39 ka), the Bølling-Allerød, and the Holocene interglacial. The Holocene is the most stable of the three, followed by the Bølling-Allerød and lastly by the Dansgaard-Oeschger event.

Another startling discovery was that ECM flickerings do not support the notion gleaned from isotopic evidence of a relatively smooth climatic transition between the Younger Dryas–Preboreal stadial–interstadial transition. The Younger Dryas termination is actually characterized by an oscillatory transition between two climate states before settling into the warm, relatively stable Preboreal and Holocene interglacial periods. Other late glacial climate transitions were equally unstable. Taylor and colleagues attributed the oscillating ECM signal to changing alkaline dust content due to "a reordering of atmospheric circulation [which] could rapidly change the surface moisture, wind speeds and air-mass routing and, hence, change the quantity of airborne loess" (1993:435). They argue that the climate system controlling ECM dust content fluctuates at a high amplitude before settling into a relatively stable state. Changing temperature gradients that would influence atmospheric circulation patterns were a favored hypothesis to account for these rapid climate shifts over Greenland.

The rapid shifts in Greenland ice core proxy indicators occurring over a century or less are manifestations of local or regional shifts in atmospheric conditions. Chronologies are not sufficiently refined to correlate centennial, decadal, or annual climate events in the North Atlantic region to events occurring elsewhere. Nonetheless, ice cores provide compelling evidence that regional climate flickers during stadials, interstadials, and the transitions between them over a century or less. These strikingly abrupt climate oscillations represent a previously unrecognized level of sensitivity of climate to shift from one mode to another within a human lifetime.

MECHANISMS OF CLIMATE CHANGE

One ultimate goal of paleoclimatology is to understand the causes of climate change. The bonanza of new information about paleo-atmospheric conditions discovered from ice cores is still being digested with respect to the causes of climate change. Nonetheless, it is useful to distill the salient features of the ice-core records as they pertain to current thinking on climate mechanisms.

As discussed in prior chapters, climate change can result from many forcing factors; each type should in theory have a distinct "fingerprint" that might be deciphered from the geographic patterns and the timing of observed climatic events. For example, orbital changes that affect insolation in the Northern Hemisphere require feedback mechanisms to produce a global climate change. Rind and Chandler (1991) and Crowley (1991) argue on the basis of general circulation climate model simulations that climatic warming due to oceanic circulation changes enhancing global heat export from low to high latitudes would be manifested mainly in high latitudes but not near the tropics. Conversely, other factors being equal, global warmth due to elevated greenhouse gases should show a more equitable warming over all latitudes. Climate changes triggered by regional events such as meltwater-driven changes in high-latitude deep-water formation (thermohaline circulation changes) might take 500–1000 years to propagate through the world's oceans. Atmospheric forcing by CO_2, CH_4, and/or water vapor might have a global signal and result in rapid climate response. Climate changes occurring over centuries or less probably reflect atmospheric processes because of the rapid response time of the atmosphere compared with that of oceans or ice sheets. Thus, both the temporal and spatial scales of past climate events are critical factors in establishing causality between pattern and process of climate change.

Many of these causal mechanisms have already been mentioned in the previous pages. Five of the most important discoveries from ice cores bearing upon the causes of climate change can be encapsulated as follows. First, CO_2 very likely plays a complex role in amplifying orbitally induced climate changes at high latitudes over the past 400 ka, as indicated by the in-phase (< 1000-yr lag) relationship between Vostok temperature and CO_2 concentrations. Lorius et al. (1990) attributed about one half of the total Quaternary glacial-interglacial climate signal (notably the Vostok temperature record) to Northern Hemisphere ice-sheet decay and one half to greenhouse gas forcing. Moreover, with respect to sea-level change due to ice volume, CO_2 leads ice volume changes by 4–7 ka and must influence ice sheets through complex oceanic feedbacks and linkages (Bender et al. 1994; Keigwin and Jones 1994). The several-thousand-year lag of CO_2 behind temperature during global cooling after interglacial warmth suggests a delayed oceanic and ice-volume response to orbital changes (or other factors) as carbon was sequestered from the atmosphere. A CO_2 increase of 25 ppmv might also exert a small impact during interstadial events 16 and 17, between 55 and 65 ka (Bender et al. 1994), translating into a global temperature change of about 0.6°C (Lorius et al. 1990).

Second, low latitude precessional influence on climate is apparent in the CH_4 record of Greenland and Antarctic ice cores, in which atmospheric and hydrological changes over the past 110 ka seem to mediate wetland CH_4 production (Chappellaz et al. 1993; Blunier et al. 1995; Brook et al. 1996). However, high-frequency (millennial-scale) Holocene CH_4 variability of 15% above background levels reflects unknown climate forcing related to tropical (early Holocene) and high-latitude (mid to late Holocene) CH_4 production (Blunier et al. 1995).

Third, the near interhemispheric synchroneity of millennial-scale Dansgaard-Oeschger events between about 110 and 50 ka (and perhaps up to 20 ka) suggest that oceanic thermohaline circulation and ice-sheet behavior are important mechanisms to transfer climate change from one hemisphere to another, although atmospheric processes may be important. During some interstadials, ocean circulation might act in concert with sea-level and ice-volume changes and/or be sometimes triggered by internal ice-sheet surges (Bender et al. 1994).

Fourth, during the last deglaciation, one school of thought holds that polar synchroneity of climate changes is evidence that atmospheric processes are causing climate to change over millennial time scales because synchronous global climate changes are too rapid to be explained by oceanic processes. Another school holds that interhemispheric asynchroneity, with the Southern Hemisphere leading the Northern Hemisphere both during the initiation of deglaciation (about 18 ka) and during climate reversals like the ACR and Younger Dryas, is more consistent with oceanic mechanisms and 1000-yr leads and lags (Jouzel et al. 1995). North Atlantic Ocean sensitivity to relatively minor perturbations to sea-surface temperature and salinity caused by ice sheet discharges suggests, however, that the initial triggering mechanism for some brief regional climate reversals (i.e., at 8.2 ka) might originate in the Northern Hemisphere (Alley et al. 1997).

Fifth, extremely rapid and widespread events such as those recorded during the deglacial interval in the polar ice cores (e.g., Alley et al. 1993) suggests that a critical threshold was reached in the climate system, triggering atmospheric temperatures to plunge, snow accumulation rate to drop, and atmospheric circulation to shift. Indeed, the Younger Dryas event ended in Greenland so abruptly—in just a few years—that some researchers believe (e.g., Alley et al. 1993; Mayewski et al. 1996) that oceanic circulation cannot solely account for it and that thresholds in atmospheric circulation were reached, causing an exceedingly sharp climatic reversal.

In closing, compelling evidence from ice cores about the existence of abrupt natural climate change occurring within a human lifetime and the critical role of trace gases in Quaternary climate must rank high among this century's most important discoveries of the earth's global environment. They reveal a heretofore unrealized dimension to humans' perception about their natural environment. The integration of paleo-atmospheric trends with climate histories from oceans and continents will be fertile research ground over the next few decades.

References

Aagaard, K. and E. C. Carmack. 1989. The role of sea ice and other fresh waters in the Arctic circulation. *Journal of Geophysical Research* 94:14,485–14,989.

Aagaard, K., J. H. Swift, and E. C. Carmack. 1985. Thermohaline circulation in the Arctic Mediterranean seas. *Journal of Geophysical Research* 90(C3): 4833–4846.

Agassiz, L. 1840. *Etudes sur les Glaciers.* Neuchatel.

Alibert, C. and M. T. McCulloch. 1997. Strontium/calcium ratios in modern *Porites* corals from the Great Barrier Reef as a proxy for sea surface temperatures: Calibration of the thermometer and monitoring of ENSO. *Paleoceanography* 12:345–363.

Allen, T. F. H. and T. W. Hoekstra. 1992. *Toward a Unified Ecology.* New York: Columbia University Press.

Alley, R. B. 1995. Resolved: the Arctic controls global climate change. In *Arctic Oceanography: Marginal Ice Zones and Continental Shelves, Coastal and Estuarine Studies,* 49:263–283. Washington, D.C.: American Geophysical Union.

Alley, R. B. and S. Anandakrishnan. 1995. Variations in melt-layer frequency in the GISP2 ice core: implications for Holocene summer temperatures in central Greenland. *Annals of Glaciology* 21:64–70.

Alley, R. B. and D. R. MacAyeal. 1994. Ice-rafted debris associated with binge/purge oscillations of the Laurentide ice sheet. *Paleoceanography* 9:503–511.

References

Alley, R. B., P. A. Mayewski, T. Sowers, M. Stuiver, K. C. Taylor, and P. U. Clark. 1997. Holocene climatic instability: A prominent widespread event at 8200 yr ago. *Geology* 25:483–486.

Alley, R. B., D. A. Meese, C. A. Shuman, et al. 1993. Abrupt increase in Greenland snow accumulation at the end of the Younger Dryas event. *Nature* 362:527–529.

Altenbach, A. V. and M. Sarnthein. 1989. Productivity record in benthic foraminifers. In W. H. Berger, V. S. Smetacek, and G. V. Wefer, eds., *Productivity of the Ocean: Present and Past,* Dahlem Workshop Report, pp. 255–270. New York: John Wiley & Sons.

Ammann, B. and A. F. Lotter. 1989. Late-glacial radiocarbon and palynostratigraphy of the Swiss Plateau. *Boreas* 18:109–126.

Anderson, R. S. and S. J. Smith. 1994. Paleoclimatic interpretations of meadow sediment and pollen stratigraphies from California. *Geology* 22:723–726.

Anderson, R. Y. 1992. Possible connection between surface winds, solar activity and the Earth's magnetic field. *Nature* 358:51–53.

Anderson, R. Y. 1993. The varve chronometer in Elk Lake: Record of climatic variability and evidence of solar–geomagnetic–^{14}C-climate connection. In J. P. Bradbury and W. E. Dean, eds., *Elk Lake, Minnesota: Evidence for Rapid Climate Change in the North-Central United States,* Geological Society of America Special Paper 276, pp. 45–67. Boulder: Geological Society of America.

Anderson, R. Y., A. Soutar, and T. C. Johnson. 1992. Long-term changes in El Niño/Southern Oscillation: Evidence from marine and lacustrine sediments. In H. F. Diaz and V. Markgraf, eds., *El Niño: Historical and Paleoclimatic Aspects of the Southern Oscillation,* pp. 419–433. Cambridge: Cambridge University Press.

Andrews, J. T., A. E. Jennings, M. Kerwin, et al. 1995a. A Heinrich-like event, H-0 (CD-0): Source(s) for detrital carbonate in the North Atlantic during the Younger Dryas chronozone. *Paleoceanography* 10:943–952.

Andrews, J. T., G. H. Miller, J. S. Vincent, and W. W. Shilts. 1986. Quaternary correlations in eastern Canada. *Quaternary Science Reviews* 5:243–249.

Andrews, J. T., A. S. Dyke, K. Tedesco, and J. W. White. 1993a. Meltwater along the Arctic margin of the Laurentide Ice Sheet (8–12 ka): Stable isotopic evidence and implications for past salinity anomalies. *Geology* 21: 881–884.

Andrews, J. T., H. Erlenkeuser, K. Tedesco, A. Aksu, and A. J. T. Jull. 1994. Late Quaternary (stage 2 and 3) meltwater and Heinrich events. *Quaternary Research* 41:26–34.

Andrews, J. T., B. Maclean, M. Kerwin, W. Manley, A. E. Jennings, and F. Hall. 1995b. Final stages in the collapse of the Laurentide Ice Sheet, Hudson Strait, Canada, NWT, ^{14}C AMS dates, seismic stratigraphy, and magnetic susceptibility logs. *Quaternary Science Reviews* 14:983–1004.

Andrews, J. T. and K. Tedesco. 1992. Detrital carbonate-rich sediments, northwest Labrador Sea: Implications for ice-sheet dynamics and iceberg rafting Heinrich events in the North Atlantic. *Geology* 20:1987–1990.

Andrews, J. T., K. Tedesco, and A. Jennings. 1993b. Heinrich events: Chronology and processes, east-central Laurentide Ice Sheet and NW Labrador Sea. In W. Peltier, ed., *Ice in the Climate System,* NATO ASI Series 1, 12:167–186. Berlin: Springer-Verlag.

Annals of Glaciology. 1994. Vol. 21.

Arrhenius, G. 1952. Sediment cores from the east Pacific. Properties of the sediment. *Report of the Swedish Deep Sea Expedition, 1947–1948.* 1:1–228. Goteborg: Elander.

Arrhenius, S. 1896. On the influence of carbonic acid in the air upon the temperature of the ground. *Philosophical Magazine* 41:237–245.

Ashworth, A. C. and J. W. Hoganson. 1993. The magnitude and rapidity of the climate change marking the end of the Pleistocene in the mid-latitudes of South America. *Palaeogeography, Palaeoclimatology, Palaeoecology* 101:263–270.

Atkinson, T. C., K. R. Briffa, and G. R. Coope. 1987. Seasonal temperatures in Britain during the past 22,000 years, reconstructed using beetle remains. *Nature* 325:587–592.

Bales, R. C. and E. W. Wolff. 1995. Interpreting natural signals in ice cores. *EOS, Transactions, American Geophysical Union* 76(47):477.

Bard, E., M. Arnold, R. G. Fairbanks, and B. Hamelin. 1993. $^{230}Th/^{234}U$ and ^{14}C ages obtained by mass spectrometry on corals. *Radiocarbon* 35:191–195.

Bard, E., R. G. Fairbanks, M. Arnold, and B. Hamelin. 1992. $^{230}Th/^{234}U$ and ^{14}C ages obtained by mass spectrometry on corals from Barbados (West Indies), Isabel (Galapagos) and Mururoa (French Polynesia). In E. Bard and W. S. Broecker, eds., *The Last Deglaciation: Absolute and Radiocarbon Chronologies,* pp. 103–110. Berlin: Springer-Verlag.

Bard, E., B. Hamelin, and R. G. Fairbanks. 1990a. U/Th ages obtained by mass spectrometry in corals from Barbados: Sea level during the past 130,000. *Nature* 346:456–458.

Bard, E., B. Hamelin, R. G. Fairbanks, and A. Zindler. 1990b. Calibration of the ^{14}C timescale over the past 30,000 years using mass spectrometric U-Th ages from Barbados corals. *Nature* 345:405–410.

Bard, E., B. Hamelin, M. Arnold, et al. 1996. Deglacial sea-level record from Tahiti corals and the timing of global meltwater discharge. *Nature* 382:241–244.

Bard, E., F. Rostek, and C. Sonzogni. 1997. Interhemispheric synchrony of the last deglaciation inferred from alkenone palaeothermometry. *Nature* 385:707–710.

Barnes, D. J. and J. M. Lough. 1992. Systematic variations in depth of skeleton occupied by coral tissue in massive colonies of *Porites* from the Great Barrier Reef. *Journal of Experimental Marine Biology and Ecology* 159:113–128.

Barnes, D. J. and J. M. Lough. 1993. On the nature and causes of density banding in massive coral skeleton. *Journal of Experimental Marine Biology and Ecology* 167:91–108.

Barnola, J. M., M. Anklin, J. Porcheron, D. Raynaud, J. Schwander, and B. Stauffer.

1995. CO_2 evolution during the last millennium as recorded by Antarctic and Greenland ice. *Tellus* 47B:264–272.

Barnola, J.-M., P. Pimienta, D. Raynaud, and Y. S. Korotkevich. 1991. CO_2-climate relationship as deduced from the Vostok ice core: A re-examination based on new measurements and on a re-evaluation of the air dating. *Tellus* 43B:83–90.

Barnola, J. M., D. Raynaud, Y. S. Korotkevich, and C. Lorius. 1987. Vostok ice core provides 160,000-year record of atmospheric CO_2. *Nature* 329: 408–414.

Barron, E. J., P. J. Fawcett, W. H. Peterson, D. Pollard, and S. L. Thompson. 1995. A "simulation" of mid-Cretaceous climate. *Paleoceanography* 10:953–962.

Barron, E. J., P. J. Fawcett, D. Pollard, and S. Thompson. 1993. Model simulations of Cretaceous climates: the role of geography and carbon dioxide. *Philosophical Transactions of the Royal Society, Series B, Biological Sciences* 341:307–316.

Barron, E. J. and W. M. Washington. 1985. Warm Cretaceous climates: High atmospheric CO_2 as a plausible mechanism. In E. T. Sundquist and W. S. Broecker, eds., *The Carbon Cycle and Atmospheric CO_2: Natural Variations Archaen to Present, American Geophysical Union Monograph* 32:546–553.

Barron, J. A., B. Larsen, and J. G. Baldauf. 1991. Evidence for late Eocene to Early Oligocene Antarctic glaciation and observations on late Neogene glacial history of Antarctica: Results from Leg 119. *Proceedings of Ocean Drilling Program Scientific Results* 119:869–891.

Bartlein, P. J., I. C. Prentice, and T. Webb III. 1986. Climatic response surfaces from pollen data for some eastern North American taxa. *Journal of Biogeography* 13:35–57.

Bartlett, K. B. and R. C. Harriss. 1993. Review and assessment of methane emissions from wetlands. *Chemosphere* 26:261–320.

Battle, M., M. Bender, T. Sowers, et al. 1996. Atmospheric gas concentrations over the past century measured in air from firn at the South Pole. *Nature* 383:231–235.

Bazzaz, F. A. 1990. The response of natural ecosystems to the rising global CO_2 levels. *Annual Review of Ecology and Systematics* 21:167–196.

Be, A. W. H. 1977. An ecological, zoogeographical, and taxonomic review of recent planktonic foraminifera. In A. T. S. Ramsey, ed., *Oceanic Micropaleontology*, pp. 1–100. London: Academic Press.

Be, A. W. H. and W. H. Hamlin. 1967. Ecology of recent planktonic foraminifera. *Micropaleontology* 13:87–106.

Beck, J. W., R. L. Edwards, E. Ito, et al. 1992. Sea-surface temperature from coral skeletal Strontium/Calcium ratios. *Science* 257:644–647.

Beck, J. W., J. Recy, F. Taylor, R. L. Edwards, and G. Cabioch. 1997. Abrupt changes in early Holocene tropical sea surface temperature derived from coral records. *Nature* 385:705–707.

Becker, B. 1993. An 11,000-year German oak and pine dendrochronology for radiocarbon calibration. *Radiocarbon* 35:201–213.

Beer, J., S. J. Johnsen, G. Bonani, et al. 1990. ^{10}Be peaks as time markers in polar ice cores. In E. Bard and W. S. Broecker, eds., *The Last Deglaciation: Absolute and Radiocarbon Chronologies,* pp. 141–153. Berlin: Springer-Verlag.

Beer, J., F. Joos, C. Lukasczyk, et al. 1988. ^{10}Be as an indicator of solar variability. In G. C. Castagnoli, ed. 1988. *Solar-Terrestrial Relationships and the Earth Environment in the Last Millennia,* pp. 221–233. New York: Elsevier/North Holland Physics Publishing.

Beer, J., F. Joos, C. Lukasczyk, et al. 1994. In E. Nesme-Ribes, ed., *The Solar Engine and Its Influence on Terrestrial Atmospheres and Climate,* NATO ASI Series 25:221–233. Berlin: Springer-Verlag.

Beer, J., W. Mende, R. Stellmacher, and O. R. White. 1996. Intercomparisons of proxies for past solar variability. In P. D. Jones, R. S. Bradley, and J. Jouzel, eds., *Climatic Variations and Forcing Mechanisms of the Last 2000 Years,* NATO ASI Series 141:501–517. Berlin: Springer-Verlag.

Behl, R. J. and J. P. Kennett. 1996. Brief interstadial events in the Santa Barbara Basin, NE Pacific, during the past 60 kyr. *Nature* 379:243–246.

Bender, M., T. Sowers, M.-L. Dickson, et al. 1994. Climate correlations between Greenland and Antarctica during the past 100,000 years. *Nature* 372:663–666.

Bender, M. L., R. G. Fairbanks, F. W. Taylor, R. K. Matthews, J. G. Goddard, and W. S. Broecker. 1979. Uranium-series dating of the Pleistocene reef tracts of Barbados, West Indies. *Geological Society of America Bulletin* 90:577–594.

Bender, M. L., D. Labeyrie, D. Raynaud, and C. Lorius. 1985. Isotopic composition of atmospheric O_2 in ice linked with deglaciation and global productivity. *Nature* 31:349–352.

Bender, M. L., F. T. Taylor, and R. K. Matthews. 1973. Helium-uranium dating of corals from middle Pleistocene Barbados reef tracts. *Quaternary Research* 3:142–146.

Bennett, K. D. 1990. Milankovitch cycles and their effects on species in ecological and evolutionary time. *Paleobiology* 16:11–16.

Bennett, K. D. 1997. *Evolution and Ecology, The Pace of Life.* Cambridge: Cambridge University Press.

Benson, L. V., J. W. Burdett, M. Kashgarian, S. P. Lund, F. M. Phillips, and R. O. Rye. 1996. Climatic and hydrologic oscillations in the Owens Lake Basin and adjacent Sierra Nevada, California. *Science* 274:746–749.

Bentley, C. R. and M. B. Giovinetto. 1991. Mass balance of Antarctica and sea level change. In G. Weller, C. L. Wilson, and B. A. B. Severin, eds., *International Conference on the Role of the Polar Regions in Global Change,* pp. 481–488. Fairbanks: University of Alaska.

Berger, A., ed. 1981. *Climatic variations and variability: facts and theories.* Dordrecht: D. Reidel.

Berger, A. 1984. Accuracy and frequency stability of the Earth's orbital elements during the Quaternary. In A. Berger, J. Imbrie, J. Hays, G. Kukla, and B. Salztman, eds., *Milankovitch and Climate: Understanding the Response to Astronomical Forcing,* NATO ASI Series, vol. 126, Parts 1, 2, pp. 3–40. Dordrecht: D. Reidel.

Berger, A., J. Imbrie, J. Hays, G. Kukla, and B. Saltzman, eds. 1984. *Milankovitch and Climate: Understanding the Response to Astronomical Forcing.* NATO ASI Series, vol. 126, Parts 1, 2. Dordrecht: D. Reidel.

Berger, A. and M. F. Loutre. 1991. Insolation values for the climate of the last 10 million years. *Quaternary Science Reviews* 10:297–317.

Berger, A. and M. F. Loutre. 1994. Precession, eccentricity, obliquity, insolation, and paleoclimates. In J.-C. Duplessy and M.-T. Spyridakis, eds., *Long-term Climatic Variations,* NATO ASI Series 1, 22:107–151. Berlin: Springer-Verlag.

Berger, A., M. F. Loutre, and J. Laskar. 1992. Stability of the astronomical frequencies over the Earth's history for paleoclimatic studies. *Science* 255:560–566.

Berger, G. W. 1995. Process in luminescence dating methods for Quaternary sediments. In N. W. Rutter and N. R. Catto, eds., *Dating Methods for Quaternary Deposits.* Geoassociation of Canada.

Berger, W. H. 1978. Oxygen-18 stratigraphy in deep-sea sediments: additional evidence for the deglacial meltwater effect. *Deep-Sea Research* 25:473–480.

Berger, W. H. 1979. Stable isotopes in foraminifera. Cushman Foundation Short Course in Foraminifera, pp. 156–198.

Berger, W. H., A. Be, and E. Vincent, eds. 1981. Oxygen and carbon isotopes in foraminifera. *Palaeogeography, Palaeoclimatology, Palaeoecology* 133(1–3):1–277.

Berger, W. H. and J. C. Crowell. 1982. *Climate in Earth History, Studies in Geophysics.* Washington, D.C.: National Academy Press.

Berger, W. H. and J. V. Gardner. 1975. On the determination of Pleistocene temperatures from planktonic foraminifers. *Journal of Foraminiferal Research* 5:102–113.

Berger, W. H., J. C. Herguera, C. B. Lange, and R. Schneider. 1994. Paleoproductivity: flux proxies versus nutrient proxies and other problems concerning the Quaternary productivity record. In R. Zahn, T. F. Pedersen, M. A. Kaminski, and L. Labeyrie, eds., *Carbon Cycling in the Glacial Ocean: Constraints on the Ocean's Role in Global Change.* NATO ASI Series 1, 17:385–412. Berlin: Springer-Verlag.

Berger, W. H. and L. D. Labeyrie. 1987. Abrupt climatic change—an introduction. In W. H. Berger and L. D. Labeyrie, eds., *Abrupt Climatic Change: Evidence and Implications,* pp. 3–22. Dordrecht: D. Reidel.

Berger, W. H., V. S. Smetacek, and G. Wefer., eds. 1989. *Productivity of the Ocean: Present and Past,* Dahlem Workshop Report. New York: John Wiley & Sons.

Berger, W. H. and G. Wefer. 1991. Productivity of the glacial ocean: Discussion of the iron hypothesis. *Limnology and Oceanography* 36:1899–1918.

Berggren, W. A. 1998. The Cenozoic Era: Lyellian (chrono)stratigraphy and nomenclatural reform at the millennium. In press.

Berggren, W. A., F. J. Hilgen, C. G. Langereis, et al. 1995. Late Neogene chronology: New perspectives in high-resolution stratigraphy. *Geological Society of America Bulletin* 107:1272–1287.

Berggren, W. A. and J. A. Van Couvering, eds. 1984. *Catastrophes and Earth History: The New Uniformitarianism.* Princeton, N.J.: Princeton Univeristy Press.

Bergthorsson, P. 1969. An estimate of drift ice and temperature in Iceland in 1000 years. *Jokull* 19:94–101.

Bernabo, J. C. and T. Webb. 1977. Changing patterns in the Holocene pollen record of northeastern North America: A mapped summary. *Quaternary Research* 8:64–96.

Berner, E. K. and R. A. Berner. 1996. *Global Environment: Water, Air, and Geochemical Cyles.* New Jersey: Prentice Hall.

Berner, R. A. 1990. Atmospheric carbon dioxide levels over Phanerozoic time. *Science* 249:1382–1386.

Berner, R. A. 1994. Geocarb II: A revised model of atmospheric CO_2 over Phanerozoic time. *American Journal of Science* 294:56–91.

Berner, R. A. 1997. The rise of plants and their effect on weathering and atmospheric CO_2. *Science* 276:544–546.

Bindschadler, R. 1997. Actively surging West Antarctic ice streams and their response characteristics. *Annals of Glaciology* 24:409–414.

Birks, H. J. B. and H. H. Birks. 1980. *Quaternary Palaeoecology.* London: Edward Arnold.

Bischof, J. F. 1994. The decay of the Barents ice sheet as documented in nordic seas' ice-rafted debris. *Marine Geology* 117:35–55.

Bjerknes, J. 1966. A possible response of the atmospheric Hadley circulation to equatorial anomalies of ocean temperature. *Tellus* 18:820–829.

Bjerknes, J. 1969. Atmospheric teleconnections from the equatorial Pacific. *Monthly Weather Review* 97:163–172.

Björck, S., B. Kromer, S. Johnsen, et al. 1996a. Synchronized terrestrial-atmospheric deglacial records around the North Atlantic. *Science* 274: 1155–1160.

Björck, S., S. Olsson, C. Ellis-Evans, H. Hakansson, O. Humlum, and J. M. de Lirio. 1996b. Late Holocene palaeoclimatic records from lake sediments on James Ross Island, Antarctic. *Palaeogeography, Palaeoclimatology, Palaeoecology* 121:195–220.

Björck, S. B. Wohlfarth, and G. Possnert. 1993. ^{14}C AMS measurements from the late Weischelian part of the Swedish timescale. *Quaternary International* 27:11–18.

Blanchon, P. and J. Shaw. 1995. Reef drowning during the last deglaciation: Evidence for catastrophic sea-level rise and ice-sheet collapse. *Geology* 23:4–8.

Bloemendal, J. and P. deMenocal. 1989. Evidence for a change in the periodicity

of tropical climate cycles at 2.4 myr from whole core magnetic susceptibility measurements. *Nature* 342:897–899.
Bloom, A. L. 1967. Pleistocene shorelines: A new test of isostasy. *Geological Society of America Bulletin* 78:1477–1494.
Bloom, A. L. 1977. *IGCP Project 61 Atlas of Sea Level Curves.* Ithaca: Cornell University.
Bloom, A. L. 1992. Sea level and coastal morphology of the United States through the Late Wisconsin glacial maximum. In H. E. Wright, Jr., ed., *Late-Quaternary Environments of the United States,* pp. 215–229. Minneapolis: University of Minnesota Press.
Bloom, A. L., W. S. Broecker, J. M. A. Chappell, R. K. Matthews, and K. J. Mesolella. 1974. Quaternary sea level fluctuations on a tectonic coast: New $^{230}Th/^{234}U$ dates from the Huon Peninsula, New Guinea. *Quaternary Reseach* 4:185–205.
Blunier, T., J. Chappellaz, J. Schwander, et al. 1993. Atmospheric methane record from a Greenland ice core over the last 1000 years. *Geophysical Research Letters* 20:2219–2222.
Blunier, T., J. Chappellaz, J. Schwander, B. Stauffer, and D. Raynaud. 1995. Variations in atmospheric methane concentrations during the Holocene Epoch. *Nature* 374:46–49.
Boesch, D. F. (ed.) 1994. Scientific assessment of coastal wetland loss, restoration and management in Louisiana. *Journal of Coastal Research Special Issue* 20:1–103.
Bond, G., W. Broecker, S. Johnsen, et al. 1993. Correlations between climate records from North Atlantic sediments and Greenland ice. *Nature* 365:143–147.
Bond, G., H. Heinrich, W. Broecker, et al. 1992. Evidence for massive discharges of icebergs into the North Atlantic ocean during the last glacial period. *Nature* 360:245–249.
Bond, G., W. Showers, M. Cheseby, et al. 1997. A pervasive millennial-scale cycle in North Atlantic Holocene and glacial climates. *Science* 278: 1257–1266.
Bond, G. C. and R. Lotti. 1995. Iceberg discharges into the North Atlantic on millennial time scales during the last glaciation. *Science* 267:1005–1010.
Boulton, G. S. and T. Payne. 1994. Mid-latitude ice sheets through the last glacial cycle: Glaciological and geological reconstructions. In J.-C. Duplessy and M.-T. Spyridakis, eds., *Long-term Climatic Variations,* NATO ASI Series 1, 22:177–212. Berlin: Springer-Verlag.
Bowen, D. Q. 1978. *Quaternary Geology: A Stratigraphic Framework for Multidisciplinary Work.* Oxford: Pergamon.
Boyle, E. A. 1988a. Cadmium: chemical tracer of deepwater paleoceanography. *Paleoceanography* 3:471–489.
Boyle, E. A. 1988b. The role of vertical chemical fractionation in controlling late Quaternary atmospheric carbon dioxide. *Journal of Geophysical Research* 93(C12):15,701–15,714.
Boyle, E. A. 1992. Cadmium and $\delta^{13}C$ paleochemical ocean distributions dur-

ing stage 2 glacial maximum. *Annual Review of Earth and Planetary Science* 20:243–287.

Boyle, E. A. 1994. A comparison of carbon isotopes and cadmium in the modern and glacial maximum ocean: Can we account for the discrepancies? In R. Zahn, T. F. Pedersen, M. A. Kaminski, and L. Labeyrie, eds., *Carbon Cycling in the Glacial Ocean: Constraints on the Ocean's Role in Global Change.* NATO ASI Series 17:167–194. Berlin: Springer-Verlag.

Boyle, E. A. and L. D. Keigwin. 1987. North Atlantic thermohaline circulation during the past 20,000 years linked to high latitude surface temperature. *Nature* 330:35–40.

Bradbury J. P. and W. E. Dean, eds. 1993. *Elk Lake, Minnesota: Evidence for Rapid Climate Change in the North-Central United States.* Geological Society of America Special Paper 276. Boulder: Geological Society of America.

Bradley, R., E. Bard, G. Farquhar, et al. 1993. Group report: Evaluating strategies for reconstructing past global changes—what and where are the gaps? In J. A. Eddy and H. Oeschger, eds., *Global Changes in the Perspective of the Past,* pp. 145–171. Chichester: John Wiley & Sons.

Bradley, R. S. 1985. *Quaternary Paleoclimatology: Methods of Paleoclimatic Reconstruction.* London: Allen and Unwin.

Bradley, R. S. 1987. The explosive volcanic eruption record in Northern Hemisphere temperature records. In W. H. Berger and L. D. Labeyrie, eds., *Abrupt Climatic Change,* pp. 59–60. Dordrecht: D. Reidel.

Bradley, R. S., ed. 1989. *Global Changes of the Past.* Boulder: University Corporation for Atmospheric Research (UCAR)/Office for Interdisciplinary Earth Studies.

Bradley, R. S. and P. D. Jones, eds. 1992. *Climate Since* A.D. *1500.* New York: Routledge.

Bradley, R. S. and P. D. Jones. 1993. "Little Ice Age" summer temperature variations: their nature and relevance to recent global warming trends. *The Holocene* 3:367–376.

Brandon, R. N. 1978. Adaptation and evolutionary theory. *Studies in the History and Philosophy of Science* 9:181–206.

Brandon, R. N. 1990. *Adaptation and Environment.* Princeton, N.J.: Princeton University Press.

Brassell, S. C., G. Eglington, I. T. Marlowe, U. Pflaumann, and M. Sarnthein. 1986. Molecular stratigraphy: a new tool for climatic assessment. *Nature* 320:129–133.

Bray, J. R. 1959. An analysis of the possible recent change in atmospheric carbon dioxide concentration. *Tellus* 11:220–230.

Bray, J. R. and J. T. Curtis. 1957. An ordination of the upland forest communities of southern Wisconsin. *Ecological Monographs* 27:325–349.

Briffa, K. R. 1994. Grasping at shadows? A selective review of the search for sunspot related variability in tree rings. In E. Nesme-Ribes, ed., *The Solar Engine and Its Influence on Terrestrial Atmospheres and Climate,* NATO ASI Series 25:417–435. Berlin: Springer-Verlag.

Briffa, K. R., P. D. Jones, T. S. Bartholin, et al. 1992. Fennoscandinavian summers from A.D 500: Temperature changes on short and long timescales. *Climate Dynamics* 7:111–119.

Briffa, K. R., P. D. Jones, F. H. Schweingruber, W. Karlen, and S. G. Shiyatov. 1996. Tree ring variables as proxy-climate indicators: Problems with low frequency signals. In P. D. Jones, R. S. Bradley, and J. Jouzel, eds., *Climate Variations and Forcing Mechanisms of the Last 2000 Years,* NATO ASI Series 1, 41:9–41. Berlin: Springer-Verlag.

Broecker, W. S. 1968. In defense of the astronomical theory of glaciation. *Meteorological Monographs* 8:139–141.

Broecker, W. S. 1982. Glacial to interglacial changes in ocean chemistry. *Progress in Oceanography* 11:151–197.

Broecker, W. S. 1994. Massive iceberg discharge triggers for global climate change. *Nature* 372:421–424.

Broecker, W. S. 1996. Chaotic Climate. *Scientific American.* November 1995:62–68.

Broecker, W. S. 1998. Paleocean circulation during the last deglaciation: A bipolar seesaw. *Paleoceanography* 13:119–121.

Broecker, W. S., M. Andree, G. Bonani, W. Wolfi, H. Oeschger, and M. Kas. 1988a. Can the Greenland climatic jumps be identified in records from ocean and land? *Quaternary Research* 30:1–6.

Broecker, W. S., M. Andree, W. Wolfli, et al. 1988b. The chronology of the last deglaciation: Implications for the cause of the Younger Dryas Event. *Paleoceanography* 3:1–19.

Broecker, W. S., G. Bond, M. Klas, G. Bonani, and W. Wolfli. 1990. A salt oscillator in the glacial Atlantic? I. The concept. *Paleoceanography* 4:469–477.

Broecker, W. S., G. Bond, M. Klas, E. Clark, and J. McManus. 1992. Origin of the northern Atlantic's Heinrich events. *Climate Dynamics* 6:265–273.

Broecker, W. S. and G. H. Denton. 1989. The role of ocean-atmosphere reorganizations in glacial cycles. *Geochimica et Cosmochimica Acta* 53: 2465–2501.

Broecker, W. S. and W. Farrand. 1963. Radiocarbon age of the Two Creeks Forest bed, Wisconsin. *Geological Society of America Bulletin* 74:795–802.

Broecker, W. S., J. P. Kennett, B. F. Flower, et al. 1989. Routing of meltwater from the Laurentide ice sheet during the Younger Dryas. *Nature* 341:318–321.

Broecker, W. S. and T.-H. Peng. 1982. *Tracers in the Sea.* New York: Eldigio Press.

Broecker, W. S. and T.-H. Peng. 1989. The cause of the glacial to interglacial atmospheric CO_2 change: A polar alkalinity hypothesis. *Global Biogeochemical Cycles* 3:215–239.

Broecker, W. S., D. M. Peteet, and D. Rind. 1985. Does the ocean-atmosphere system have more than one stable mode of operation. *Nature* 315:21–25.

Broecker, W. S. and D. L. Thurber. 1965. Uranium series dating of corals and oolites from Bahaman and Florida Key limestones. *Science* 149:58–60.

Broecker, W. S., D. L. Thurber, J. Goddard, T.-L. Ku, R. K. Matthews, and K. L. Mesolella. 1968. Milankovitch supported by precise dating of coral reefs and deep-sea sediments. *Science* 159:297–300.

Broecker, W. S. and J. van Donk. 1970. Insolation changes, ice volumes, and the O-18 record in deep-sea sediments. *Reviews in Geophysics and Space Physics* 8:169–198.

Brook, E. J., T. Sowers, and J. Orchado. 1996. Rapid variations in atmospheric methane concentration during the past 110,000 years. *Science* 273: 1087–1091.

Brown, J. H. 1995. *Macroecology.* Chicago: University of Chicago Press.

Buckland, W. 1823. Relinquiae Diluvianae; or observations on the organic remains contained in caves, fissures and diluvial gravel, and on other geological phenomena attesting the action of a universal deluge. London: John Murray. (Cited in Dott 1992.)

Budd, W. F. and I. N. Smith. 1985. The state of balance of the Antarctic Ice Sheet, an updated assessment 1984. In *Glaciers, Ice Sheets, and Sea Level: Effects of a CO-2 Induced Climatic Change,* pp. 172–177. Washington, D.C.: National Academy Press.

Buddmeier, R. W. J. E. Maragas, and D. W. Knutson. 1974. Radiograph studies of reef coral exoskeletons: rates and patterns of coral growth. *Journal of Experimental Biology and Marine Ecology* 14:179–200.

Budyko, M. I. 1982. *The Earth's Climate: Past and Future.* New York: Academic Press.

Bunkers, M. J., J. R. Miller, Jr., and A. T. DeGaetana. 1996. An examination of El Niño–La Niña–related precipitation and temperature anomalies across the Great Plains. *Journal of Climate* 9:147–160.

Burian, R. 1984. Adaptation. In M. Grene, ed., *Dimensions of Darwinism,* pp. 287–314. Cambridge: Cambridge University Press.

Campbell, I. D. and J. H. McAndrews. 1993. Forest disequilibrium caused by rapid Little Ice Age cooling. *Nature* 366:336–338.

Cande, S. C. and D. V. Kent. 1992. A new geomagnetic polarity time scale for the late Cretaceous and Cenozoic. *Journal of Geophysical Research* 97:13,917–14,951.

Cane, M. A. 1986. El Niño. *Annual Review of Earth and Planetary Science* 14:43–70.

Carozzi, A. V. 1992. De Maillet's Telliamed (1748): The discrimination of the sea or the fall portion of a complete cosmic eustatic cycle. In R. H. Dott, Jr., ed. *Eustasy: The Historical Ups and Downs of a Major Geological Concept,* pp. 17–24. Boulder: Geological Society of America.

Carriquiry, J. D., M. J. Risk, and H. P. Schwarcz. 1994. Stable isotope geochemistry of corals from Costa Rica as a proxy indicator of the El Niño/Southern Oscillation (ENSO). *Geochimica et Cosmochimica Acta* 58:335–351.

Carrasco, S. and H. Santander. 1987. The El Niño event and its influence on the zooplankton off Peru. *Journal of Geophysical Research* 92(C13): 14,405–14,410.

Carstens, J., D. Hebbeln, and G. Wefer. 1997. Distribution of planktic foraminifera at the ice margin in the Arctic (Fram Strait). *Marine Micropaleontology* 29:257–269.

Casanova, J. and C. Hillaire-Marcel. 1993. Carbon and oxygen isotopes in

Africa lacustrine stromatolites: Paleohydrological interpretations. In P. K. Swart, K. C. Lohman, J. McKenzie, and S. Savin, eds., *Climate Change in Continental Isotope Records, American Geophysical Union Monograph* 78:123–133.

Castagnoli, G. C., ed. 1988. *Solar-Terrestrial Relationships and the Earth Environment in the Last Millennia.* New York: Elsevier/North Holland Physics Publishing.

Cathles, L. M. 1975. *The Viscosity of the Earth's Mantle.* Princeton, N.J.: Princeton University Press.

Cathles, L. M. and A. Hallam. 1991. Stess-induced changes in plate density, Vail sequences, epeirogeny and short-lived global sea-level fluctuations. *Tectonics* 10:659–671.

Cayan, D. R. and R. H. Webb. 1992. El Niño/Southern Oscillation and streamflow in the western United States. In H. F. Diaz and V. Markgraf, eds., *El Niño: Historical and Paleoclimatic Aspects of the Southern Oscillation,* pp. 29–68. Cambridge: Cambridge University Press.

Cerling, T. 1991. Carbon dioxide in the atmosphere: Evidence from Cenozoic and Mesozoic paleosols. *American Journal of Science* 291:377–400.

Cerling, T. 1992. Use of carbon isotopes as an indicator of the P(CO_2) of the paleoatmosphere. *Global Biogeochemical Cycles* 6:307–314.

Cerling, T., Y. Wang, and J. Quade. 1993. Expansion of C4 ecosystems as an indicator of global ecological change in the late Miocene. *Nature* 361: 344–345.

Cerling, T. E. and J. Quade. 1993. Stable carbon and oxygen isotopes in soil carbonates. In P. K. Swart, K. C. Lohman, J. McKenzie, and S. Savin, eds., *Climate Change in Continental Isotope Records, American Geophysical Union Monograph* 78:217–231.

Cess, R. D., G. Potter, M.-H. Zhang, J.-P. Blanchet, et al. 1991. Interpretation of snow-climate feedback as produced by 17 general circulation models. *Science* 253:888–892.

Cess, R. D., M.-H. Zhang, G. L. Potter, H. W. Barker, et al. 1993. Uncertainties in carbon dioxide radiative forcing in atmospheric general circulation models. *Science* 262:1252–1255.

Chamberlin, T. C. 1897. A group of hypotheses bearing on climate changes. *Journal of Geology* 5:653–683.

Chamberlin, T. C. 1898. The ulterior basis of time divisions and the classification of geologic history. *Journal of Geology* 6:449–462.

Chamberlin, T. C. 1899. An attempt to frame a working hypothesis on the cause of glacial periods on an atmospheric basis. *Journal of Geology* 7:545–584, 667–685, 751–787.

Chamberlin, T. C. 1916. *The Origin of the Earth.* Chicago: University of Chicago.

Chao, B. F. 1996. "Concrete" testimony to Milankovitch cycle in Earth's changing obliquity. *EOS, Transactions, American Geophysical Union* 77:433.

Chappell, J. 1974. Geology of coral terraces, Huon Peninsula, New Guinea: A study of Quaternary movements and sea level. *Geological Society of America Bulletin* 85:553–570.

Chappell, J., A. Omura, T. Esat, et al. 1996. Reconciliation of late Quaternary sea levels derived from coral terraces at Huon Peninsula with deep-sea isotope records. *Earth and Planetary Science Letters* 141:227–236.

Chappell, J. and H. Polach. 1991. Post-glacial sea-level rise from a coral record at Huon Peninsula, Papua New Guinea. *Nature* 349:147–149.

Chappell, J. and N. J. Shackleton. 1986. Oxygen isotopes and sea level. *Nature* 324:137–140.

Chappellaz, J., J. M. Barnola, D. Raynaud, Y. S. Korotkevich, and C. Lorius. 1990. Ice-core record of atmospheric methane over the past 160,000 years. *Nature* 345:127–131.

Chappellaz, J., T. Blunier, D. Raynaud, J. M. Barnola, J. Schwander, and R. Stauffer. 1993. Synchronous changes in atmospheric CH_4 and Greenland climate between 40 and 8 kyr BP. *Nature* 366:443–445.

Charles, C. D. and R. G. Fairbanks. 1990. Glacial to interglacial changes in the isotopic gradients of Southern Ocean surface water. In U. Bleil and J. Thiede, eds., *Geological History of the Polar Oceans: Arctic versus Antarctic,* pp. 519–538. Norwell, Mass.: Kluwer Academic.

Charles, C. D., D. E. Hunter, and R. G. Fairbanks. 1997. Interaction between the ENSO and the Asian monsoon in a coral record of tropical climate. *Science* 277:925–928.

Charles, C. D., J. Lynch-Steiglitz, U. S. Ninneman, and R. G. Fairbanks. 1996. Climate connections between the hemispheres revealed by deep-sea sediment core/ice core correlations. *Earth and Planetary Science Letters* 142:19–27.

Charles, C. D., D. Rind, J. Jouzel, R. D. Koster, and R. G. Fairbanks. 1995. Seasonal precipitation timing and ice core records. *Science* 269:247–248.

Charleson, R. J., J. E. Lovelock, M. O. Andreae, and S. G. Warren. 1987. Oceanic phytoplankton, atmospheric sulphur, cloud albedo and climate. *Nature* 326:655–661.

Chavez, F. P. and R. T. Barber. 1987. An estimate of new production in the equatorial Pacific. *Deep Sea Research* 34:1229–1243.

Chelton, D. B. and M. G. Schlax. 1996. Global observations of oceanic Rossby waves. *Science* 272:234–238.

Chen, D., S. E. Zebiak, A. J. Busalacchi, and M. A. Cane. 1995a. An improved procedure for El Niño forecasting: implications for predictability. *Science* 269:1699–1702.

Chen, J., J. W. Farrell, D. W. Murray, and W. L. Prell. 1995b. Timescale and paleoceanographic implications of a 3.6 m.y. oxygen isotope record from the northeast Indian Ocean (Ocean Drilling Program site 758). *Paleoceanography* 10:21–47.

Chen, J. H., H. A. Curran, B. White, and G. J. Wasserburg. 1991. Precise chronology of the last interglacial period: ^{234}U-^{230}Th data from fossil coral reefs in the Bahamas. *Geological Society of America Bulletin* 103:82–97.

Cheney, R. E., L. Miller, R. W. Agreen, and N. S. Doyle. 1994. TOPEX/Poseidon: The 2-cm solution. *Journal of Geophysical Research* 99(C2): 24,555–24,563.

Chisolm, S. W. and F. M. M. Morel, eds. 1991. What controls phytoplankton production in nutrient-rich areas of the open sea? *Limnology and Oceanography* 36(8):1507–1965.

Chivas, A. R., P. De Deckker, J. A. Cali, A. Chapman, E. Kiss, and J. M. G. Shelley. 1993. Coupled stable-isotope and trace-element measurements of lacustrine carbonates as paleoclimatic indicators. In P. K. Swart, K. C. Lohman, J. McKenzie, and S. Savin, eds., *Climate Change in Continental Isotope Records, American Geophysical Union Monograph* 78:113–121.

Chivas, A. R., P. De Deckker, and J. M. G. Shelley. 1986. Magnesium content of non-marine ostracode shells: A new palaeosalinometer and palaeothermometer. *Palaeogeography, Palaeoclimatology, Palaeoecology* 54:43–61.

Christie-Blick, N., G. S. Mountain, and K. G. Miller. 1988. Technical comments: Sea level history. *Science* 241:596.

Ciais, P., P. P. Tans, M. Troiler, J. W. C. White, and R. J. Francey. 1995. A large northern hemisphere terrestrial CO_2 sink indicated by the $^{13}C/^{12}C$ ratio of atmospheric CO_2. *Science* 269:1098–1102.

Clapperton, C. M. 1990. Quaternary glaciations in the southern hemisphere: An overview. *Quaternary Science Reviews* 9:299–304.

Clark, J. A., W. E. Farrell, and W. R. Peltier. 1978. Global changes in post-glacial sea level: A numerical simulation. *Quaternary Research* 9:265–287.

Clark, P. U. 1994. Unstable behavior of the Laurentide Ice Sheet over deforming sediments and its implications for climate change. *Quaternary Research* 41:19–25.

Clark, P. U., R. B. Alley, L. D. Keigwin, J. M. Licciardi, S. J. Johnsen, and H. Wang. 1996a. Origin of the first global meltwater pulse following the last glacial maximum. *Paleoceanography* 11:563–577.

Clark, P. U. and P. J. Bartlein. 1995. Correlation of late Pleistocene glaciation in the western United States with North Atlantic Heinrich events. *Geology* 23:483–486.

Clark, P. U., J. M. Liccardi, D. R. MacAyeal, and J. W. Jenson. 1996b. Numerical reconstruction of a soft-bedded Laurentide Ice Sheet during the last glacial maximum. *Geology* 24:679–682.

Clausen, H. B., N. S. Gundestrup, H. Shoji, and O. Watanabe. 1996. Scientific research collaboration efforts for Greenland ice core studies. *Memoirs National Institute Polar Research, Special Issue* 51:337–342.

Clausen, H. B. and C. U. Hammer. 1988. The Laki and Tambora eruptions as revealed in Greenland ice cores from 11 locations. *Annals of Glaciology* 10:16–22.

Clausen, H. B., C. U. Hammer, J. Christensen, et al. 1995. 1250 years of global volcanism as revealed by central Greenland ice cores. In R. J. Delmas, ed., *Ice Core Studies of Global Biogeochemical Cycles*, pp. 175–194. Berlin: Springer-Verlag.

Clausen, H. B., C. Hammer, C. S. Huidberg, D. Dahl-Jensen, J. P. Steffensen, J. Kipfstuhl, M. Legrand. 1997. A comparison of the volcanic records over the past 4000 years from the Greenland ice core project and Dye 3 Greenland ice cores. *Journal of Geophysical Research* 102, no. C12: 26, 707–26,723.

Clemens, S. C. and Prell, W. L. 1990. Large Pleistocene variability of Arabian Sea Summer Monsoon Winds and Continental Aridity: Eolian Records from the lithogenic component of deep-sea sediments. *Paleoceanography* 5:109–146.

Clemens, S. C. and W. L. Prell. 1991. Late Quaternary forcing of the Indian Ocean summer monsoon winds: A comparison of Fourier and general circulation model results. *Journal of Geophysical Research* 96:22,683–22,700.

Clemens, S. C. and R. Tiedemann. 1997. Eccentricity forcing of Pliocene–early Pleistocene climate revealed in a marine oxygen-isotope record. *Nature* 385:801–804.

Clements, F. E. 1916. Plant succession: An analysis of the development of vegetation. Publication 242. Washington, D.C.: Carnegie Institution of Washington.

CLIMAP Project Member. 1976. The surface of the ice-age Earth. *Science* 191:1131–1137.

CLIMAP Project Member. 1981. Seasonal reconstruction of the earth's surface during the last glacial maximum. Map and Chart Series No. 36. Boulder: Geological Society of America.

CLIMAP Project Members. 1984. The last interglacial ocean. *Quaternary Research* 21:123–224.

Cline, R. M. and J. D. Hays, eds. 1976. *Investigation of Late Quaternary Paleoceanography and Paleoclimatology.* Geological Society of America Memoir 145. Boulder, Colo.: Geological Society of America.

Codispoti, L. A. 1989. Phosphorous versus nitrogen of new and export production. In W. H. Berger, V. S. Smetacek, and G. Wefer, eds., *Productivity of the Ocean: Present and Past,* Dahlem Workshop Report, pp. 377–394. New York: John Wiley & Sons.

Cody, M. L. and J. M. Diamond, eds. 1975. *Ecology and Evolution of Communities.* Cambridge, Mass.: Harvard University Press.

COHMAP Members. 1988. Climatic changes of the last 18,000 years: Observations and model simulations. *Science* 241:1043–1052.

Cole, J. 1996. Coral records of climate change: understanding past variability in the tropical ocean-atmosphere. In P. D. Jones, R. S. Bradley, and J. Jouzel, eds., *Climatic Variations and Forcing Mechanisms of the Last 2000 Years,* NATO ASI Series 41:331–353. Berlin: Springer-Verlag.

Cole, J. E. and R. G. Fairbanks. 1990. The Southern Oscillation recorded in the $\delta^{18}O$ of corals from Tarawa Atoll. *Paleoceanography* 5:669–683.

Cole, J. E., R. G. Fairbanks, and G. T. Shen. 1993a. The spectrum of recent variability in the Southern Oscillation: results from a Tarawa Atoll coral. *Science* 260:1790–1793.

Cole, J. E., D. Rind, and R. G. Fairbanks. 1993b. Isotopic response to interannual climatic variability simulated by an atmospheric general circulation model. *Quaternary Science Reviews* 12:387–406.

References

Cole-Dai, J., L. G. Thompson, E. Mosley-Thompson. 1995. A 485-year record of atmospheric chloride, nitrate and sulfate: results of chemical analyses of ice cores from Dyer Plateau, Antarctic Peninsula. *Annals of Glaciology* 21:182–188.

Colinvaux, P. A. 1996. Quaternary environmental history and forest diversity in the Neotropics. In J. B. C. Jackson, A. F. Budd, and A. C. Coates, eds., *Evolution and Environment in Tropical America,* pp. 359–405. Chicago: The University of Chicago Press.

Colman, S. M. and R. B. Mixon. 1987. The record of major Quaternary sea-level changes in a large coastal plain estuary, Chesapeake Bay, eastern United States. *Palaeogeography, Palaeoclimatology, Palaeoecology* 68: 99–116.

Colman, S. M., J. A. Peck, E. B. Karabanov, et al. 1995. Continental climate response to orbital forcing from biogenic silica records in lake Baikal. *Nature* 378:769–771.

Colman, S. M. and K. L. Pierce. 1981. Weathering rinds on andesitic and basaltic stones as a Quaternary age indicator, western United States. *U.S. Geological Survey Professional Paper* 1210.

Connell, J. H. 1961. Effects of competition, predation by *Thais lapillus* and other factors on natural populations of the barnacle *Balanus balanoides. Ecological Monographs* 31:61–104.

Cook, E. R. 1992. Using tree rings to study past El Niño/Southern Oscillation influences on climate. In H. F. Diaz and V. Markgraf, eds., *El Niño: Historical and Paleoclimatic Aspects of the Southern Oscillation,* pp. 203–214. Cambridge: Cambridge University Press.

Cook, E. R. 1995. Temperature history from tree rings and corals. *Climate Dynamics* 11:211–222.

Cook, E. R., T. Bird, M. Peterson, et al. 1991. Climatic change in Tasmania inferred from a 1089-year tree-ring chronology of subalpine Huon pine. *Science* 253:1266–1268.

Cook, E. R., B. M. Buckley, R. D. D'Arrigo. 1996. Inter-decadal climate oscillations in the Tasmanian sector of the southern Hemisphere: Evidence from tree rings over the past three millennia. In P. D. Jones, R. S. Bradley, and J. Jouzel, eds., *Climate Variations and Forcing Mechanisms of the Last 2000 Years,* NATO ASI Series 1, 41:141–160. Berlin: Springer-Verlag.

Cook, E. R. and G. C. Jacoby, Jr. 1979. Evidence for quasi-periodic July drought in the Hudson Valley, New York. *Nature* 282:390–392.

Cook, E. R. and G. C. Jacoby, Jr. 1983. Potomac River streamflow since 1730 as reconstructed by tree rings. *Journal of Climate and Applied Meteorology* 32:1659–1672.

Cook, E. R. and P. Mayes. 1987. Decadal-scale patterns of climatic change over eastern North America inferred from tree rings. In W. H. Berger and L. D. Labeyrie, eds., *Abrupt Climatic Change,* pp. 61–66. Dordrecht: D. Reidel.

Coope, G. R. 1977. Fossil coleopteran assemblages as sensitive indicators of climatic changes during the Devensian (last) cold stage. *Philosophical Transactions of the Royal Society, Series B, Biological Sciences* 280:313–340.

Coplen, T. B., I. J. Winograd, J. M. Landwehr, and A. C. Riggs. 1994. 500,000-year stable carbon isotopic record from Devils Hole, Nevada. *Science* 263:361–365.

Corliss, B. H. 1985. Microhabitats of benthic foraminifera within deep-sea sediments. *Nature* 314:435–438.

Corliss, B. H. 1991. Morphology and microhabitat preferences of benthic foraminifers from the northeast Atlantic Ocean. *Marine Micropaleontology* 17:195–226.

Corliss, B. H., A. S. Hunt, and L. D. Keigwin. 1982. Benthonic foraminiferal fauna and isotopic data for the postglacial evolution of the Champlain Sea. *Quaternary Research* 17:325–338.

Cortijo, E., L. Labeyrie, L. Vidal, et al. 1997. Changes in sea surface hydrology associated with Heinrich event 4 in the North Atlantic Ocean between 40 and 60°N. *Earth and Planetary Science Letters* 146:29–45.

Covey, C. 1995. Using paleoclimates to predict future climate: How far can analogy go? *Climatic Change* 29:403–407.

Covey, C., L. C. Sloan, and M. I. Hoffert. 1996. Paleoclimatic data constraints on climate sensitivity: the paleocalibration method. *Climatic Change* 32:165–184.

Craig, H. 1965. The measurement of oxygen isotope paleotemperature. In E. Tongiori, ed., *Second Conference on Oceanographic Studies and Paleotemperatures,* pp. 161–182. Speleto, Italy: Consiglio Naz della Richerche.

Craig, H. and L. I. Gordon. 1965. Deuterium and oxygen-18 variations in the ocean and the marine atmosphere. *Proceedings of the Spoleto Conference on Stable Isotopes in Oceanographic Studies and Palaeotemperatures* 2:1–87.

Craig, H., Y. Horibe, and T. Sowers. 1988. Gravitational separation of gases and isotopes in polar ice cores. *Science* 242:1675–1678.

Croll, J. 1864. On the physical cause of the change of climate during geological epochs. *Philosophical Magazine* 28:121–137.

Croll, J. 1867a. On the eccentricity of the earth's orbit, and its physical relations to the glacial epoch. *Philosophical Magazine* 33:119–151.

Croll, J. 1867b. On the change in the obliquity of the ecliptik, its influence on climate and the polar regions and on the level of the sea. *Philosophical Magazine* 33:426–445.

Croll, J. 1875. *Climate and Time.* New York: Appleton & Co.

Cronin, T. M. 1977. Late Wisconsin marine environments of the Champlain Valley (New York, Quebec). *Quaternary Research* 7:238–253.

Cronin, T. M. 1982. Rapid sea level and climate change: evidence from continental and island margins. *Quaternary Science Reviews* 1:177–214.

Cronin, T. M. 1985. Speciation and stasis in marine Ostracoda: Climatic modulation of evolution. *Science* 227:60–63.

Cronin, T. M. 1987. Evolution, paleobiogeography, and systematics of *Puriana:* Evolution and speciation in Ostracoda III. *Journal of Paleontology Memoir* 21:1–71.

Cronin, T. M. 1988. Evolution of marine climates of the U.S. Atlantic coastal

plain during the last 4 million years. *Philosophical Transactions of the Royal Society, Series B, Biological Sciences* 318(1191):661–678.
Cronin, T. M. and H. J. Dowsett. 1990. A quantitative micropaleontologic method for shallow marine paleoclimatology: Application to Pliocene deposits of the western North Atlantic. *Marine Micropaleontology* 16:117–147.
Cronin, T. M., G. A. Dwyer, P. A. Baker, J. Rodriguez-Lazaro, and W. M. Briggs, Jr. 1996. Deep-sea ostracode shell chemistry (Mg:Ca ratios) and late Quaternary Arctic Ocean history. In J. T. Andrews, W. E. Austin, H. Bergsten, and A. E. Jennings, eds., *The Late Quaternary Paleoceanography of the North Atlantic Margins,* Geological Society of London Special Publication 111, pp. 117–134. London: Geological Society of London.
Cronin, T. M., T. R. Holtz, Jr., R. Stein, R. Spielhagen, D. Fütterer, and J. Wollenberg. 1995. Late Quaternary paleoceanography of the Eurasian Basin, Arctic Ocean. *Paleoceanography* 10:259–281.
Cronin, T. M. and N. Ikeya. 1992. Tectonic events and climatic change: Opportunities for speciation in Cenozoic marine ostracodes. In R. M. Ross and W. Allmon, eds., *Causes of Evolution: A Paleontological Perspective,* pp. 210–248. Chicago: University of Chicago Press.
Cronin, T. M., A. Kitamura, N. Ikeya, and T. Kamiya. 1994. Mid-Pliocene paleoceanography of the Sea of Japan. *Palaeogeography, Palaeoclimatology, Palaeoecology* 108:437–455.
Cronin, T. M. and M. E. Raymo. 1997. Orbital forcing of deep-sea benthic species diversity. *Nature* 385:624–627.
Cronin, T. M., B. J. Szabo, T. A. Ager, J. E. Hazel, and J. P. Owens. 1981. Quaternary climates and sea levels of the U.S. Atlantic Coastal Plain. *Science* 211:233–240.
Crowley, T. J. 1991. Modeling Pliocene warmth. *Quaternary Science Reviews* 10:275–282.
Crowley, T. J. 1994. Potential reconciliation of Devils Hole and deep-sea Pleistocene chronologies. *Paleoceanography* 9:1–5.
Crowley, T. J., T. A. Criste, and N. R. Smith. 1993. Reassessment of Crete (Greenland) ice core acidity/volcanism link to climate change. *Geophysical Research Letters* 20:209–212.
Crowley, T. J. and K.-Y. Kim. 1993. Towards development of a strategy for determining the origin of decadal-centennial scale climate variability. *Quaternary Science Reviews* 12:375–387.
Crowley, T. J. and K.-Y. Kim. 1996. Comparison of proxy records of climate change and solar forcing. *Geophysical Research Letters* 23:359–362.
Crowley, T. J. and G. R. North. 1991. *Paleoclimatology.* New York: Oxford University Press.
Crowley, T. J., T. M. Quinn, F. W. Taylor, C. Henin, and P. Joannot. 1997. Evidence for a volcanic cooling signal in a 335-year coral record from New Caledonia. *Paleoceanography* 12:633–639.
Cubasch, U., G. C. Hegerl, A. Hellbach, et al. 1995. A climate change simulation starting from 1935. *Climate Dynamics* 11:71–84.
Cuffey, K. M., G. D. Clow, R. B. Alley, M. Stuiver, E. D. Waddington, and R. W.

Saltus. 1995. Large Arctic temperature change at the Wisconsin-Holocene glacial transition. *Science* 270:455–458.

Curry, B. B. and L. R. Follmer. 1992. The last glacial/interglacial transition in Illinois: 122–125 ka. In P. U. Clark and P. D. Lea, eds., *The Last Interglacial/Glacial Transition in North America,* Geological Society of America Special Paper 270, pp. 71–88. Boulder, Colo.: Geological Society of America.

Curry, B. B. and M. J. Pavich. 1996. Absence of glaciation in Illinois during marine isotope stages 3 through 5. *Quaternary Research* 46:19–26.

Curry, W. B., J. C. Duplessy, L. D. Labeyrie, and N. J. Shackleton. 1988. Changes in the distribution of $\delta\ ^{13}C$ of deep water and $\Sigma\ CO_2$ between the last glaciation and the Holocene. *Paleoceanography* 3:317–341.

Curry, W. B. and R. K. Matthews. 1981. Paleoceanographic utility of oxygen isotopic measurements on planktonic foraminifera: Indian Ocean core top evidence. *Palaeogeography, Palaeoclimatology, Palaeoecology* 33:173–192.

Curry, W. B. and D. W. Oppo. 1997. Synchronous high-frequency oscillations in tropical sea surface temperatures and North Atlantic deep water production during the last glacial cycle. *Paleoceanography* 12:1–14.

Curry, W. B., R. C. Thunnell, and S. Honjo. 1983. Seasonal changes in the isotopic composition of planktonic foraminifera collected in the Panama Basin sediment traps. *Earth and Planetary Science Letters* 64:33–43.

Curtis, J. H. and D. A. Hodell. 1993. An isotopic and trace element study of ostracodes from Lake Miragone, Haiti: A 10,500 year record of paleosalinity and paleotemperature changes in the Caribbean. In P. K. Swart, K. C. Lohmann, J. McKenzie, and S. Savin, eds., *Climate Change in Continental Isotopic Records, American Geophysical Union Monograph* 78:135–152.

Cwynar, L. C. and A. J. Levesque. 1995. Chironomid evidence for late glacial climatic reversals in Maine. *Quaternary Research* 43:405–413.

Daly, R. A. 1910. Pleistocene glaciation and the coral reef problem. *American Journal of Science* 30:297–308.

Damuth, J. 1985. Selection among "species": A formulation in terms of natural functional units. *Evolution* 39:1132–1146.

Dana, J. D. 1853. *Coral Reefs and Islands.* New York: G. Putnam & Co.

Dansgaard, W., H. B. Clausen, N. Gundestrup, et al. 1982. A new Greenland deep ice core. *Science* 218:1273–1277.

Dansgaard, W., S. J. Johnsen, H. B. Clausen, et al. 1993. Evidence for general instability of climate from a 250-kyr ice-core record. *Nature* 364:218–220.

Dansgaard, W., S. J. Johnsen, H. B. Clausen, and N. Gundestrup. 1973. Stable isotope glaciology. *Meddeleleser om Grønland* 197:1–53.

Dansgaard, W., S. J. Johnsen, H. B. Clausen, and C. C. Langway. 1971. Climatic record revealed by the Camp Century ice core. In K. K. Turekian, ed., *The Late Cenozoic Glacial Ages,* pp. 37–56. New Haven, Conn.: Yale University Press.

Dansgaard, W., S. J. Johnsen, I. Moller, and C. C. Langway, Jr. 1969. One thousand centuries of climatic record from Camp Century on the Greenland Ice Sheet. *Science* 166:377–381.

Dansgaard, W., S. J. Johnsen, N. Reeh, N. Gundestrup, H. B. Clausen, and C. U. Hammer. 1975. Climate changes, Norseman, and modern man. *Nature* 255:24–28.

Dansgaard, W. and H. Oeschger. 1989. Past environmental long-term records from the Arctic. In H. Oeschger and C. C. Langway, Jr., eds., *The Environmental Record in Glaciers and Ice Sheets*, pp. 287–318. Chichester: John Wiley & Sons.

Dansgaard, W. and H. Tauber. 1969. Glacier oxygen-18 content and Pleistocene ocean temperatures. *Science* 166:499–502.

Dansgaard, W., J. W. C. White, and S. J. Johnsen. 1989. The abrupt termination of the Younger Dryas climatic event. *Nature* 339:532–534.

D'Arrigo, R. D. and G. C. Jacoby 1993. Secular trends in high northern latitude temperature reconstructions based on tree rings. *Climatic Change* 25:163–177.

Darwin, C. 1859. *On the Origin of Species*. London: John Murray.

Davis, M. B. 1967. Late-glacial climate in Northern United States: A comparison of New England and the Great Lakes region. In E. J. Cushing and H. E. Wright, eds., *Quaternary Paleoecology*, pp. 11–43. New Haven, Conn.: Yale University Press.

Davis, M. B. 1981. Quaternary history and the stability of forest communities. In D. C. West, H. H. Shugart, and D. B. Botkin, eds. *Forest Succession, Concepts and Applications*, pp. 132–153. New York: Springer-Verlag.

Davis, M. B. and D. B. Botkin. 1985. The sensitivity of cool-temperate forests and their fossil pollen to rapid temperature change. *Quaternary Research* 23:327–340.

Davis, M. B., K. D. Woods, S. L. Webb, and R. P. Futuyma. 1986. Dispersal versus climate: expansion of *Fagus* and *Tsuga* into the upper Great Lakes region. *Vegetatio* 67:93–103.

Davis, W. M. 1928. *The Coral Reef Problem*. New York: American Geographical Society.

Dawkins, R. 1976. *The Selfish Gene*. Oxford: Oxford University Press.

Dawkins, R. 1996. *Climbing Mount Improbable*. New York: W. W. Norton & Co.

Dean, J. S. 1994. The Medieval Warm Period on the southern Colorado Plateau. In M. K. Hughes and H. F. Diaz, eds., *The Medieval Warm Period*, pp. 225–241. Dordrecht: Kluwer Academic.

Dean, W. E. 1997. Rates, timing, and cyclicity of Holocene eolian activity in north-central United States: Evidence from varved lake sediments. *Geology* 25:331–334.

de Beaulieu, J.-L and M. Reille. 1984. A long upper Pleistocene pollen record from Les Echets, near Lyon, France. *Boreas* 13:111–132.

de Beaulieu, J.-L. and M. Reille. 1992. The last climate cycle at La Grande Pile (Vosges, France): A new pollen profile. *Quaternary Science Reviews* 11:431–438.

de Boer, P. L and D. G. Smith, eds. 1994. *Orbital Forcing and Cyclic Sequences*. International Association of Sedimentologist Special Publication No. 19. Oxford: Blackwell Scientific Publications.

De Deckker, P., A. R. Chivas, J. M. G. Shelley, and T. Torgersen. 1988. Ostracode shell chemistry: A new palaeoenvironmental indicator applied to a regressive/transgressive record from the Gulf of Carpentaria, Australia. *Palaeogeography, Palaeoclimatology, Palaeoecology* 66:231–241.

Deevey, E. S. and R. F. Flint. 1957. Postglacial hypsithermal interval. *Science* 125:182–184.

De Geer, G. 1888–1890. Om Skandinaviens nivaforandringar under Quatarperioden. *Geologiska Foüreningen Stockholm Forlandlinder* 10:366–379 and 12:61–110.

De Geer, G. 1892. On Pleistocene changes of level in eastern North America. *Proceedings of the Boston Society of Natural History* 25:454–477.

De Geer, G. 1912. A geochronology of the last 12,000 years. *11th International Geological Congress Stockholm 1910, Compte Rendu* 1:241–258.

Delaney, M. L., J. L. Linn, and E. R. M. Druffel. 1993. Seasonal cycles of manganese and cadmium in coral from the Galapagos Islands. *Geochimica et Cosmochimica Acta* 57:347–354.

Delcourt, H. R. and P. A. Delcourt. 1991. *Quaternary Ecology: A Paleoecological Perspective.* London: Chapman & Hall.

Delcourt, P. A. and H. R. Delcourt. 1987. Long-term forest dynamics of the temperate zone. *Ecological Studies 63.* New York: Springer Verlag.

Delmas, R. J. 1992. Environmental records from ice cores. *Reviews of Geophysics* 30:1–21.

Delmas, R. J. 1993. A natural artefact in Greenland ice core CO_2 measurements. *Tellus* 45B:391–396.

Delmas, R., ed. 1995. *Ice Core Studies of Global Biogeochemical Cycles.* NATO ASI Series, vol. 30. Berlin: Springer-Verlag.

Delmas, R. J., J.-M. Ascencio, and M. Legrand. 1980. Polar ice evidence that atmospheric CO_2 20,000 yr B.P. was 50% of present. *Nature* 284:155–157.

Delmas, R. J., M. Briat, and M. Legrand. 1982. Chemistry of South Polar snow. *Journal of Geophysical Research* 87:4314–4318.

Delmas, R. J., M. Legrand, A. J. Aristarain, and F. Zanolini. 1985. Volcanic deposits in Antarctic snow and ice. *Journal of Geophysical Research* 90:901–920.

deMenocal, P. B. 1995. Plio-Pleistocene African climate. *Science* 270:53–59.

deMenocal, P. B., W. F. Ruddiman, and E. M. Pokras. 1993. Influences of high- and low-latitude processes on African terrestrial climate: Pleistocene eolian records from equatorial Atlantic Ocean drilling program Site 663. *Paleoceanography* 8:209–242.

Denton, G. H. and C. H. Hendy. 1994. Younger Dryas-age advance of Franz Josef Glacier in the southern Alps of New Zealand. *Science* 264:1434–1437.

Denton, G. H. and T. J. Hughes. 1981. *The Last Great Ice Sheets.* New York: John Wiley & Sons.

Denton, G. H. and W. Karlén. 1973. Holocene climatic variations—their pattern and possible causes. *Quaternary Research* 3:155–205.

Denton, G. H. and W. Karlén. 1976. Holocene glacial and tree-line variations in the White River Valley and Skolai Pass, Alaska and Yukon Territory, Canada. *Quaternary Research* 7:63–111.

Denton, G. H., D. E. Sugden, D. R. Marchant, B. L. Hall, and T. I. Wilch. 1993. East Antarctic Ice Sheet sensitivity to Pliocene climatic change from a Dry Valleys perspective. *Geografiska Annaler* 75A:155–204.

Deuser, W. G., E. H. Ross, C. Hemleben, and M. Spindler. 1981. Seasonal changes in species composition, number, mass, size and isotopic composition of planktonic foraminifera settling into the deep Sargasso Sea. *Palaeogeography, Palaeoclimatology, Palaeoecology* 33:103–127.

de Vernal, A., C. Hillaire-Marcel, and G. Bilodeau. 1996. Reduced meltwater outflow from the Laurentide ice margin during the Younger Dryas. *Nature* 381:774–777.

de Villiers, S., B. K. Nelson, and A. R. Chivas. 1995. Biological controls on coral Sr/Ca and $\delta^{18}O$ reconstructions of sea surface temperatures. *Science* 269:1247–1249.

De Visser, J. P., J. H. J. Ebbing, L. Gudjonsson, et al. 1989. The origin of rhythmic bedding in the Pliocene Trubi Formation of Sicily, southern Italy. *Palaeogeography, Palaeoclimatology, Palaeoecology* 69:45–66.

de Vries, H. 1958. Variation in the concentration of radiocarbon with time and location on Earth. *Proceedings Koninklijk Nederlands Akademie van Weterschappen, Series B* 61:94.

DeVries, T. J., L. Ortleib, A. Diaz, L. Wells, and C. Hillaire-Marcel. 1997. Determining the early history of El Niño. *Science* 276:996.

de Wolde, J. R., R. Bintania, and J. Oerlemans. 1995. On thermal expansion over the last hundred years. *Journal of Climate* 8:2881–2891.

Diaz, H. F. 1996. Temperature changes on long time and large spatial scales: Inferences from instrumental and proxy records. In P. D. Jones, R. S. Bradley, and J. Jouzel, eds., *Climate Variations and Forcing Mechanisms of the last 2000 Years,* NATO ASI Series 1, 41:587–601. Berlin: Springer-Verlag.

Diaz, H. F. and G. N. Kiladis. 1992. Atmospheric teleconnections associated with the extreme phases on the Southern Oscillation. In H. F. Diaz and V. Markgraf, eds., *El Niño: Historical and Paleoclimatic Aspects of the Southern Oscillation,* pp. 7–28. Cambridge: Cambridge University Press.

Diaz, H. F. and V. Markgraf, eds. 1992. *El Niño: Historical and Paleoclimatic Aspects of the Southern Oscillation.* Cambridge: Cambridge University Press.

Diaz, H. F. and R. S. Pulwarty. 1994. An analysis of the time scales of variability in centuries-long ENSO-sensitive records in the last 1000 years. *Climatic Change* 26:317–342.

Dickinson, R. E., G. A. Meehl, and W. M. Washington. 1987. Ice-albedo feedback in a CO_2-doubling simulation. *Climatic Change* 10:241–248.

Dickson, R. R. and J. Brown. 1994. The production of North Atlantic deep water: Sources, rates and pathways. *Journal of Geophysical Research* 99:12,319–12,341.

Dickson, R. R., H. H. Lamb, S.-A. Malberg, and J. M. Colebrook. 1975. Climatic reversal in the northern North Atlantic. *Nature* 256:479–482.

Dickson, R. R., Lazier, J. Meincke, P. Rhines, and J. Swift. 1996. *Progress in Oceanography* 38:241–295.

Dickson, R. R., J. Meincke, S.-A. Malberg, and A. J. Lee. 1988. The "great salinity anomaly" in the northern North Atlantic: 1968–1982. *Progress in Oceanography* 20:103–151.

DiMichele, W. A. 1994. Ecological patterns in time and space. *Paleobiology* 20:89–92.

Ding, Z., Z. Yu, N. W. Rutter, and T. Liu. 1994. Towards an orbital timescale for Chinese loess deposits. *Quaternary Science Reviews* 13:39–70.

Dixon, R. K., S. Brown, R. A. Houghton, A. M. Solomon, M. C. Trexler, and J. Wisniewski. 1994. Carbon pools and flux of global forest ecosystems. *Science* 263:185–190.

Dlugokencky, E. J., K. A. Masaire, P. P. Tans, L. P. Steele, and E. G. Nisbet. 1994. A dramatic decrease in the growth rate of atmospheric methane in the northern hemisphere during 1992. *Geophysical Research Letters* 21:45–48.

Dodge, R. E., R. G. Fairbanks, L. K. Benninger, and F. Maurrasse. 1983. Pleistocene sea levels from raised coral reefs of Haiti. *Science* 219:1423–1425.

Dolan, R., B. Hayden, and S. May. 1983. Erosion of US shorelines. In *Handbook of Coastal Processes and Erosion Control,* pp. 285–299. Boca Raton, Fla.: CRC Press.

Domack, E. W., S. E. Ishman, A. B. Stein, C. E. McClennen, and A. J. T. Jull. 1995. Late Holocene advance of the Müller Ice Shelf, Antarctic Peninsula: sedimentological, geochemical and palaeontological evidence. *Antarctic Science* 7:159–170.

Domack, E. W., T. A. Mashiotta, and L. A. Burkley. 1993. 300-year cyclicity in organic matter preservation in Antarctic fjord sediments. *The Antarctic Paleoenvironment: A Perspective on Global Change. Antarctic Research Series (American Geophysical Union)* 60:265–272.

Donovan, A. D. and E. J. W. Jones. 1979. Causes of world-wide changes in sea level. *Journal of the Geological Society of London* 136:187–192.

Dott, R. H. Jr., ed. 1992. *Eustasy: The Historical Ups and Downs of a Major Geological Concept.* Boulder: Geological Society of America.

Douglas, A. V. and P. J. Englehart. 1981. On a statistical relationship between Autumn rainfall in the central equatorial Pacific and subsequent winter precipitation in Florida. *Monthly Weather Review* 105:2377–2382.

Douglas, B. C. 1991. Global sea level rise. *Journal of Geophysical Research* 96:6981–6992.

Douglas, B. C. 1992. Global sea level acceleration. *Journal of Geophysical Research* 97(C8):12,699–12,706.

Dowdeswell, J. A., M. A. Maslin, J. T. Andrews, and I. N. McCave. 1995. Iceberg production, debris, rafting, the extent and thickness of Heinrich layers (H-1, H-2) in North Atlantic sediments. *Geology* 23:301–304.

Dowsett, H. J. and T. M. Cronin. 1990. High eustatic sea level during the middle Pliocene: Evidence from the southeastern U. S. Atlantic Coastal Plain. *Geology* 18:435–438.

Dowsett, H. J., J. Barron, and R. Poore. 1996. Middle Pliocene sea surface temperatures: A global reconstruction. *Marine Micropaleontology* 27:13–25.

Dowsett, H. J., T. M. Cronin, R. Z. Poore, R. C. Thompson, R. C. Whatley, and A. M. Wood. 1992. Micropaleontological evidence for increased meridional heat transport in the north Atlantic Ocean during the Pliocene. *Science* 258:1133–1135.

Dowsett, H., R. Thompson, J. Barron, et al. 1994. Joint investigations of the middle Pliocene climate I: PRISM paleoenvironmental reconstructions. *Global and Planetary Change* 9:169–195.

Dragan, J. C. and S. Airinei. 1989. Geoclimate and History. Rome: NAGARD. (Translated from Romanian.)

Dreimanis, A. and P. F. Karrow. 1972. Glacial history of the Great Lakes–St. Lawrence region, the classification of the Wisconsin (an) Stage, and its correlatives. *24th International Geological Congress,* Montreal. pp. 5–15.

Druffel, E. M. 1982. Banded corals: Changes in oceanic carbon-14 during the Little Ice Age. *Science* 218:13–19.

Druffel, E. M. 1985. Detection of El Niño and decade time scale variations of sea surface temperature from banded coral records: implications for the carbon dioxide cycle. In E. T. Sundquist and W. S. Broecker, eds., *The Carbon Cycle and Atmospheric* CO_2*: Natural Variations Archean to Present, American Geophysical Union Monograph* 32:111–121.

Druffel, E. M., R. B. Dunbar, G. M. Wellington, and S. A. Minnis. 1990. Reef-building corals and identification of ENSO warming episodes. In P. W. Glynn, ed., *Global Ecological Consequences of the 1982–83 El Niño–Southern Oscillation,* pp. 233–254. Amsterdam: Elsevier.

Druffel, E. M. and S. Griffin. 1993. Large variations of surface ocean radiocarbon: Evidence of circulation changes in the southwestern Pacific. *Journal of Geophysical Research* 218:20,249–20,259.

Dunbar, R. B. 1983. Stable isotope record of upwelling and climate from Santa Barbara Basin, California. In J. Thiede and E. Suess, eds., *Coastal Upwelling: Its Sediment Record,* Part B, pp. 217–246. New York: Plenum Press.

Dunbar, R. B. and J. E. Cole. 1993. *Coral Records of Ocean-Atmosphere Variability.* National Oceanographic and Atmosphere Administration Climate and Global Change Program Special Report No. 10, pp. 1–38. Boulder: University Corporation for Atmospheric Research.

Dunbar, R. B., B. K. Linsley, and G. M. Wellington. 1996. Eastern Pacific corals monitor El Niño/Southern Oscillation, precipitation, and sea surface temperature variability over the past 3 centuries. In P. D. Jones, R. S. Bradley, and J. Jouzel, eds., *Climate Variations and Forcing Mechanisms of the last 2000 Years,* NATO ASI Series 1, 41:373–405. Berlin: Springer-Verlag.

Dunbar, R. B. and G. M. Wellington. 1981. Stable isotopes in a branching coral monitor seasonal temperature variations. *Nature* 293:453–455.

Dunbar, R. B., G. M. Wellington, M. W. Colgan, and P. W. Glynn. 1994. Eastern Pacific sea surface temperature since 1600 A.D: The $\delta\ ^{18}O$ record of climate variability in Galapagos corals. *Paleoceanography* 9:291–315.

Duplessy, J.-C., G. Delibrias, J. L. Touron, C. Pujol, and J. Duprat. 1981. Deglacial warming of the northeastern Atlantic Ocean: Correlation with the paleoclimatic evolution of the European continent. *Palaeogeography, Palaeoclimatology, Palaeoecology* 35:121–144.

Duplessy, J.-C., L. Labeyrie, A. Julliet-Leclerc, F. Maitre, J. Dupont, and M. Sarnthein. 1991. Surface salinity reconstruction of the North Atlantic Ocean during the last glacial maximum. *Oceanologica Acta* 14:311–324.

Duplessy, J.-C., N. J. Shackleton, R. G. Fairbanks, L. Labeyrie, D. Oppo, and N. Kallel. 1988. Deep water source variations during the last climatic cycle and their impact on the global deep water circulation. *Paleoceanography* 3:343–360.

Duplessy, J.-C. and M.-T. Spyridakis, eds. 1994. *Long-term Climatic Variations: Data and Modelling,* NATO ASI Series, vol. 22. Berlin: Springer-Verlag.

Dupont, L. M., H.-J. Beug, H. Stalling, and R. Tiedemann. 1989. First palynological results from site 658 at 21°N off northwest Africa: Pollen as climate indicators. *Proceedings of the Ocean Drilling Program Scientific Results* 108:93–111.

Dutton, C. E. 1871. The cause of regional elevations and subsidences. *Proceedings of the American Philosophical Society* 7:70–72.

Dwyer, G. S., T. M. Cronin, P. A. Baker, M. E. Raymo, J. S. Buzas, and T. Correge. 1995. North Atlantic deepwater temperature change during late Pliocene and late Quaternary climatic cycles. *Science* 270:1347–1351.

Dwyer, T. R., H. T. Mullins, and S. C. Good. 1996. Paleoclimatic implictions of Holocene lake-level fluctuation: Oswego Lake, New York. *Geology* 24:519–522.

Dyke, A. S. and V. K. Prest. 1987. Late Wisconsin and Holocene history of the Laurentide Ice Sheet. *Geographie Physique et Quaternaire* 41:237–263.

Eddy, J. A. 1976. The Maunder minimum. *Science* 192:1189–1202.

Eddy, J. A. 1983. An historical review of solar variability, weather, and climate. In B. M. McCormac, ed., *Weather and Climate Responses to Solar Variations,* pp. 1–15. Boulder: Colorado Associated University Press.

Eddy, J. A. and H. Oeschger, eds. 1993. *Global Changes in the Perspective of the Past.* Chichester: John Wiley & Sons.

Edwards, L., P. J. Mudie, and A. de Vernal. 1991. Pliocene paleoclimatic reconstruction using dinoflagellate cysts: Comparison of methods. *Quaternary Science Reviews* 10:259–274.

Edwards, R. L. 1995. Paleotopography of glacial-age ice sheets. *Science* 267:536.

Edwards, R. L., J. H. Chen, and G. J. Wasserburg. 1987. ^{238}U-^{234}U-^{230}Th-^{232}Th systematics and the precise measurement of time over the past 500,000 years. *Earth and Planetary Science Letters* 81:175–191.

Edwards, R. L., H. Cheng, M. T. Murrell, and S. J. Goldstein. 1997. Protactinium-231 dating of carbonates by thermal ionization mass spectrometry: Implications for Quaternary climate changes. *Science* 276:782–786.

Edwards, R. L. and C. D. Gallup. 1993. Dating of the Devils Hole calcite vein. *Science* 259:1626–1627.

Eisenhauer, A., G. J. Wasserburg, J. H. Chen, et al. 1993. Holocene sea-level determination relative to the Australian continent: U/Th (TIMS) and ^{14}C (AMS) dating of coral cores from the Abrolhos Islands. *Earth and Planetary Science Letters* 114:529–547.

Elderfield, H. 1990. Tracers of ocean paleoproductivity and paleochemistry: An introduction. *Paleoceanography* 5:711–717.

Eldredge, N., ed. 1992. *Systematics, Ecology, and the Biodiversity Crisis.* New York: Columbia University Press.

Eldredge, N. and M. Grene. 1992. *Interactions: The Biological Context of Social Systems.* New York: Columbia University Press.

Elias, S. A. 1994. *Quaternary Insects and Their Environments.* Washington, D.C.: Smithsonian Institution Press.

Elson, J. A. 1969. Radiocarbon dates: *Mya arenaria* phase of the Champlain Sea. *Canadian Journal of Earth Science* 6:367–372.

Elverhoi, A., W. Fjeldskaar, A. Solheim, M. Nyland-Berg, and L. Russwurm. 1993. The Barents Sea Ice Sheet—a model of its growth and decay during the last ice maximum. *Quaternary Science Reviews* 12:863–873.

Emery, K. O. and D. G. Aubrey. 1991. *Sea Levels, Land Levels, and Tide Gauges.* New York: Springer-Verlag.

Emiliani, C. 1955. Pleistocene temperatures. *Journal of Geology* 63:538–578.

Emiliani, C. 1993. Milankovitch theory verified. *Nature* 364:583–584.

Epstein, S., R. Buchsbaum, H. A. Lowenstam, and H. C. Urey. 1953. Revised carbonate-water isotopic temperature scale. *Geological Society of America Bulletin* 64:1315–1326.

Epstein, S., R. P. Sharp, and A. J. Gow. 1970. Antarctic ice sheet: stable isotope analyses of Byrd Station cores and interhemispheric climatic implications. *Science* 168:1570–1572.

Erez, J. and S. Honjo. 1981. Comparison of isotopic composition of planktonic foraminifera in plankton tows, sediment traps, and sediments. *Palaeogeography, Palaeoclimatology, Palaeoecology* 33:129–156.

Ericson, D. B. and G. Wollin. 1968. Pleistocene climates and chronology in deep-sea sediments. *Science* 162:1227-1229.

Etheridge, D. M., G. I. Pearman, and P. J. Fraser. 1992. Changes in tropospheric methane between 1841 and 1978. *Tellus* 44B:282–294.

Etheridge, D. M., L. P. Steele, R. L. Langenfields, R. J. Francey, J.-M. Barnola, and V. I. Morgan. 1996. Natural and anthropogenic changes in atmospheric CO_2 over the last 1,000 years from air in Antarctic ice and firn. *Journal of Geophysical Research* 101:4115–4128.

Fairbanks, R. G. 1989. A 17,000-year glacio-eustatic sea level record: influence of glacial melting rates on the Younger Dryas event and deep-ocean circulation. *Nature* 342:637–642.

Fairbanks, R. G. and R. E. Dodge. 1979. Annual periodicity of the $^{18}O/^{16}O$ and $^{13}C/^{12}C$ ratios in the coral *Montastrea annularis. Geochimica et Cosmochimica Acta* 43:1009–1020.

Fairbanks, R. G., and R. K. Matthews. 1978. The marine oxygen isotope record in Pleistocene coral, Barbados, West Indies. *Quaternary Research* 10:181–196.

Fairbridge, R. W. 1961. Eustatic changes in sea-level. *Physics and Chemistry of the Earth* 4:99–185.

Farrell, J. W., D. W. Murray, V. S. McKenna, and A. C. Ravelo. 1995. Upper ocean temperature and nutrient contrasts inferred from Pleistocene planktonic foraminifer $\delta^{18}O$ and $\delta^{13}C$ in the eastern equatorial Pacific. *Proceedings of the Ocean Drilling Program Scientific Results* 138:289–319.

Field, M. E., E. P. Meisburger, E. A. Stanley, and S. J. Williams. 1979. Upper Quaternary peat deposits on the Atlantic inner shelf of the United States. *Geological Society of America Bulletin* 90:618–628.

Field, M., H. B. Huntley, and H. Müller. 1994. Eemian climate fluctuations observed in a European pollen record. *Nature* 371:779–783.

Fischer, A. F. 1981. Climatic oscillations in the biosphere. In M. H. Nitecki, ed., *Biotic Crises in Ecological and Evolutionary Time,* pp. 103–131. New Yord: Academic Press.

Fischer, A. G., T. D. Herbert, G. Napoleone, I. Premoli Silva, and M. Ripepe. 1991. Albian pelagic rhythms (Piobbico core). *Journal of Sedimentary Petrology* 61:1164–1172.

Fisher, D. A. and R. M. Koerner. 1994. Signal and noise in four ice-core records from the Agassiz Ice Cap, Ellesmere Island, Canada: Details of the last millennium for stable isotopes, melt and solid conductivity. *The Holocene* 4:113–120.

Fisher, D. A., R. M. Koerner, K. Kuivinen, et al. 1996. Inter-comparison of ice core del ^{18}O and precipitation records from sites in Canada and Greenland over the past 3500 years and over the last few centuries in detail using EOF techniques. In P. D. Jones, R. S. Bradley, and J. Jouzel, eds., *Climate Variations and Forcing Mechanisms of the Last 2000 Years,* NATO ASI Series 1, 41:297–328. Berlin: Springer-Verlag.

Fitt, W. K., H. Spero, J. Halas, M. W. White, and J. W. Porter. 1993. Recovery of the coral *Monastrea annularis* in the Florida Keys after the 1987 "bleaching event." *Coral Reefs* 12:57–64.

Fletcher, C. H. III. 1992. Sea-level trends and physical consequences: applications to the U.S. shore. *Earth-Science Reviews* 33:73–109.

Fletcher, C. H. III and J. F. Wehmiller, eds. 1992. *Quaternary Coasts of the United States: Marine and Lacustrine Systems. SEPM Special Publication* 48.

Flint, R. F. 1971. *Quaternary and Glacial Geology.* New York: John Wiley & Sons.

Flower, B. P. and J. P. Kennett. 1990. The Younger Dryas cool episode in the Gulf of Mexico. *Paleoceanography* 5:949–961.

Follmer, L. R. 1983. Sangamonian and Wisconsinan pedogenesis in the midwestern United States. In S. C. Porter, ed., *Late Quaternary Environments of the United States,* Vol. 1, *The Late Pleistocene,* pp. 138–144. Minneapolis: University of Minnesota Press.

Fortuin J. P. G. and J. Oerlemans. 1990. Parameterization of the annual surface temperature and mass balance of Antarctica. *Annals of Glaciology* 14:78–84.

Foukal, P. and J. Lean. 1990. An empirical model of total solar irradiance between 1874 and 1988. *Science* 254:698–700.

Frakes, L. A. 1979. *Climates Throughout Geologic Time.* Amsterdam: Elsevier.

Frakes, L. A., J. E. Francis, and J. I. Syktus. 1992. *Climate Modes of the Phanerozoic.* New York: Cambridge University Press.

Freeze, R. A. and J. A. Cherry. 1979. *Groundwater Hydrology.* New York: Prentice-Hall.

Friedli, H., H. Loetscher, H. Oeschger, U. Siegenthaler, and B. Stauffer. 1986. Ice core record of the $^{13}C/^{12}C$ ratio of atmospheric CO_2 in the past two centuries. *Nature* 324:237–238.

Friis-Christensen, E. and K. Lassen. 1991. Length of the solar cycle: An indicator of solar activity closely associated with climate. *Science* 254:698–700.

Fritts, H. C. 1976. *Tree Rings and Climate.* London: Academic Press.

Froelich, P. N., R. A. Mortlock, and A. Shemesh. 1989. Inorganic germanium and silica in the Indian Ocean: biological fractionation during (Ge/Si) opal formation. *Global Biogeochemical Cycles* 3:79–88.

Fronval, T. and E. Jansen. 1996. Rapid changes in ocean circulation and heat flux in the Nordic seas during the last interglacial period. *Nature* 383:806–810.

Fronval, T. and E. Jansen. 1997. Eemian and early Weischelian (140–60 ka) paleoceanography and paleoclimate in the Nordic seas with comparisons to Holocene conditions. *Paleoceanography* 12:443–462.

Frost, B. W. 1991. The role of grazing in nutrient-rich areas of the open sea. *Limnology and Oceanography* 36:1616–1630.

Fry, B. 1996. $^{13}C/^{12}C$ fractionation by marine diatoms. *Marine Ecology Progress Series* 134:283–294.

Fry, W. E. and S. B. Goodwin. 1997. Resurgence of the Irish potato famine fungus. *Bioscience* 47:363–371.

Fu, L.-L., C. J. Koblinsky, J.-F. Minster, and J. Picaut. 1996. Reflecting on the first three years of TOPEX/Poseidon. *EOS* March 19:109.

Gagan, M. K., A. R. Chivas, and P. J. Isdale. 1994. High-resolution isotope records from corals using ocean temperature and mass-spawning chronometers. *Earth and Planetary Science Letters* 121:549–558.

Gallup, C. G., R. L. Edwards, and R. G. Johnson. 1994. The timing of high sea levels over the past 200,000 years. *Science* 263:796–800.

Gard, G. 1993. Late Quaternary coccoliths at the North Pole: Evidence of ice-free conditions and rapid sedimentation in the central Arctic. *Geology* 21:227–230.

Gayes, P. T., D. B. Scott, E. S. Collins, and D. D. Nelson. 1992. A late Holocene sea-level fluctuation in South Carolina. In C. H. Fletcher, III and J. F. Wehmiller, eds., *Quaternary Coasts of the United States: Marine and Lacustrine Systems. SEPM Special Publication* 48:155–160.

Genthon, C., J. M. Barnola, D. Raynaud, et al. 1987. Vostok ice core: Climatic response to CO_2 and orbital forcing changes over the last climatic cycle. *Nature* 329:414–418.

Ghil, M. and S. Childress. 1987. Topics in geophysical fluid dynamics: Atmospheric dynamics, dynamo theory and climate dynamics. New York: Springer-Verlag.

Ghiselin, M. T. 1969. *The Triumph of the Darwinian Method.* Chicago: University of Chicago Press.

Gilbert, G. K. 1890. Lake Bonneville. *U. S. Geological Survey Memoir* 1:1–438.

Gilbert, G. K. 1895. Sedimentary measurement of geologic time. *Journal of Geology* 3:121–127.

Gillespie, A. and P. Molnar. 1995. Asynchronous maximum advances of mountain and continental glaciers. *Reviews of Geophysics* 33:311–364.

Gleason, H. A. 1922. The vegetational history of the middle west. *Annals of the Association of American Geographers.* 12:39–85.

Gleason, H. A. 1926. The individualistic concept of plant association. *Bulletin of Torrey Biological Club* 53:7–26.

Gleason, H. A. 1939. The individualistic concept of the plant association. *American Midland Naturalist* 21:92–108.

Gleissberg, W. 1966. Ascent and descent in the eighty-year cycles of solar activity. *Journal of the British Astronomical Association* 76:265–270.

Glynn, P. W., ed. 1990. *Global Ecological Consequences of the 1982–83 El Niño–Southern Oscillation.* Amsterdam: Elsevier.

Golley, F. B. 1993. *A History of the Ecosystem Concept in Ecology.* New Haven, Conn.: Yale University Press.

Gooday, A. J. 1986. Meiofaunal foraminiferans from the bathyal Porcupine Seabight: Size, structure, taxonomic composition, species diversity, and vertical distribution in the sediment. *Deep-sea Research* 33:1345–1373.

Gooday, A. J. 1988. A response by benthic foraminifera to the deposition of phytodetritus in the deep sea. *Nature* 332:70–73.

Gooday, A. J. 1993. Deep-sea benthic foraminiferal species which exploit phytodetritus: Characteristic features and controls on distribution. *Marine Micropaleontology* 22:187–205.

Gooday, A. J., L. A. Levin, P. Linke, and T. Heeger. 1992. The role of benthic foraminifera in deep-sea food webs and carbon cycling. In G. T. Rowe and V. Pariente, eds., *Deep-sea Food Chains and the Global Carbon Cycle,* pp. 63–91. Amsterdam: Kluwer Academic Publishers.

Gooday, A. J. and C. M. Turley. 1990. Responses by benthic organisms to inputs of organic material to the ocean floor: a review. *Philosophical Transactions of the Royal Society, Series A* 331:119–138.

Gornitz, V. 1995a. Monitoring sea level changes. *Climate Change* 31:515–544.

Gornitz, V. 1995b. Sea-level rise: A review of recent past and near-future trends. *Earth Surface Processes and Landforms* 20:7–20.

Gornitz, V. and S. Lebedeff. 1987. Global sea level changes during the past century. In D. Nummendal, O. H. Pilkey, and J. D. Howard, eds., *Sea Level Fluctuations and Coastal Evolution, SEPM Special Publication* 41:3–16.

Gornitz, V., C. Rosenzweig, and D. Hillel. 1994. Is sea level rising or falling? *Nature* 371:481.
Goslar, T., M. Arnold, M. F. Pazdur. 1995. The Younger Dryas cold event: Was it synchronous over the North Atlantic region? *Radiocarbon* 37:63–70.
Gosse, J. C., E. B. Evenson, J. Klein, B. Lawn, and R. Middleton. 1995. Precise cosmogenic ^{10}Be measurements in western North America: Support for a global Younger Dryas cooling event. *Geology* 23:877–880.
Gould, S. J. 1985. Paradox of the first tier: An agenda for paleobiology. *Paleobiology* 11:2–12.
Gould, S. J. 1987. *Times's Arrow, Time's Cycle: Myth and Metaphor in the Discovery of Geological Time.* Cambridge, Mass.: Harvard Univ. Press.
Gould, S. J. and R. C. Lewontin. 1979. The spandrels of San Marco and the Panglossian paradigm: A critique of the adapationist programme. *Proceedings of the Royal Society of London* 205:581–598.
Gould, S. J. and E. Vrba. 1982. Exaptation—a missing term in the science of form. *Paleobiology* 8:4–15.
Grabau, A. W. 1940. *The Rhythm of the Ages.* Peking: Henri Vetch.
Graedel, T. E. and P. J. Crutzen. 1993. *Atmospheric Change: An Earth System Perspective.* New York: W. H. Freeman and Company.
Graham, N. E. 1995. Simulation of recent global temperature trends. *Science* 267:666–671.
Graumlich, L. J. 1993. A 1000-year record of temperature and precipitation in the Sierra Nevada. *Quaternary Research* 39:249–255.
Graybill, D. A. and S. G. Shiyatov. 1992. Dendroclimatic evidence from the northern Soviet Union. In R. S. Bradley and P. D. Jones, eds., *Climate Since AD 1500,* pp. 393–414. London: Routledge.
Greenland Ice-core Project (GRIP) Members. 1993. Climate instability during the last interglacial period recorded in the GRIP ice core. *Nature* 364:203–207.
Greenland Summit Ice Cores. 1997. *Journal of Geophysical Research* 102, no. C12, pp. 26,315–26,886.
Grene, M. ed. 1984. *Dimensions of Darwinism.* Cambridge: Cambridge University Press.
Grimm, E. C., G. L. Jacobson, Jr., W. A. Watts, B. C. S. Hansen, and K. A. Maasch. 1993. A 50,000-year record of climate oscillations from Florida and its temporal correlation with the Heinrich Events. *Science* 261:198–200.
Grinnell, J. 1914. An account of mammoth and birds of the lower Colorado River Valley. *University of California Publications in Zoology* 12:51–294.
Grootes, P. M. 1993. Interpreting continental oxygen isotope records. In P. K. Swart, K. C. Lohman, J. McKenzie, and S. Savin, eds., *Climate Change in Continental Isotope Records. American Geophysical Union Monograph* 78:37–46.
Grootes, P. M. 1995. Ice cores as archives of decade-to-century-scale climate variability. In. D. G. Martinson, K. Bryan, M. Ghil, et al., eds., *Natural Climate Variability on Decade-to-Century Time Scales,* pp. 544–554. Washington, D. C.: National Research Council, National Academy of Science Press.

Grootes, P. M. and M. Stuiver. 1987. Ice sheet elevation changes from isotope profiles. In E. D. Waddington and J. S. Walder, eds., *The Physical Basis of Ice Sheet Modeling,* International Association of Hydrological Sciences No. 70, pp. 269–281.

Grootes, P. M., M. Stuiver, T. L. Saling, et al. 1990. A 1400-year oxygen isotope history from the Ross Sea area, Antarctica. *Annals of Glaciology* 14:94–98

Grootes, P., M. Stuiver, J. W. C. White, S. Johnsen, and J. Jouzel. 1993. Comparison of oxygen isotope records from the GISP2 and GRIP Greenland ice cores. *Nature* 366:552–554.

Grosswald, M. G. 1993. Extent and melting history of the late Weischelian ice sheet: the Barents-Kara continental margin. In W. R. Peltier, ed., *Ice in the Climate System,* pp. 1–20. New York: Springer-Verlag.

Grotch, S. L. and M. C. MacCracken. 1991. The use of general circulation models to predict regional climatic changes. *Journal of Climate* 4:286–303.

Grousset, F. E., L. Labeyrie, J. A. Sinko, et al. 1993. Patterns of ice-rafted detritus in the glacial North Atlantic (40°–55°N). *Paleoceanography* 8:175–192.

Grove, J. M. 1988. *The Little Ice Age.* London: Methuen.

Guilderson, T. P., R. G. Fairbanks, and J. L. Rubenstone. 1994. Tropical temperature variations since 20,000 years ago: Modulating interhemispheric climate change. *Science* 263:663–665.

Guiot, J. 1990. Methodology of the last climatic cycle reconstruction in France from pollen data. *Palaeogeography, Palaeoclimatology, Palaeoecology* 80:49–69.

Guiot, J., J.-L. de Beaulieu, R. R. Cheddadi, F. David, P. Ponel, and M. Reille. 1993. The climate in western Europe during the last Glacial/Interglacial cycle derived from pollen and insect remains. *Palaeogeography, Palaeoclimatology, Palaeoecology* 103:73–93.

Guiot, J., A. Pons, J.-L. de Beaulieu, and M. Reille. 1989. A 140,000-year continental climate reconstruction from two European pollen records. *Nature* 338:309–313.

Guiot, J., M. Reille, J.-L. de Beaulieu, and A. Pons. 1992. Calibration of the climatic signal in a new pollen sequence from La Grande Pile. *Climate Dynamics* 6:259–264.

Gwiazda, R. H., S. R. Hemming, and W. S. Broecker. 1996. Provenance of icebergs during Heinrich event 3 and the contrast to their sources during other Heinrich episodes. *Paleoceanography* 11:371–378.

Haake, F.-W. and U. Pflaumann. 1989. Late Pleistocene foraminiferal stratigraphy on the Vøring Plateau, Norwegian Sea. *Boreas* 18:343–356.

Haflidason, H., H. P. Sejrup, D. K. Kristiansen, and S. Johnsen. 1995. Coupled response of the late glacial climatic shifts of northwest Europe reflected in Greenland ice cores: evidence from the northern North Sea. *Geology* 23:1059–1062.

Hagadorn, J. W., L. D. Stott, A. Sinha, and M. Rincon. 1995. Geochemical and sedimentological variations in inter-annually laminated sediments from Santa Monica Basin. *Marine Geology* 125:111–131.

Hagelberg, T. K., G. Bond, and P. deMenocal. 1994. Milankovitch band forcing of sub-Milankovitch climate variability during the Pleistocene. *Paleoceanography* 9:545–558.

Hagelberg, T. K. and N. Pisias. 1990. Nonlinear response of Pliocene climate to orbital forcing: Evidence from the eastern equatorial Pacific. *Paleoceanography* 5:595–617.

Hajdas, I., S. Ivy, L. Beer, et al. 1993. AMS radiocarbon dating and varve chronology of Lake Soppensee: 6000–12,000 ^{14}C years BP. *Climate Dynamics* 9:107–116.

Hajdas, I., B. Zolitschka, S. D. Ivy-Ochs, et al. 1995. AMS radiocarbon dating of annually laminated sediments from Lake Holzmaar, Germany. *Quaternary Science Reviews* 14:137–143.

Hallam, A. 1963. Major epeirogenic and eustatic changes since the Cretaceous and their possible relationship to crustal structure. *American Journal of Science* 261:397–423.

Hallam, A. 1977. Secular changes in marine inundation of USSR and North America through the Phanerozoic. *Nature* 269:769–772.

Hallam, A. 1992. *Phanerozoic Sea-Level Changes.* New York: Columbia University Press.

Halpert, M. S. and C. F. Ropelewski. 1992. Surface temperature patterns associated with the Southern Oscillation. *Journal of Climate* 5:577–593.

Halpert, M. S. and T. M. Smith. 1994. The global climate for March-May 1993: Mature ENSO conditions persist and a blizzard blankets the Eastern United States. *Journal of Climate* 7:1772–1793.

Hamilton, K. and R. R. Garcia. 1986. El Niño/Southern Oscillation events and their associated midlatitude teleconnections. *Bulletin of the American Meteorological Society* 67:1354–1361.

Hammer, C. U. 1977. Past volcanism revealed by Greenland ice sheet impurity. *Nature* 270:482–486.

Hammer, C. U. 1984. Traces of Icelandic eruptions in the Greenland Ice Sheet. *Jokull* 34:51–65.

Hammer, C. U. 1989. Dating by physical and chemical seasonal variation and reference horizons. In H. Oeschger and C. C. Langway, Jr., eds., *The Environmental Record in Glaciers and Ice Sheets,* pp. 99–121. Chichester: John Wiley & Sons.

Hammer, C. U., H. B. Clausen, and W. Dansgaard. 1980. Greenland ice sheet evidence of post-glacial volcanism and its climatic impact. *Nature* 288:230–235.

Haq, B. U., J. Hardenbol, and P. R. Vail. 1987. The chronology of fluctuating sea level since the Triassic. *Science* 235:1156–1167.

Haq, B. U., J. Hardenbol, and P. R. Vail. 1988. Mesozoic and Cenozoic chronostratigraphy and cycles of sea-level change. *SEPM Special Publication* 42:71–108.

Harmon, R. S., R. M. Mitterer, N. Kriausakul, et al. 1983. U-series and amino-acid racemization geochronology of Bermuda: Implications for eustatic sea-level fluctuation over the past 250,000 years. *Palaeogeography, Palaeoclimatology, Palaeoecology* 44:41–70.

Hay, W. W. 1992. The cause of the late Cenozoic northern hemisphere glaciations: a climate change enigma. *Terra Nova* 4:305–311.

Hays, J. D., J. Imbrie, and N. J. Shackleton. 1976. Variations in the Earth's orbit: Pacemaker of the ice ages. *Science* 194:1121–1132.

Hazel, J. E. 1970. Atlantic continental shelf and slope of the United States—ostracode zoogeography in the southern Nova Scotian and northern Virginian faunal provinces. *U.S. Geological Survey Professional Paper* 529-E.

Hearty, P. J. 1987. New data on the Pleistocene of Mallorca. *Quaternary Science Reviews* 6:245–257.

Hearty, P. J. and P. Kindler. 1995. Sea-level highstand chronology from stable carbonate platforms (Bermuda and the Bahamas). *Journal of Coastal Research* 11:675–689.

Hecht, A. D., ed. 1985. *Paleoclimate Analyses and Modeling.* New York: John Wiley & Sons.

Heinrich, H. 1988. Origin and consequence of cyclic ice rafting in the northeast Atlantic Ocean during the past 130,000 years. *Quaternary Research* 29:142–152.

Helmans, K. F., and T. van der Hammen. 1994. The Pliocene and Quaternary of the high plains of Bogota (Colombia): A history of tectonic uplift, basin development and climatic changes. *Quaternary International* 21:41–61.

Hemleben, Ch. and J. Bijma. 1994. Foraminiferal population dynamics and stable carbon isotopes. In R. Zahn, T. F. Pedersen, M. A. Kaminski, and L. Labeyrie, eds., *Carbon Cycling in the Glacial Ocean: Constraints on the Ocean's Role in Global Change,* NATO ASI Series 17, pp. 145–166. Berlin: Springer-Verlag.

Hemleben, Ch., M. Spindler, and O. R. Anderson. 1989. *Modern Planktonic Foraminifera.* Berlin: Springer-Verlag.

Henrich, R., H. Kassens, E. Vogelsang, and J. Thiede. 1989. Sedimentary facies of glacial-interglacial cycles in the Norwegian Sea during the last 350 ka. *Marine Geology* 86:283–319.

Herguera, J. C. and W. H. Berger. 1994. Glacial to postglacial drop in productivity in the western equatorial Pacific: Mixing rate vs. nutrient concentrations. *Geology* 22:629–632.

Herschel, W. 1801. Observations tending to investigate the nature of the sun in order to find the causes or symptoms of its variable emission of light and heat. *Philosophical Transactions of the Royal Society London,* vol. 265 (cited in Hoyt and Schatten 1997).

Hicks, S. D. and L. E. Hickman. 1988. United States sea level variations through 1986. *Shore and Beach* 56:3–7.

Hilgen, F. J. 1987. Sedimentary rhythms and high-resolution chronostratigraphic correlations in the Mediterranean Pliocene. *Newsletters in Stratigraphy* 17:109–127.

Hilgen, F. J. 1991a. Astronomical calibration of Gauss to Matuyama sapropels in the Mediterranean and implication for the geomagnetic polarity time scale. *Earth and Planetary Science Letters* 104:226–244.

Hilgen, F. J. 1991b. Extension of the astonomically calibrated (polarity) time scale to the Miocene/Pliocene boundary. *Earth and Planetary Science Letters* 107:349–368.

Hilgen, F. J. and C. G. Langereis. 1989. Periodicities of $CaCO_3$ cycles in the Pliocene of Sicily: Discrepancies with the quasi-periods of the Earth's orbital cycles. *Terra Nova* 1:409–415.

Hilgen, F. J. and C. G. Langereis. 1993. A critical re-evaluation of the Miocene/Pliocene boundary as defined in the Mediterranean. *Earth and Planetary Science Letters* 118:167–179.

Hillaire-Marcel, C., A. de Vernal, G. Bilodeau, and G. Wu. 1993. Isotope stratigraphy, sedimentation rates, deep circulation, and carbonate events in the Labrador Sea during the last 200 ka. *Canadian Journal of Earth Sciences* 31:63–89.

Hillaire-Marcel, C., G. Gariepy, B. Ghaleb, J.-L. Goy, C. Zazo, and J. C. Barcelo. 1996. U-series measurements in Tyrrhenian deposits from Mallorca—further evidence for two last-interglacial high sea levels in the Balearic Islands. *Quaternary Science Reviews* 15:63–75.

Hillaire-Marcel, C. and S. Occhietti. 1980. Chronology, paleogeography, and paleoclimatic significance of the late and post-glacial events in eastern Canada. *Zeitschrift fur Geomorphologie* 24:373–392.

Hodell, D. A., J. H. Curtis, and M. Brenner. 1995. Possible role of climate in the collapse of Classic Maya civilization. *Nature* 375:391–394.

Hodell, D. A. and K. Venz. 1992. Towards a high-resolution stable isotopic record of the Southern Ocean during the Pliocene-Pleistocene (4.8–0.8 Ma). In J. P. Kennett and D. A. Warnke, eds., *The Antarctic Paleoenvironment: A Perspective on Global Change,* Part 1, Antarctic Research Series 56:265–310. Washington, D.C.: American Geophysical Union.

Hoffert, M. I. and C. Covey. 1992. Deriving global climate sensitivity from palaeoclimate reconstructions. *Nature* 360:573–576.

Hogbom, A. G. 1921. Nivaforandringarna I Norden. *Goteborgs Kungl. Vetebsjaps-och Vitterhats Saümhalle Handlungen* 4th f., 21, 3:1–160. (Cited in Morner 1979a.)

Holland, H. D. 1984. *The Chemical Evolution of the Atmosphere and Oceans.* Princeton, N.J.: Princeton University Press.

Holmes, J. 1994. Nonmarine ostracodes as Quaternary palaeoenvironmental indicators. *Progress in Physical Geography* 16:405–431.

Hooghiemstra, H. 1995. Environmental and paleoclimatic evolution in Late Pliocene-Quaternary Colombia. In E. S. Vrba, G. H. Denton, T. C. Partridge, and L. H. Burckle, eds., *Paleoclimate and Evolution, With Emphasis on Human Origins,* pp. 249–261. New Haven, Conn.: Yale University Press.

Hooghiemstra, H., C. O. C. Agwu, and H.-J. Beug. 1986. Pollen and spore distribution in recent marine sediments: a record of NW African seasonal wind patterns and vegetation belts. *Meteor Forschungen-Ergebnisse. C* 40:87–135.

Hooghiemstra, H. and A. M. Cleef. 1995. Pleistocene climatic change and environmental and generic dynamics in the North Andean montane forest and Paramo. In S. P. Churchill , H. Balslev, E. Farero, J. T. Luteyn, eds., *Bio-*

diversity and Conservation of Neotropical Montane Forests, pp. 35–49. New York: New York Botanical Garden.

Hooghiemstra, H., J. I. Melice, A. Berger, and N. J. Shackleton. 1993. Frequency spectra and paleoclimatic variability of the high-resolution 30–1450 ka Funza I pollen record (eastern Cordillera, Colombia). *Quaternary Science Reviews* 12:141–156.

Hooghiemstra, H. and E. T. H. Ran. 1994. Late Pliocene-Pleistocene high resolution pollen sequence of Colombia: An overview of climatic change. *Quaternary International* 21:63–80.

Hooghiemstra, H. and G. Sarmiento. 1991. Long continental pollen record from a tropical intermontane basin: Late Pliocene and Pleistocene history from a 540-meter core. *Episodes* 14:107–115.

Hopley, D. 1982. *The Geomorphology of the Great Barrier Reef.* New York: John Wiley & Sons.

Houghton, J. T., G. J. Jenkins, and J. J. Ephraums, eds. 1990. *Climate Change: the IPCC Scientific Assessment.* Cambridge: Cambridge University Press.

Houghton, J. T., L. G. Meira Filho, B. A. Callender, N. Harris, A. Kattenberg, and K. Maskell, eds. 1996. *Climate Change 1995: The Science of Climate Change.* Cambridge: Cambridge University Press.

Hovan, S. A., D. K. Rea, N. G. Pisias, and N. J. Shackleton. 1989. A direct link between the China loess and marine $\delta\ ^{18}O$ records: Eolian flux to the north Pacific. *Nature* 340:296–298.

Hoyt, D. V. and K. H. Schatten. 1993. A discussion of plausible solar irradiance variations, 1700–1992. *Journal of Geophysical Research* 98:18,895–18,906.

Hoyt, D. V. and K. H. Schatten. 1997. *The Role of the Sun in Climate Change.* New York: Oxford University Press.

Hudson, J. H., J. V. D. Powell, M. B. Robblee, and T. J. Smith. 1989. A 107-year old coral from Florida Bay: barometer of natural and man-induced catastrophes? *Bulletin of Marine Science* 44:283–291.

Hughen, K., J. T. Overpeck, L. C. Peterson, and S. Trumbore. 1996. Rapid climate changes in the tropical Atlantic regions during the last deglaciation. *Nature* 380:51–54.

Hughes, M. K. and P. M. Brown. 1992. Drought frequency in central California since 101 B.C. recorded in giant sequoia tree rings. *Climate Dynamics* 6:161–167.

Hughes, M. K. and H. F. Diaz. 1994a. *The Medieval Warm Period.* Dordrecht: Kluwer Academic.

Hughes, M. K. and H. F. Diaz, eds. 1994b. Was there a "Medieval Warm Period," and if so where and when? *Climatic Change* 26:109–142.

Hughes, M. K. and L. J. Graumlich. 1996. Multimillennial dendroclimatologic studies from the western United States. In P. D. Jones, R. S. Bradley, and J. Jouzel, eds., *Climate Variations and Forcing Mechanisms of the Last 2000 Years,* NATO ASI Series 1, 41:109–124. Berlin: Springer-Verlag.

Hughes, T. J., G. H. Denton, B. G. Andersen, D. H. Schilling, J. L. Fastook, and C. S. Lingle. 1981. Numerical reconstruction of paleo-ice sheets. In G. H.

Denton and T. J. Hughes, eds., *The Last Great Ice Sheets*, pp. 263–317. New York: John Wiley & Sons.

Hull, D. 1973. *Darwin and His Critics*. Cambridge, Mass.: Harvard University Press.

Huntley, B. and I. C. Prentice. 1988. July temperatures in Europe from pollen data 6000 years before present. *Science* 241:687–690.

Hurrell, J. W. 1995. Decadal trends in the North Atlantic Oscillation: Regional temperatures and precipitation. *Science* 269:676–679.

Huston, M. A. 1994. *Biological Diversity*. Cambridge: Cambridge University Press.

Hutchinson, G. E. 1957. Concluding Remarks. *Cold Spring Harbor Symposium on Quantitative Biology* 22:415–427.

Hutchinson, G. E. 1959. Homage to Santa Rosalia or why are there so many kinds of animals? *American Naturalist* 93:145–159.

Hutchinson, G. E. 1978. *An Introduction to Population Ecology*. New Haven, Conn.: Yale University Press.

Huybrechts, P. 1990. A 3-D model for the Antarctic ice sheet: a sensitivity study on the glacial interglacial contrast. *Climate Dynamics* 5:79–92.

Huybrechts, P. 1993. Glaciological modeling of the late Cenozoic East Antarctic ice sheet: stability or dynamism. *Geografiska Annaler* 75A:221–238.

Huybrechts, P. 1994. The present evolution of the Greenland ice sheet: an assessment by modeling. *Global and Planetary Change* 9:39–51.

Ikeya, N. and T. M. Cronin. 1993. Quantitative analysis of Ostracoda and water masses around Japan: Application to Neogene paleoceanography. *Micropaleontology* 39:263–281.

Imbrie, J., A. McIntyre, and A. Mix. 1989. Oceanic response to orbital forcing in the late Quaternary: Observational and experimental strategies. In A. Berger, ed., *Climate and Geo-sciences* 285:121–164. Kluwer: Norwell, Massachusetts.

Imbrie, J., A. Berger, E. Boyle, et al. 1993a. On the structure and origin of major glaciation cycles. 2. The 100,000-year cycle. *Paleoceanography* 8:699–735.

Imbrie, J., E. Boyle, S. Clemens, et al. 1992. On the structure and origin of major glaciation cycles. 1. Linear responses to Milankovitch forcing. *Paleoceanography* 7:701–738.

Imbrie, J., J. D. Hays, D. G. Martinson, et al. 1984. The orbital theory of Pleistocene climate: support from a revised chronology of the marine del ^{18}O record. In A. Berger, J. Imbrie, J. Hays, G. Kukla, and B. Saltzman, eds., *Milankovitch and Climate: Understanding the Response to Astronomical Forcing*, NATO ASI Series, vol. 126, Parts 1, 2, pp. 269–305. Dordrecht: D. Reidel.

Imbrie, J. and K. P. Imbrie. 1979. *Ice Ages, Solving the Mystery*. Cambridge: Harvard University Press.

Imbrie, J. and J. Z. Imbrie. 1980. Modeling the climatic response to orbital variations. *Science* 207:943–953.

Imbrie, J. and N. G. Kipp. 1971. A new micropaleontological method for quantitative paleoclimatology: Application to a late Pleistocene Caribbean core.

In K. K. Turekian, ed., *Late Cenozoic Glacial* Ages, pp. 71–181. New Haven, Conn.: Yale University Press.

Imbrie, J., A. C. Mix, and D. G. Martinson. 1993b. Milankovitch theory viewed from Devils Hole. *Nature* 363:531–533.

Iversen, J. 1954. The late-glacial flora of Denmark and its relationship to climate and soil. *Danmarks Geologische Undersogelse Series II* 75:1–175.

Jablonski, D. and J. J. Sepkoski. 1996. Paleobiology, community ecology, and scales of ecological pattern. *Ecology* 77:1367–1378.

Jacobs, S. S. 1992. Is the Antarctic ice sheet growing? *Nature* 360:29–33.

Jacoby, G. C. and R. D'Arrigo. 1989. Reconstructed northern hemisphere annual temperatures since 1671 based on high-latitude tree-ring data from North America. *Climatic Change* 14:39–59.

Jacoby, G. C., R. D. D'Arrigo, and B. Luckman. 1996a. Millennial and near-millennial scale dendroclimatic studies in northern North America. In P. D. Jones, R. S. Bradley, and J. Jouzel, eds., *Climate Variations and Forcing Mechanisms of the Last 2000 Years,* NATO ASI Series 1, 41:67–84. Berlin: Springer-Verlag.

Jacoby, G. C., R. D. D'Arrigo, T. Devaajamts. 1996b. Mongolian tree rings and 20th-century warming. *Science* 273:771–773.

Jacoby, R. and N. Glauberman, eds. 1995. *The Bell Curve Debate.* New York: Times Books.

Jamieson, T. E. 1865. On the history of the last geological changes in Scotland. *Quarterly Journal of the Geological Society of London* 21:161–203.

Janacek, T. R. and D. K. Rea. 1985. Quaternary fluctuation in the Northern Hemisphere trade winds and westerlies. *Quaternary Research* 24:645–672.

Jansen, E. 1987. Rapid changes in the inflow of Atlantic water into the Norwegian Sea at the end of the last glaciation. In W. H. Berger and L. D. Labeyrie, eds., *Abrupt Climatic Change,* pp. 299–310. Dordrecht: D. Reidel.

Jansen, E. and K. R. Bjørklund. 1985. Surface Ocean circulation in the Norwegian Sea 15,000 B.P. to present. *Boreas* 14:243–257.

Jansen, E. and J. Sjøholm. 1991. Reconstruction of glaciation over the past 6 Myr from ice-borne deposits in the Norwegian Sea. *Nature* 349:600–603.

Jansen, E., J. Sjøholm, U. Bleil, and J. A. Erichsen. 1990. Neogene and Pleistocene glaciation in the northern hemisphere and late Miocene-Pliocene global ice volume fluctuations: evidence from the Norwegian Sea. In U. Bleil and J. Thiede, eds., *Geological History of the Polar Oceans: Arctic versus Antarctic,* pp. 677–705. Netherlands: Kluwer.

Jelgersma, S. 1996. Land subsidence in coastal lowlands. In J. D. Milliman and B. U. Haq, eds. Sea-Level Rise and Coastal Subsidence, pp. 47–62. Dordrecht: Kluwer Academic.

Jenkins, G. M. and D. G. Watts. 1968. *Spectral Analysis and Its Applications.* San Francisco: Holden-Day.

Jensen, K. 1935. Archaeological dating in the history of North Jutland's vegetation. *Acta Archaeologica 5:185–214.*

Jirikowic, J. L. and P. E. Damon. 1994. The Medieval solar activity maximum. *Climatic Change* 26:309–316.

REFERENCES

Jirikowic, J. L., R. M. Kalin, and O. K. Davis. 1993. Tree-ring ^{14}C as a possible indicator of climate change. In P. K.Swart, K. C. Lohman, J. McKenzie, and S. Savin, eds., *Climate Change in Continental Isotope Records, American Geophysical Union Monograph* 78:353–366.

Johnsen, S. J., H. B. Clausen, W. Dansgaard, et al. 1992. Irregular glacial interstadials recorded in a new Greenland ice core. *Nature* 359:311–313.

Johnsen, S. J., H. B. Clausen, W. Dansgaard, et al. 1997. The $\delta^{18}O$ record along the Greenland Ice Core Project deep ice core and the problem of possible Eemian climatic instability. *Journal of Geophysical Research* 102, no. C12: 26, 397–26,410.

Johnsen, S. J., D. Dahl-Jensen, W. Dansgaard, and N. Gundstrup. 1995. Greenland palaeotemperature derived from GRIP borehole temperatures and ice core isotope profiles. *Tellus* 45B:624–630.

Johnsen, S. J., W. Dansgaard, and J. W. C. White. 1989. The origin of Arctic precipitation under present and glacial conditions. *Tellus* 41B:452–468.

Jones, G. A. and L. D. Keigwin. 1988. Evidence from Fram Strait (78°N) for early deglaciation. *Nature* 336:56–59.

Jones, G. A. and W. F. Ruddiman. 1982. Assessing the global meltwater spike. *Quaternary Research* 17:148–172.

Jones, P. D. 1994. Hemispheric surface air temperature variations: A reanalysis and an update to 1993. *Journal of Climate* 7:1794–1802.

Jones, P. D., R. S. Bradley, and J. Jouzel, eds., 1996. *Climate Variations and Forcing Mechanisms of the Last 2000 Years,* NATO ASI Series 1, Vol. 41. Berlin: Springer-Verlag.

Jones, P. D. and K. R. Briffa. 1996. What can the instrumental record tell us about longer timescale paleoclimate reconstructions? In P. D. Jones, R. S. Bradley, and J. Jouzel, eds., *Climate Variations and Forcing Mechanisms of the Last 2000 Years,* NATO ASI Series 1, 41:625–644. Berlin: Springer-Verlag.

Jouzel, J., N. I. Barkov, J.-M. Barnola, et al. 1993. Extending the Vostok ice-core record of palaeoclimate to the penultimate glacial period. *Nature* 364:407–412.

Jouzel, J., N. I. Barkov, J. M. Barnola, et al. 1989. Global change over the last climatic cycle from the Vostok ice core record (Antarctica). *Quaternary International* 2:15–24.

Jouzel, J., C. Lorius, J. R. Petit, et al. 1987. Vostok ice core: A continous isotopic temperature record over the last climatic cycle (160,000 years). *Nature* 329:403–408.

Jouzel, J., L. Merlivat, and C. Lorius. 1982. Deuterium excess in an east Antarctic ice core suggests higher relative humidity at the oceanic surface during the last glacial maximum. *Nature* 299:688–691.

Jouzel, J., J. R. Petit, J. M Barnola, et al. 1992. The last deglaciation in Antarctica: Further evidence for a "Younger Dryas" type climatic event. In E. Bard and W. S. Broecker, eds., *The Last Deglaciation: Absolute and Radiocarbon Chronologies,* pp. 229–266. Berlin: Springer-Verlag.

Jouzel, J., R. Vaikmae, J. R. Petit, et al. 1995. The two-step shape and timing of the last deglaciation in Antarctica. *Climate Dynamics* 11:151–161.

Kaiser, K. F. 1994. Two Creeks interstade dated through dendrochronology and AMS. *Quaternary Research* 42:288–298.

Kapsner, W. R., R. B. Alley, C. A. Shuman, S. Anandakrishnan, and P. M. Grootes. 1995. Dominant influence of atmospheric circulation on snow accumulation in Greenland over the past 18,000 years. *Nature* 373:52–54.

Kareiva, P. M., J. G. Kingsolver, and R. B. Huey. 1993. *Biotic Interactions and Global Change.* Sunderland, Mass.: Sinauer Associates.

Kasting, J. F. 1993. Earth's early atmosphere. *Science* 259:920–926.

Kaufman, A., W. S. Broecker, T.-L. Ku, and D. I. Thurber. 1971. The status of U-series methods of mollusc dating. *Geochimica et Cosmochimica Acta* 35:1115–1183.

Kaufman, D. S., G. H. Miller, J. A. Stravers, and J. T. Andrews. 1993. Abrupt early Holocene (9.9–9.6 ka) ice-stream advance at the mouth of Hudson Strait, Arctic, Canada. *Geology* 21:1063–1066.

Keeling, C. D. 1973. Industrial production of carbon dioxide from fossil fuels and limestone. *Tellus* 5:174–198.

Keeling, C. D., R. B. Bacastow, A. F. Carter, et al. 1989. A three-dimensional model of atmospheric CO_2 transport based on observed winds. I. Analysis and observational data. In D. H. Peterson, ed., *Aspects of Climatic Variability in the Pacific and Western Americas, American Geophysical Union Monograph* 55:165–236.

Keigwin, L. D. 1996. The Little Ice Age and Medieval Warm Period in the Sargasso Sea. *Science* 274:1504–1508.

Keigwin, L. D., W. B. Curry, S. J. Lehman, and S. Johnsen. 1994. The role of the deep ocean in North Atlantic climate change between 70 and 130 kyr ago. *Nature* 371:323–326.

Keigwin, L. D. and G. A. Jones. 1994. Western North Atlantic evidence for millennial-scale changes in ocean circulation and climate. *Journal Geophysical Research* 99:12,397–12,410.

Keigwin, L. D., G. A. Jones, and S. J. Lehman. 1991. Deglacial meltwater discharge, North Atlantic deep circulation, and abrupt climate change. *Journal of Geophysical Research* 96(C9):16,811–16,826.

Kennedy, J. A. and S. Brassell. 1992. Molecular records of twentieth-century El Niño events in laminated sediments from the Santa Barbara Basin. *Nature* 357:62–64.

Kennett, J. P., K. Elmstrom, and N. L. Penrose. 1985. The deglaciation in the Orca Basin Gulf of Mexico: High-resolution planktonic foraminifera. *Palaeogeography, Palaeclimatology, Palaeoecology* 50:189–216.

Kennett, J. P. 1990. The Younger Dryas cooling event: An introduction. *Paleoceanography* 5:891–895.

Kennett, J. P. and D. A. Hodell. 1993. Evidence for relative climatic stability of Antarctica during the early Pliocene: A marine perspective. *Geografiska Annaler* 75A:205–220.

Kennett, J. P. and B. L. Ingram. 1995. A 20,000-year record of ocean circulation and climate change from the Santa Barbara Basin. *Nature* 377:510–514.

Kennett, J. P. and N. J. Shackleton. 1975. Laurentide ice sheet meltwater recorded in the Gulf of Mexico. *Science* 188:147–150.

Kier, R. S. and W. H. Berger. 1984. Atmospheric CO_2 content in the last 120,000 years: The phosphate extraction model. *Journal of Geophysical Research* 88:6027–6038.

Kinealy, C. 1994. *This Great Calamity.* Boulder: Roberts Rinehart Publishers.

Knox, F. and M. B. McElroy. 1984. Changes in atmospheric CO_2: Influence of the marine biota at high latitudes. *Journal of Geophysical Research* 89:4629–4637.

Knudson, D. W., R. W. Buddmeier, and S. V. Smith. 1972. Coral chronologies: seasonal growth bands in reef corals. *Science* 177:270–272.

Knutson, T. R. and S. Manabe. 1994. Impact of increased CO_2 on simulated ENSO-like phenomena. *Geophysical Research Letters* 21:2295–2298.

Koç-Karpuz, N. and E. Jansen. 1992. A high resolution diatom record of the last deglaciation from the southeastern Norwegian Sea: Documentaion of rapid climatic changes. *Paleoceanography* 5:557–580.

Koç, N. and E. Jansen. 1994. Response of the high-latitude Northern Hemisphere to orbital climatic forcing: Evidence from the Nordic Seas. *Geology* 22:523–526.

Koch, G. W. and H. A. Mooney, eds. 1996. *Carbon Dioxide and Terrestrial Ecosystems.* New York: Academic Press.

Kohfeld, K. E., R. G. Fairbanks, S. L. Smith, and I. D. Walsh. 1996. *Neogloboquadrina pachyderma* (sinistral coiling) as paleoceanographic tracers in polar oceans: Evidence from Northeast water polynya plankton tows, sediment traps, and surface sediments. *Paleoceanography* 11:679–699.

Kominz, M. 1984. Oceanic ridge volumes and sea-level change—an error analysis. *American Association of Petroleum Geologists Memoir* 36: 109–127.

Konishi, K., A. Omura, and O. Nakamichi. 1974. Radiometric coral ages and sea level records from the late Quaternary reef complexes of the Ryukyu Islands. In *Proceedings of the 2nd International Coral Reef Symposium,* 2:595–613. Brisbane: Great Barrier Reef Commission.

Korner, C. and F. A. Bazzaz, eds. 1996. *Carbon Dioxide, Populations, and Communities.* New York: Academic Press.

Kousky, V. E., M. T. Kagano, and I. F. A. Cavalcanti. 1984. A review of the Southern Oscillation: Oceanic-atmospheric circulation changes and related rainfall anomalies. *Tellus* 36A:490–504.

Kreutz, K. J., P. A. Mayewski, L. D. Meeker, M. S. Twickler, S. I. Whitlow, I. I. Pittalwala. 1997. Bipolar changes in atmospheric circulation during the little ice age. *Science* 277:1294–1296.

Krijgsman, W., F. J. Hilgen, C. G. Langereis, A. Santarelli, and W. J. Zachariasse. 1995. Late Miocene magnetostratigraphy, biostratigraphy and cyclostratigraphy in the Mediterranean. *Earth and Planetary Science Letters* 136:475–494.

Kromer, B. and B. Becker. 1992. Tree ring ^{14}C calibration at 10,000 BP. In E. Bard and W. S. Broecker, eds., *The Last Deglaciation: Absolute and Radiocarbon Chronologies,* NATO ASI Series 1, 2:3–12. Berlin: Springer-Verlag.

Kromer, B. and B. Becker. 1993. German oak and pine ^{14}C calibration, 7200–9439 B.C. *Radiocarbon* 35:125–135.

Kromer, B., B. Becker, M. Spurk, and P. Trimborn. 1994. Radiocarbon timescale in early Holocene and isotope time series based on tree ring chronologies. *Terra Nostra* 1:31–33.

Ku, T.-L., M. A. Kimmel, W. H. Easton, T. J. O'Neil. 1974. Eustatic sea level 120,000 years ago on Oahu, Hawaii. *Science* 183:959–961.

Kuhn, T. S. 1962. *The Structure of Scientific Revolutions.* Chicago: The University of Chicago Press.

Kukla, G. 1975. Loess stratigraphy in Europe. In K. W. Butzer and G. L. Isaac, eds., *After the Australopithecines,* pp. 99–188. The Hague: Mouton.

Kukla, G. 1987. Loess stratigraphy in central China. *Quaternary Science Reviews* 6:191–219.

Kukla, G., F. Heller, X. M. Liu, T. C. Xu, T. S. Kiu, and Z. S. An. 1988. Pleistocene climates in China dated by magnetic susceptibility. *Geology* 16:811–814.

Kukla, G., J. F. McManus, D.-D. Rousseau, and I. Chiune. 1997. How long and how stable was the last interglacial. *Quaternary Science Reviews* 16:605–612.

Kutzbach, J. E. and P. J. Guetter. 1986. The influence of changing orbital parameters and surface boundary conditions on climatic simulations for the past 18,000 years. *Journal of Atmospheric Science* 43:1726–1759.

Labeyrie, L. D., J.-C. Duplessy, and P. L. Blanc. 1987. Variations in mode of formation and temperature of oceanic deep waters over the past 125,000 years. *Nature* 327:477–482.

Ladurie, E. L. 1971. *Times of Feast, Times of Famine.* New York: Doubleday.

Laird, K. R., S. C. Fritz, E. C. Grimm, and P. G. Mueller. 1996. Century-scale paleoclimate reconstruction from Moon Lake, a closed basin lake in the northern Great Plains. *Limnology and Oceanography* 41:890–902.

Laird, K. R., S. C. Fritz, K. A. Maasch, and B. F. Cumming. 1997. Greater drought intensity and frequency before AD 1200 in the northern Great Plains, USA. *Nature* 384:552–554.

Lamb, H. H. 1965. The early Medieval Warm Epoch and its sequel. *Palaeogeography, Palaeoclimatology, Palaeoecology* 1:13–27.

Lamb, H. H. 1977. *Climate History and the Future,* Vol. 2, *Climate Present, Past, and Future.* London: Methuen & Co.

Lamb, H. H. 1984. Climate history in northern Europe and elsewhere. In N.-A. Morner and W. Karlen, eds., *Climatic Changes on a Yearly to Millennial Basis,* pp. 225–240. Dordrecht: D. Reidel.

Lamb, H. H. 1995. *Climate History and the Modern World,* 2nd ed. London: Routledge.

Landsberg, H. E. 1985. Historic weather data and early meteorological observations. In A. D. Hecht ed., *Paleoclimatic Analysis and Modeling,* pp. 27–69. New York: Wiley.

Landwehr, J. M., I. J. Winograd, and T. B. Coplen. 1994. No verification for Milankovitch. *Nature* 368:594.

Langway, C. C. Jr., H. B. Clausen, and C. U. Hammer. 1988. An inter-hemispheric volcanic time-marker in ice cores from Greenland and Antarctica. *Annals of Glaciology* 10:102–108.

Langway, C. C. Jr., H. Oeschger, and W. Dansgaard, eds. 1985. *Greenland Ice Core: Geophysics, Geochemistry, and the Environment, American Geophysical Union Monograph* 33.

Lara, A. and R. Villalba. 1993. A 3620-year temperature record from *Fitzroya cupressoides* tree rings in southern South America. *Science* 260:1104–1106.

Larsen, E., F. Eide, O. Lonva, and J. Mangerud. 1984. Allerød-Younger Dryas climatic inferences from cirque glaciers and vegetational development in the Nordfjord area, western Norway. *Arctic and Alpine Research* 16:137–160.

Larsen, E., H. P. Sejrup, S. J. Johnsen, and K. L. Knudsen. 1995. Do Greenland ice cores reflect NW European interglacial climate variations? *Quaternary Research* 43:125–132.

Larsen, H. C., A. D. Saunders, P. D. Clift, et al. 1994. Seven million years of glaciation in Greenland. *Science* 264:952–955.

LaSalle, P. and J. A. Elson. 1975. Emplacement of the St. Narcisse Moraine as a climatic event in eastern Canada. *Quaternary Research* 5:621–625.

LaSalle, P. and W. W. Shilts. 1993. Younger Dryas–age readvance of Laurentide ice into the Champlain Sea. *Boreas* 22:25–37.

Latif, M., T. P. Barnett, M. A. Cane, et al. 1994. A review of ENSO prediction studies. *Climate Dynamics* 9:167–179.

Lazier, J. R. N. 1995. The salinity decrease in the Labrador sea over the past thirty years. In D. G. Martinson, K. Bryan, M. Ghil, et al., eds., *Natural Climate Variability on Decade-to-Century Timescales,* pp. 295–304. Washington, D.C.: National Academy of Sciences.

Lea, D. W. and E. A. Boyle. 1990. Foraminiferal reconstruction of barium distributions in water masses of the glacial oceans. *Paleoceanography* 5:719–742.

Lea, D. W., G. T. Shen, and E. A. Boyle. 1989. Coralline barium records temporal variability in equatorial Pacific upwelling. *Nature* 340:373–376.

Lean, J., J. Beer, and R. Bradley. 1995. Reconstruction of solar irradiance since 1610: Implications for climate change. *Geophysical Research Letters* 22:3195–3198.

Lean, J., A. Skumanich, and O. White. 1992. Estimating the sun's radiative output during the Maunder Minimum. *Geophysical Research Letters* 19:1591–1594.

Leder, J. J., P. K. Swart, A. Szmant, and R. E. Dodge. 1996. The origin of variations in the isotopic record of scleractinian corals: I. Oxygen. *Geochimica et Cosmochimica Acta* 60:2857–2870.

Legrand, M. 1995. Sulfur-derived species in polar ice: A review. In R. Delmas, ed., *Ice Core Studies of Global Biogeochemical Cycles,* NATO ASI Series 30:91–119. Berlin: Springer-Verlag.

Lehman, S. 1993. Ice sheets, wayward winds and sea change. *Nature* 365: 108–109.

Lehman, S. J. and L. D. Keigwin. 1992. Sudden changes in North Atlantic circulation during the last deglaciation. *Nature* 356:757–762.

Leinen, M., D. Cwienk, G. R. Heath, et al. 1986. Distribution of biogenic silica and quartz in recent deep-sea sediments. *Geology* 14:199–203.

Leinen, M. and M. Sarnthein, eds. 1989. *Paleoclimatology and Paleometeorology: Modern and Past Patterns of Global Atmospheric Transport.* Norwell, Mass.: Kluwer Academic.

Leroy, S. A. G. and L. M. Dupont. 1994. Development of vegetation and continental aridity in northwestern Africa during the upper Pliocene: the pollen record of ODP 658. *Palaeogeography, Palaeoclimatology, Palaeoecology* 109:295–316.

Le Treut, H. and M. Ghil. 1983. Orbital forcing, climatic interactions, and glaciation cycles. *Journal of Geophysical Research* 88:5167–5190.

Leuenberger, M., U. Siegenthaler, and C. C. Langway, Jr. 1992. Carbon isotope composition of atmospheric CO_2 during the last ice age from an Antarctic ice core. *Nature* 357:488–490.

Leventer, A., D. E. Williams, and J. P. Kennett. 1983. Relationship between anoxia, glacial meltwater, and microfossil preservation in the Orca Basin, Gulf of Mexico, *Marine Geology* 53:23–40.

Leventer, A., E. W. Domack, S. E. Ishman, S. Brachfeld, C. E. McClennen, and P. Manley. 1996. Productivity cycles of 200–300 years in the Antarctic Peninsula region: Understanding linkages among the sun, atmosphere, oceans, sea ice and biota. *Geological Society of America Bulletin* 108:1626–1644.

Levesque, A. J., L. C. Cwynar, and I. R. Walker. 1994. A multiproxy investigation of late-glacial climate and vegetation change at Pine Ridge park, southwest New Brunswick, Canada. *Quaternary Research* 42:316–327.

Levesque, A. J., L. C. Cwynar, and I. R. Walker. 1997. Exceptionally steep north-south gradients in lake temperatures during the last deglaciation. *Nature* 385:423–426.

Levesque, A., F. E. Mayle, I. R. Walker, and L. C. Cwynar. 1993. A previously unrecognized late-glacial cold event in eastern North America. *Nature* 361:623–626.

Levin, I. 1994. The recent state of carbon cycling through the atmosphere. In R. Zahn, T. F. Pedersen, M. A. Kaminski, and L. Labeyrie, eds., *Carbon Cycling in the Glacial Ocean: Constraints on the Ocean's Role in Global Change,* NATO ASI Series, Global Environmental Change 17:3–13. Berlin: Springer-Verlag.

Levin, S. I. 1992. The problem of pattern and scale in ecology. *Ecology* 73:1943–1967.

Lewontin, R. C. 1978. Adaptation. *Scientific American* 239:156–169.

Lighty, R. G., I. G. Macintyre, and R. Stuckenrath. 1982. *Acropora palmata* reef framework: A reliable indicator of sea level in the western Atlantic for the past 10,000 years. *Coral Reefs* 1:125–130.

Lindzen, R. S. and W. Pan. 1994. A note on orbital control of equator-pole heat-fluxes. *Climate Dynamics* 10:49–57.

Linn, L. J., M. J. Delaney, and E. R. M. Druffel. 1990. Trace metals in contemporary and seventeenth-century Galapagos coral: records of seasonal and annual variations. *Geochimica et Cosmochimica Acta* 54:387–394.

Linsley, B. K., R. B. Dunbar, G. M. Wellington, and D. A. Mucciarone. 1994. A coral-based reconstruction of intertropical convergence zone variability over Central America since 1707. *Journal of Geophysical Research* 99(C5):9977–9994.

Lockwood, J. G. 1979. *Causes of Climate.* New York: John Wiley & Sons.

Longhurst, A. R. 1991. Role of marine biosphere in the global carbon cycle. *Limnology and Oceanography* 36:1507–1526.

Lorius, C., J. Jouzel, D. Raynaud, J. Hansen, and H. Le Treut. 1990. The ice-core record: Climate sensitivity and future greenhouse warming. *Nature* 347:139–145.

Lorius, C., J. Jouzel, C. Ritz, et al. 1985. A 150,000-year climatic record from Antarctic ice. *Nature* 316:591–596.

Lorius, C., L. Merlivat, J. Jouzel, and M. Pourchet. 1979. A 30,000-yr isotope climatic record from Antarctic ice. *Nature* 280:644–648.

Lough, J. M. 1992. An index of Southern Oscillation reconstructed from North American tree-ring chronologies. In H. F. Diaz and V. Markgraf, eds., *El Niño: Historical and Paleoclimatic Aspects of the Southern Oscillation,* pp. 215–226. Cambridge: Cambridge University Press.

Lough, J. M. and D. J. Barnes. 1989. Possible relationships between environmental variables and skeletal density in a coral colony from the central Great Barrier Reef. *Journal of Experimental Marine Biology and Ecology* 134:221–241.

Lough, J. M. and D. J. Barnes. 1990. Intra-annual timing of density band formation of *Porites* coral from the central Great Barrier Reef. *Journal of Experimental Marine Biology and Ecology* 135:35–47.

Lough, J. M. and D. J. Barnes. 1992. Comparisons of skeletal density variations in *Porites* from the Great Barrier Reef. *Journal of Experimental Marine Biology and Ecology* 155:1–25.

Lourens, L. J., A. Antonarakau, F. J. Hilgen, A. A. M. Van Hoof, C. Vergnaud-Grazzini, and W. J. Zachariasse. 1996. Evaluation of the Plio-Pleistocene astronomical timescale. *Paleoceanography* 11:391–413.

Lovelock, J. E. 1972. Gaia as seen through the atmosphere. *Atmospheric Environment* 6:579–580.

Lovelock, J. E. 1989. *The Ages of Gaia.* New York: W. W. Norton.

Lovelock J. E. and L. R. Kump. 1994. Failure of climate regulation in a geophysiological model. *Nature* 369:732–734.

Lovelock, J. E. and L. Margulis. 1974. Atmospheric homeostasis by and for the biosphere. *Tellus* 26:1–10.

Lowell, T. V., C. J. Heusser, B. G. Andersen, et al. 1995. Interhemispheric correlation of late Pleistocene glacial events. *Science* 269:1541–1549.

Lozano, J. A. and J. D. Hays. 1976. Relationship of radiolarian assemblages to sediment types and physical oceanography in the Atlantic and western Indian Ocean sectors of the Antarctic Ocean. Investigations of late Quaternary paleoceanography and paleoclimatology. *Geological Society of America Memoir* 145:303–336.

Luckman, B. H. 1996. Reconciling the glacial and dendroclimatological records for the last millennium in the Canadian Rockies. In P. D. Jones, R. S. Bradley, and J. Jouzel, eds., *Climate Variations and Forcing Mechanisms of the Last 2000 Years*, NATO ASI Series 1, 41:85–108. Berlin: Springer-Verlag.

Ludwig, K. R., K. R. Simmons, I. J. Winograd, B. J. Szabo, and A. C. Riggs. 1993. Dating of the Devils Hole calcite vein: Response to Edwards and Gallup. *Science* 259:1626–1627.

Lyell, C. 1830–1833. *Principles of Geology*, Vols. 1–3. London: Murray.

Ma, T. Y. H. 1937. On the growth rate of reef corals and its relation to seawater temperature. *Paleontologica Sinica Series B* 16:1–426. (Cited in Dunbar and Cole 1993.)

MacArthur, R. H. 1972. *Geographical Ecology*. New York: Harper and Row.

MacArthur, R. H. and E. O. Wilson. 1967. *The Theory of Island Biogeography*. Princeton, N.J.: Princeton University Press.

MacAyeal, D. R. 1993. Binge/Purge oscillations of the Laurentide Ice Sheet as a cause of the North Atlantic's Heinrich Events. *Paleoceanography* 8:775–784.

Machida, T. T. Nakazawa, Y. Fujii, S. Aoke, and O. Watanabe. 1995. Increase in atmospheric nitrous oxide concentrations during the last 250 years. *Geophysical Research Letters* 22:2921–2924.

Mackensen, A., H. Grobe, H.-W. Hubberton, and G. Kuhn. 1994. Benthic foraminiferal assemblages and the del ^{13}C signal in the Atlantic sector of the Southern Ocean: glacial to interglacial contrasts. In R. Zahn, T. F. Pedersen, M. A. Kaminski, and L. Labeyrie, eds., *Carbon Cycling in the Glacial Ocean: Constraints on the Ocean's Role in Global Change*. NATO ASI Series 17, pp. 145–166. Berlin: Springer-Verlag.

Mackensen, A., H.-W. Hubberton, T. Bickert, G. Fischer, and D. Fütterer. 1993. The $\delta^{13}C$ in benthic foraminiferal tests of *Fontbotia wuellerstorfi* (Schwager) relative to the $\delta^{13}C$ of dissolved inorganic carbon in Southern Ocean circulation models. *Paleoceanography* 8:587–610.

Maclaren, C. 1842. The glacial theory of Professor Agassiz of Neuchatel. *American Journal of Science* 42:346–365.

Manabe, S. and T. Broccoli. 1985. The influence of ice sheets on the climate of an ice age. *Journal of Geophysical Research* 90(C2):2167–2190.

Manabe, S. and R. J. Stouffer. 1988. Two stable equilibria of a coupled ocean-atmosphere model. *Journal of Climate* 1:841–866.

Manabe, S. and R. J. Stouffer. 1994. Multiple-century response of a coupled ocean-atmosphere model to an increase of atmospheric carbon dioxide. *Journal of Climate* 7:5–23.

Mangerud, J. 1987. The Allerød/Younger Dryas Boundary. In W. H. Berger and L. D. Labeyrie, eds., *Abrupt Climatic Change*, pp. 163–171. Dordrecht: D. Reidel.

Mangerud, J., E. Jansen, and J. Y. Landvik. 1996. Late Cenozoic history of the Scandinavian and Barents Sea ice sheets. *Global and Planetary Change* 12:11–26.

Mangerud, J., S. T. Anderson, B. E. Birklund, and J. J. Donner. 1974. Quaternary stratigraphy of Norden, a proposal for terminology and classification. *Boreas* 3:109–128.

Mangerud, J., E. Sonstegaard, and H. P. Sejrup. 1979. Correlation of the Eemian (interglacial) Stage and the deep-sea oxygen isotope stratigraphy. *Nature* 277:189–192.

Mann, K. H. and J. R. N. Lazier. 1996. *Dynamics of Marine Ecosystems*. Cambridge: Blackwell Science, Inc.

Mann, M. E., J. Park, and R. S. Bradley. 1995. Global interdecadal and century-scale climate oscillations during the past five centuries. *Nature* 378: 266–270.

Mantua, N. J., S. R. Hare, Y. Zhang, et al. 1997. A Pacific interdecadal climate oscillation with impacts on salmon production. *Bulletin American Meteorological Society* 78:1069–1079.

Markgraf, V. 1991. Younger Dryas in southern South America. *Boreas* 20:63–69.

Markgraf, V. 1993. Younger Dryas in southernmost South America—an update. *Quaternary Science Reviews* 12:351–355.

Martinson, D. G., et al. ed. 1995. *Natural Climate Variability on Decade- to Century-Time Scales*. Washington, D.C.: National Academy Press; National Research Council (U.S.) Climate Research Committee.

Martinson, D. G., N. G. Pisias, J. D. Hays, J. Imbrie, T. C. Moore, Jr., and N. J. Shackleton. 1987. Age dating and the orbital theory of the Ice Ages: Development of a high-resolution 0 to 300,000-year chronostratigraphy. *Quaternary Research* 27:1–29.

Matthes, F. E. 1939. Report of Committee on Glaciers, April 1939. *Transactions American Geophysical Union* 20:518–523.

Matthews, R. K. 1988. Comment on Haq et al. 1987. *Science* 241:597–599.

Matthews, R. K. and R. Z. Poore. 1980. Tertiary $\delta\ ^{18}O$ and glacio-eustatic sea level fluctuations. *Geology* 8:501–504.

Maul, G. A. and D. M. Martin. 1993. Sea level rise at Key West, Florida, 1846–1992: America's longest instrument record? *Geophysical Research Letters* 20:1955–1958.

Maunder, E. W. 1922. The prolonged sunspot minimum, 1645–1715. *The British Astronomical Association Journal* 32:140–145.

May, R. M., ed. 1981. *Theoretical Ecology: Principles and Applications*. Oxford: Blackwell Scientific.

Mayewski, P. A., L. D. Meeker, S. Whitlow, et al. 1993. The atmosphere during the Younger Dryas. *Science* 261:195–197.

Mayewski, P. A., M. S. Twickler, S. I. Whitlow, et al. 1996. Climate change during the last deglaciation in Antarctica. *Science* 272:1636–1638.

Mayle, F. E., A. J. Levesque, and L. C. Cwynar. 1993. Accelerator mass spectrometer ages for the Younger Dryas event in Atlantic Canada. *Quaternary Research* 39:355–360.

Mayr, E. 1982. *The Growth of Biological Thought.* Cambridge: Harvard University Press.

McCann, M. P., A. J. Semtner, Jr., and R. M. Chervin. 1994. Transports and budgets of volume, heat, and salt from a global eddy-resolving ocean model. *Climate Dynamics* 10:59–80.

McCartney, M. S. 1992. Recirculating components to the deep boundary current of the northern North Atlantic. *Progress in Oceanography* 29:283–383.

McConnaughey, T. A. 1989. C-13 and O-18 isotopic disequilibria in biological carbonates: I. Patterns. *Geochimica et Cosmochimica Acta* 53:151–163.

McCormac, B. M., ed. 1983. *Weather and Climate Responses to Solar Variations.* Boulder: Colorado Associated University Press.

McCorkle, D. C., L. D. Keigwin, B. C. Corliss, and S. R. Emerson. 1990. The influence of microhabitats on the carbon isotopic composition of deep-sea sediments. *Paleoceanography* 5:161–186.

McCorkle, D. C., P. A. Martin, D. W. Lea, and G. P. Klinkhammer. 1995. Evidence of a dissolution effect on benthic foraminiferal shell chemistry: δ^{13}C, Cd/Ca, Ba/Ca, Sr/Ca results from the Ontong Java Plateau. *Paleoceanography* 10:699–714.

McCorkle, D. C., H. H. Veeh, and D. T. Heggie. 1994. Glacial-Holocene paleoproductivity off western Australia: a comparison of proxy records. In R. Zahn, T. F. Pedersen, M. A. Kaminski, and L. Labeyrie, eds., *Carbon Cycling in the Glacial Ocean: Constraints on the Ocean's Role in Global Change,* pp. 443–479. Berlin: Springer-Verlag.

McElwain, J. C. and W. C. Chaloner. 1996. The fossil cuticle as a skeletal record of environmental change. *Palaios* 11:376–388.

McGinnis, N. W. Driscoll, G. D. Karner, W. D. Brumbaugh, and N. Cameron. 1993. Flexural response of passive margins to deep-sea erosion and slope retreat: Implications for relative sea-level change. *Geology* 21:893–896.

McGowan, J. A. 1989. Pelagic ecology and Pacific climate. In D. G. Peterson, ed., *Aspects of Climate Variability in the Pacific and the Western Americas, American Geophysical Union Monograph* 55:141–150.

McGowan, J. A. 1990. Climate and change in oceanic ecosystems: the value of time-series data. *Trends in Ecology and Evolution* 5:293–299.

McGuffie, K. and A. Henderson-Sellers, eds. 1997. *A Climate Modeling Primer.* New York: Chichester.

McIntosh, R. P. 1981. Succession in ecological theory. In D. C. West, H. H. Shugart, and D. B. Botkin, eds., *Forest Succession: Concepts and Applications,* pp. 10–23. New York: Springer-Verlag.

McIntyre, A. and B. Molfino. 1996. Forcing of Atlantic equatorial and subpolar millennial cycles by precession. *Science* 274:1867–1870.

McIntyre, A., W. F. Ruddiman, K. Karlin, and A. C. Mix. 1989. Surface water response of the equatorial Atlantic Ocean to Orbital Forcing. *Paleoceanography* 4:19–56.

McKenna, V. S., J. W. Farrell, D. W. Murray, and S. C. Clemens. 1995. The foraminifer record at site 847: Paleoceanographic response to late Pleistocene climate variability. In N. G. Pisias, L. A. Mayer, T. R. Janacek, A. Palmer-Julson, and T. H. van Andel, eds., *Proceedings of the Ocean Drilling Program Scientific Results* 138:695–714.

McKenzie, J. A. and G. P. Eberli. 1987. Indication for abrupt Holocene climate change: Late Holocene oxygen isotope stratigraphy of the Great Salt Lake. In W. H. Berger and L. D. Labeyrie, eds., *Abrupt Climatic Change,* NATO ASI Series C 216:127–136. Dordrecht: D. Reidel.

McManus, J. F., G. C. Bond, W. S. Broecker, S. Johnsen, L. Labeyrie, and S. Higgins. 1994. High-resolution climate records from the North Atlantic during the last interglacial. *Nature* 371:326–329.

Meehl, G. A. and W. M. Washington. 1990. CO_2 climate sensitivity and snow-sea-ice albedo parameterization in an atmospheric CGM coupled to a mixed-layer ocean model. *Climatic Change* 16:283–306.

Meese, D. A., A. J. Gow, P. Grootes, et al. 1994. The accumulation record from the GISP2 core as an indicator of climate change throughout the Holocene. *Science* 266:1680–1685.

Meier, M. F. 1984. Contribution of small glaciers to global sea level. *Science* 226:1418–1421.

Meier, M. F. 1993. Ice, climate, and sea level: Do we really know what is happening? In W. R. Peltier, ed., *Ice in the Climate System,* NATO ASI Series 112:142–160. Berlin: Springer-Verlag.

Meko, D. M. 1992. Spectral properties of tree-ring data in the United States southwest as related to El Niño/Southern Oscillation. In H. F. Diaz and V. Markgraf, eds., *El Niño: Historical and Paleoclimatic Aspects of the Southern Oscillation,* pp. 227–241. Cambridge: Cambridge University Press.

Mercer, J. H. 1969. The Allerød oscillation: A European climatic anomaly. *Arctic and Alpine Research* 1:227–234.

Mercer, J. H. 1981. West Antarctic ice volume: the interplay of sea level and temperature, a strandline test for absence of the ice sheet during the last interglacial. In I. Allison, ed., *Sea Level, Ice and Climate,* International Association of Hydrological Science Publication 131, pp. 323–330. Oxford: Wellingford/International Association of Hydrological Science.

Mesolella, K. J., R. K. Matthews, W. S. Broecker, and D. L. Thurber. 1969. The astronomical theory of climatic change: Barbados data. *Journal of Geology* 77:250–274.

Miall, A. D. 1997. *The Geology of Stratigraphic Sequences.* Berlin: Springer.

Mikolajewicz, U. and E. Maier-Reimer. 1990. Internal secular variability in an ocean general circulation model. *Climate Dynamics* 4:145–156.

Milankovitch, M. 1930. Mathematische Klimalehre und astronomische Theorie der Klimaschwankungen. In W. Koppen and R. Geiger, eds., *Handbuch der Klimatologie 1 A,* pp. 1–176. Berlin: Gebruder Borntraeger.

Milankovitch, M. 1938. Astronomische Mittel zur Erforschung der erdgeschichtlichen Klimate. *Handbuch der Geophysik* 9:593–698.

Milankovitch, M. 1941. Kanon der Erdbestrahlung und seine Andwendung auf das Eiszeitenproblem. *Royal Serbian Academy Special Publication* 133:1–633; Transl. 1969. Israel Program for Scientific Translation, U.S. Department of Commerce.

Miller, G. H., J. W. McGee, and A. J. T. Jull. 1997. Low latitude glacial cooling in the Southern Hemisphere from amino-acid racemization in emu eggshells. *Nature* 385:241–244.

Miller, K. G., G. S. Mountain, the Leg 150 Shipboard Party, and Members of the New Jersey Coastal Plain Drilling Project. 1996. Drilling and dating New Jersey Oligocene-Miocene sequences: Ice volume, global sea level, and Exxon records. *Science* 271:1092–1094.

Miller, K. G., J. D. Wright, and R. G. Fairbanks. 1991. Unlocking the icehouse: Oligocene-Miocene oxygen isotopes, eustasy, and margin erosion. *Journal of Geophysical Research* 96:6829–6848.

Milliman, J. D. and B. U. Haq. 1996. *Sea-level Rise and Coastal Subsidence: Causes Consequences, and Strategies.* Dordrecht: Kluwer Academic.

Mitchell, G. F., L. F. Penny, F. W. Shotten, and R. G. West. 1973. A correlation of Quaternary deposits in the British Isles. *Geological Society London Special Report* 4.

Mitrovica, J. X. and W. R. Peltier. 1991. *Journal of Geophysical Research* 96:20,053.

Mix, A. C. 1989. Influence of productivity variations on long-term atmospheric CO_2. *Nature* 337:541–543.

Mix, A. C. 1992. The marine oxygen isotope record: Constraints on timing and extent of ice-growth events (120–65 ka). In P. U. Clark and P. D. Lea, eds., *The Late Interglacial-Glacial Transition in North America,* Geological Society of America Special Paper 270, pp. 19–30. Boulder: Geological Society of America.

Molfino, B., N. G. Kipp, and J. J. Morley. 1982. Comparison of foraminiferal, coccolithophorid, and radiolarian paleotemperature equations: Assemblage coherency and estimate concordancy. *Quaternary Research* 17:279–313.

Molfino, B. and A. McIntyre. 1990. Nutricline variation in the equatorial Atlantic coincident with the Younger Dryas. *Paleoceanography* 5:997–1008.

Mommersteeg, H. J. P. M., T. A. Wijmstra, H. Hooghiemstra, R. Young, M. F. Loutre, and A. Berger. 1995. Orbital forced frequencies in the 975,000 year pollen record from Tenagi Philippon (Greece). *Climate Dynamics* 11:4–24.

Montaggioni, L. F., G. Cabioch, G. F. Camoinau, et al. 1997. Continuous record of reef growth over the past 14 k.y. on the mid-Pacific island of Tahiti. *Geology* 25:555–558.

Moore, T. C. 1973. Late Pleistocene-Holocene radiolarian assemblages and their relationship to oceanographic parameters. *Quaternary Research* 3:73–88.

Mora, C. I., S. G. Driese, and L. A. Colarusso. 1996. Middle to late Paleozoic atmospheric CO_2 levels from soil carbonate and organic matter. *Science* 271:1105–1107.

Morley, J. J. 1989. Radiolarian-based transfer functions for estimating paleoceanographic conditions in the South Indian Ocean. *Marine Micropaleontology* 13:293–307.
Morner, N.-A. 1979a. The Fennoscandinavian uplift and late Cenozoic geodynamics: Geological evidence. *Geojournal* 3.3:287–318.
Morner, N.-A. 1979b. The Fennoscandinavian uplift: geological data and their geodynamical implication. In N.-A. Morner, ed., *Earth Rheology, Isostasy, and Eustasy.* New York: John Wiley & Sons.
Morner, N.-A. 1981. Revolution in Cretaceous sea-level analysis. *Geology* 9:344–346.
Morner, N.-A. and W. Karlen, eds. 1984. *Climatic Changes on a Yearly to Millennial Basis.* Dordrecht: D. Reidel.
Morrison, R. B. 1969. The Pleistocene-Holocene boundary. *Geol. en Mijnb.* 48:363–372.
Mosley-Thompson, E. 1996. Holocene climate changes recorded in an East Antarctica ice core. In P. D. Jones, R. S. Bradley, and J. Jouzel, eds., *Climate Variations and Forcing Mechanisms of the Last 2000 Years,* NATO ASI Series 1, 41:pp. 263–279. Berlin: Springer-Verlag.
Mosley-Thompson, E., J. Dai, L. G. Thompson, P. M. Grootes, J. K. Arbogast, and J. F. Paskievich. 1991. Glaciological studies at Siple Station (Antarctica): Potential ice-core paleoclimatic record. *Journal of Glaciology* 37:11–22.
Mosley-Thompson, E., L. G. Thompson, J. Dai, M. Davis, and P. N. Lin. 1993. Climate of the past 500 years: High resolution ice core records. *Quaternary Science Reviews* 12:419–430.
Mosley-Thompson, E., L. G. Thompson, P. M. Grootes, and N. Gundestrup. 1990. Little Ice Age (Neoglacial) paleoenvironmental conditions at Siple Station, Antarctica. *Annals of Glaciology* 14:199–204.
Mott, R. J., D. R. Grant, R. Stea, and S. Occhietti. 1986. Late-glacial climatic oscillation in Atlantic Canada equivalent to the Allerød/Younger Dryas event. *Nature* 323:247–250.
Mott, R. J. and R. R. Stea. 1993. Late-glacial (Allerød/Younger Dryas) buried organic deposits, Nova Scotia, Canada. *Quaternary Science Reviews* 12:645–657.
Muhs, D. R. 1992. The last interglacial-glacial transition in North America: Evidence from uranium-series dating of coastal deposits. In P. U. Clark and P. D. Lea, eds., *The Last Interglacial-Glacial Transition in North America,* Geological Society of America Special Paper 270, pp. 31–51. Boulder: Geological Society of America.
Muhs, D. R., G. L. Kennedy, and T. K. Rockwell. 1994. Uranium-series ages of marine terrace corals from the Pacific coast of North America and implications for last-interglacial sea level history. *Quaternary Research* 42:72–87.
Muhs, D. R. and B. J. Szabo. 1994. New uranium-series ages of the Waimanalo Limestone, Oahu, Hawaii: Implications for sea level during the last interglacial period. *Marine Geology* 118:315–326.

Müller, H. 1974. Pollenanalytische Untersuchungen und Jahresschichtenzahlungen an der eem-zeitlichen Kieselgur von Bispingen/Luhe. *Geologisches Jahrbuch* A21:149–169.

Murray, D. W., J. W. Farrell, and V. McKenna. 1995. Biogenic sedimentation at site 847, eastern equatorial Pacific Ocean, during the past 3 M.Y. In N. G. Pisias, L. A. Mayer, T. R. Janacek, A. Palmer-Julson, and T. H. van Andel, eds., *Proceedings of the Ocean Drilling Program Scientific Results* 138:429–459.

Murray, J. W., R. T. Barber, M. R. Roman, M. P. Bacon, and R. A. Feeley. 1994. Physical and biological controls on carbon cycling in the equatorial Pacific. *Science* 266:58–65.

Nairn, A. E. M. N., ed. 1961. *Descriptive Palaeoclimatology.* New York: Interscience Publishers, Inc.

Naish, T. 1997. Constraints on the amplitude of late Pliocene eustatic sea-level fluctuations: New evidence from the New Zealand shallow-marine sediment record. *Geology* 25:1139–1142.

Nakazawa, T., Machida, M. Tanaka, Y. Fujii, S. Aoki, and O. Watanabe. 1993. Differences of the atmospheric CH_4 concentration between the Arctic and Antarctic regions in pre-industrial/agricultural era. *Geophysical Research Letters* 20:943–946.

Nansen, F. 1922. The strandflat and isostasy: Vitenskapsselskapets skrifter, I Matematisk-Naturevitenskapelig Klasse 1921, No. 11. Kristiana I, Kommission hos Jacob Dybwad. (Cited in Dott 1992.)

Neftel, A., E. Moore, H. Oeschger, and B. Stauffer. 1985. Evidence from polar ice cores for the increase in atmospheric CO_2 in the past two centuries. *Nature* 315:45–47.

Neftel, A. E., H. Oeschger, J. Schwander, B. Stauffer, and R. Zumbrunn. 1982. Ice core measurements give atmospheric CO_2 content during the past 40,000 years. *Nature* 295:220–223.

Neftel, A., H. Oeschger, T. Staffelbach, and B. Stauffer. 1988. CO_2 record in the Byrd ice core 50,000–5,000 years BP. *Nature* 331:609–611.

Nerem, R. S. 1995. Measuring global sea level variations using TOPEX/Poseidon altimeter data. *Journal of Geophysical Research* 100:25,135-25,151.

Nesme-Ribes, E., ed. 1994. *The Solar Engine and Its Influence on Terrestrial Atmospheres and Climate,* NATO ASI Series, vol. 25. Berlin: Springer-Verlag.

Neumann, A. C. and P. J. Hearty. 1996. Rapid sea-level changes at the close of the last interglacial (substage 5e) recorded in Bahamian island geology. *Geology* 24:775–778.

Neumann, A. C. and W. S. Moore. 1975. Sea level events and Pleistocene coral ages in the northern Bahamas. *Quaternary Research* 5:215–224.

Newman, W. S., L. J. Cinquemani, R. R. Pardi, and L. F. Marcus. 1980. Holocene deleveling of the United States' east coast. In N.-A. Morner, ed., *Earth Rheology, Isostasy, and Eustasy,* pp. 449–463. New York: Wiley.

Nicholls, N., G. V. Gruza, J. Jouzel, T. R. Karl, L. A. Ogallo, and D. E. Parker.

1996. Observed climate variability and change. In J. T. Houghton, L. G. Meira Filho, B. A. Callender, N. Harris, A. Kattenberg, and K. Maskell, eds., *Climate Change 1995: The Science of Climate Change,* pp. 137–192. Cambridge: Cambridge University Press.

Nielson, R. and D. Marks. 1995. A global perspective of regional vegetation and hydrologic sensitivities from climate change. *Journal of Vegetative Science* 5:715–730.

Nummendal, D., O. H. Pilkey, and J. D. Howard, eds. 1987. *Sea Level Fluctuations and Coastal Evolution, SEPM Special Publication* 41:3–16.

Oba, T. 1969. Biostratigraphy and isotope paleotemperatures of some deep-sea cores from the Indian Ocean. *Tohoku University Science Reports, 2nd Series (Geology)* 41:129–195.

O'Brien, J. J., T. S. Richards, and A. C. Davis. 1996. The effects of El Niño on U.S. landfalling hurricanes. *Bulletin Meteorological Society* 77:773–774.

O'Brien, S. R., P. A. Mayewski, L. D. Mecker, D. A. Meese, M. S. Twickler, and S. I. Whitlow. 1995. Complexity of Holocene climate as reconstructed from a Greenland ice core. *Science* 270:1962–1964.

Ochoa, N. and O. Gomez. 1987. Dinoflagellates as indicators of water masses during El Niño 1982–1983. *Journal of Geophysical Research* 92(C14): 14,355–14,367.

Oerlemans, J. 1982. Glacial cycles and ice-sheet modeling. *Climatic Change* 4:353–374.

Oeschger, H. and C. C. Langway, Jr., eds. 1989. *The Environmental Record in Glaciers and Ice Sheets.* Chichester: John Wiley & Sons.

Oeschger, H., U. Siegenthaler, U. Schotterer, and A. Gugelmann. 1975. A box diffusion model to study carbon dioxide exchange in nature. *Tellus* 27:168–192.

Oeschger, H., B. Stauffer, R. Finkel, and C. C. Langway, Jr. 1985. Variations of the CO_2 concentration of occluded air and of anions and dust in polar ice cores. In E. T. Sundquist and W. S. Broecker, eds., *The Carbon Cycle and Atmospheric CO_2: Natural Variations Archean to Present, American Geophysical Union Monograph* 32:132–142.

Olausson, E. 1965. Evidence of climatic changes in North Atlantic deep-sea cores with remarks on isotopic palaeotemperature analysis. *Progress Oceanography* 3:221–252.

O'Neill, R. V., D. L. DeAngelis, J. B. Waile, and T. F. H. Allen. 1989. *A Hierarchical Concept of Ecosystems.* Princeton, N.J.: Princeton University Press.

Oppo, D. W. and R. G. Fairbanks. 1987. Variability in the deep and intermediate water circulation of the Atlantic Ocean: northern hemisphere modulation of the Southern Ocean. *Earth and Planetary Science Letters* 86:1–15.

Oppo, D. W., M. Horowitz, and S. J. Lehman. 1997. Marine core evidence for reduced deep water production during Termination II followed by a relatively stable substage 5e (Eemian). *Paleoceanography* 12:51–63.

Oppo, D. W. and S. J. Lehman. 1995. Suborbital timescale variability of North Atlantic deep water during the past 200,000 years. *Paleoceanography* 10:901–910.

Oppo, D. W., M. E. Raymo, G. P. Lohman, A. C. Mix, J. D. Wright, and W. L. Prell. 1995. A $\delta^{13}C$ record of upper North Atlantic deep water during the past 2.6 million years. *Paleoceanography* 10:373–394.

Osmond, J. K., J. R. Carpenter, and H. L. Windom. 1965. Th^{230}/U^{234} age of the Pleistocene corals and oolites of Florida. *Journal of Geophysical Research* 70:1843–1847.

Oster, G. F. and E. O. Wilson. 1978. *Caste and Ecology in the Social Insects.* Princeton, N.J.: Princeton University Press.

Overpeck, J., K. Hughen, D. Hardy, et al. 1997. Arctic environmental change of the last four centuries. *Science* 278:1251–1256.

Overpeck, J., D. Rind, A. Lacis, and R. Healy. 1996. Possible role of dust-induced regional warming in abrupt climate change during the last glacial period. *Nature* 384:447–449.

Overpeck, J. T., T. Webb III, and I. C. Prentice. 1985. Quantitative interpretaion of fossil pollen spectra: Dissimilarity coefficients and the method of modern analogs. *Quaternary Research* 23:87–108.

Paine, R. T. 1966. Food web complexity and species diversity. *American Naturalist* 100:65–75.

Pandolfi, J. M. 1996. Limited membership in Pleistocene reef coral assemblages from the Huon Peninsula, Papua, New Guinea: constancy during global change. *Paleobiology* 22:152–176.

Parrish, J. T. 1998. *Interpreting Pre-Quaternary Climate from the Geological Record.* New York: Columbia University Press.

Paterson, W. S. B. 1978. *The Physics of Glaciers,* 2nd ed. Oxford: Pergamon.

Paterson, W. S. B. 1993. World sea level and the present mass balance of the Antarctic Ice Sheet. In W. R. Peltier, ed., *Ice in the Climate System,* NATO ASI Series 112:1131–140. Berlin: Springer-Verlag.

Patzold, J. 1984. Growth rhythms recorded in stable isotopes and density bands in the reef coral *Porites lobata* (Cebu, Philippines). *Coral Reefs* 3:87–90.

Paytan, A., M. Kastner, and F. P. Chavez. 1996. Glacial to interglacial fluctuations in productivity in the equatorial Pacific as indicated by marine barite. *Science* 274:1355–1357.

Paytan, A., M. Kastner, E. E. Martin, J. D. Macdougall, and T. Herbert. 1993. Marine barite as a monitor of seawater strontium isotope composition. *Nature* 366:445–449.

Pecker, J.-C. and K. Runcorn, eds. 1990. The Earth's climate and variability of the Sun over recent millennia: geophysical, astronomical, and archaeological aspects. *Philosophical Transactions of the Royal Society of London* 330:395–697.

Pederson, T. F., M. Pickering, J. S. Vogel, J. N. Southon, and D. E. Nelson. 1988. The response of benthic foraminifera to productivity cycles in the eastern equatorial Pacific: Faunal and geochemical constraints on glacial bottom water oxygen. *Paleoceanography* 3:157–168.

Peel, D. A., R. Mulvaney, and E. C. Pasteur. 1996. Climate changes in the Atlantic sector of Antarctica over the past 500 years from ice-core and other evidence. In P. D. Jones, R. S. Bradley, and J. Jouzel, eds., *Climatic Variations and Forcing Mechanisms of the Last 2000 Years,* pp. 243–262. Berlin: Springer-Verlag.

Pekar, S. and K. G. Miller. 1996. New Jersey Oligocene "Icehouse" sequences (ODP Leg 150) correlated with global del ^{18}O and Exxon eustatic records. *Geology* 24:567–570.

Peltier, W. R. 1974. The impulse response of a Maxwell Earth. *Reviews of Geophysics and Space Physics* 12:649–705.

Peltier, W. R. 1988. Global sea level and Earth rotation. *Science* 240:895–901.

Peltier, W. R., ed. 1993. *Ice in the Climate System,* NATO ASI Series, vol. 12. Berlin: Springer-Verlag.

Peltier, W. R. 1994. Ice age Paleotopography. *Science* 265:195–201.

Peltier, W. R. 1995. Technical Comments: Paleotopography of glacial-age ice sheets. *Science* 267:536–538.

Peltier, W. R. and J. T. Andrews. 1976. Glacial isostatic adjustment II: the inverse problem. *Geophysics of Journal of the Royal Astronomical Society* 46:605–646.

Peltier, W. R., W. E. Farrell, J. A. Clark. 1978. Glacial isostasy and relative sea level: A global finite element model. *Tectonophysics* 50:81–110.

Peltier, W. R. and A. M. Tushingham. 1989. Global Sea level and the Greenhouse effect: might they be connected. *Science* 244:806–810.

Peltier, W. R. and A. M. Tushingham. 1991. Influence of glacial isostatic adjustments on tide gauge measurements of secular sea level change. *Journal of Geophysical Research* 96:6779–6796.

Penck, A. 1882. Schwankungen des Meeresspiegel. *Geographica Gesellschaftlichen der Munchen* 7:1–70.

Peng, T.-H., W. S. Broecker, H. D. Freyer, and S. Trumbore. 1983. A deconvolution of the tree ring based $\delta^{13}C$ record. *Journal of Geophysical Research* 88:3609–3620.

Pestiaux, P., I. van der Mersch, and A. Berger. 1988. Paleoclimatic variability at frequencies ranging from 1 cycle per 10,000 years to 1 cycle per 1000 years: Evidence for nonlinear behaviour of the climate system. *Climatic Change* 12:9–37.

Peteet, D. M. 1987. Younger Dryas in North America: modeling, data analysis and re-evaluation. In W. H. Berger and L. D. Labeyrie, eds., *Abrupt Climatic Change,* pp. 185–193. Dordrecht: D. Reidel.

Peteet, D. M., ed. 1993. Global Younger Dryas? *Quaternary Science Reviews* 12(5):277–355.

Peteet, D. M. ed. 1995. Global Younger Dryas. *Quaternary Science Reviews* 14(2).

Peters, R. L. and T. E. Lovejoy, eds. 1992. *Global Warming and the Biodiversity Crisis.* New Haven: Yale University Press.

Peterson, I. 1993. *Newton's Clock: Chaos in the Solar System.* New York: W. H. Freeman & Co.

Petersen, K. L. 1994. A warm and wet little climatic optimum and a cold and dry little ice age in the southern Rocky Mountains, U.S.A. *Climatic Change* 26:243–269.

Peterson, L. C., J. T. Overpeck, N. G. Kipp, and J. Imbrie. 1991. A high-resolution late Quaternary upwelling record from the anoxic Cariaco Basin, Venezuela. *Paleoceanography* 6:99–119.

Petit, J. R., I. Basile, A. Leruyuet, et al. 1997. Four climate cycles in Vostok ice core. *Nature* 387:359.

Petit, J. R., L. Mournier, J. Jouzel, V. S. Korotkevich, V. I. Kotlyakov, and C. Lorius. 1990. Palaeoclimatological and chronological implications of the Vostok core dust record. *Nature* 343:56–58.

Philander, S. G. H. 1983. El Niño Southern Oscillation phenomena. *Nature* 302:295–301.

Philander, S. G. H. 1990. *El Niño, La Niña, and the Southern Oscillation.* San Diego: Academic Press.

Phillips, F. M., M. G. Zreda, L. V. Benson, M. A. Plummer, D. Elmore, and P. Sharma. 1996. Chronology for fluctuations in late Pleistocene Sierra Nevada glaciers and lakes. *Science* 274:749–751.

Phillips, F. M., M. G. Zreda, S. S. Smith, D. Elmore, P. W. Kubik, and P. Sharma. 1990. Cosmogenic chlorine-36 chronology for glacial deposits at Bloody Canyon, eastern Sierra Nevada. *Science* 248:1529–1532.

Phleger, F. B. 1976. Interpretations of late Quaternary foraminifera in deep-sea cores. *Progress in Micropaleontology* pp. 263–276.

Phleger, F. B., F. L. Parker, and J. F. Pierson. 1953. North Atlantic foraminifera. In H. Pettersson, ed., *Reports of the Swedish Deep-Sea Expedition 1947–1948,* 7:1–122. Goteborg: Elanders.

Pianka, E. R. 1975. Niche relations of desert lizards. In M. I. Cody and J. M. Diamond, eds., *Ecology and Evolution of Communities,* pp. 292–314. Cambridge: Harvard University Press.

Pike, J. and A. E. S. Kemp. 1996. Records of seasonal flux in Holocene laminated sediments, Gulf of California. In A. E. S. Kemp, ed., *Paleoceanography from Laminated Sediments,* Geological Society of America Special Publication 116, pp. 157–169. Boulder, Colo.: Geological Society of America.

Pike, J. and A. E. S. Kemp. 1997. Early Holocene decadal-scale ocean variability recorded in Gulf of California laminated sediments. *Paleoceanography* 12:227–238.

Pilgrim, L. 1904. Versuch einer rechnerischen Behandlung des Eiszeitenproblems. *Jahreschefte fur Vaterlandische Naturkunde in Wurttemberg* 60.

Pilkey, O. H., R. A. Morton, J. T. Kelley, and S. Penland. 1989. Coastal Land Loss. *American Geophysical Union Short Course in Geology,* Vol. 2. Washington, D.C.: American Geophysical Union.

Pirazzoli, P. A. 1991. *World Atlas of Holocene Sea-Level Changes.* Amsterdam: Elsevier.

Pirazzoli, P. A., D. R. Grant, and P. Woodworth. 1989. Trends of relative sea-level change: Past, present and future. *Quaternary International* 2: 63–71.

Pisias, N. and D. K. Rea. 1988. Late Pleistocene paleoclimatology of the central equatorial Pacific: Sea surface response to the southeast trade winds. *Paleoceanography* 3:21–38.

Pisias, N. G. and J. Imbrie. 1986/87. Orbital geometry, CO_2 and Pleistocene climate. *Oceanus* 29:43–49.

Pisias, N. G., D. G. Martinson, T. C. Moore, Jr., et al. 1984. High resolution stratigraphic correlation of benthic oxygen isotopic records spanning the last 300,000 years. *Marine Geology* 56:119–136.

Pisias, N. G., L. A. Mayer, T. R. Janacek, A. Palmer-Julson, and T. H. Van Andel, eds. 1995. *Proceedings of Ocean Drilling Program Scientific Results Leg 138.*

Pisias, N. G., A. Mix, and R. Zahn. 1990. Nonlinear response in the global climate system: Evidence from benthic oxygen isotopic record in core RC13–110. *Paleoceanography* 5:147–160.

Pisias, N. G. and N. J. Shackleton. 1984. Modeling the global climate response to orbital forcing and atmospheric carbon dioxide changes. *Nature* 310:757–759.

Pitman, W. C. 1978. Relationship between eustasy and stratigraphic sequences of passive margins. *Geological Society of America Bulletin* 89:1389–1403.

Plummer, L. N. 1992. Stable isotope enrichment in paleowaters of the southeastern Atlantic Coastal Plain, United States. *Science* 262:2016–2020.

Pokras, W. M. and A. C. Mix. 1987. Earth's precession cycle and Quaternary climatic changes in tropical Africa. *Nature* 326:486–487.

Pollard, D. 1978. An investigation into the astronomical theory of the ice ages using a simple climate-ice sheet model. *Nature* 272:233–235.

Pollock, D. E. 1997. The role of diatoms, dissolved silicate and Antarctic glaciation in glacial/interglacial climatic change: a hypothesis. *Global and Planetary Change* 14:113–125.

Polyak, L., S. J. Lehman, V. Gataulin, and A. J. T. Jull. 1995. Two-step deglaciation of the southeastern Barents Sea. *Geology* 23:567–571.

Poore, R. Z., R. L. Phillips, and H. J. Reick. 1993. Paleoclimate record for northwind ridge, Western Arctic Ocean. *Paleoceanography* 8:149–159.

Porter, S. C. 1986. Pattern and forcing of northern hemisphere glacier variations during the last millennium. *Quaternary Research* 26:27–48.

Porter, S. C. and G. H. Denton. 1967. Chronology of neoglaciation in the North American cordillera. *American Journal of Science* 265:177–210.

Porter, S. C., K. L. Pierce, and T. D. Hamilton. 1983. Late Wisconsin mountain glaciation in the western United States. In S. C. Porter, ed., *Late Quaternary Environments of the United States.* Vol. 1, *The Late Pleistocene,* pp. 71–114. Minneapolis: University of Minnesota Press.

Porter, S. C. and A. Zisheng. 1995. Correlation between climate events in the North Atlantic and China during the last glaciation. *Nature* 375:305–308.

Prahl, F. G. and S. G. Wakeham. 1987. Calibration of unsaturation patterns in long chain keytone compositions for paleotemperature assessment. *Nature* 330:367–369.

Prather, M., R. Derwent, D. Ehhalt, P. Fraser, E. Sanhueza, and X. Zhoi. 1995. Other trace gases and atmospheric chemistry. In J. T. Houghton, L. G.

Meira Filho, J. Bruce, et al., eds., *Climate Change 1994*, pp. 77–126. Cambridge: Cambridge University Press.

Prell, W. 1984. Covariance patterns of foraminiferal $\delta^{18}O$; An evaluation of Pliocene ice volume changes near 3.2 million years ago. *Science* 226:692–695.

Prell, W. L. and W. B. Curry. 1981. Faunal and isotopic indices of monsoonal upwelling: western Arabian Sea. *Oceanologica Acta* 4:91–98.

Prell, W. L. and J. E. Kutzbach. 1987. Monsoon variability over the past 150,000 years. *Journal of Geophysical Research* 92:8411–8425.

Prentice, I. C., P. J. Bartlein, and T. W. Webb, III. 1991. Vegetation and climate in eastern North America since the last glacial maximum. *Ecology* 72:2038–2056.

Prentice, I. C. and M. Sarnthein. 1993. Self-regulatory processes in the biosphere in the face of climate change. In J. A. Eddy and H. Oeschger, eds., *Global Climate Changes in the Perspective of the Past*, pp. 29–38. Chichester: John Wiley & Sons.

Prentice, M. L. and R. K. Matthews. 1988. Cenozoic ice-volume history: Development of a composite oxygen isotope record. *Geology* 16:963–966.

Prentice, M. L. and R. K. Matthews. 1991. Tertiary ice sheet dynamics: The snow gun hypothesis. *Journal of Geophysical Research* 96(B4):6811–6827.

PRISM Project Members. 1995. Middle Pliocene paleoenvironments of the Northern Hemisphere. In E. Vrba, G. H. Denton, T. C. Partridge, and L. H. Burckle, eds., *Paleoclimate and Evolution with Emphasis on Human Origins*, pp. 197–212. New Haven: Yale University Press.

Pye, K. 1987. *Aeolian Dust and Dust Deposits*. San Diego: Academic.

Quinn, T. M., T. J. Crowley, F. W. Taylor, C. Henin, P. Joannot, and Y. Join. 1998. A multicentury stable isotope record from a New Caledonia coral: Interannual and decadal SST variability in the southwest Pacific since 1657 A.D. *Paleoceanography* 13:412–426.

Quinn, T. M. and R. K. Matthews. 1990. Post-Miocene diagenetic and eustatic history of Enewetak Atoll: Model and data comparison. *Geology* 18: 942–945.

Quinn, T. M., F. W. Taylor, T. J. Crowley, and S. M. Link. 1996. Evaluation of sampling resolution in coral stable isotope records: A case study using records from New Caledonia and Tarawa. *Paleoceanography* 11:529–542.

Quinn, W. H. 1992. A study of Southern Oscillation-related climatic activity for A.D. 622–1990 incorporating Nile River flood data. In H. F. Diaz and V. Markgraf, eds. 1992. *El Niño: Historical and Paleoclimatic Aspects of the Southern Oscillation*, pp. 119–149. Cambridge: Cambridge University Press.

Quinn, W. H. and V. T. Neal. 1992. The historical record of El Niño events. In R. S. Bradley and P. D. Jones, eds., *Climate Since A.D. 1500*, pp. 623–648. London: Routledge.

Quinn, W. H., V. T. Neal, and S. E. Antunez de Mayolo. 1987. El Niño occurrences over the past four and a half centuries. *Journal of Geophysical Research* 92:14,449–14,461.

Quinn, W. H., D. O. Zopf, K. S. Short, and R. T. Kuoyang. 1978. Historical trends and statistics of the Southern Oscillation, El Niño, Indonesian droughts. *Fisheries Bulletin, U. S.* 76:663–678.

Rahmstorf, S. 1994. Rapid climate transitions in a coupled ocean-atmosphere model. *Nature* 372:82–85.

Raisbeck, G. M., F. Yiou, D. Bourles, C. Lorius, J. Jouzel, and N. I. Barkov. 1987. Evidence for two intervals of enhanced ^{10}Be deposition in Antarctic ice during the last glacial period. *Nature* 326:273–277.

Raisbeck, G. M., F. Yiou, J. Jouzel, and J. R. Petit. 1990. ^{10}Be and $\delta\,^{2}$H in polar ice cores as a probe of the solar variability's influence on climate. *The Earth's climate and variability of the Sun over recent millennia: geophysical, astronomical, and archaeological aspects. Philosophical Transactions of the Royal Society of London* 330:463–469.

Raisbeck, G. M., F. Yiou, J. Jouzel, J. R. Petit, N. I. Barkov, and E. Bard. 1992. ^{10}Be deposition at Vostok, Antarctica during the last 50,000 years and its relationship to possible cosmogenic production variations during this period. In E. Bard and W. S. Broecker, eds., *The Last Deglaciation: Absolute and Radiocarbon Chronologies,* pp. 125–139. Berlin: Springer-Verlag.

Ramanathan, V., B. Subasilar, G. J. Zhang, et al. 1995. Warm pool heat budget and shortwave cold forcing: A missing physics. *Science* 267:499–503.

Raper, S. C. B., T. M. L. Wigley, and R. A. Warrick. 1996. Global sea level rise: past and future. In J. D. Milliman and B. U. Haq, eds., *Sea-level Rise and Coastal Subsidence: Causes, Consequences, and Strategies,* pp. 11–45. Dordrecht: Kluwer Academic.

Rasmusson, E. M. and T. H. Carpenter. 1982. Variations in tropical sea surface temperature and surface wind fields associated with Southern Oscillation/El Niño. *Monthly Weather Review* 110:354–383.

Rasmusson, E. M. and J. M. Wallace. 1983. Meteorological aspects of El Niño/Southern Oscillation. *Science* 222:1195–1202.

Rasmusson, E. M., X. Wang, and C. F. Ropelewski. 1990. The biennial component of ENSO variability. *Journal of Marine Systems* 1:71–96.

Rasmusson, T. L., E. Thomsen, T. C. E. van Weering, and L. Labeyrie. 1996. Rapid changes in surface and deep water conditions at the Faeroe Margin during the last 58,000 years. *Paleoceanography* 11:757–771.

Rau, G. H., P. N. Froelich, T. Takahashi, and D. J. Des Marais. 1991. Does sedimentary organic $\delta\,^{13}$C record variations in Quaternary ocean [CO_2(aq)]? *Paleoceanography* 6:335–347.

Rau, G. H., T. Takahashi, and D. J. Des Marais. 1989. Latitudinal variations in plankton $\delta\,^{13}$C: Implications for CO_2 and productivity in past oceans. *Nature* 341:516–518.

Rau, G. H., T. Takahashi, D. J. Des Marais, D. J. Repeta, and J. H. Martin. 1992. The relationship between δ^{13}C of organic matter and [CO_2 (aq)] in ocean surface water: Data from a JGOFS site in the northeast Atlantic Ocean and a model. *Geochimica et Cosmochimica Acta* 56:1413–1419.

Ravelo, A. C. and N. J. Shackleton. 1995. Evidence for surface water circulation changes at Site 851 in the eastern tropical Pacific Ocean. In N. G. Pisias,

L. A. Mayer, T. R. Janacek, A. Palmer-Julson, and T. H. van Andel, eds., *Proceedings of Ocean Drilling Program Scientific Results* 138:503–514.

Raymo, M. E. 1991. Geochemical evidence supporting T. C. Chamberlin's theory of glaciation. *Geology* 19:344–347.

Raymo, M. E. 1994. The initiation of Northern Hemisphere glaciation. *Annual Review of Earth and Planetary Science* 22:353–383.

Raymo, M. E., D. Hodell, and E. Jansen. 1992. Response of deep ocean circulation to initiation of Northern Hemisphere glaciation. *Paleoceanography* 7:645–672.

Raymo, M. E., W. F. Ruddiman, J. Backman, S. M. Clemens, and D. G. Martinson. 1989. Late Pliocene variation in northern hemisphere ice sheets and North Atlantic deep water circulation. *Paleoceanography* 4:413–446.

Raymo, M. E., W. F. Ruddiman, and P. N. Froelich. 1988. Influence of late Cenozoic mountain building on ocean geochemical cycles. *Geology* 16:649–653.

Raymo, M. E., W. F. Ruddiman, N. J. Shackleton, and D. W. Oppo. 1990. Evolution of Atlantic-Pacific $\delta^{13}C$ gradients over the last 2.5 m.y. *Earth and Planetary Science Letters* 97:353–368.

Raynaud, D., and J.-M. Barnola. 1985. An Antarctic ice core reveals atmospheric CO_2 variations over the past few centuries. *Nature* 315:309–311.

Raynaud, D., J.-M. Barnola, J. Chappellaz, and P. Martinerie. 1996. Changes in trace gas concentrations during the last 2000 years and more generally, the Holocene. In P. D. Jones, R. S. Bradley, and J. Jouzel, eds., *Climate Variations and Forcing Mechanisms of the Last 2000 Years*, NATO ASI Series 1, 41:547–561. Berlin: Springer-Verlag.

Raynaud, D., J. Chappellaz, J. M. Barnola, Y. S. Korotkevich, and C. Lorius. 1988. Climatic and CH_4 cycle, implications of glacial-interglacial CH_4 change in the Vostok ice core. *Nature* 333:655–657.

Raynaud, D., J. Jouzel, J. M. Barnola, J. Chappellaz, R. J. Delmas, and C. Lorius. 1993. The ice record of greenhouse gases. *Science* 259:926–934.

Rea, D. K. 1994. The paleoclimatic record provided by eolian deposition in the deep sea: the geologic history of wind. *Reviews of Geophysics* 32:159–195.

Reid, G. C. 1993. Do solar variations change climate? *EOS, Transactions of the American Geophysical Union* Jan. 12:23.

Retallack, G. J. 1990. *Soils of the Past: An Introduction to Paleopedology*. London: Harper Collins Academic.

Revelle, R. 1985. Introduction: the scientific history of carbon dioxide. In E. T. Sundquist and W. S. Broecker, eds., *The Carbon Cycle and Atmospheric CO_2: Natural Variations Archean to Present, American Geophysical Union Monograph* 32:1–4.

Revelle, R. and H. E. Suess. 1957. Carbon dioxide exchange between atmosphere and ocean and the question of an increase of atmospheric CO_2 during past decades. *Tellus* 9:18–27.

Rind, D. 1996. The potential for modelling the effects of different forcing factors on climate during the past 2000 years. In P. D. Jones, R. S. Bradley, and

J. Jouzel, eds., *Climate Variations and Forcing Mechanisms of the Last 2000 Years,* NATO ASI Series 1, 41:563–581. Berlin: Springer-Verlag.

Rind, D. and M. Chandler. 1991. Increased ocean heat transports and warmer climate. *Journal of Geophysical Research* 96:7437–7461.

Rind, D., G. Kukla, and D. Peteet. 1989. Can Milankovitch orbital variations initiate the growth of ice sheets in a general circulation model? *Journal of Geophysical Research* 94:12,851–12,871.

Rind, D. and J. Overpeck. 1993. Hypothesized causes of decade-to-century-scale climatic variability: Climate model results. *Quaternary Science Reviews* 12:357–374.

Rind, D. and D. Peteet. 1985. Terrestrial conditions at the last glacial maximum and CLIMAP sea-surface temperature estimates: Are they consistent? *Quaternary Research* 24:1–22.

Rind, D., D. Peteet, W. Broecker, A. McIntyre, and W. Ruddiman. 1986. The impact of cold North Atlantic sea surface temperatures on climate: Implications for the Younger Dryas cooling (11–10 k). *Climate Dynamics* 1:3–33.

Roberts, N. 1989. *The Holocene: An Environmental History.* Oxford: Blackwell, Ltd.

Robin G. de Q., ed. 1983. *The Climatic Record in Polar Ice Sheets.* Cambridge: Cambridge University Press.

Rodrigues, C. G. and G. Vilks. 1994. The impact of glacial lake runoff on the Goldthwait and Champlain Seas: The relationship between Glacial Lake Agassiz runoff and the Younger Dryas. *Quaternary Science Reviews* 13:923–944.

Roemmich, D. 1992. Ocean warming and sea level rise along the southwest U.S. coast. *Science* 257:373–375.

Ropelewski, C. F. and M. S. Halpert. 1986. North American precipitation and temperature patterns associated with the El Niño/Southern Oscillation (ENSO). *Monthly Weather Review* 114:2352–2362.

Ropelewski, C. F. and M. S. Halpert. 1987. Global and regional scale precipitation patterns associated with the El Niño/Southern Oscillation. *Monthly Weather Review* 115:1606–1626.

Rose, M. R. and G. V. Lauder, eds. 1996. *Adaptation.* San Diego: Academic Press.

Rothlisberger, F. 1986. *10,000 Jahre Gletschergeschichte der Erde.* Aarau: Verlag Sauerlander.

Roulier, L. M. and T. M. Quinn. 1995. Seasonal- to decadal-scale climatic variability in southwest Florida during the Middle Pliocene: Inferences from a coralline stable isotope record. *Paleoceanography* 10:429–443.

Roy, K., J. W. Valentine, D. Jablonski, and S. Kidwell. 1996. Scales of climatic variability and time averaging in Pleistocene biotas: implications for ecology and evolution. *Trends in Ecology and Evolution* 11:458–463.

Rozanski, K., L. Araguas-Araguas, and R. Gonfiantini. 1993. Isotopic patterns in modern global precipitation. In P. K. Swart, K. C. Lohman, J. McKenzie, and S. Savin, eds., *Climate Change in Continental Isotope Records, American Geophysical Union Monograph* 78:1–36.

Ruddiman, W. F. 1977. Late Quaternary deposition of ice-rafted sand in the subpolar North Atlantic (lat 40° to 65°N). *Geological Society of America Bulletin* 88:1813–1827.

Ruddiman, W. F. and J. E. Kutzbach. 1989. Forcing of late Cenozoic northern hemisphere climate by plateau uplift in southeast Asia and the American southwest. *Journal of Geophysical Research* 94(D15):18,409–18,427.

Ruddiman, W. F. and A. McIntyre. 1973. Time-transgressive deglacial retreat of polar waters from the North Atlantic. *Quaternary Research* 3:117–130.

Ruddiman, W. F. and A. McIntyre. 1981. Oceanic mechanisms for amplification of the 23,000-year ice-volume cycle. *Science* 212:617–627.

Ruddiman, W. F. and M. E. Raymo. 1988. Northern Hemisphere climate regimes during the past 3 Ma: possible tectonic connections. *Philosophical Transactions of the Royal Society of London, Series B, Biological Sciences* 318:411–430.

Ruddiman, W. F., M. E. Raymo, and A. McIntyre. 1986. Matuyama 41,000-year cycles: North Atlantic Ocean and northern hemisphere ice sheets. *Earth and Planetary Science Letters* 80:117–129.

Ruddiman, W. F., M. E. Raymo, D. G. Martinson, B. M. Clement, and J. Backman. 1989. Pleistocene evolution: northern hemisphere ice sheets and North Atlantic Ocean. *Paleoceanography* 4:353–412.

Ruddiman, W. F., C. D. Sancetta, and A. McIntyre. 1977. Glacial/interglacial response rate of subpolar North Atlantic waters to climatic change: the record in oceanic sediments. *Philosophical Transactions of the Royal Society of London, Series B, Biological Sciences* 280:119–142.

Ruddiman, W. F. and H. E. Wright, Jr. 1987. *North America and Adjacent Oceans During the Last Deglaciation. The Geology of North America*, Vol. K-3. Boulder: Geological Society of America.

Rudels, B. and D. Quadfasel. 1991. Convection and deep water formation in the Arctic Ocean-Greenland Sea system. *Journal of Marine Systems* 2:435–450.

Ruse, M. 1981. What kind of revolution occurred in geology? In P. Asquith and I. Hackring, eds., *Philosophy of Science Association* 2: 240–273. East Lansing, Michigan: Philosophy of Science Association.

Rutter, N., Z. Ding, and T. Liu. 1996. Long paleoclimate records from China. *Geophysica* 32:7–34.

Rutter, N. W., N. R. Catto, eds. 1995. *Dating Methods for Quaternary Deposits.* Geoassociation of Canada.

Sahagian, D. L., F. W. Schwartz, and D. K. Jacobs. 1994. Direct anthropogenic contributions to sea-level rise in the twentieth century. *Nature* 367:54–57.

Saltzman, B. and A. Sutera. 1987. The mid-Quaternary climatic transition as the free response of a three-variable dynamical model. *Journal of Atmospheric Science* 44:236–241.

Sancetta, C. 1992. Primary production in the glacial North Atlantic and North Pacific oceans. *Nature* 360:249–251.

Sancetta, C. and S. M. Silvestri. 1986. Pliocene-Pleistocene evolution of the

North Pacific ocean-atmosphere system interpreted from fossil diatoms. *Paleoceanography* 1:163–180.
Sandweiss, D. H., J. B. Richardson III, E. J. Reitz, H. B. Rollins, and K. A. Maasch. 1996. Geoarchaeological evidence from Peru for a 5000 years B.P. onset of El Niño. *Science* 273:1531–1533.
Santer, B. D., A. Berger, J. A. Eddy, et al. 1993. Group Report: How can paleodata be used to evaluate the forcing mechanisms responsible for past climate changes? In J. A. Eddy and H. Oeschger, eds., *Global Changes in the Perspective of the Past*, pp. 344–367. Chichester: John Wiley & Sons.
Sarachik, E. S., ed. 1996. *Learning to Predict Climate Variations Associated With El Niño and the Southern Oscillation.* Washington, D.C.: National Academy Press.
Sarmiento, J. L. and C. Le Quére. 1996. Oceanic carbon dioxide uptake in a model of century-scale global warming. *Science* 274:1346–1350.
Sarmiento, J. and J. R. Toggweiler. 1984. A new model for the role of the oceans in determining atmospheric P_{CO_2} *Nature* 308:621–624.
Sarnthein, M. and A. V. Altenbach. 1995. Late Quaternary changes in surface water and deep water masses of the Nordic Seas and north-eastern North Atlantic: a review. *Geological Rundschau* 84:89–107.
Sarnthein, M. and R. Tiedemann. 1989. Towards a high-resolution stable isotope stratigraphy of the last 3.4 million years: Sites 658 and 659 off northwest Africa. *Proceedings of the Ocean Drilling Program Scientific Results* 108:167–185.
Sarnthein, M. and R. Tiedemann. 1990. Younger Dryas-style cooling events at glacial terminations I–IV at ODP Site 658: Associated benthic $\delta^{13}C$ anomalies constrain meltwater hypothesis. *Paleoceanography* 5:1041–1055.
Sarnthein, M., K. Winn, S. J. A. Jung, et al. 1994. Changes in east Atlantic deepwater circulation over the last 30,000 years: Eight time slice reconstructions. *Paleoceanography* 9:209–267.
Scherer, R. P., A. Aldahar, S. Tulaczyk, G. Possnert, H. Engelhardt, B. Kamb. 1998. Pleistocene collapse of the West Antarctic Ice Sheet. *Science* 281:82–85.
Schimel, D., I. G. Enting, M. Heimann, et al. 1995. CO_2 and the carbon cycle. In J. T. Houghton, L. G. Meira Filho, J. Bruce, et al., eds., *Climate Change 1994*, pp. 39–71. Cambridge: Cambridge University Press.
Schlesinger, W. 1997. *Biogeochemistry: A Study of Global Change.* San Diego: Academic Press.
Schneider, S. H. 1993. Can paleoclimatic and paleoecological analyses validate future global climate and ecological change projections? In J. A. Eddy and H. Oeschger, eds., *Global Climate Changes in the Perspective of the Past*, pp. 317–340. Chichester: John Wiley & Sons.
Schneider, S. H. 1994. Detecting climatic change signals: are there any fingerprints? *Science* 263:341–347.
Schneider, S. H. and P. J. Boston. 1991. *Scientists on Gaia.* Cambridge, Mass.: MIT Press.
Schnitker, D. 1979. The deep waters of the western North Atlantic during the

past 24,000 years, and the re-initiation of the western boundary undercurrent. *Marine Micropaleontology* 4:265–280.

Schoener, T. W. 1988. The ecological niche. In J. M. Cherret, ed., *Ecological Concepts,* pp. 79–113. Oxford: Blackwell, Ltd.

Schott, W. 1935. Die Foraminiferen in dem aquatorialen Teil des Atlantischen Ozeans. *Deutsch Atlantic Expedition Meteor 1925–1927, Wissenschaften Ergebnisse* 3:43–134. (Cited in Imbrie and Imbrie 1979.)

Schove, D. J., ed. 1983. *Sunspot Cycles.* Stroudsburg, Penn.: Hutchinson Ross.

Schrader H. and T. Baumgartner 1983. Decadal variation of upwelling in the central Gulf of California. In J. Thiede and E. Suess, eds. *Coastal Upwelling: Its Sediment Record,* Part B, pp. 247–276. New York: Plenum.

Schrag, D. P., G. Hampt, and D. W. Murray. 1996. Pore fluid constraints on the temperature and oxygen isotopic composition of the glacial ocean. *Science* 272:1930–1932.

Schwabe, A. N. 1844. Sonnen-Beobachtungen in Jahr 1843. *Astronomische Nachrichten* 21:233. (Cited in Hoyt and Schatten 1997.)

Schwander, J. 1989. The transformation of snow to ice and the occlusion of gases. In H. Oeschger and C. C. Langway, Jr., eds., *The Environmental Record in Glaciers and Ice Sheets,* pp. 53–67. Chichester: John Wiley & Sons.

Schwander, J. and B. Stauffer. 1984. Age differences between polar ice and air trapped in its bubbles. *Nature* 311:45–47.

Schwarcz, H. P. 1989. Uranium series dating of Quaternary deposits. *Quaternary International* 1:7–17.

Schwarzacher, W. and A. G. Fischer. 1982. Limestone-shale bedding and perturbations in the earth's orbit. In G. Einsele and A. Seilacher, eds., *Cyclic Event Stratification,* pp. 72–95. Berlin: Springer-Verlag.

Schweingruber, F. H. 1988. *Tree Rings: Basics and Applications of Dendrochronology.* Dordrecht: D. Reidel.

Schweingruber, F. H. and K. R. Briffa. 1996. Tree-ring density networks for climate reconstruction. In P. D. Jones, R. S. Bradley, and J. Jouzel, eds., *Climate Variations and Forcing Mechanisms of the Last 2000 Years,* NATO ASI Series 1, 41:43–66. Berlin: Springer-Verlag.

Scott, D. B. 1978. Vertical zonations of marsh foraminifera as accurate indicators of former sea-levels. *Nature* 272:528–531.

Scott, D. B., K. Brown, E. S. Collins, and F. S. Medioli. 1995a. A new sea-level curve from Nova Scotia: Evidence for a rapid acceleration of sea-level rise in the late mid-Holocene. *Canadian Journal of Earth Sciences* 32: 2071–2080.

Scott, D. B. and E. S. Collins. 1996. Late Mid-Holocene sea-level oscillation: a possible cause. *Quaternary Science Reviews* 15:851–856.

Scott, D. B., P. T. Gayes, and E. S. Collins. 1995b. Mid-Holocene precedent for a future rise in sea level along the Atlantic coast of North America. *Journal of Coastal Research* 11:615–622.

Scott, D. B. and G. Vilks. 1991. Benthonic foraminifera in the surface sediments of the deep-sea Arctic Ocean. *Journal of Foraminiferal Research* 21:20–38.

Seidenkrantz, M.-S., L. Bornmalm, S. J. Johnsen, et al. 1996. Two-step deglaciation at the oxygen isotope stage 6/5e transition: the Zeifen-Kattegat climate oscillation. *Quaternary Science Reviews* 15:77–90.

Seidenkrantz, M.-S., P. Kristensen, and K. L. Knudsen. 1995. Marine evidence for climatic instability during the last interglacial in shelf records from northwest Europe. *Journal of Quaternary Science* 10:77–82.

Selle, W. 1962. Geologische und vegetationskundliche Untersuchungen an einigen wichtigen Vorkommen des letzen Interglazials in Nordwestdeutschland. *Geologisches Jahrbuch* 79:295–352.

Semtner, A. J. 1995. Modeling ocean circulation. *Science* 269:1379.

Senum, G. I. and J. S. Gaffney. 1985. A reexamination of the tropospheric methane cycle: Geophysical implications. In W. T. Sundquist and W. S. Broecker, eds., *The Carbon Cycle and Atmospheric CO_2: Natural Variations Archean to Present, American Geophysical Union Monograph* 32:61–69.

Seret, G., J. Guiot, G. Wansard, J. L. de Beaulieu, and M. Reille. 1992. Tentative palaeoclimatic reconstruction linking pollen and sedimentology in La Grande Pile (Vosges, France). *Quaternary Science Reviews* 11:425–430.

Serre-Bachet, F. and J. Guiot. 1987. Summer temperature changes from tree rings in the Mediterranean area during the last 800 years. In W. H. Berger and L. D. Labeyrie, eds., *Abrupt Climatic Change,* pp. 89–97. Dordrecht: D. Reidel.

Severinghaus, J. P., T. Sowers, E. J. Brook, R. B. Alley, and M. L. Bender. 1998. Timing of abrupt climate change at the end of the Younger Dryas interval from the thermally fractioned gases in polar ice. *Nature* 391:141–146.

Shackleton, N. J. 1967. Oxygen isotope analyses and Pleistocene temperatures re-assessed. *Nature* 215:15–17.

Shackleton, N. J. 1969. The last interglacial in the marine and terrestrial records. *Proceedings of the Royal Society of London, Series B, Biological Sciences* 174:135–154.

Shackleton, N. J. 1993. Last interglacial in Devils Hole. *Nature* 362:596.

Shackleton, N. J., A. Berger, and W. R. Peltier. 1990. An alternative astronomical calibration of the lower Pleistocene timescale based on ODP Site 677. *Transactions of the Royal Society of Edinburgh* 81:251–261.

Shackleton, N. J., S. Crowhurst, T. Hagelberg, N. G. Pisias, and D. A. Schneider. 1995a. A new late Neogene time scale: application to Leg 138 sites. *Proceedings of Ocean Drilling Program Scientific Results* 138:73–101.

Shackleton, N. J., T. K. Hagelberg, and S. J. Crowhurst. 1995b. Evaluating the success of astronomical tuning: pitfalls of using coherence as a criterion for assessing pre-Pleistocene timescales. *Paleoceanography* 10:693–698.

Shackleton, N. J. and N. D. Opdyke. 1973. Oxygen isotope and paleomagnetic stratigraphy of equatorial Pacific core V28–238: Oxygen isotope temperatures and ice volumes on a 10^5 and 10^6 year scale. *Quaternary Research* 3:39–55.

Shackleton, N. J. and N. D. Opdyke. 1976. Oxygen isotopes and paleomagnetic stratigraphy of Pacific core V28–239: Late Pliocene to latest Pleistocene: An investigation of late Quaternary paleoceanography and paleoclimatology. *Geological Society of America Memoir* 145:449–464.

Shackleton, N. J. and N. Pisias. 1985. Atmospheric carbon dioxide, orbital forcing, and climate. In E. T. Sundquist and W. S. Broecker, eds., *The Carbon Cycle and Atmospheric CO_2: Natural Variations Archean to Present, American Geophysical Union Monograph* 32:303–317.

Shemesh, A., C. D. Charles, and R. G. Fairbanks. 1992. Oxygen isotopes in biogenic silica: Global changes in ocean temperature and isotopic composition. *Science* 256:1434–1436.

Shemesh, A., R. A. Mortlock, and P. N. Froelich. 1989. Late Cenozoic Ge/Si record of marine biogenic opal: Implications for variations of riverine fluxes to the ocean. *Paleoceanography* 4:221–234.

Shen, G. T. and E. A. Boyle. 1988. Determination of lead, cadmium, and other trace metals in annually-banded corals. *Chemical Geology* 67:47–62.

Shen, G. T., J. E. Cole, D. W. Lea, L. J. Lin, T. A. McConnaughey, and R. G. Fairbanks. 1992a. Surface ocean variability at Galapagos from 1936–1982: calibration of geochemical tracers in corals. *Paleoceanography* 7:563–583.

Shen, G. T. and R. B. Dunbar. 1995. Environmental controls on uranium in reef corals. *Geochimica et Cosmochimica Acta* 59:2009–2024.

Shen, G. T., R. B. Dunbar, G. M. Wellington, M. W. Colgan, and P. W. Glynn. 1991. Paleochemistry of manganese in corals from the Galapagos Islands. *Coral Reefs* 10:91–101.

Shen, G. T., L. J. Linn, T. M. Campbell, J. E. Cole, and R. G. Fairbanks. 1992b. A chemical indicator of trade wind reversal in corals from the western tropical Pacific. *Journal of Geophysical Research* 97:12,689–12,698.

Shen, G. T. and Sanford, C. L. 1990. Trace element indicators of climate variability in reef-building corals. In P. W. Glynn, ed., *Global Ecological Consequences of the 1982–83 El Niño–Southern Oscillation*, pp. 255–283. Amsterdam: Elsevier.

Sheppard, P. A. 1966. Preface. *Proceedings of the International Symposium on World Climate: 8000–0 B.C.*, p. 1. London: Imperial College.

Sibrava, V., D. Q. Bowen, and G. M. Richmond. 1986. Quaternary Glaciations in the Northern Hemisphere. *Quaternary Science Reviews* 5:1–514.

Siegenthaler, U. and J. L. Sarmiento. 1993 Atmospheric carbon dioxide and the ocean. *Nature* 365:119–125.

Siegenthaler, U., H. Friedli, H. Loeetscher, et al. 1988. Stable isotope ratios and concentration of CO_2 in air from polar ice cores. *Annals of Glaciology* 10:151–156.

Sikes, E. L., J. W. Farrington, and L. D. Keigwin. 1991. Use of alkenone unsaturation ratio U^k_{37} to determine past sea surface temperature: coretop SST calibrations and methodological considerations. *Earth and Planetary Science Letters* 104:36–47.

Sikes, E. L. and L. D. Keigwin. 1994. Equatorial Atlantic sea surface temperature for the last 30 kyr: A comparison of U^k_{37}, $\delta^{18}O$ and foraminiferal assemblage temperature estimates. *Paleoceanography* 9:31–45.

Singer, C., J. Shulmeister, B. McLea. 1998. Evidence against a significant Younger Dryas cooling event in New Zealand. *Science* 281: 812–814.

Skinner, B. J. and S. C. Porter. 1995. *The Blue Planet.* New York: John Wiley & Sons.

Sloan, L. C. and D. K. Rea. 1995. Atmospheric carbon dioxide and early Eocene climate: A general circulation modeling sensitivity study. *Palaeogeography, Palaeoclimatology, Palaeoecology* 119:275–292.

Sloan, L. C., J. C. G. Walker, and T. C. Moore, Jr. 1995. Possible role of oceanic heat transport in early Eocene climate. *Paleoceanography* 10:347–356.

Sloss, L. L. 1991. The tectonic factor in sea level change: the countervailing view. *Journal of Geophysical Research* 96:6609–6617.

Slowey, N. C., G. M. Henderson, and W. B. Curry. 1996. Direct U-Th dating of marine sediments from the two most recent interglacial periods. *Nature* 383:242–244.

Smith, S. V., R. W. Buddmeier, R. C. Redalje, and J. E. Houck. 1979. Strontium-calcium thermometry in coral skeletons. *Science* 204:404–407.

Smith, T. J., J. H. Hudson, M. B. Robblee, G. V. N. Powell, and P. J. Isdale. 1989. Freshwater flow from the Everglades to Florida Bay: A historical reconstruction based on fluorescent banding in the coral *Solenastrea bournoni. Bulletin of Marine Science* 44:274–282.

Sober, E. ed. 1984. *Conceptual Issues in Evolutionary Biology: An Anthology.* Cambridge: MIT Press.

Sober, E. 1993. *Philosophy of Biology.* Boulder: Westview Press.

Sonnett C. P., M. S. Giampapa, and M. S. Matthews, eds. 1992. *The Sun in Time.* Tucson: University of Arizona Press.

Souchez, R., M. Lemmens, and J. Chappellaz. 1995. Flow-induced mixing in the GRIP basal ice deduced from the CO_2 and CH_4 records. *Geophysical Research Letters* 22:41–44.

Sowers, T., E. Brook, D. Etheridge, et al. 1997. An inter-laboratory comparison of techniques for extracting and analyzing gases in ice cores. *Journal of Geophysical Research* 102:26,527–26,539.

Sowers, T. and M. Bender. 1995. Climate records covering the last deglaciation. *Science* 269:210–214.

Sowers, T., M. Bender, L. Labeyrie, et al. 1993. A 135,000-year Vostok-SPECMAP common temporal framework. *Paleoceanography* 8:737–766.

Sowers, T., M. Bender, D. Raynaud, and Y. S. Korotkevich. 1992. The $\delta^{15}N$ of N_2 in air trapped in polar ice: a tracer of gas transport in the firn and a possible constraint on ice age–gas age differences. *Journal of Geophysical Research* 97:15,683–15,697.

Sowers, T. M. Bender, D. Raynaud, Y. S. Korotkevich, and J. Orchado. 1991. The $\delta\ ^{18}O$ of atmospheric O_2 from air inclusions in the Vostok ice core: Timing of CO_2 and ice volume changes during the penultimate deglaciation. *Paleoceanography* 6:679–696.

Spero, H. J. and D. W. Lea. 1993. Intraspecific stable isotope variability in the planktonic foraminifer *Globigerinoides sacculifer:* Results from laboratory study. *Marine Micropaleontology* 22:221–234.

Spero, H. J., I. Lerche, and D. F. Williams. 1991. Opening the carbon isotope "Vital Effect" black box. 2. Quantitative model for interpreting foraminiferal carbon isotope data. *Paleoceanography* 6:639–655.

Spoerer, G. 1889. Uber die Periodicitat de Sonnenflecken seit dem Jahr 1618. *Nova Acta der Ksl. Leop.-Carol. Deutschen Akademie der Naturforscher* 53:283–324. (Cited in Hoyt and Schatten 1997.)

Stager, J. C. and P. A. Mayewski. 1997. Abrupt early to mid-Holocene climatic transition registered at the equator and the poles. *Science* 276:1834–1836.

Stahle, D. W. and M. K. Cleaveland. 1994. Tree-ring reconstructed rainfall over the southeastern U.S.A. during the Medieval Warm period and Little Ice Age. *Climatic Change* 26:199–212.

Stahle, D. W. and M. K. Cleaveland. 1996. Large-scale climatic influences on baldcypress tree growth across the southeastern United States. In P. D. Jones, R. S. Bradley, and J. Jouzel, eds., *Climatic Variations and Forcing Mechanisms of the Last 2000 Years.* NATO ASI Series 1, 41:125–140. Berlin: Springer-Verlag.

Stahle, D. W., M. K. Cleaveland, and J. G. Hehr. 1988. North Carolina climate changes reconstructed from tree rings: A.D. 372–1985. *Science* 240:1517–1519.

Stanley, S. M. 1979. *Macroevolution.* San Francisco: W. H. Freeman.

Starkel, L. 1991. Environmental change at the Younger Dryas–Preboreal transition and during the early Holocene: some distinctive aspects in central Europe. *The Holocene* 1:234–242.

Stauffer, B., H. Hofer, H. Oeschger, J. Schwander, and U. Siegenthaler. 1984. Atmospheric CO_2 concentration during the last glaciation. *Annals of Glaciology* 5:160–164.

Stauffer, B., E. Lochbronner, H. Oeschger, J. Schwander. 1988. Methane concentration in the glacial atmosphere was only half that of preindustrial Holocene. *Nature* 332: 812–814.

Stea, R. and R. J. Mott. 1989. Deglaciation environments and evidence for glaciers of Younger Dryas age in Nova Scotia, Canada. *Boreas* 18:169–187.

Steig, E. J., P. M. Grootes, and M. Stuiver. 1994. Seasonal precipitation timing and ice core records. *Science* 266:1885–1886.

Stein, M., G. J. Wasserburg, J. H. Chen, Z. R. Zhu, A. Bloom, and J. Chappell. 1993. TIMS U-series dating and stable isotopes of the last interglacial event in Papua New Guinea. *Geochimica et Cosmochimica Acta* 57: 2541–2554.

Stein, M., G. J. Wasserburg, K. R. Lajoie, and J. H. Chen. 1991. U-series ages of solitary coral from the California coast by mass spectrometry. *Geochimica et Cosmochimica Acta* 55:3709–3722.

Stein, R., S.-I. Nam, C. Schubert, C. Vogt, D. Fütterer, and J. Heinemeier. 1994. The last deglaciation event in the eastern central Arctic Ocean. *Science* 264:692–696.

Stille, H. 1924. *Grundfagen der vergleichenden Tectonik.* Berlin: Borntraeger.

Stirling, C. H., T. M. Esat, M. T. McCulloch, and K. Lambeck. 1995. High-precision U-series dating of corals from western Australia and implications for the timing and duration of the last interglacial. *Earth and Planetary Science Letters* 135:115–130.

Stokes, M. A. and T. L. Smiley. 1996. *An Introduction to Tree-ring Dating.* Tucson: University of Arizona Press.

Stoll, H. M. and D. P. Schrag. 1996. Evidence for glacial control of rapid sea level changes in the Early Cretaceous. *Science* 272:1771–1774.

Stoner, J. S., J. E. T. Channell, and C. Hillaire-Marcel. 1996. The magnetic signature of rapidly deposited detrital layers from the deep Labrador Sea: relationship to North Atlantic Heinrich layers. *Paleoceanography* 11:309–325.

Strahler, A. N. 1987. *Science and Earth History: The Evolution/Creation Controversy.* Buffalo, N.Y.: Promethean Books.

Street-Perrot, F. A. and S. P. Harrison. 1985. Lake levels and climate reconstruction. In A. D. Hecht, ed., *Paleoclimate Analyses and Modeling,* pp. 291–340. New York: John Wiley & Sons.

Stromberg, B. 1994. Younger Dryas deglaciation at Mt. Billinen and clay varve dating of the Younger Dryas/Preboreal transition. *Boreas* 23:177–193.

Stuiver, M. 1965. Carbon-14 content of the 18th and 19th century wood: Variations correlated with sunspot activity. *Science* 149:533–537.

Stuiver, M. 1993. A note on single-year calibration of the AD radiocarbon timescale. *Radiocarbon* 35:67–72.

Stuiver, M. 1994. In E. Nesme-Ribes, ed., *The Solar Engine and Its influence on Terrestrial Atmospheres and Climate,* NATO ASI Series 25:202–220. Berlin: Springer-Verlag.

Stuiver, M. and T. F. Braziunas. 1987. Tree cellulose $^{13}C/^{12}C$ isotope ratios and climatic change. *Nature* 328:58–60.

Stuiver, M. and T. F. Braziunas. 1989. Atmospheric ^{14}C and century-scale solar oscillations. *Nature* 388:405–408.

Stuiver, M. and T. F. Braziunas. 1993. Sun, ocean, climate and atmospheric $^{14}CO_2$: an evaluation of causal and spectral relationships. *The Holocene* 3:289–305.

Stuiver, M., T. F. Braziunas, and P. M. Grootes. 1997. Is there evidence for solar forcing of climate in the GISP2 oxygen isotope record. *Quaternary Research* 48:259–266.

Stuiver, M., P. M. Grootes, and T. F. Braziunas. 1995. The GISP2 $\delta\ ^{18}O$ climate record of the past 16,500 years and the role of the sun, ocean, and volcanos. *Quaternary Research* 44:341–354.

Stuiver, M. and H. A. Polach. 1977. Discussion: Reporting of ^{14}C data. *Radiocarbon* 19:355–363.

Stuiver, M. and P. D. Quay. 1980. Changes in atmospheric carbon-14 attributed to a variable sun. *Science* 207:11–19.

Stuiver, M. and P. J. Reimer. 1993. Extended ^{14}C data base and revised CALIB 3.0 ^{14}C age calibration program. *Radiocarbon* 35:215–230.

Stute, M. and P. Schlosser. 1993. Principles and applications of the noble gas paleothermometer. In P. K. Swart, K. C. Lohman, J. McKenzie, and S. Savin, eds., *Climate Change in Continental Isotope Records, American Geophysical Union Monograph* 78:89–100.

Suess, E. 1885–1909. *Das Anlitz der Erde.* Vol. 2, *Die Meere der Erde.* Prague: F. Tempsky.

Suess, H. 1965. Secular variations of the cosmic ray produced carbon-14 in the atmosphere and their interpretations. *Journal of Geophysical Research* 70:5935-5952.

Suess, H. 1968. Climatic changes, solar activity and the cosmic-ray production rate of natural radiocarbon. *Meteorological Monographs* 8:146–150.

Sugden, D. E., D. R. Marchant, and G. H. Denton, eds. 1993. The case for a stable East Antarctic ice sheet. *Geografiska Annaler* 75A:151–351.

Suggate, R. P. 1990. Late Pliocene and Quaternary glaciations of New Zealand. *Quaternary Science Reviews* 9:175–197.

Sundquist, E. T. 1985. Geological perspectives on carbon dioxide and the carbon cycle. In E. T. Sundquist and W. S. Broecker, eds., *The Carbon Cycle and Atmospheric* CO_2*: Natural Variations Archean to Present, American Geophysical Union Monograph* 32:5–60.

Sundquist, E. T. and W. S. Broecker, eds. 1985. *The Carbon Cycle and Atmospheric* CO_2*: Natural Variations Archaen to Present, American Geophysical Union Monograph* 32.

Swart, P. K. 1983. Carbon and oxygen isotope fractionation in scleractinian corals: A review. *Earth Science Reviews* 19:51–80.

Swart, P. K., G. F. Healy, R. E. Dodge, et al. 1996a. The stable oxygen and carbon isotopic record from a coral growing in Florida Bay: A 160 year record of climatic and anthropogenic influence. *Palaeogeography, Palaeoclimatology, Palaeoecology* 123:219–237.

Swart, P. K., J. Leder, A. M. Szmant, and R. E. Dodge. 1996b. The origin of variations in the isotopic record of scleractinian corals. II. Carbon. *Geochimica et Cosmochimica Acta* 60:2871–2885.

Swart, P. K., K. C. Lohman, J. McKenzie, and S. Savin, eds. 1993. *Climate Change in Continental Isotopic Records, American Geophysical Union Monograph* 78.

Swetnam, T. W. 1993. Fire history and climate change in giant sequoia groves. *Science* 262:685–689.

Swetnam, T. W. and J. L. Betancourt. 1992. Temporal patterns of El Niño–Southern Oscillation—wildfire teleconnections in the southwestern United States. In H. F. Diaz and V. Markgraf, eds., *El Niño: Historical and Paleoclimatic Aspects of the Southern Oscillation,* pp. 259–270. Cambridge: Cambridge University Press.

Swithinbank, C. 1988. Antarctica. In R. Williams and J. Ferrigno, eds., Satellite Image Atlas of Glaciers of the World. U. S. Geological Survey Professional Paper 1386-B-Antarctica.

Sy, A., M. Rhein, J. R. N. Lazier, et al. 1997. Surprisingly rapid spreading of newly formed intermediate waters across the North Atlantic Ocean. *Nature* 386:675–679.

Szabo, B. J. 1979. Uranium-series age of coral reef growth on Rottnest Island, Western Australia. *Marine Geology* 29:11–15.

Szabo, B. J. 1985. Uranium-series dating of fossil corals from marine sediments of southeastern United States Atlantic Coastal Plain. *Geological Society of America Bulletin* 96:398–406.

Szabo, B. J., K. R. Ludwig, D. R. Muhs, and K. R. Simmons. 1994. Thorium-230 ages of corals and duration of the last interglacial sea-level high stand on Oahu, Hawaii. *Science* 266:93–96.

Szabo, B. J., J. I. Tracey, and E. R. Goter. 1985. Ages of subsurface stratigraphic intervals in the Quaternary of Enewetak Atoll, Marshall Islands. *Quaternary Research* 23:54–61.

Szabo, B. J., W. C. Ward, A. E. Weidie, and M. J. Brady. 1978. Age and magnitude of the late Pleistocene sea-level rise on the eastern Yucatan Peninsula. *Geology* 9:451–457.

Tanner, W. F. 1992. 3000 years of sea level change. *Bulletin American Meteorological Society* 73:297–303.

Tans, P. P., I. Y. Fung, and T. Takahashi. 1990. Observational constraints on the global atmospheric CO_2 budget. *Science* 247:1431–1438.

Tansley, A. G. 1935. The use and abuse of vegetational terms and concepts. *Ecology* 16:284–307.

Taylor, K. C. 1994. Climate models for the study of paleoclimates. In J.-C. Duplessy and M.-T. Spyridakis, eds., *Long-term Climatic Variations: Data and Modelling,* pp. 21–41. Berlin: Springer-Verlag.

Taylor, K. C., G. W. Lamorey, G. A. Doyle, et al. 1993a. The "flickering switch" of late Pleistocene climate variability. *Nature* 361:432–436.

Taylor, K. C., P. A. Mayewski, R. B. Alley, et al. 1997. The Holocene-Younger Dryas transition recorded at Summit, Greenland. *Science* 278:825–827.

Taylor, K. C., P. A. Mayewski, M. S. Twickler, and S. I. Whitlow. 1996. Biomass burning recorded in GISP2 ice core: A record from eastern Canada. *The Holocene* 6:1–6.

Taylor, R. B., D. J. Barnes, and J. M. Lough. 1993b. Simple models of density band formation in massive corals. *Journal of Experimental Marine Biology and Ecology* 167:109–125.

Tchernia, P. 1980. *Descriptive Regional Oceanography.* Oxford: Pergamon.

Teller, J. T. 1990. Volume and routing of late-glacial runoff from the southern Laurentide Ice Sheet. *Quaternary Research* 34:12–23.

Teller, J. T. and A. E. Kehew. 1994. Introduction to the last glacial history of large proglacial lakes and meltwater runoff along the Laurentide Ice Sheet. *Quaternary Science Reviews* 13:795–799.

Thiede, J., A. M. Myhre, J. V. Firth, G. L. Johnson, and W. F. Ruddiman, eds. 1996. *Proceedings Ocean Drilling Program Scientific Results,* Vol. 151.

Thiede, J. and Suess, E. 1981. *Coastal Upwelling: Its Sediment Record.* New York: Plenum.

Thomas E., L. Booth, M. Maslin, and N. J. Shackleton. 1995. Northeastern Atlantic benthic foraminifers during the last 45,000 years: Changes in productivity seen from the bottom up. *Paleoceanography* 10:545–562.

Thompson, L. G. 1996. Climate change for the last 2000 years inferred from ice-core evidence in tropical ice cores. In P. D. Jones, R. S. Bradley, and J.

Jouzel, eds., *Climate Variations and Forcing Mechanisms of the Last 2000 Years*, NATO ASI Series 1, 41:281–295. Berlin: Springer-Verlag.

Thompson, L. G., M. E. Davis, E. Mosley-Thompson, and K. Liu. 1988. Pre-Incan agricultural activity recorded in dust layers in two tropical ice cores. *Nature* 336:763–765.

Thompson, L. G., S. Hastenrath, and B. Morales Arnao. 1979. Climatic ice core records from the tropical Quelccaya ice cap. *Science* 203:1240–1243.

Thompson, L. G., E. Mosley-Thompson, J. F. Bolzan, and B. R. Koci. 1985. A 1500-year record of tropical precipitation in ice cores from Quelccaya ice cap, Peru. *Science* 229:971–973.

Thompson, L. G., E. Mosley-Thompson, W. Dansgaard, and P. M. Grootes. 1986. The Little Ice Age as recorded in the stratigraphy of the tropical Quelccaya Ice Cap. *Science* 234:361–364.

Thompson, L. G., E. Mosley-Thompson, M. E. Davis, et al. 1989. Holocene-Late Pleistocene climatic ice core records from Qinghai-Tibetan Plateau. *Science* 246:474–477.

Thompson, L. G., E. Mosley-Thompson, M. E. Davis, et al. 1995a. A 1000 year climatic ice-core record from the Guliya ice cap, China: Its relationship to global climate variability. *Annals of Glaciology* 21:175–181.

Thompson, L. G., E. Mosley-Thompson, M. E. Davis, et al. 1995b. Late-glacial stage and Holocene tropical ice core records from Huascarán, Peru. *Science* 269:46–48.

Thompson, L. G., E. Mosley-Thompson, and B. Morales Arnao. 1984. El Niño-Southern Oscillation events recorded in the stratigraphy of the tropical Quelccaya ice cap, Peru. *Science* 226:50–53.

Thompson, L. G., E. Mosley-Thompson, and P. A. Thompson. 1992. Reconstructing interannual climate variability from tropical and subtropical ice cores. In H. F. Diaz and V. Markgraf, eds. *El Niño: Historical and Paleoclimatic Aspects of the Southern Oscillation*, pp. 295–322. Cambridge: Cambridge University Press.

Thompson, L. G., T. Yao, M. E. Davis, et al. 1997. Tropical climate instability: the last glacial cycle from a Qinghai-Tibetan ice core. *Science* 276: 1821–1825.

Thunnell, R., C. Pride, E. Tappa, and F. Muller-Karger. 1993. Varve formation in the Gulf of California: Insights from time series sediment trap sampling and remote sensing. *Quaternary Science Reviews* 12:451–464.

Thurber, D. L., W. S. Broecker, R. L. Blanchard, and H. A. Potratz. 1965. Uranium-series ages of Pacific atoll corals. *Science* 149:55–58.

Tiedemann, R., M. Sarnthein, and N. J. Shackleton. 1994. Astronomic timescale for the Pliocene Atlantic $\delta^{18}O$ and dust flux records of Ocean Drilling Program site 659. *Paleoceanography* 9:619–638.

Tiedemann, R., M. Sarnthein, and R. Stein. 1989. Climatic changes in the western Sahara: Aeolo-marine sediment record of the last 8 million years. *Proceedings of the Ocean Drilling Program Scientific Results* 180:241–178.

Tolderlund, D. S. and A. W. H. Be. 1971. Seasonal distribution of planktonic foraminifera in the thermocline of the North Atlantic. *Micropaleontology* 17:297–329.

Tooley, M. J. and I. Shennan. 1987. *Sea-Level Changes.* Oxford: Blackwell.

Tomczak, M. and J. S. Godfrey. 1994. *Regional Oceanography: An Introduction.* Oxford: Elsevier.

Trenberth, K. E. 1994. *Climate Modeling.* Cambridge: Cambridge Univ. Press.

Trenberth, K. E., G. W. Branstator, and P. A. Arkin. 1988. Origins of the 1988 North American drought. *Science* 242:1640–1645.

Trenberth, K. E. and J. W. Hurrell. 1994. Decadal atmosphere-ocean variations in the Pacific. *Climate Dynamics* 9:303–319.

Trenberth, K. E. and D. J. Shea. 1987. On the evolution of the Southern Oscillation. *Monthly Weather Review* 115:3078–3096.

Trupin, A. S., M. F. Meier, and J. M. Wahr. 1992. Effects of melting glaciers on the Earth's rotation and gravitational field: 1965–1984. *Geophysical Journal International* 108:1–15.

Turekian, K. K., ed. 1971. *Late Cenozoic Glacial Ages.* New Haven, Conn.: Yale University Press.

Tushingham, A. M. and W. R. Peltier. 1991. ICE-3G: a new global model of late Pleistocene deglaciation based upon geophysical predictions of late-glacial relative sea level change. *Journal of Geophysical Research* 96:4497–4523.

Umbgrove, J. H. F. 1939. On rhythms in the history of the Earth. *Geological Magazine* 76:116–129.

Urey, H. C. 1947. The thermodynamic properties of isotopic substances. *Journal of the Chemical Society (London)* April 1947:562–581.

Urey, H. C., H. A. Lowenstam, S. Epstein, and C. R. McKinney. 1951. Measurement of paleotemperatures and temperatures of the upper Cretaceous of England, Denmark, and the southeastern United States. *Geological Society of America Bulletin* 62:399–416.

Vail, P. R. 1992. The evolution of seismic stratigraphy and the global sea-level curve. In R. H. Dott, Jr., ed., *Eustasy: The Historical Ups and Downs of a Major Geological Concept,* pp. 83–91. Boulder: Geological Society of America.

Vail, P. R. and J. Hardenbol. 1979. Sea level changes during the Tertiary. *Oceanus* 22:71–79.

Vail, P. R., R. M. Mitchum, Jr., R. G. Todd, et al. 1977. Seismic stratigraphy and global changes of sea level. *American Association of Petroleum Geologists Memoir* 26:49–212.

van de Plassche, O., ed. 1986. *Sea-Level Research: A Manual for the Collection and Evaluation of Data.* Norwich: Geo Books.

Van der Burgh, J., H. Visscher, D. Dilcher, and W. M. Kurschner. 1993. Paleoatmospheric signatures in Neogene fossil leaves. *Science* 260:1788–1790.

van der Hammen, T., J. H. Werner, and H. van Dommelen. 1973. Palynological record of the upheaval of the northern Andes: a study of the Pliocene and lower Quaternary of the Colombian eastern Cordillera and the early evolution of its high-Andean biota. *Review of Paleobotany and Palynology* 16:1–122.

van der Hammen, T., T. A. Wijmstra, and W. H. Zagwijn. 1971. The floral record of late Cenozoic Europe. In K. K. Turekian, ed., *Late Cenozoic Glacial Ages*, pp. 391–424. New Haven, Conn.: Yale University Press.

Varekamp, J. C. and E. Thomas. 1998. Climate change and the rise and fall of sea level over the millennium. *EOS, Transactions of the American Geophysical Union* 79:69, 74–75.

Varekamp, J. C., E. Thomas, and O. van de Plassche. 1992. Relative sea-level rise and climate change over the last 1500 years. *Terra Nova* 4:293–304.

Vaughan, D. G. and C. S. M. Doake. 1995. Recent atmospheric warming and retreat of ice shelves on the Antarctic Peninsula. *Nature* 379:328–331.

Veeh, H. H. 1966. $^{230}Th/^{238}U$ and $^{234}U/^{238}U$ ages of Pleistocene high sea level stand. *Journal of Geophysical Research* 71:3379–3386.

Veeh, H. H. and J. Chappell. 1970. Astronomical theory of climate change: Support from New Guinea. *Science* 167:862–865.

Verardo, D. V., P. N. Froelich, and A. McIntyre. 1990. Determination of organic carbon and nitrogen in marine sediments using the Carbo Erba NA-1500 analyzer. *Deep-Sea Research* 37:157–165.

Verbitsky, M. and B. Saltzman. 1995. Behavior of the East Antarctic Ice Sheet as deduced from a coupled GCM/ice sheet model. *Geophysical Research Letters* 22:2913–2916.

Versteeg, G. J. M. 1994. Recognition of cyclic and non-cyclic environmental changes in the Mediterranean Pliocene: A palynological approach. *Marine Micropaleontology* 23:147–183.

Versteeg, G. J. M. 1996. The onset of major Northern Hemisphere glaciations and their impact on dinoflagellate cysts and acritshapearchs from the Singa section, Calabria (southern Italy), and DSDP Holes 607/607A (North Atlantic). *Marine Micropaleontology* 30:319–343.

Vilks, G. 1989. Ecology of recent foraminifera of the Canadian Continental Shelf of the Arctic Ocean. In Y. Herman, ed., *The Arctic Seas*, pp. 497–569. New York: Van Nostrand Reinholt.

Villalba, R. 1990. Climatic fluctuations in Northern Patagonia in the last 1000 years as inferred from tree-ring records. *Quaternary Research* 34:346–360.

Villalba, R. 1994. Tree-ring and glacial evidence for the Medieval Warm Epoch and the Little Ice Age in southern South America. *Climatic Change* 26:183–197.

Villalba, R., J. A. Boninsegna, A. Lara, et al. 1996. Interdecadal climatic variations in millennial temperature reconstruction from southern South America. In P. D. Jones, R. S. Bradley, and J. Jouzel, eds., *Climatic Variations and Forcing Mechanisms of the Last 2000 Years*, NATO ASI Series 141: 161–189. Berlin: Springer-Verlag.

Vitousek, P. M. 1994. Beyond global warming: ecology and global change. *Ecology* 75:1861–1876.

Vogt, P. R., K. Crane, and E. Sundvor. 1994. Deep Pleistocene iceberg plowmarks on the Yermak Plateau: Sidescan and 3.5 kHz evidence for thick calving ice fronts and a possible marine ice sheet in the Arctic Ocean. *Geology* 22:403–406.

Vostok Project Members. 1995. International effort helps decipher mysteries of paleoclimate from Antarctic ice cores. *EOS, Transactions of the American Geophysical Union* 76:172.

Vrba, E. S. and N. Eldredge. 1984. Individuals, hierarchies, and processes: Towards a more complete evolutionary theory. *Paleobiology* 10:146–171.

Walcott, R. I. 1972. Past sea levels, eustasy, and deformation of the Earth. *Quaternary Research* 2:1–14.

Walker, G. T. 1924. Correlation in seasonal variations of weather. IX. A further study of world weather. *Memoir Indian Meteorological Department* 24:275–332. (Cited in Philander 1990.)

Walker, G. T. and E. W. Bliss. 1932. World weather V. *Memoirs of the Royal Meteorological Society* 4:53–84. (Cited in Philander 1990.)

Wallace, J. M. and D. S. Gutzler. 1981. Teleconnections in the geopotential height field during the Northern Hemisphere winter. *Monthly Weather Review* 109:784–811.

Wang, B. 1995. Interdecadal changes in El Niño onset in the last four decades. *Journal of Climate* 8:267–285.

Wanless, H. R. and F. P. Shepard. 1935. Permo-carboniferous coal series related to Southern Hemisphere glaciation. *Science* 81:521–522.

Wardlaw, B. R. and T. M. Quinn. 1991. The record of Pliocene sea level change at Enewetak Atoll. *Quaternary Science Reviews* 10:247–258.

Warrick, R. and J. Oerlemans. 1990. Sea level rise. In J. T. Houghton, G. J. Jenkins, and J. J. Ephraums, eds., *Climate Change: The IPCC Scientific Assessment*, pp. 257–281. Cambridge: Cambridge University Press.

Warrick, R. A., C. Le Provost, M. F. Meier, J. Oerlemans, and P. L. Woodward. 1996. Changes in sea level. In J. T. Houghton, L. G. Meira Filho, B. A. Callender, N. Harris, A. Kattenberg, and K. Maskell, eds., *Climate Change 1995: The Science of Climate Change*. Cambridge: Cambridge University Press.

Washington, W. M. and G. A. Meehl. 1996. High latitude climate change in a global coupled ocean–atmosphere–sea ice model with increased atmospheric CO_2. *Journal of Geophysical Research* 101:12,795-12,801

Washington, W. M., G. A. Meehl, L. VerPlank, and T. W. Bettge. 1994. A world ocean model for greenhouse sensitivity studies: resolution intercomparison and the role of diagnostic forcing. *Climate Dynamics* 9:321–344.

Washington, W. M. and C. L. Parkinson. 1986. *An Introduction to Three-Dimensional Climate Modelling*. Mill Valley, Calif.: University Science Books/Oxford: Oxford Univ. Press.

Webb, P.-N. and D. M. Harwood. 1991. Late Cenozoic glacial history of the Ross Embayment, Antarctica. *Quaternary Science Reviews* 10:215–223.

Webb, P.-N., D. M. Harwood, B. C. McKelvey, J. H. Mercer, and L. D. Stott. 1984. Cenozoic marine sedimentation and ice-volume variation on the east Antarctic craton. *Geology* 12:287–291.

Webb, T. III, T. J. Crowley, B. Frenzel, et al. 1993. Group Report: Use of paleoclimatic data as analogs for understanding future global changes. In J. A. Eddy and H. Oeschger, eds., *Global Climate Changes in the Perspective of the Past*, pp. 51–70. Chichester: John Wiley & Sons.

Webb, T. III, and T. M. L. Wigley. 1985. What past climates can indicate about a warmer world. In M. C. MacCracken and F. M. Luther, eds., *Detecting the Climatic Effects of Increasing Carbon Dioxide*. Department of Energy Report ER-0237, pp. 239–257. Washington D.C.

Weber, J. N., E. W. White, and P. H. Weber. 1975. Correlation of density banding in reef coral skeletons with environmental parameters: the basis for interpretation of chronological records preserved in coralla of corals. *Paleobiology* 1:137–149.

Weertman, J. 1976. Milankovitch solar radiation variations and ice age ice sheet sizes. *Nature* 261:17–20.

Wehmiller, J. F. 1982. A review of amino acid racemization studies in Quaternary mollusks: stratigraphic and chronologic applications in coastal interglacial sites, Pacific and Atlantic coasts of the United States, United Kingdom, Baffin Island, and tropical islands. *Quaternary Science Reviews* 1:83–120.

Wellington, G. M. and R. B. Dunbar. 1995. Stable isotopic signature of El Niño–Southern Oscillation events in eastern tropical Pacific reef corals. *Coral Reefs* 14:5–25.

Wellington, G. M., R. B. Dunbar, and G. Merlen. 1996. Calibration of stable oxygen isotope signatures in Galapagos coral. *Paleoceanography* 11: 467–480.

Wellington, G. M. and P. W. Glynn. 1983. Environmental influences on skeletal banding in eastern Pacific (Panama) corals. *Coral Reefs* 1:215–222.

Wenk, T. and U. Siegenthaler. 1985. The high-latitude ocean as a control of atmospheric CO_2. In E. T. Sundquist and W. S. Broecker, eds., *The Carbon Cycle and Atmospheric CO_2: Natural Variations Archean to Present, American Geophysical Union Monograph* 32:185–194.

Whetton, P., R. Allan, and I. Rutherford. 1996. Historical ENSO teleconnections in the eastern hemisphere: a comparison with latest El Niño series of Quinn. *Climatic Change* 32:103–109.

Whetton, P. and I. Rutherford. 1994. Historical ENSO teleconnections in the eastern hemisphere. *Climatic Change* 28:221–253.

White, J., B. Molfino, L. Labeyrie, B. Stauffer, and G. Farquhar. 1993. How reliable and consistent are paleodata from continents, oceans and ice? In J. A. Eddy and H. Oeschger, eds., *Global Climate Changes in the Perspective of the Past*, pp. 73–102. Chichester: John Wiley & Sons.

White, J. W. C., P. Ciais, R. A. Figge, R. Kenny, and V. Markgraf. 1994. A high-resolution record of atmospheric CO_2 content from carbon isotopes in peat. *Nature* 367:153–156.

Whittlesey, C. 1868. Depression of the ocean during the ice period. *American Association for the Advancement of Science Proceedings* 16:92–97.

Wigley, T. M. L. 1995. Global-mean temperature and sea level consequences of greenhouse gas concentration stabilization. *Geophysical Research Letters* 22:45–48.

Wigley, T. M. L. and S. C. B. Raper. 1987. Thermal expansion of sea level associated with global warming. *Nature* 330:127–131.

Williams, D. F., J. Peck, E. B. Karabanov, et al. 1997. Lake Baikal record of continental climate response to orbital insolation during the past 5 million years. *Science* 278:1114–1117.

Williams, R. and J. Ferrigno, eds. 1988. Satellite Image Atlas of Glaciers of the World. U. S. Geological Survey Professional Paper 1386-B-Antarctica.

Williams, R. and J. Ferrigno, eds. 1993. Satellite Image Atlas of Glaciers of the World. U. S. Geological Survey Professional Paper 1386-E-Europe.

Willson, R. C. 1997. Total solar irradiance trend during solar cycles 21 and 22. *Science* 277:1963–1965.

Wilson, G. S. 1995. The Neogene East Antarctic Ice Sheet: A dynamic or stable feature? *Quaternary Science Reviews* 14:101–123.

Winograd, I. J., T. B. Coplen, J. M. Landwehr, et al. 1992. Continuous 500,000-year climate record from vein calcite in Devils Hole, Nevada. *Science* 258:255–260.

Winograd, I. J. and J. M. Landwehr. 1993. A response to "Milankovitch theory viewed from Devils Hole" by J. Imbrie, A. C. Mix, and D. G. Martinson. U.S. Geological Survey Open-File Report 93–357, pp. 1–9.

Winograd, I. J., J. M. Landwehr, K. R. Ludwig, T. B. Coplen, and A. C. Riggs. 1997. Duration and structure of the last four interglaciations. *Quaternary Research* 48: 141–154.

Winograd, I. J., B. J. Szabo, T. B. Coplen, and A. C. Riggs. 1988. A 250,000-year climatic record from Great Basin vein calcite: Implications for Milankovitch theory. *Science* 242:1275–1280.

Wintle, A. G. 1990. A review of current research on TL dating of loess. *Quaternary Science Reviews* 9:385–397.

Wohlfarth, B. 1996. The chronology of the last termination: A review of radiocarbon-dated, high-resolution terrestrial stratigraphies. *Quaternary Science Reviews* 15:267–284.

Woillard, G. M. 1978. Grand Pile Peat bog: A continuous pollen record for the last 140,000 years. *Quaternary Research* 9:1–21.

Woillard, G. M. and W. G. Mook. 1982. Carbon-14 dates at Grande Pile: Correlation of land and sea chronologies. *Science* 215:159–161.

Wolf, R. 1868. In *Astronomische Mittheilungen.* (Cited in Hoyt and Schatten 1997.)

Wolff, T., S. Mulitza, H. Arz, J. Patzold, G. Wefer. 1998. Oxygen isotopes versus CLIMAP (18 ka) temperatures: a comparison from the tropical Atlantic. *Geology* 26: 675–678.

Wright, H. E., J. E. Kutzbach, T. Webb III, W. E. Ruddiman, F. A. Street-Perrott, and P. J. Bartlein, eds. 1993. *Global Climates Since the Last Glacial Maximum.* Minneapolis: University of Minnesota.

Wright, J. D. and K. G. Miller. 1992. *Proceedings of the Ocean Drilling Program Leg 120,* p. 855.

Wyrtki, K. 1973. Teleconnections in the equatorial Pacific. *Science* 180:66–68.

Wyrtki, K. 1975. El Niño—the dynamic response of the equatorial Pacific Ocean to atmospheric forcing. *Journal of Physical Oceanography* 5: 572–584.

Yiou, F., G. M. Raisbeck, C. Lorius, and N. L. Barkov. 1985. ^{10}Be in ice at Vostok, Antarctica during the last climatic cycle. *Nature* 316:616–617.

Yiou, P., M. Ghil, J. Jouzel, D. Paillard, and R. Vautard. 1994. Nonlinear variability of the climatic system from singular and power spectra of Late Quaternary records. *Climate Dynamics* 9:371–389.

Zachariasse, W. J., L. Gudjonsson, F. J. Hilgen, et al. 1990. Late Gauss to early Matuyama invasions of *Neogloboquadrina atlantica* in the Mediterranean and associated records of climatic change. *Paleoceanography* 5:239–252.

Zachariasse, W. J., J. D. A. Zijderveld, C. G. Langereis, F. J. Hilgen, and P. J. J. M. Verhallen. 1989. Early late Pliocene biochronology and surface water temperature variations in the Mediterranean. *Marine Micropaleontology* 14:339–355.

Zagwijn, W. H. 1961. Vegetation, climate and radiocarbon datings in the late Pleistocene of the Netherlands. I. Eemian and Early Weischelian. *Mededelingen Geologische Stichtung N. S.* 14:15–45.

Zahn, R., T. F. Pedersen, M. A. Kaminski, and L. Labeyrie, eds. 1994. *Carbon Cycling in the Glacial Ocean: Constraints on the Ocean's Role in Global Change,* NATO ASI Series 17. Berlin: Springer-Verlag.

Zbinden, H., M. Andree, H. Oeschger, et al. 1989. Atmospheric radiocarbon at the end of the Last Glacial: An estimate based on AMS radiocarbon dates on terrestrial macrofossils from lake sediments. *Radiocarbon* 31:795–804.

Zebiak, S. E. and M. A. Cane. 1987. A model El Niño/Southern Oscillation. *Monthly Weather Review* 115:2262–2278.

Zhisheng, A. and S. C. Porter. 1997. Millennial-scale climatic oscillations during the last interglacial in central China. *Geology* 25:603–606.

Zhu, Z. R., K.-H. Wyrwoll, L. B. Collins, J. H. Chen, G. J. Wasserburg, and A. Eisenhauer. 1993. High-precision U-series dating of Last Interglacial events by mass spectrometry: Houtman Abrolhos Islands, Western Australia. *Earth and Planetary Science Letters* 118:281–293.

Zielinski, G. A., P. A. Mayewski, L. D. Meeker, et al. 1997. Volcanic aerosol records and tephrachronology of the summit Greenland ice cores. *Journal of Geophysical Research* 102, no. C125: 26,625–26,640.

Zijderveld, J. D. A., F. J. Hilgen, C. G. Langereis, P. J. J. M. Verhallen, and W. J. Zachariasse. 1991. Integrated magnetostratigraphy and biostratigraphy of the upper Pliocene-lower Pleistocene from the Monte Singa and Crotone areas in Calabria. *Earth and Planetary Science Letters* 107:697–714.

Zijderveld, J. D. A., W. J. Zachariasse, P. J. J. M. Verhallen, and F. J. Hilgen. 1986. The age of the Miocene/Pliocene boundary. *Newsletters in Stratigraphy* 16:169–181.

Zubakov, V. A. and I. I. Borzenkova. 1990. *Global Paleoclimate of the Late Cenozoic. Developments in Paleontology and Stratigraphy,* No. 12. Amsterdam: Elsevier.

Zwally, H. J., A. C. Brenner, J. A. Major, R. A. Bindschadler, and J. G. Marsh. 1989. Growth of Greenland Ice Sheet: Measurement. *Science* 246:1587–1592.

INDEX